MORAL ASPECTS OF HEALTH CARE

Vincent Barry
Bakersfield College

Wadsworth Publishing Company
Belmont, California
A division of Wadsworth, Inc.

For Dave Nowell, who has prevailed.

Philosophy Editor: Kenneth King
Production: Del Mar Associates
Copy Editors: Rebecca Smith and Magda Grant
Cover design: Neiheisel•Slick

Printed in the United States of America

1 2 3 4 5 6 7 8 9 10—86 85 84 83 82

Library of Congress Cataloging in Publication Data

Barry, Vincent E.
  Moral aspects of health care.

  Includes bibliographies and index.
  1. Medical ethics.   I. Title. [DNLM: 1. Ethics, Medical. W 50 B281m]
R724.B33   174'.2   81-21854
ISBN 0-534-01090-3 (pbk.)   AACR2

CONTENTS

PART II: THE HEALTH CARE PROFESSIONAL WITHIN THE ORGANIZATION   93

4. THE HEALTH CARE SYSTEM   94

5. THE HEALTH CARE SETTING   128

PART III: THE HEALTH CARE PROFESSIONAL AND THE PATIENT   163

6. PATIENT RIGHTS   164

13. HUMAN EXPERIMENTATION   431

14. ALLOCATION OF RESOURCES   467

APPENDIXES

PREFACE

In recent years, bioethics has emerged as an area of intense social and academic interest. Abortion, euthanasia, human experimentation, the allocation of scarce resources—the list of bioethical issues goes on. Whatever the topic, one thing seems clear: Embedded in such issues are the most important and complex moral questions we can ask.

A spate of textbooks has greeted this burgeoning interest in bioethics. These books are commendable for crystallizing the issues, introducing health care professionals to them, and elevating the quality of moral debate. At their best they have dramatized the need for social discussion and policy.

At the same time, many of these textbooks share decided shortcomings. For one thing, they generally focus on the more "glamorous" bioethical issues, such as those mentioned above. Although the social and moral importance of these issues is indisputable, other issues also challenge health care professionals and society as a whole. One is the treatment of special categories of patients—children, the elderly, the dying, the mentally ill, the poor. Another is the relationship between the health care professional and the health care organization. These topics rarely get the coverage warranted by their centrality to the lives of health care professionals.

Another shortcoming of many bioethics textbooks is that they largely ignore the fundamental role of values in moral decision making. They do not try to help students examine, define, and order their values. And yet values and valuations help shape the health care professional's interactions with organization, institution, patient, colleagues, and society. Any study of morality in health care that ignores the central role of values is therefore incomplete.

Still another deficiency of these textbooks is that they rarely frame health care issues within traditional moral terms. To be sure, they inevitably speak of *utilitarianism* or *the categorical imperative* or *natural law ethics*. But very few seriously attempt to apply these conceptual models to the moral problems that confront health care professionals. This oversight can result in an unfortunate gap between theory and practice: Students are disconnected from the theoretical bases they need for resolving concrete health care problems.

Clearly, then, there is a need for a book designed specifically for health care students that not only treats some standard bioethical issues but also covers values and priorities, the position of health care professionals within the organization, and their relationships with special types of patients. Such a book must also strike a balance between theory and practice. *Moral Aspects of Health Care* is intended to be such a book.

The organization and coverage of this book are readily apparent in the table of contents. However, the following list of subjects that are covered will help indicate the scope of *Moral Aspects of Health Care:*

The patient as consumer

The influence of culture, science, and religion on values

The transmission of values, including values clarification

The inadequacy of law, oaths, and codes in moral decision making

The influence of medical schools, professional associations, and drug suppliers on the health care system

The impact of free enterprise and technology on the health care system

The extension of medical influence into nonhealth areas

Conflicts between the bureaucratic and professional modes of organization

Tensions between physicians and nurses

Sexism in the medical professions

The impact of concepts of health on the delivery of care

Paternalistic acts of health care professionals, which meet some patient needs but violate other rights and needs

Advocacy by health care professionals

The impact of the relationship among parents, children, health care professionals, and the institutional environment on the health care of children

The child's right to truth and information, consent, and autonomy

Myths about old age and the elderly

The effects of ageism on health care for the elderly

Paternalistic acts by health care professionals directed at protecting the elderly from external and internal dangers

Moral aspects of the social organization of old-age institutions

The restructuring of old-age institutions

The philosophical, physiological, and methodical categories in discussions of death

The dangers of compelling the dying to conform to a "script for dying"

The hospice movement

The connection between values and mental health care

Mental health professionals as agents of the state

The mental health professional's relationship to the law

The role of third parties in health care

Emergency and nonemergency involuntary commitments

Competency and privacy considerations in mental health care

The medical profession's traditional antiabortion view

Nurses' attitudes toward abortion

The connection between the feminist movement and abortion

The relationship between concepts of death and decisions about death

The psychosocial climate in which medical research is done

Therapeutic versus nontherapeutic experimentation

Hospital patients as experimental subjects

The use of fetuses in research

Moral aspects of the double-blind method

The role of peer review committees in research

Crisis medicine versus preventive medicine

The relationship between economic privation and health problems

The inequitable distribution of health care

Complex criteria versus random selection systems in allocating health care resources

Moral considerations in organ acquisition

In addition, several pedagogical features of this book are worth noting. For instance, each chapter opens with a vignette about a health care professional who is caught in a moral dilemma. A Values and Priorities section follows in most chapters, consisting of questions designed to stimulate introspection about personal attitudes, feelings, and values related to the subject of the chapter. Each chapter ends with a summary, list of readings, and exercises designed to help readers analyze their own moral values. Beginning with Part II, each chapter also contains a case presentation and a reading selection. For the most part, these reading selections have been written by people in health care and allied fields. Finally, beginning with Part III, each chapter has a section applying the major ethical theories to an issue raised in the chapter.

I would like to take this opportunity to acknowledge those who helped shape *Moral Aspects of Health Care*. One of them was the Wadsworth philosophy editor, Ken King. This is the fourth book we have done together, but Ken approached it with the same infectious enthusiasm as he did the first. Others were professor of philosophy Richard T. Hull of the State University of New York at Buffalo, whose detailed reviews were extraordinarily useful, and Jeannine Cox, who provided many insightful reflections on the problems facing mental health professionals. Still other shapers were the following nurses and educators: Dixalene Bahleda, University of Wisconsin, Oshkosh; Eleanor M. Crowder, University of Texas, Austin; Lucille H. Davis, Saint Xavier College; Michael Finn, Saint Mary's Junior College; Marsha D. Fowler, Ph.D. Candidate, University of Southern California; Sylvia C. Gendrop, Boston State College; Ernest Marshall, East Carolina University; Sally Moroney, Nursing Instructor, University of Lowell; Theodore Scharle, Bradley University; Barbara C. Scholz, Columbus Technical Institute; and James L. Smith, East Carolina University.

Part I

Decisions, Values,
and Theories

1
THE MORAL CLIMATE OF HEALTH CARE

One evening on ambulance duty, my fellow paramedic and I were summoned to the home of a very ill man. While we were getting our equipment ready to transport him to the hospital, his wife filled us in on his history.

Her husband, she said, was in his early seventies; three years earlier he had been diagnosed as having terminal cancer. Recently his condition had become "much worse." At the time of her call to us, he seemed to be having difficulty in breathing.

Upon initial examination, I found the patient to be experiencing Cheyne-Stokes respiration, which indicates impending death. Within moments he went into respiratory arrest; his heart continued to beat, albeit irregularly. Immediately I readied the resuscitation equipment and applied it to the patient.

But before I could start, the wife said to both of us, "No, don't. He has been in such misery for so long, please don't do that. Let him go." The driver and I looked at each other, our minds working a mile a minute, then looked at her. She said, "Please. He wouldn't want this."

I asked, "You're sure you feel that way? We should start now if we're going to."

Without hesitation she said, "Don't start."

How would you react to such a request? Would you follow it? Or would you ignore it and try to keep the man alive? Do you think a health care professional is ever justified in letting a life expire? The following situation provokes similar questions.

In an effort to provide a more homelike atmosphere in the convalescent hospital where I worked at one time, aides were assigned to the same group of patients for an entire month. Of course, we still had to take days off, so occasionally some aides had to work other sections. During the first few months, this system caused some scheduling problems. One of these was that I ended up with the same section two months in a row. This became a crucial element in a very trying dilemma.

Rachel was seventy-eight, nearly deaf, and very independent. Her diagnosis was cerebrovascular accident—stroke. Her prognosis: uncertain. Rachel didn't take kindly to strangers; in fact, she had the reputation of being the most difficult patient in the hospital. She also had a very devoted and concerned family.

It took about a week for Rachel and I to get used to each other. I discovered that she had a very poor appetite and some broken skin areas on her heels and coccyx. I also noticed that the evening shift was leaving her hearing aid in at night, which caused a large sore on her ear. These things worried me, because they were not being tended to when I was not there. By the third week, Rachel decided that I was a saint and that she didn't want any other nurse. So she proceeded to be as difficult as possible with the other personnel. To add to the problem, Rachel was apparently coming down with a case of the flu.

One day during this time, Rachel's daughter-in-law cornered me. She wanted some answers. Why was Rachel getting worse? Why hadn't the doctor been in to see her? Did she have any reason to be angry at the staff? The daughter-in-law said that the doctor had painted a very rosy picture: just a little rest and some physical therapy and Rachel would be as good as new. But I knew that, in fact, when the charge nurse had called the doctor for medication and treatment orders, he had instructed her simply to have us make Rachel as comfortable as possible because she wasn't going to last very long anyway.

By the sixth week of my working this section the doctor still hadn't come to see Rachel; in fact, he hadn't come in since admitting her two months earlier. The bout with the flu had left her too weak to take physical therapy. Worse, she could only tolerate about half an hour of sitting up and had practically stopped eating. And the daughter-in-law was still shooting questions at me and complaining that nobody would give her straight answers. I knew, of course, that professionally I should keep my mouth shut and not make anybody look bad. But I really felt sorry for Rachel, and I thought her daughter-in-law was getting the runaround.

If you were the aide, what would you do? Would you tell the daughter-in-law what you know and think? Would you "stonewall" her? Or would you send her along to your superiors? Do you think it is ever right for one health care professional to say anything derogatory about another or about an insitution?

The people in these situations are not imaginary. They are real health care professionals (HCPs) who had to decide what would be the right thing to do. Such situations represent only a smattering of the multitude of moral dilemmas that arise daily in health care. You need only recall your own or someone else's experiences in the field, read a professional journal, or skim the pages of the daily newspaper to multiply such incidents beyond count. Some situations involve split-second, life-or-death decisions. Others are mundane. But all raise questions that are important to the study of morality in health care.

Here are some of the questions that HCPs may be forced to consider:

Should heroic measures always be taken to preserve a person's life?

How much information should a patient and his or her family receive about the patient's condition, prognosis, and treatment?

Is it ever justifiable to perform a medical procedure on a patient without the patient's consent?

Should a nonphysician HCP always follow a physician's orders?

Are HCPs obliged to relinquish their personal philosophies and to support the philosophies of others, such as patients?

Are the individual rights of patients more important than the rights of society at large?

Should HCPs support all the positions on ethical questions taken by their professional organizations?

Which patients should receive the benefit of scarce medical resources?

Is it ever right to violate a patient's confidentiality?

Should infants with severely disabling and irreparable handicaps be left to die?

Does a woman have a right to an elective abortion?

Is research on fetuses wrong?

Should life-support systems be discontinued if a patient's electroencephalogram shows no brain response for several days?

Are HCPs ever justified in violating health team loyalty?

Should HCPs speak publicly on health issues, even at the risk of injuring the reputation or image of their profession?

These questions are far-ranging and involve the HCP's relationships with patients, colleagues, institutions, professional societies, and society as a whole. Such questions not only typify health care issues with clearly moral overtones but also raise the central concern of this chapter: the need to study health care ethics in general.

As you shall see, one reason for studying health care ethics is the increasing ethical content of health care decisions. More and more people in the health field must decide not just what is legally and professionally the right thing to do but also what is morally right.

Another reason is to understand the dilemma that HCPs often experience when trying to separate personal and professional values from bureaucratic responsibilities. The two are not easily divorced.

A third reason is the quandary professionals face in making moral health care decisions. This quandary develops because (1) individuals must make moral decisions without clear-cut standards to guide them; (2) the health care professions are often deficient in spelling out these standards and assigning rights and responsibilities to the proper parties; and (3) HCPs, like most people, seldom have well-formed ideas about moral accountability for policies, decisions, and actions.

This chapter focuses on these three reasons for studying health care ethics. But before discussing these subjects, it is important to define some basic terms, such as *ethics, health care ethics,* and *morality.* With a clear understanding of what these words mean, you will be able to understand the issues raised in subsequent chapters.

Ethics and Health Care Ethics

The following chapters consider problems and issues that raise questions about good and bad human conduct. *Ethics is the study of what constitutes good and bad human conduct, incuding related actions and values.*

Occasionally the term *ethics* is used interchangeably with *morals*. Although this usage is acceptable, it is more accurate to restrict the terms *moral* and *morality* to the conduct itself and to use *ethics* and *ethical* for the study of moral conduct or the code a person follows. The term *moral,* then, refers to an action or a person insofar as either is considered right or good; *immoral* refers to an action or person insofar as either one is considered wrong or bad. *Morality, therefore, refers to the property of an action by which it conforms to a standard or norm of human conduct.*

Sometimes in the study of ethics the term *nonmoral* arises. This word refers to something outside the sphere of moral concern. For example, whether a community health center is built on Elm or Spruce Street generally is a nonmoral question; it is simply not a question of right or wrong. But whether the facility is accessible to those most in need of its services certainly could be a moral question.

Ethics is concerned with questions of right and wrong, of duty and obligation, of moral responsibility. When ethicists and moralists use words such as *good* or *right* to describe a person or action, they usually mean that the person or action conforms to some moral standards. A "good" person or action has desirable qualities. Although moralists and ethicists often disagree about the nature of these qualities, they do agree that having determined the right thing to do, one should do it and should avoid doing wrong.

This book is concerned with ethics as it applies to health care. *Health care ethics—also often termed medical ethics, bioethics, or biomedical ethics—is the study of what constitutes good and bad human conduct, including related actions and values, in the context of health science and allied fields.* Health care ethics, then, is concerned with what is right and wrong, good and bad, and with what ought to be done or avoided in a health science situation that calls for a moral decision. Such situations, as we have suggested, are omnipresent and far-reaching. They can arise in any health care context and can involve any number of health care professionals—including electroencephalograph, electrocardiograph, laboratory, and x-ray technicians; intravenous, occupational, physical, and respiratory therapists; home care and in-service coordinators; nurse-practitioners, nursing aides, nursing students, nursing supervisors, licensed vocational nurses, and mental health nurses; and an array of physicians.

Moral Aspects of Health Care, then, is concerned with what constitutes good and bad human conduct, including related actions and values, in the context of health care practice, policy, and relationships. It deals with the values and behavior of HCPs in health care situations, the impact of their actions on other persons within the health care professions and outside the field, as individuals or as members of groups.

Since the study of health care is part of the general study of ethics, this book must also discuss general ethical concepts. For example, if we are to discover whether there are guidelines to help the ambulance unit and the aide (or any other HCP) make moral decisions, we must first explore whether there are guidelines for making *any* moral decisions. Similarly, in order to help HCPs determine and order their values, we must find out whether any methodology is available for doing so. Therefore, in addition to examining specific health care matters, later chapters discuss the general ethical theory of value and obligation.

Although many current situations signal the need to investigate health care ethics, three stand out. One relates to the ethical content of health care decisions; a second to the human element, the HCPs themselves; and a third to the moral quandary that the health system faces.

The Ethical Content of Health Care

A number of phenomena explain the increasing ethical content of health care today. Five are particularly noteworthy: (1) advances in medical science; (2) the recognition of the patient as a consumer of health services; (3) medical malpractice case law; (4) court-ordered treatment; and (5) the Nuremberg Code and related guidelines.

Advances in Medical Science

Mrs. Huff is forty-one years old and pregnant for the first time. Since the risk of birth defects is known to rise with increased maternal age, Mrs. Huff's physician suggests that she have amniocentesis, the procedure whereby a sample of amniotic fluid is obtained for analysis. The Huffs agree. Amniocentesis shows that the fetus has Down's syndrome, a chromosomal abnormality. The couple is left wondering whether or not to have an abortion. Inevitably, the health care team is drawn into their decision.

Bill Wright is twenty-three years old. His wife is five months pregnant with their second child. The first child is three. Bill was brought to the intensive care unit of a local community hospital after an automobile accident three months ago. He has been in a coma and on a respirator ever since. As it happened, Bill was just starting up his own paint company when the accident occurred. His wife does not work, and neither her family nor Bill's is able to underwrite the young couple's quickly mounting medical bills. The health care team is convinced that even if he comes out of the coma, which is unlikely, Bill will be no more than a "vegetable," due to severe brain damage. Several members of the team are concerned about the morality of subjecting Mrs. Wright to further physical, emotional, and financial strain by using extraordinary measures to keep Bill alive. They wish Mrs. Wright would request that they "pull the plug." But evidently she cannot bring herself to do that. They wonder whether they ought to broach the subject and perhaps attempt to influence her.

Not too long ago, the Huffs and Wrights and the health care teams ministering to them would not have been faced with these dilemmas. Procedures like amniocentesis and equipment like respirators simply were not available. Thus there would have been no way for the Huffs or their health care team to know the condition of the baby before birth, and so the issue of aborting a seriously defective fetus would not have arisen. Similarly, in the not too distant past, Bill Wright undoubtedly would have died shortly after being hospitalized. But now, he could be kept alive indefinitely, while Mrs. Wright and HCPs wrestle with such concepts as the right to life, the right to die, extraordinary measures, and the quality of life.

Today advanced medical procedures, respirators, dialysis machines, pacemakers, implants and transplants, new drugs, and advances in the treatment of chronic, long-term illnesses are all part of a breathtaking expansion of medical knowledge that has increased the options available to patients and HCPs. As in the two cases just cited, increased options often mean moral choices.

In addition, expanded medical knowledge and the accompanying technological advances have increased the need for medical specialization. Specialization, in turn, has made cooperation among HCPs vital. As a result, the fullest use of today's medical wonders is creating radical changes in patterns of health care and in the organization of the health professions. A good example is the emergence of the "health care team," which typically consists of a physician, nurse, social worker, various therapists (for example, physical, occupational, respiratory), a psychologist or psychiatrist, and sometimes a public health nurse.

Inherent in any team arrangement are a number of problems.[1] One of these is the tendency to diffuse responsibility: Where everyone is responsible, no single individual is. A second is the competence of the group as a team: As individuals, members may be superb HCPs; but as team members who are required to work in harmony with others, they may be inefficient. A third concerns supervising patterns: The more team members, the greater the management problems. A fourth relates to confidentiality: The threat to patient confidentiality increases with the number of team members. Finally, the advent of teams in many instances has meant the end of person-to-person, one-to-one relationships. Such problems can readily lead to moral concerns about the adequacy of health care.

So, increases in medical knowledge and advances in scientific technology go a long way toward explaining the increase in the ethical content of health care decisions. Advances in medical science have made more options available, which, in turn, call for moral decision making. These same advances have necessitated specialization and professional cooperation, which require a heightened sense of professional rights and responsibilities and of patient rights.

There is another reason, apart from advances in medical science, that helps account for the increasing ethical content of health care decisions. This is the changing perception we have of patients and their own changing perceptions of themselves.

The Patient as Consumer

Certainly one of the most important legal and social movements in the last three decades has been the recognition of individual rights. The initial thrust of this movement was in racial equality, but in recent years we have seen the movement enlarged to embrace women's rights, gay rights, children's rights, the rights of the elderly, and so forth. An integral part of this phenomenon has been the consumer rights movement, which aims to protect consumers from exploitation. The literature of consumerism outlines broad areas of consumer rights, including (1) the right to have certain needs met in the marketplace; (2) the right to clear, accurate, and adequate information about products and ser-

---

1. E. D. Pellegrino, "Ethical Implications in Changing Practice," American Journal of Nursing, September 1964, p. 111.

vices; (3) the right to safe products and services; and (4) the right to high-quality products and services. Needs, information, safety, and quality—these are the areas that have been carved out for discussion of consumer rights and concomitant business responsibilities.

The recognition of consumer rights affects the health care field as well. Increasingly, the patient is being looked upon as a consumer of health care services. As is true of any consumer, the consumer of health care services is regarded as having specific rights. As you shall see in later chapters, widespread attention is being given to the rights of consumer-patients generally and also to the rights of special consumer-patients, such as children, the dying, the elderly, the mentally ill, and the disadvantaged. Indeed, the growing recognition of the patient as a consumer of medical services may partly explain the tendency in some health fields to view patients as clients. Perhaps the term *client*, which denotes a person who engages the services of a professional, captures our current concept of those who consume medical services, and of the relationship between them and HCPs, better than *patient* does. In any event, use of the term *client* underscores the increasing attention being given to the rights of those seeking health care and the responsibilities of those providing it.

The growing emphasis on the role of the patient as consumer has produced several phenomena of moral note. First, patients generally have a more sharply developed sense of their rights and expect more of health care and of those providing it. Second, in becoming sensitized to patient rights, HCPs have had to reassess their own concepts of duties and responsibilities to patients. This heightened awareness on the part of patients and HCPs has increased, even intensified, the moral content of health care decisions. In many instances it has forced HCPs to ask themselves serious questions about the propriety of their loyalties. Thus an x-ray technician who is convinced that a patient is receiving too many x-ray treatments may now wonder whether to inform the patient and maybe risk the physician's wrath or to say nothing. A nurse may wonder whether to honor a dying patient's incessant pleas to be allowed to die. A psychiatrist may wonder whether to inform authorities after reading about a gruesome murder that fits exactly the bizarre fears and secret wishes expressed earlier by a patient. Whether or not the ultimate decision comes down on the side of the patient's rights is not so important here as the fact that the formation and high visibility of these rights largely accounts for the debates that now occur among individuals and within institutions.

Medical Malpractice Case Law

A third factor that helps account for the increasing ethical content of health care decisions is medical malpractice case law, the source of our current understanding of basic patient rights. Take, for example, the patient's right to informed consent. How much information must HCPs disclose to ensure that a patient has been informed before consenting to treatment or therapy? When is a patient competent to give consent? Under what conditions is consent voluntary? In recent years these and other important issues relating to patient rights have been placed in a legal frame that has crystallized the concerns. At the same time, medical malpractice case law has had the effect of highlighting the moral dilemmas in which HCPs can find themselves in attempting to meet divergent patient rights. For example, what should HCPs do when fully informing a patient may mean jeopardizing adequate health care? Ought they to conceal information, thereby maximizing the care, or reveal the information, thereby jeopardizing the care? The point is that medical malpractice case law, rather than resolving dilemmas like these, raises them. Just as important, in many instances it has left HCPs more concerned with avoiding malpractice suits than with delivering health care, a phenomenon honeycombed with the most serious kind of moral implications.

Court-ordered Treatment

A fourth factor influencing the ethical content of health care decisions is the intervention of the courts in the treatment of various cases. Albeit infrequently, the courts have on occasion enunciated principles concerning the treatment of incompetents, adherents of religious sects, and others. In the celebrated Karen Ann Quinlan case, for example, a court in effect permitted authorities to withdraw life-support systems from a patient. In other cases, courts have allowed HCPs to provide life-sustaining treatment for children whose parents or guardians refused such treatment on grounds of conscience or religious principle. By contrast, where the minor's life has not been threatened, the courts generally have supported the parents' right to refuse treatment for their children. These legal interventions, like medical malpractice cases, have flushed out the moral problems embedded in such situations, although they have not resolved them. The pages ahead say considerably more about the interplay between law and health care ethics.

The Nuremberg Code and Related Guidelines

Finally, the Nuremberg war crimes trials and the resultant Nuremberg Code have been important sources of increased ethical content and concern in health care decisions. The Nuremberg Code, developed by the Allies after World War II, specifically provided the standards for judging Nazis who practiced human experimentation. Since then, numerous related guidelines and codes have been formulated, some of which are considered in Chapters 2 and 13. Here it is sufficient to note that such documents not only reflect but increase the importance of moral decision making in the provision of health care.

There are, then, at least five major factors that account for the increasing ethical content of health care decisions. Taken together, these factors have increased the frequency and complexity of moral decision making for HCPs; have altered organizational patterns in the delivery of health care and have raised serious questions about the adequacy of and responsibility for the care of patients; and have increased awareness of patient rights and provider responsibilities.

The Health Care Professional

A second reason for studying health care ethics is that HCPs increasingly find themselves stuck between the professional ideals and values that they have learned in training and the bureaucratic values that they are expected to pursue on the job. For example, consider a public health nurse who, in trying to show the sensitivity that she (or he) has been trained to value, chats with a patient while changing his bed. As a result, she takes longer than usual with the chore and thus displeases her supervisor. Was the nurse right or wrong in her conduct? Situations like this reflect the discrepancy between the professional world view and the work world view: The operational values of the work world often clash with the conceived values of the academic world.

As a result of this clash, HCPs frequently find themselves in a double bind. On the one hand, they are saddled with lofty professional ideals and values, which they have been educated to aspire to and which are focal points in their effort to provide health care. On the other hand, they are expected to conform to bureaucratic objectives in order to get along, to receive good evaluations, even to keep their jobs. In some instances, HCPs are expected to do what they think is actually wrong. If they refuse or fail to do it, they are not commended for their courage or virtue but are reprimanded for stubbornness and unbridled individualism.

For the individual who is ill equipped to handle it, this discrepancy between the professional and work world views can produce fear, anger, hate, self-pity, outrage, and withdrawal. These reactions cannot help but influence the HCP's behavior with co-workers and patients. Undoubtedly HCPs in the throes of this kind of reality shock give inferior care.

HCPs who cannot resolve conflicts between their own value systems and the prevailing value system in the work world are handicapped in effecting positive, lasting changes in the work situation. Rather, they try to resolve such conflicts in various self-defeating ways. Some decide to go along by suppressing their own values and ideals and accepting the status quo. But when people are preoccupied with going along, needed changes in the health care system often go begging, and gross inequities can go uncorrected. Others simply refuse to incorporate any work values into their own system, or they adopt only a few, preferring instead to maintain what they perceive as their professional values. But for such stalwarts, the occupational road can be long and difficult. They are reminded daily, sometimes in unforgiving ways, of their disloyalty to the bureaucracy. Still other HCPs wish a plague on both houses. Unable to reconcile conflicts, they disavow allegiance to either value system. If this becomes a permanent pattern of adjustment, it can seriously undermine patient care. Health care can become just a job that is given minimal effort, with no thought to improvement.

These and other adaptive behavioral patterns raise serious moral questions about the quality of health care, the rights of patients, and the treatment of HCPs within the institution. It would be presumptuous to suggest that the study of health care ethics would resolve an HCP's work conflicts. But a grounding in values, ethical theory, and the moral aspects of health care is bound to sensitize those in the field to the problems, enlarge the debate, and provide some bases for resolving the conflicts.

Health Care's Moral Quandary

Although many factors characterize health care's "moral quandary," three are of major ethical importance: (1) the absence of a clear moral standard; (2) the uncertainty in defining and assigning rights and responsibilities; and (3) an ill-defined concept of moral accountability.

The Absence of a Clear Moral Standard

At a time when the moral content of health care decisions is increasing, HCPs do not agree on an ethical standard or norm. Thus there is no universal measuring rod against which HCPs can judge their actions as right or wrong. As a result, some HCPs make decisions strictly on the basis of self-interest, and others act rigidly in accord with institutional policies. Still others look primarily to the patient's interest or to what the law prescribes or proscribes. And others still attempt to apply some maxim of moral behavior, such as the Golden Rule. There are even those who operate strictly on whim—for example, honoring patient self-determination in one instance but violating it in another. Of course, no group of professionals has ever been in complete agreement about how to determine moral behavior. But the proliferation of ethical issues in health care seems to have divided HCPs more than ever.

Two of the most popular sources of health care norms are professional codes and the law, both of which are examined in Chapter 2. Now it is sufficient to say that neither is an adequate substitute for a moral standard.

Rights and Responsibilities

On many occasions this book uses the terms *rights* and *responsibilities*. Although there are numerous definitions of what constitutes a right, we will use the following definition: *A right is a claim or sphere of decision that is, or ought to be, respected by other individuals and protected by society. A responsibility is a sphere of duty or obligation assigned to a person by the nature of that person's position, function, or work.*[2] There is a connection between rights and responsibilities: Rights imply responsibilities, and responsibilities imply rights.

Clearly, HCPs have responsibilities to patients, employers, co-workers, and society at large. For example, HCPs have the responsibility to provide their patients with adequate health care and to preserve their self-determination, and health team members have a responsibility to work together cooperatively in the interest of optimum health care for the patient. Conversely, patients have the

---

2. C. Wellman, *Morals and Ethics (Glenview, Ill.: Scott, Foresman, 1975), p. 252.*

right to expect those treating them to provide adequate health care and to help them preserve their autonomy; members of a health care team have a right to expect other members to work cooperatively with them. Patients have responsibilities to those caring for them: to be cooperative, to follow instructions, to keep appointments, and so on. HCPs, in turn, have the right to expect patients to meet these responsibilities.

It is easy enough to indicate which parties have rights and responsibilities in health care, but it is quite difficult to sort out the rights and responsibilities themselves and to resolve conflicts when they arise. For example, an HCP has a right to professional autonomy, but a patient also has the right to self-determination. But what happens when the HCP's right to exercise professional judgment in the best interests of the patient and in keeping with the noblest ideals of the profession conflicts with the patient's right to determine the nature and extent of the treatment that he or she will be subjected to? The situation in which the ambulance unit found itself poses just such a dilemma. Then the area of shared responsibilities in the delivery of health care must be considered. Although it is often easy to pinpoint responsibility, there are times when it is not. This is especially true when teams are involved and when HCPs work within institutions and are therefore constrained by rules and procedures.

The point is that defining and assigning rights and responsibilities often poses a difficult problem. HCPs have duties and obligations to various claimants: patients, supervisors, employers, institutions, professions, society at large. But defining these responsibilities, and assigning them to the proper claimants in the proper degree, can be troublesome—especially for those who find themselves operating within a bureaucracy.

The structure of a health care bureaucracy can be as complex as any other business corporation. Generally speaking, the larger the bureaucracy, the more likely it is that responsibilities will get blurred. Individuals can easily lose sight of their own responsibilities or pass the buck, especially on tough matters. The danger of not clarifying one's own responsibilities and those of others is that one can act immorally, often through omission.

Perhaps the only way to heighten one's awareness of rights and responsibilities is to scrutinize individual cases and to identify specific persons to whom rights and responsibilities can properly be assigned. Even in a bureaucratic maze, rights and responsibilities can usually be assigned to specific individuals. For example, a health care facility's responsibilities to provide proper health care for its patients can be sorted out into components: responsibilities of administrators, physicians, nurses, technicians, dieticians, and others to provide what patients have, in effect, contracted for. No doubt, in many cases of bureaucratic decisions and actions, not one but many HCPs share the responsibility. Although it is true that one person may merit responsibility more than another, we are not necessarily justified in placing all blame or praise on that person.

Even where responsibility has specifically been assigned, we can still ask whether the individual is blameworthy or praiseworthy in this instance. Can we rightly hold this person morally accountable? Rather than asking what the responsibilities are and who has them, we should determine the conditions under which we hold people morally accountable for their actions. Not only is this question fundamental to any study of ethics, it is critically important in any bureaucratic organization where individuals define or justify questionable actions and transactions, and avoid decision making altogether, by appealing to severe restrictions on their personal freedom of choice. This problem looms large in any context where rights and responsibilities are not always easy to define and assign. In short, moral accountability is a third issue we should consider in examining health care's quandary and in explaining the need to study health care ethics.

Moral Accountability

Even when rights and responsibilities are defined and properly assigned, it is not always easy to assign moral accountability. Moral accountability is blaming or praising someone for an action. When we hold someone morally accountable for an action, we mean that the person should be blamed or praised for it.

The nature of a health care institution is such that on many occasions just about everyone can be said to share the accountability for an action. Thus everyone is to be blamed or praised. In practice this usually means that no one is held morally accountable; or in the case of a malpractice suit, only the HCP that is sued bears the moral accountability. Nothing threatens real morality in health care more than the doctrine of *non*accountability, or the belief that liability is the exclusive test of moral accountability.

Awareness of the conditions under which we hold people morally accountable is as important as sorting out specific cases of rights and responsibilities. If we are not aware of the importance of the conditions, then we are likely to founder on the shoals of the "Bormann defense." The Bormann defense takes its name from Martin Bormann, third deputy under the Nazi regime, whose defense for wartime atrocities was that he was just following orders. When individuals have no sense of moral accountability, or an underdeveloped sense, they ordinarily try to justify their behavior as Bormann did: "I was told to do it"; "I was expected to do it"; "I was just doing my job"; "That's how things operate around here"; and so forth.

There are standard features in the structure of any organization that encourage just such moral rationalizations. One is their generally autocratic nature. When confronted by a tight chain of command, in which decisions are made at the top and then passed down, most people working within organizations survive by not questioning their superiors. In addition, when the organizational structure is complex, moral accountability is diluted. So many people, even institutions, can get involved at so many levels that moral buck-passing can become the order of the day. The blind pursuit of prestige and profit also blurs moral accountability. And most important, the intense pressure to keep one's job and to secure promotions can be used to justify almost anything. The point is that working within an organization provides easy excuses for abdicating personal moral accountability for decisions and actions. This observation is no less true in health care institutions than in business, government, or education—although particular organizational features that encourage moral rationalizations vary in intensity from one to another.

Under what conditions, then, do we hold people morally accountable? A convenient way of answering this question is to isolate the circumstances under which we ordinarily excuse people—that is, hold them blameless or not morally accountable. Professor of philosophy Paul Taylor suggests four of these circumstances:[3]

1. *Excusable ignorance of consequences.* We generally excuse people when we do not believe that they were aware of the unfavorable consequences their actions would produce or because they could not reasonably be expected to have known how to prevent the consequences. For example, today we are extremely cautious about subjecting patients to x-rays, because we now know that they can be damaging. Years ago, physicians, radiologists, and technicians were not aware of the serious potential danger. Therefore, barring cases of egregiously excessive and unnecessary x-ray exposure, we probably would not hold such individuals morally accountable for damage from x-rays administered during that time.

   In other cases we excuse people because, in our estimation, they could not reasonably be expected to have known how to prevent the actual consequences of their actions. Thus a physician may prescribe sleeping pills for a middle-aged man who complains of insomnia. The physician is always very cautious in prescribing sleeping pills, but she feels warranted in this case, because she has treated the patient for years and has always found him to be rational, stable, and levelheaded. The pharmacist fills the prescription, the man goes home, intentionally overdoses, and dies. Clearly, the physician is not morally accountable. Although she realizes that patients can and do overdose, she could not reasonably be expected to have known that it would occur in this case.

2. *Constraints.* We usually excuse people when we think that they could not help what they did, that they had little or no choice in a matter. A person's capacity to choose can be limited by external constraints or by some inner compulsion.

   In the case of external constraints—that is, outside factors or forces—we speak of people acting against their wills. Thus a health care team cannot be held morally accountable for not treating a rational adult patient who simply refuses the treatment.

   In the second form of constraint, the compelling element comes from within, not from without. We speak of people who act in certain ways because they feel an overwhelming inner urge, desire, craving, or impulse to do so. Thus we would hardly hold patients morally accountable for the damage they cause during a period of postoperative psychosis.

3. *Uncontrollable circumstances.* When, in our estimation, the circumstances of an act were beyond the person's control, we generally excuse the behavior. There are numerous circumstantial excuses that we readily accept as legitimate excuses. In each instance the circumstances must truly be beyond the

---

3. *P. Taylor,* Problems of Moral Philosophy, *2nd ed. (Belmont, Calif. Dickenson, 1973), p. 277.*

person's control and not have resulted from voluntary actions in the past. For example, we would not hold a nurse morally accountable for being late for work if, through no fault of her own, she had been involved in an automobile accident. But we would hold her accountable if the accident resulted from her reckless driving.

4. *Lack of alternatives.* Ordinarily we excuse actions when we think that people lacked either the ability or the opportunity to do the right act. For example, a young man with no identification is rushed into the hospital after lapsing into unconsciousness on a street corner. His appendix has ruptured. The surgical resident operates and saves his life. Upon recovery, the young man tells the surgeon that he never would have consented to the appendectomy had he been conscious. It turns out that the man is a member of a religious sect opposed to the shedding of blood. Surely we would not hold the surgeon morally accountable for violating the patient's right to informed consent in this matter, because there simply was no way for the surgeon to honor this principle short of letting the man die. Indeed, had the surgeon allowed him to die, we would probably hold the physician morally accountable, because he could reasonably be expected to have known how to prevent the consequences. But as it happened, the surgeon had no way of gaining informed consent, nor did he have the time to wait. Since he lacked any other alternative to surgery, we would not hold him morally accountable.

In health care situations, and in everyday life, there are occasions when we excuse people and occasions when we hold them responsible. When any of the above conditions accompany a person's action, whether in health care or outside it, we do not hold the person accountable. But when none of these conditions are present, we hold the person morally accountable; we do not excuse. Keeping these conditions in mind can help HCPs deal with at least this part of the moral quandary they face.

Summary

The need to study health care ethics arises from a consideration of several factors. The first is the ethical content of health care decisions. Advances in medical science and the emergence of the patient as consumer have heightened the moral dimension of health care decisions. As a result, HCPs need the moral grounding that the study of health care ethics provides.

A second factor is the plight of individual HCPs, who too often find their allegiance divided between the profession, with its values and goals, and the work organization, with its objectives and expectations. HCPs who experience this conflict can benefit from the on-the-job moral direction that health care ethics provides.

A third reason for health care ethics is what has been termed health care's moral quandary. Although HCPs are asked to make moral decisions in the context of health care, they often lack the ability to decide. They have no clear-cut moral standard to guide them. Furthermore, their decisions depend on clearly defined and assigned rights and responsibilities, which they rarely have. This quandary is intensified by the erosion of moral accountability, stemming in part from the complex nature and structure of health care bureaucracies and from the tendency to practice "defensive medicine." The study of health care ethics is helpful in dealing with this quandary because it airs standards of moral conduct and values, investigates rights and responsibilities, and refines one's ability to recognize and act with moral accountability.

For Further Reading

Alford, R. R. *Health Care Politics: Ideological Interest Group Barriers to Reform.* Chicago: University of Chicago Press, 1975.

Beauchamp, T. L., and Walters, L., eds. *Contemporary Issues in Bioethics.* Encino, Calif.: Dickenson, 1978.

Dedek, J. F. *Contemporary Medical Ethics.* Mission, Kans.: Sheed, Andrews, and McNeel, 1975.

Edelstein, L. *Ancient Medicine.* Baltimore: Johns Hopkins University Press, 1967.

Edwards, M. H. *Hazardous to Your Health: A New Look at the "Health Crisis" in America.* New York: Arlington House, 1972.

Ehrenreich, B., and Ehrenreich, J. *The American Health Empire: Power, Profits, and Politics.* New York: Vintage Books, 1971.

Ginzborg, E., and Ostow, M. *Men, Money, and Medicine.* New York: Columbia University Press, 1969.

Krizay, J., and Wilson, A. *The Patient as Consumer in Health Care Financing in the United States.* Lexington, Mass.: D. C. Heath, 1974.

Reiser, S. J., Dyck, A. J., and Curran, W. J. *Ethics in Medicine: Historical Perspectives and Contemporary Concerns.* Cambridge, Mass.: MIT Press, 1977.

Shannon, T. A., ed. *Bioethics.* New York: Paulist Press, 1976.

Stevens, R. *American Medicine and the Public Interest.* New Haven, Conn.: Yale University Press, 1971.

Vaux, K. *Biomedical Ethics: Morality for the New Medicine.* New York: Harper & Row, 1974.

2

VALUES AND PRIORITIES

Jim Drury had been in the intensive care unit for fifteen days and comatose for the last six. As his condition deteriorated, intensive efforts were undertaken to reverse his course. His wife Rita was permitted to see him only for short periods of time, since the medical team was working on him almost continuously.

On the sixteenth day, Jim's vital signs faltered. Treatment measures were stepped up, but to no avail. After a few hours, Jim Drury died.

A few minutes later, Rita Drury arrived. An intern told her of her husband's death and asked her permission for an autopsy. Rita refused. The chief resident was summoned, and she too pressed for permission, arguing that the autopsy would help medical science. Again Rita refused. Finally, the physician in charge stepped in and requested permission. But Rita stood her ground, insisting that she wanted her husband to have the peace and dignity she felt he was denied in the last days of his life.

Suppose you were a member of the health care team who had cared for Jim Drury and you overheard these conversations. If you could do only one of the following, which would it be?

1. Do nothing.

2. Intervene on Rita's behalf.

3. Support the health care team.

4. Do nothing at the time, but report the entire incident.

Whatever your choice, one thing is sure: Your decision reflects a value you hold and the priority you give to that value.

The foregoing situation involves at least three values: the value of patient advocacy, the value of team loyalty, and the value of professional health care action. What makes a decision like this one so difficult is that the values conflict. Today's HCP frequently experiences similar conflict in the choices to be made. The problem is acute for newcomers to the health care field. Typical areas where conflict arises are

Aging

Euthanasia

Suicide

Death

Institutional rules

Authority

Informed consent

Drugs and drug addiction

Privacy

Sex

Psychosurgery

Genetic testing and screening

In vitro fertilization

Patient autonomy

Cloning

Genetic intervention

Some of the reasons for confusion about values are social and cultural; others are professional. In the last thirty years profound, sometimes violent, changes have occurred in society's attitudes, feelings, and values in many areas: politics, religion, work, leisure time, education, family, health, aging, and death, to mention just a few. Nobody is immune to the fallout from these changes.

Accompanying these changes, even advancing them, has been a so-called information explosion. In some instances this development has forced us to relinquish the easy "truths" of the past and to formulate new viewpoints; in others it has led us to reserve judgment until more evidence is in. Whatever the full and precise impact of this onslaught of new information, we now have more factors to consider before deciding on a course of action. Compounding the problem is the fact that we live in an age in which absolutes of any sort are spurned as hallmarks of a former, less enlightened time, a time when the credibility of political, educational, religious, and health care institutions was not seriously questioned. Lacking institutions we can still believe in, we are left without clear-cut directives and are left to our own devices.

On the professional side, dramatic changes in health care have widened the range of problems and choices for HCPs. One obvious change is the impressive advance in medical technology, which was mentioned in Chapter 1. The relatively recent proliferation of life-sustaining apparatus has raised questions about the extent to which HCPs should sustain life artificially. Not too long ago, such questions would not have arisen.

In the health care professions, there is an accelerating drive toward professionalization among nonphysician HCPs. But that phenomenon, and the concomitant shift in attitudes, values, and responsibilities, at times brings nonphysician HCPs into conflict with physicians. The conflict is understandable, for historically physicians have borne the ultimate responsibility for patient care and hospital administration. What is more, they continue to influence the work other HCPs do, as well as the conditions under which they do it and the compensation they get for doing it. As a result, today's nonphysician HCP must frequently choose between loyalty to the physician-administrator and loyalty to a specific health care profession. Such choices are not easy.

In the midst of all this change, HCPs are often not clear about the values they hold in professional areas. Equally as troublesome, they often do not know or understand their patients' values. And yet the values of patients influence the nature and morality of the treatment HCPs provide them. When HCPs are sure of their own and others' values, they are frequently still not sure of the priority to give them in value conflicts. And like most of us, they rarely know how to go about clarifying values and assigning priorities among them. The result is that HCPs today experience considerable inner tension as well as feelings of helplessness and even despair.

Throughout our study, we will have many occasions to analyze particular values and to consider value content, such as the three values involved in the preceding case. But before doing so, we need to consider the whole concept of value, including the sources of values and how values are transmitted. We particularly need to examine the valuing process and how HCPs can apply it to beliefs and patterns of behavior already formed, as well as to those emerging. This chapter is designed to do that. It aims to help you reflect on your own particular values and to help you use the valuing process in health care. It also shows why the discovery of values needs to be supplemented with criteria grounded in moral theory.

Values

All week long you have been eagerly anticipating Mr. Byron's release from the hospital. Not that Mr. Byron has been a bad patient. You wish that all your patients recovering from stroke would behave as well as Mr. Byron has during his long and difficult period of rehabilitation. Now that he has finally learned to walk with the aid of a cane and is ready to go home, you could not be happier. It is little wonder that Mr. Byron would like to show off his newly mastered skill to his wife when she comes to take him home. Who could blame him?

Unfortunately, it is a hospital rule that all patients must be discharged in a wheelchair. When you mention this to Mr. Byron, he shows uncharacteristic pique. "Damn it," he says, "I came in here on my back, I'd like to leave on my feet!"

You are sympathetic to his feelings and are quite sure that he can walk the distance to the hospital entrance. But rules are rules.

"Look," he says softly, his hand covering yours. "It means a lot to me. Please don't rob me of the chance to walk out of here."

You feel terrible, torn between fidelity to hospital rules and compassion for Mr. Byron. What to do?

The decision that any HCP would make in such a situation is largely determined by values. *A value is an assessment of worth.* When we value something, we view it as having worth. In this case, hospital rules constitute a value; so do compassion for the patient and patient autonomy.

In choosing personal and professional goals, in selecting the means to attain them, and in resolving conflicts, we are deeply influenced by our conceptions of what is desirable, preferable, important, and appropriate—by what we see as having value. The quality of our relationships with family, friends, and co-workers; the way we conduct ourselves professionally; the degree of respect we have for others and ourselves; the social, political, religious, and leisure activities we undertake; the books we read, films we see, food we eat—all are influenced by our system of values.

This does not mean that values are the only determinant of behavior. Any choice or act reflects a vast range of factors: our concept of reality, our perceptions of the possible, our immediate motivational patterns, and so forth. But it is our key choices that affect the kind of person we become, the quality of our personal and professional lives. And key choices are largely determined by our basic values.

Since the time of the ancient Greeks, philosophers have written about the theoretical side of values. Today the word *axiology* (from the Greek *axios,* meaning "worthy") *is used to refer to the study of the general theory of values, including the origin, nature, classification, and place of values in the world.* One specialized field within axiology is ethics, the study of values in human conduct, which is the concern of this book.

Axiology is far too broad a study to consider fully here. But we should consider some recurring axiological concerns, because they relate to value content and the priority we give specific values. One of these concerns deals with the distinction between facts and values; a second deals with the subjective or objective nature of values; and a third deals with value selection.

Facts and Values

At the outset, a crucial philosophical distinction must be made between facts and values, specifically between factual judgments and value judgments. *A factual judgment describes an empirical relationship or quality.* The word *empirical* refers to whatever can be derived from or depends on experience or experiment alone. Thus a statement expresses a matter of empirical fact when its truth can be determined through sensory observations. "Mrs. Willingham has a temperature of 102°," "St. Joseph's is a hospital operated by the Sisters of Mercy," and "Mr. Rodriguez has a melanoma on his chest" are factual empirical judgments.

In contrast to such factual judgments, *value judgments are statements about the worth of objects, acts, feelings, attitudes, and even people.* For example: "Mrs. Willingham is a <u>model</u> patient," "St. Joseph's is the <u>best</u> hospital in the city," "Mr. Rodriguez is a <u>stubborn</u> man," "Dr. Johnson is a <u>brilliant</u> surgeon," "I <u>should</u> stick up for the rights of my patients," "You were <u>wrong</u> to misinform the patient about her condition." Notice the underlined words in these examples.

Inevitably, value judgments contain words like these—words that imply appraisals of worth. *Good, bad, right, wrong, should, ought,* and the like are *value words* that express moral judgments. Sometimes a statement implies a value judgment even though it does not include a value word. For example, someone may say, "What! You've discontinued Mrs. Willingham's I.V.!" The tone and context of the exclamation are enough to signal that discontinuing the I.V. was ill-advised—a value judgment.

Although it is relatively easy to distinguish factual judgments from value judgments, it is difficult, if not impossible, to separate the two, because of the interplay between facts and values. For example, the answer to the question of whether an HCP *ought* to support Rita Drury or *should* allow Mr. Byron to walk to the hospital entrance depends in part on facts concerning all the people involved, the HCP's standing in the hospital, the institution's standard operating procedure, the HCP's relationship with other health team members, and so on. Similarly, just how much an HCP *ought* to tell patients about their conditions depends in part on the emotional maturity and stability of the patient, preferences of the patient's family, the certainty of the diagnosis, the tact of the HCP, and similar considerations. The observable characteristics of things—that is, facts—enter into our assessments of worth. Value judgments, then, are fact-dependent. When conditions and facts change, our value judgments may change also.

So, value judgments are not factual judgments, but the two cannot be divorced. When I assert "Abortion is wrong (or right)" or "Euthanasia is bad (or good)," I am not expressing a factual judgment, as I am when I say "Abortion is legal" or "Euthanasia is practiced by some physicians." But I am probably basing my value judgments about abortion and euthanasia on some facts that I believe support my assessment.

Subjectivity versus Objectivity

Some philosophers view value judgments as strictly subjective—that is, expressions of one's own feelings and desires. Thus, when I say "HCPs act properly when they respect patient privacy," I am expressing personal approval. Similarly, when I say "HCPs act improperly when they don't respect patient privacy," I am expressing personal disapproval.

Other philosophers claim that a value judgment identifies a quality within an object or situation that is there independently of the individual's response and is therefore objective. Thus there is some quality in the way HCPs behave toward patients with respect to patient privacy that is good or bad independently of anyone's response of approval or disapproval.

Still other philosophers interpret value judgments as having both objective and subjective elements. In this interpretation, when I say "HCPs act properly when they respect patient privacy," I am recognizing the objective quality of confidentiality in the HCP-patient relationship, which I subjectively approve of. This interpretation seems most useful in health care matters. It permits one to judge facts on the basis of criteria that are independent of prevailing attitudes and then to use such judgments to evaluate subjective attitudes and preferences critically, to determine whether they conform to the objective criteria.

Value Selection

One of the most important axiological questions we can ask is what we should value. There is general agreement that certain groups of values exist, such as moral, political, esthetic, religious, and intellectual. And there is agreement that genetic, biological, and cultural influences produce many of these values. But there is little agreement about their nature, their relative importance, or their relationship to one another. Nevertheless, most philosophers use the following principles in discussing value issues:

1. *We should prefer what is of intrinsic value to what is of extrinsic value.* A thing has intrinsic value when it is valued for its own sake. For example, some believe that pleasure has intrinsic value; that is, pleasure is worthwhile in itself, not because it can yield something else. On the other hand, a thing has extrinsic value when it is valued as a means to something else. A triple bypass operation could be said to have extrinsic value: It is not a value in itself but can yield a value, such as improved cardiac function or good health. Intrinsic and extrinsic values are not necessarily mutually exclusive. What is of value in itself may also be a means to something else, as in the case of good health. Good health may be considered worthwhile in itself, but it is also important as a means to greater enjoyment of life.

2. *We should prefer values that are productive and lasting to ones that are not.* Physical and material values are generally less productive and long-lived than social, artistic, intellectual, and religious values. For example, long after a fortune has been spent, good health or personal integrity persists.

3. *We should choose values according to our own goals and ideals.* When we allow values to be thrust upon us, we end up living others' lives, not our own. Our values should be consistent with one another and responsive to our own circumstances.

4. *In choosing between two values, we should prefer the greater,* which will be determined largely by the previous three criteria. Conversely, when choosing between two evils, we should choose the lesser, again looking to these criteria to determine our choice.

These principles are themselves expressions of fundamental values. They are grounded primarily in Western culture, which is one of several sources from which our values spring.

There are several reasons why we should understand the sources of our values. First, in exploring the wellsprings we also develop insights into the process by which we acquire the values we hold. Thus we are better able to examine our values critically, with an eye toward refining, modifying, or abandoning them. Second, we develop insights into the value bases of patients and consequently can interact with them more effectively. Third, as professionals, we develop insights into how and why the health care system functions as it does; institutions share value sources with individuals. Furthermore, given its prominence in society, an institution can reinforce basic values and can exert pressure on those within it to implement those values. This is particularly true of the health care institution. Examining the sources of values, then, promises personal and professional dividends for HCPs.

Sources of Values

Most of our personal and professional attitudes probably stem from a handful of values. How do values arise? Where do they come from? Why does one person see beauty in an ocean, while another stands unmoved? Why does one person risk life and limb to ensure justice, while another remains detached and indifferent? Why does one HCP advocate patient rights, even at great personal expense, while another opts for team loyalty?

There are four main sources from which we get our values: experience, culture, science, and religion. Each plays a formative role in the value systems of HCPs.

Experience

Our values are significantly shaped and formed by experience. Thus the ocean holds little beauty for a person who has watched a loved one drown in it. Those who have felt the sting of racial and sexual discrimination understandably may become champions of fair and just treatment. HCPs who have been patients themselves can become highly sensitized to patient rights. In brief, the values we hold as individuals and as groups are inseparable from the endlessly changing experiences of our lives.

Much of experience is vicarious: We learn by watching and listening to other people. Indeed, vicarious experience and the sensitization that comes from thinking about what others have experienced are powerful methodological tools for shaping and ordering values.

Culture

Each of us has a value system that differs somewhat from anyone else's. Nonetheless, our individual values spring from the core values of our culture. Some social scientists have suggested that these core values reflect a particular culture's orientation to five recurring human problems: human nature, the environment, time, activity, and human relationships.[1] Orientations to these problems affect the HCP's interactions with patients, colleagues, institutions, and professions. Furthermore, orientations in these five areas can conflict, thus producing personal turmoil.

HUMAN NATURE. The culture of each social group is based on expressed and implied positions about human nature. All cultures develop answers to such

---

1. *J. C. Coleman*, Contemporary Psychology and Effective Behavior *(Glenview, Ill.: Scott, Foresman, 1974), pp. 490–496.*

questions as the following: Is human nature essentially good, bad, or neutral? Do humans have inherent worth, or are they little more than assembled collections of physical parts to be used? Are humans basically rational or irrational?

Our own culture offers a variety of positions on these issues, which can affect one's delivery of health care. For example, believing that humans lack intrinsic worth, or that they are a bundle of atoms in the physical world, directly influences the HCP's relationship to patients, including her or his commitment to principles such as confidentiality, informed consent, and truth and views on problems such as paternalism, human experimentation, and the allocation of scarce medical resources. It is much easier to justify experimenting on humans without informed consent if one does not see people as having intrinsic worth. And one who believes that people generally "get what they deserve" is likely to be unfeeling and unsympathetic toward patients.

ENVIRONMENT. A culture's orientation toward the environment addresses the issue of the human's place in nature. Are human beings powerless to affect their destiny, or do they have a measure of free choice? Is the relationship between humans and nature basically friendly and symbiotic or hostile and exploitive? Is the natural world something to be adapted to or a resource to be used and conquered? Clearly, our own culture has viewed the environment as a resource to be harnessed and used for our survival and prosperity. Indeed, this has been an energizing force behind science, medical science included.

In recent years, a trend has developed toward a less adversarial, more harmonious relationship with nature. We observe not only a growing sensitivity to the outer world but also to the inner environment: a greater awareness of what goes on within our bodies, of the complementary relationship between physical and mental modes, and of homeostasis. As a result, many individuals, in the field and outside health care, are reassessing traditional attitudes toward health and well-being and to what is needed to promote them. More attention is given to the concept of the individual as an organic whole and to the proposition that health care requires caring for the whole person, not just the diseased part of the body.

TIME. Cultures inevitably provide answers to such questions as: To what degree should custom and tradition be preserved? To what extent should new standards and patterns be encouraged and emphasized? Should a person live for the present or the future? There is little question that the institution of medicine in the United States is tradition-bound. Standard medical practice, operational methods, and sundry policies not only die hard but resist even token change. This can produce conflict in the HCP, particularly in the newcomer. For the new HCP, fresh out of a "future-oriented" educational program, the confrontation with a monolithic medical establishment, often a bureaucratic maze of rules and regulations, can produce serious "reality shock." A later chapter says much more about this problem.

ACTIVITY. Still another basic human dimension for which cultures provide an orientation deals with how people ought to spend their time. What activity is most valued? Making money? Serving others? Acquiring knowledge? Our culture provides several orientations on activity but has traditionally emphasized doing, accomplishing, getting things done. Some have termed this a time-pressure orientation, a compulsion to use time efficiently for useful purpose. The evidence of this compulsion in health care institutions needs little documentation, although the conflicts it creates for HCPs often goes uncharted.

Time and activity orientations, of course, go hand in hand. In health care they can combine to produce wrenching conflicts for HCPs. Take as one example the experience of a nurse on a twenty-bed postcardiac surgical ward in a large city hospital, reported by Marlene Kramer in her book *Reality Shock*. The facts and the order in which they are presented may seem confused, but this illustration does highlight the conflicts between time and activity orientations that may arise.

> It is the evening shift, about 4:30 p.m. The nurse is alone because her one aide has gone to dinner. Things are quiet with the patients, who are pre- or post-cardiac surgery. Many recently have been transferred out of the cardiac ICU which is adjacent to the ward.
>
> While making her rounds to see the patients, the nurse comes upon fifty-one-year-old Mrs. Swape, a post pump patient and volunteer at the hospital. The nurse expects Mrs. Swape to be her cheerful, talkative self. But she is not. The nurse detects tears welling up in her eyes. During the post surgery, Mrs. Swape has never complained or cried, even during painful procedures such as suctioning and turning her.
>
> Thinking that Mrs. Swape is ready to "open up" and talk, the nurse draws the curtain around her bed and seats herself on the edge. She takes Mrs. Swape's hand in her own, and after a few minutes Mrs. Swape begins to talk, to reveal her fears, not about the surgery but about incurring the dislike of the nurses on the ward.
>
> It seems that a few years ago Mrs. Swape had a post-surgical psychosis that resulted in her "coming unglued" for a few days: pulling out IV's and NG tubes, throwing things, screaming at the nurses. She remembered all this afterward and now was fearful that she might do it again with this surgery, thus offending the nurses so much that they wouldn't let her work on the ward as a volunteer.[2]

Now let us permit the nurse to recount the rest of her experience, which underscores how a particular time-activity orientation can produce conflicts in an HCP.

> Mrs. Swape is in the middle of telling you this. Her face is pinched and anxious and she looks as though she will "let go" any minute now. Suddenly you hear the food cart clanging through the doors of the ward. At this hospital the trays do not come up from the kitchen prepared; there are just pots of food, and the staff has to serve it on plates, construct the special low-salt diets, etc. Furthermore, the cart has to be returned to the kitchen for use on the other floors. Most of your patients are fairly recent open-heart surgery patients. They need their food, particularly potassium. If you stay with Mrs. Swape, there's no one there to serve them food; if you don't serve them the food and return the cart, the supervisor will call or come to the ward, and you know from past experience that she takes a dim view of this kind of "inefficiency." If you leave

---

2. *M. Kramer,* Reality Shock: Why Nurses Leave Nursing *(St. Louis, Mo.: C. V. Mosby, 1974), p. 1.*

even for a minute, your intuition and judgment tell you that the climate will be broken and Mrs. Swape will clam up again.[3]

What should the nurse do? What is the *right* thing to do? Answering this question involves choosing between efficiency and patient care, between a time orientation and an activity orientation.

INTERPERSONAL RELATIONSHIPS.   Our culture also influences the way we interact with others. What should the prevailing relationship be among members of a group? Competitive or cooperative? Friendly or hostile? Human relationships vary from culture to culture. Indeed, they often vary within subgroups of the same culture. Our own culture orients individuals in both competitive and cooperative ways. For example, the education of health care professionals is highly competitive. Fierce competition for admission to programs and for job placement often spills over to job performance later. Once on the job, however, the HCP is expected to function cooperatively, even selflessly, as a member of a team devoted to patient welfare. In reality, competitive relationships—often destructively competitive—infect health care practice. HCPs have been known to guard their medical territories zealously, watchful of interlopers, resentful of intrusions, mindful of protocol—often at patient expense.

The point is that our culture predisposes us to certain widely divergent values with respect to human nature, the environment, time, activity, and human relationships. These values profoundly influence the HCP's interactions with patients, colleagues, institutions, other professions, and society at large. When these value orientations conflict, as they often do in delivering health care, HCPs can experience considerable inner turmoil. Although understanding the cultural roots of our values will not in itself resolve such conflicts, it will help us get a handle on them and help us establish priorities.

Science

Like culture, science influences value selection. This is especially true of those in health care whose professional training is dominated by science and scientific method. Indeed, it is not uncommon for HCPs to turn almost exclusively to science as a guide for action.

Although science does have the advantage of providing information that has been verified, it deals exclusively in facts, and facts alone do not provide meaning or direction. Only interpretation of fact does that. For example, genetic engineering is a scientific fact, as manifested in such techniques as cloning and artificial insemination and ovulation. But to what extent we should actually engage in scientific engineering is a value judgment that can be made only on the basis of a cluster of value assumptions about nature, human nature, and the kind of society we want. So, although science does provide us with information about ourselves and the world and with practical possibilities, it helps us make value judgments only insofar as we can and are willing to relate scientific information to value assumptions made on other bases.

---

3. *Kramer, p. 1.*

Another point about science. Because of their training, HCPs often value scientific method as the only reliable source of knowledge. *Scientific method is a way of investigating phenomena by collecting, analyzing, and interpreting evidence to determine the most probable explanation.* This method has proved invaluable in advancing knowledge about ourselves and the world, but it has distinct limitations, which those who use it daily are apt to overlook.

First, scientific method can grasp only as much as its tools allow. The more limited the scientific technique, the more speculative the theory. Thus medical theories, diagnoses, and treatment that rely on scientific method are only as effective as the medical technology that is used in determining them.

Then there is a second limitation: We can learn only what our scientific tools are designed to show. For example, if you thought you had a fever, an accurate thermometer would confirm or disprove your belief. But no matter how accurate the thermometer, it would not disclose whether you had pneumonia. Other instruments must do this. What is more, we have no tools that measure the will to live, including one's sense of purpose, meaning, or freedom—all of which may play a formidable role in the anatomy of an illness.

A third limitation concerns classification, a vital part of the scientific process. Classification refers to the process of grouping like things into subcategories. The rule of classification is to group by similar defining characteristics, so in effect, scientific laws gain validity only by ignoring some properties. In other words, scientific statements about phenomena, and the procedures for arriving at such statements, must be as isolated from any context as possible. Individually, however, entities consist of many more traits than those that determine grouping or laws. To ignore these differences is to ignore part of reality, perhaps a substantial part. Yet, there has been a tendency in medicine, albeit an understandable one, to classify patients according to case, syndrome, ailment, or some other handy pigeonhole. The person who is hospitalized may be referred to as "the cardiac in 423" or "the gallbladder in 108." The disease label can be so pronounced as to deny effectively the patient's personhood. The patient is no longer a person with a disease but a diseased person.

In some instances, classification through labeling can even function to deny patients desperately needed services. Take, for example, the experience of an HCP in public health who was trying to get a woman with a prenatal problem into a prenatal clinic. This woman really needed attention, but already she had missed a couple of appointments because she could not get any transportation to the clinic. At the woman's request, the HCP attempted to arrange for the transportation. This is what occurred, by the HCP's own account:

> [I found out] that this patient didn't fit into the right classification. They didn't have transportation services for those classified as prenatal. The poor woman just couldn't get transportation. Buses cost a lot and [she] lived a long ways away. We have a driver at the Health Department who drives TB patients, so I thought I would get this driver to go out and pick up this prenatal woman. Oftentimes I would see the driver just sitting on the bench with nothing to do. I made all arrangements and filled out all the forms and went and talked to the community worker who drives patients back and forth; and she said, "Sorry, I can't do it; I can only transport (a particular classification of) patients." The

driver was sitting there knitting, not doing a thing; but still she couldn't go out and pick up this woman. I explained that she really needs the transportation and "if you aren't busy" . . . "I'm sorry; they have to be in a certain category." I talked to everybody to see if I could get some help for this woman. The woman didn't wear the right label, so her need was just pushed aside and ignored.[4]

A fourth limitation of scientific method concerns science's ideological tendency not only to explain but to explain away. This results in the "nothing but" attitude of reductionism. *Reductionism* refers to the idea that a whole can be understood completely by analyzing its parts or that a developing process can be explained as the result of earlier, simpler stages. Thus color is "nothing but" light waves; heat is "nothing but" molecular motion; pain is "nothing but" a brain state. This tendency to reduce the whole to its parts encourages us to overlook characteristics of the whole that the parts alone may not reveal. For example, taken by themselves, hydrogen and oxygen gases do not show the same characteristics as when they are combined in the liquid state as water. Similarly, the heart, liver, and lungs taken alone do not show the characteristics of life that they show when functioning together in a body. As Gestalt psychology argues, and as high-energy physics recognizes, the whole is often greater than the sum of its parts. Missing this point, we might come to consider the parts more real than the unit, the condition of a patient more real than the individual patient.

Finally, the fifth limitation follows from the fact that neither science nor scientific method occurs in a vacuum. No matter how refined the instruments, no matter how advanced the computer, some person is still needed to make the observations, formulate the problem, collect the relevant data, establish a hypothesis, draw deductions from it, and finally to verify those deductions. And where there is a person, there is a viewpoint—a value bias bound to influence the investigations and findings in some way—and thus objective science can no longer be considered totally objective.

A good current example can be seen in the controversial May 27, 1980 report issued by the Food and Nutrition Board of the National Academy of Sciences, which found no definitive evidence that heart disease could be prevented by a reduction of cholesterol in the blood. The report also said that reducing fat intake is unnecessary, except for weight loss. Needless to say, such conclusions startled members of the medical community, who for years had linked heart disease to high levels of cholesterol and fat.

The effect was such that a group of nutritionists and consumer representatives, formed in 1974 to advise the Academy's board, severed its relationship with the board, claiming that the board's report was totally misleading. Dr. Kent Peterson, one of the group's members and executive vice-president of the American College of Preventive Medicine, said the board's report seemed "almost . . . cavalier about how healthy the American public is."[5] He further

---

4. *Kramer, p. 2.*

5. Los Angeles Times, *June 12, 1980, p. 11.*

pointed out that two-thirds of the people in the United States die from heart trouble and related illnesses. Peterson and other panel members cited numerous studies, including the monumental Framingham Heart Study, to support the claim that heart disease and fat intake are linked. The chairperson of the panel said that the cholesterol report was one of a series of projects in which panel input had been ignored.

Did the Food and Nutrition Board ignore relevant input? Was it biased in its interpretation of available data? Was it trying too hard to reassure a consuming public that is rapidly growing paranoid about what it breathes, drinks, eats, and does? Maybe so. On the other hand, maybe the panel was clinging tenaciously to a traditional theory, unwilling to relinquish an hypothesis even in the face of contradictory evidence.

The point is that scientific investigators are limited by their own physical and intellectual capabilities, by their presuppositions, and by their cultural biases. Moreover, while investigators shape a hypothesis, they are being shaped by it; the hypothesis determines further research as much as a road determines the path of a driver.

All of this is not intended as an indictment of science and its method. Rather, it is a caution that prudent scientists and practitioners of scientific method bring to their endeavors. Keeping these limitations in mind helps those who work in science to keep scientific knowledge in perspective and to keep themselves from slipping into scientism—the view that what science discloses comprises the total reality.

Religion

As customarily thought of in its institutionalized form, religion is based on "revelation" believed to be from God, as recorded in tradition and sacred literature. Religion is unquestionably a powerful source of values. Indeed, in the Western world there has probably been no influence greater than religion on people's view of the universe and humankind's place in it. Specifically, the Jewish and Christian traditions, to name just two, offer a view of humans as unique products of divine intervention, because they possess consciousness and an ability to love. Both of these traditions claim that we are creatures who stand midway between nature and spirit. On the one hand, we are finite, bound to earth, and capable of sin. On the other, we are able to transcend nature and achieve infinite possibilities.

Primarily because of the influence of Western religion, many in our culture view themselves as beings with a supernatural destiny, as possessing a life after death, as being immortal. One's purpose in life is found in serving and loving God. For the Christian, the way to serve and love God is by emulating the life of Jesus of Nazareth. In the life of Jesus, Christians find an expression of the highest virtue—love. They love when they perform selfless acts, develop a keen social-mindedness, and realize that people are creatures of God and are therefore worthwhile. For the Jew, one serves and loves God primarily through expressions of justice and righteousness. Jews also develop a sense of honor derived from a commitment to the ideals of truth, humility, fidelity, and kindness. This commitment produces a sharp sense of responsibility to family and community.

But religion has fostered beliefs, attitudes, feelings, and values not only about a supernatural dimension but about this world as well. Official religious positions concerning political, educational, economic, and medical questions are usually made known. These positions are influential in molding public opinion—as evidenced, for example, in the issues of abortion and euthanasia and in broad philosophies of health care.

Typically, then, religion involves not only a formal system of worship but prescriptions for social relationships. A good example of such a prescription is the mandate "Do unto others as you would have them do unto you." Termed the Golden Rule, this injunction represents one of humankind's highest value aspirations and can be found in essence in all the great religions of the world.

> Good people proceed while considering that what is best for others is best for themselves. (*Hitopadesa*, Hinduism)

> Thou shalt love thy neighbor as thyself. (*Leviticus* 19:18, Judaism)

> Therefore all things whatsoever ye would that men should do to you, do ye even so to them. (*Matthew* 7:12, Christianity)

> Hurt not others with that which pains yourself. (*Udanavarga* 5:18, Buddhism)

> What you do not want done to yourself, do not do to others. (*Analects* 15:23, Confucianism)

> No one of you is a believer until he loves for his brother what he loves for himself. (*Traditions*, Islam)

It is hard to imagine a profession that gives greater witness to the impact of religious beliefs on individuals than health care does. Faced with illness and grief and with loss of control, many people—patients and HCPs—understandably turn to religion for stability, comfort, and hope. Indeed, the mythical aspects of religion provide many with the meaning they seek in the otherwise inexplicable, even cruel realities of the world.

In religious anthropology, the term *myth* carries none of the popular connotations of something that is false or merely imagined. Rather, *myth* refers to a story that is circulated and accepted within some social group (for example, a clan, religious community, tribe, or nation) that gives significance and meaning to an event or explains some ultimate problem.[6] For example, the Genesis account of the Tower of Babel explains the multiplicity of languages that the Hebrews encountered in their travels. In a similar way, faced with serious health crises, religious people frequently rely on religious myths to explain the event's ultimate significance. Whether or not one agrees with the mythic formulations does not alter the fact that myths represent a continuing effort to interpret the

---

6. *J. L. Christian*, Philosophy: An Introduction to the Art of Wondering (*New York: Holt, Rinehart & Winston, 1977*), *pp. 601–604.*

world around us in meaningful ways. Myths underscore a simple but profound truth: Without meaning, humans cannot survive for long.

A moving example of the influence of religious myths is provided by anthropologist Hope Isaacs, who teaches in the graduate nursing program at the State University of New York at Buffalo. Isaacs recounts the case of a Tonawanda Indian woman with diabetes who was admitted to a local hospital with a gangrenous foot. Physicians explained to the tribal chief the need for amputating her lower leg. The chief interpreted the procedure for the woman, who did not speak English, and the woman's consent for amputation was obtained. The surgeons then removed the woman's foot. The amputation went smoothly, the stump healed well, and the woman was fitted with a prosthetic device that she learned to use. Things could not have proceeded better. But when ready for discharge, the patient asked for her leg. When it was explained that, as usual, the leg had been incinerated, the woman burst into tears and went into a deep depression. Why? Because she was of the religious belief that one must be buried with all one's parts in order to be resurrected. Little did the surgeons know that, in removing her foot, they were depriving the woman of her ultimate reward.

By overlooking or underestimating the influence of religion on patients and on themselves, HCPs ignore a vital component of the psychology of the ill and of those caring for the ill. Furthermore, if they remain ignorant of a patient's religious background and predilections, HCPs foreclose rich opportunities for providing care and may even risk injuring patients. Psychotherapist Victor Frankl, in his inspiring book *Man's Search for Meaning*, demonstrates graphically how understanding of and sensitivity to the religious main currents in a patient's life enable an HCP to provide hope and meaning within the context of health care. The conversation that follows is between narrator Frankl and a rabbi:

> He had lost his wife and their six children in the concentration camp of Auschwitz where they were gassed, and now it turned out that his second wife was sterile. I observed that procreation is not the only meaning of life, for then life in itself would become meaningless, and something that in itself is meaningless cannot be rendered meaningful merely by its perpetuation. However, the rabbi evaluated his plight as an orthodox Jew in terms of despair that there was no one of his own who would ever say *Kaddish* [prayer for the dead] for him after his death.
>
> But I would not give up. I made a last attempt to help him by inquiring whether he did not hope to see his children again in Heaven. However, my question was followed by an outburst of tears, and now the true reason of his despair came to the fore: he explained that his children, since they died as innocent martyrs, were found worthy of the highest place in Heaven, but as for himself he could not expect, as an old sinful man, to be assigned the same place. I did not give up but retorted, "Is it not conceivable, Rabbi, that precisely this was the meaning of your surviving your children; that you may be purified through these years of suffering, that finally you, too, though not innocent like your children, may *become* worthy of joining them in Heaven? Is it not written in the Psalms that God preserves all your tears? So, perhaps none of your

sufferings were in vain." For the first time in many years he found relief from his suffering, through the new point of view that I was able to open up to him.[7]

In discussing these four core sources of values, we have tried to suggest some ways that they connect to health care. Experience, culture, science, and religion are wellsprings from which flow the values that directly affect the attitudes we hold and act on in our personal and professional lives.

The subject of sources naturally leads to another: transmission of values. If we accept that experience, culture, science, and religion are powerful value sources, we must ask how their values are imparted. By what process do people internalize the values inherent in experience, culture, science, and religion?

Transmission of Values

We have noted that HCPs today must direct their lives through a professional world that can be full of confusion and conflict. How do they do this? To answer this question, it is necessary to ask a larger one: How does any person learn to direct his or her life through a world full of confusion and conflict? To answer this question, we need to examine the various ways that values are generally transmitted to individuals and groups. Sidney Simon, who has done considerable groundwork in the area of finding and setting value priorities, identifies three common ways in which values are transmitted: moralizing, modeling, and taking a laissez-faire attitude.

Moralizing

In *Values Clarification*, Simon describes moralizing as "the direct, although sometimes subtle, inculcation of the adult's values in the young."[8] The assumption behind this approach is simple enough: Experience is the best teacher. Thus a parent might say, "Experience has taught me a few things, a set of values, that would be right for you. To save you the trouble and pain of having to go through what I have, I'm going to transfer my own values to you."

Moralizing is often part of professional training. The reason is that the maintenance of any profession depends to a large extent on a solid, coherent set of values. In fact, one of the hallmarks of a professional organization—medical, educational, legal, political, or religious—is a value system that gives the group identity and that members are more or less expected to embrace.

Simon points out some problems with this approach to value transmission. One is that direct inculcation of values works best when there is complete consistency about what constitutes a desirable value. But moralizers often disagree about what is of value. It is not uncommon, for example, for parents to offer one set of values, religion another, schools still another, and peers and professions yet another. Then there are the individual moralizers we meet along

---

7. *V. E. Frankl,* Man's Search for Meaning *(New York: Beacon Press, 1963), pp. 189–190.*

8. *S. Simon,* Values Clarification *(New York: Hart Publishing, 1972), pp. 15–16.*

the road of value development: the individual religionists, teachers, counselors, friends, and professional associates. It is no wonder that many young people today are ultimately left to make their own choices. But when they have been raised by moralizing adults, they are in no position to make responsible choices on their own. As Simon puts it, "They have not learned a process for selecting the best and rejecting the worst elements contained in the various value systems which others have been urging them to follow."[9] As a result, they often make the important choices of their lives on the basis of peer pressure, blind acceptance of authority, or the power of propaganda.

Another problem with the direct inculcation of values is that this technique often results in a dichotomy between conceived and operational values. Conceived values are conceptions of the ideal. In general, they are the values the culture teaches. Some of our culture's conceived values are honesty, freedom, equality, justice, and service to others. An example of a conceived value in health care would be compassion for patients.

Conceived values may be espoused with intellectual conviction, but they frequently have little practical influence on our behavior. Thus HCPs who think they believe in showing compassion for patients may not be at all guided by this ideal in reality, even when it would be relatively easy to practice. An example would be the physicians who hurry patients through office visits in order to see as many as possible. In such cases a conflict arises between conceived values and operational values, which are the criteria or value assumptions according to which action choices are *actually* made.[10] What people actually do always reflects their operational values—but not necessarily their conceived ones. Physicians who "run patients through" reveal a commitment to efficiency. There are numerous examples of this phenomenon in the health care professions. The following poem illustrates the inconsistencies between conceived and operational values in nursing:

> I was hungry and could not feed myself.
> You left my food tray out of reach
>   on my bedside table,
> Then you discussed my nutritional needs
>   during a nursing conference.
>
> I was thirsty and helpless, but you forgot
>   to ask the attendant to
>   refill my water pitcher.
> You later charted that I refused fluids.
>
> I was lonely and afraid, but you left me alone
>   because I was so cooperative
>   and never asked for anything.

9.   *Simon, p. 16.*

10.   *Coleman, p. 489.*

I was in financial difficulties
  and in your mind I became an object
  of annoyance.

I was a nursing problem and you discussed
  the theoretical basis of my illness.
And you do not even see me.

I was thought to be dying and,
  thinking I could not hear, you said you hoped
  I would not die
  before it was time to finish your day
  because you had an appointment at the beauty parlor
  before your evening date.

You seem so well educated, well spoken,
  and so very neat in your spotless
  unwrinkled uniform.
But when I speak you seem to listen
  but do not hear me.

Help me, care about what happens to me,
I am so tired, so lonely and so very afraid.
Talk to me—reach out to me—take my hand.
Let what happens to me matter to you.

Please, nurse, listen.[11]

Sometimes the discrepancy between the individual's conceived and operational values suggests an alarming schism between the "ideal" and "real" self. Sensitive HCPs can be torn by the inner conflict that results from recognizing that what they evidently *must* do deviates from what they believe they *should* do. The personal torment that results cannot be understated, and remaining ignorant of this phenomenon courts hypocrisy and risks damage to patients and the profession, as the poem above illustrates.

Modeling

The modeling approach to instilling values sets up the adult as the paradigm of behavior. As Simon points out, the rationale here is "I will present myself as an attractive model who lives by a certain set of values. The young people with whom I come in contact will be duly impressed by me and by my values and will want to adopt and emulate my attitudes and behavior."[12] This approach recognizes the importance of a living example in learning and the need for actions to match words in teaching values.

---

11. From R. Johnston, "Listen, Nurse." Copyright ©1971, *The American Journal of Nursing Company. Reproduced, with permission, from* American Journal of Nursing, *February 1971, Vol. 71, No. 2.*

12. *Simon, p. 18.*

But as with moralizing, young people are exposed to many different models. Parents, teachers, religionists, professional athletes, celebrities—all present different models. Which one, if any, is the young person to emulate? Frequently one model's values clash with those of another. Whose example should the person follow? Answering this question can be particularly difficult for HCPs fresh out of training, who must often reconcile the model of conceived values with the new model of operational values—the "ivory tower" model pitted against the "real world" model.

For example, consider one young nurse's response to the case of Mrs. Swape which was presented earlier:

> I know what my instructor would tell me to do! Stay with Mrs. Swape. Obviously! Your concern at this moment is about her and her total needs, so stay with her and do a good job of listening to her and seeing what you can do to get her to air how she's feeling and perhaps begin to relax and mend. That would be how my instructor would see it, and me too. That's how I was when I graduated and really how I'd like to be able to do it now. . . . The thing that really burns me up is that when you take a problem like this to your supervisor, or even when I took it to the Director of Nursing, they just shrug. They don't see it; they can't seem to understand why I'm so upset about it. They see this just as something that happens only every once in awhile; they happen over and over again every day. But these little things are what's really nursing, and the constant frustration over them really eats at me. What would I do? I hate to say it, but I've been worn down. When I graduated, I would have stayed and listened to Mrs. Swape. Now? I'd probably go ahead and serve the trays. That's what's important in terms of keeping your job and getting good performance ratings. Those are really the only two choices I see.[13]

Taking a Laissez-faire Attitude

The term *laissez-faire* is synonymous with little or no interference. The laissez-faire approach to value transmission holds that adults should not intervene with the value development of young people. The rationale is that no one value system is right for everyone, that individuals have to formulate their own values. So let young people think and do essentially what they want, and eventually they will discover what is "right" for them.

But the laissez-faire approach has the very same weaknesses as moralizing and modeling. In assuming a "hands-off" position, the practitioner of the laissez-faire attitude offers nothing to help individuals sort out, select, or reject the values of the models and the moralizers that they will inevitably encounter.

Actually, all three approaches assume that there are only two ways to teach values: either to inculcate predetermined values or do nothing. But why not a middle road that helps people to formulate their own values? Simon believes that such an alternative can be found in an approach termed *values clarification*. Because this technique for finding and setting priorities has already received attention in many health fields, it deserves close examination.

---

13. *Kramer, p. 3.*

Values Clarification and the Process of Valuing

Values clarification is an approach to values transmission that attempts to help individuals sort through, analyze, and set the priorities of their own values. It also tries to help people begin to act in a way that is consistent with those values.

Certainly values clarification is not entirely a new approach. Over two thousand years ago, Socrates (469–399 BC) practiced a version of it in his dialectical method—the *Socratic method* of teaching—which used probing questions and incisive argumentation to elucidate the obvious. Through the ages, there have always been those who have tried to teach people to think on their own. The approach known as values clarification was formulated rather recently by Louis Raths, who in turn built upon the thought of the American philosopher John Dewey (1859–1952).

Unlike the other approaches to values, especially moralizing and modeling, values clarification is not concerned with the context of people's values but with the process of valuing. Because valuing as a process figures so prominently in values clarification, it deserves close inspection. Although it is impossible to say precisely what valuing consists of, much of the research in the area indicates that valuing is composed of several subprocesses. Raths, Simon, and Merrill provide a useful description: *"When we value something we choose freely from alternatives after consideration of the consequences of each alternative; are proud of and happy with the choice; are willing to affirm the choice publicly; and make the choice part of our behavior and repeat the choice."*[14] The description recognizes seven subprocesses of valuing that can be grouped under three main headings, as follows:

Choosing

1. Choosing freely
2. Selecting from alternatives
3. Choosing after careful consideration of the consequences of each alternative

Prizing

4. Being proud of and happy with the choice
5. Being willing to affirm the choice publicly

Acting

6. Making the choice part of one's behavior
7. Repeating the choice

According to Raths, Simon, and Merrill, only if all seven steps of this valuing process are satisfied does a belief or feeling become a value.

---

14. *L. Raths, S. Simon, and H. Merrill,* Values and Teaching *(Columbus, Ohio: Charles E. Merrill, 1966), p. 27.*

Choosing

Nobody would dispute that a value must be chosen freely. Self-determination in value selection is a point on which all value theorists agree. Assessments of worth that are imposed on us or that we accept blindly are not our own values but someone else's. Accepting other people's values, we end up living as they would want us to live but not necessarily as we would choose to live if we gave the matter thought. And yet, this is precisely what we sometimes do; we unquestioningly accept others' values, only to find that they are inappropriate to our own circumstances and life goals.

Psychotherapist Carl Rogers has found that many of those who seek therapy have, consciously or unconsciously, followed external value judgments, ignoring or denying their own perceptions of those values. As a result, most of these individuals' values are learned from other people or groups significant to them. Nevertheless, they regard these values as their own. In addition, the locus of evaluation is outside these individuals. In effect, they are "programmed"—they follow uncritically the input they have received from others. Worse yet, many of the accepted values and assumptions are questionable, even unsound. They are often contradictory or conflict with evidence supplied by one's own experience. Thus an HCP may calmly discuss the obligation to keep patients alive by extraordinary measures if necessary but agonize over the actual suffering of a terminally ill patient.

Also, when value assumptions have been accepted uncritically, they tend to be asserted "come hell or high water." They are not readily subject to reassessment or change. When discrepancies arise, they tend to be denied or rationalized away.

Finally, when individuals learn to accept external values unthinkingly, they generally end up feeling insecure and easily threatened, because they do not trust their own valuing process to resolve discrepancies and formulate a coherent, meaningful value system. Not seeing any reliable alternatives, they cling to the contradictions and confusing value systems. And they pay the price: the loss of contact with their own inner process of evaluation, together with the loss of potential wisdom. The result can be estrangement, not only from themselves but from their professions.

Implied in the process of choosing values are certain cognitive skills—skills that relate to the consciousness function, termed *thinking*. To choose values in the context of freedom, alternatives, and consequences, one must be able and willing to think on many levels: the critical, the creative, the logical, and the moral. These dimensions of thinking are crucial to choosing values and must be cultivated.

Prizing

Prizing or cherishing is a continuous process that represents an ever-increasing recognition of one's feelings. It can be associated with what some psychologists have termed "becoming an authentic person." For example, one psychologist describes the authentic person as the individual whose example is perhaps beyond the reach of most of us: the individual who is free and who knows it, who knows that every deed and word is a choice and hence an act of value creation, and, finally and perhaps decisively, who knows that he is the author of his own life and must be held personally responsible for the values on behalf of which he has chosen to live it, and that these values can never be justified by referring to something or somebody outside himself.[15]

Similarly, others, such as humanist psychologist Abraham Maslow, have spoken of the authentic person as one who has integrity, who has thought through his or her own values, and who cherishes and lives by them.

Although the basic theme in prizing is commitment to being true to oneself, a second theme is commitment to others. Thus the fifth subprocess is a willingness to affirm one's choice publicly.

Commitment to others follows almost automatically from commitment to oneself, for there seems to be a basic unity to humanity such that learning how to live constructively entails involvement with, and obligations and commitments to, one's fellow human beings. Rogers puts it this way:

> I believe that when the human being is inwardly free to choose whatever he deeply values, he tends to value the objects, experiences and goals which make for his own survival, growth and development, and for the survival and development of others.[16]

The commitment to others, which follows from a commitment to self, seems to be the thrust of Raths' "willingness to affirm the choice publicly"—that is, to share it with others, to recognize when it is appropriate to reveal ourselves, and then to have the courage to do so.

Acting

The final two subprocesses in Raths' scheme are action-oriented. They recognize that choosing and prizing do not of themselves constitute valuing. Choosing and prizing do provide and deepen insight, but without action they are incomplete; without behavioral change, any insight is useless. After all, what is the point of choosing and prizing the values of adequate health care, correct health care practice, patient rights, and professional autonomy if these choices are not reflected in behavior? Action means that we have incorporated the value by doing something tangible about what we say we value.

Corollary to this is that we act on values consistently and repeatedly. This does not mean that we follow a foolish consistency, that we become rigid and calcified in our values. Ultimately values emerge from experience, and experience is intrinsically dynamic. But it does mean that our actions should consistently be compatible with perceptions of worth.

---

15. *V. C. Morris,* Existentialism in Education *(New York: Harper & Row, 1966), p. 48.*

16. *C. Rogers, "Toward a Modern Approach to Values: The Valuing Process in the Mature Person,"* Journal of Abnormal and Social Psychology *68 (1964): 166.*

Values Clarification and the
Health Care Professional

Undoubtedly many of our personal and professional beliefs, feelings, aspirations, and goals do not accord with the seven subprocesses of valuing. Raths terms such beliefs *value indicators*. It is important that we distinguish value indicators from values if we are to (1) preclude self-deception, (2) determine our genuine values, (3) rank values, and (4) identify what is required of us before we can correctly be said to value something. Used conscientiously, the seven subprocesses can direct this course. One HCP puts it this way: "For me the steps of the valuing process serve as a guide for change. I check an attitude or thought with these seven steps and ask myself what steps I need to consider more carefully so that this belief can become a value for my behavior."[17] The strengths of one's beliefs and values, then, can be analyzed in terms of the seven subprocesses of valuing by using one of a number of values clarification strategies. "Experience with these strategies," writes another HCP, "can lead us to self-examination and personal growth. Values clarification allows us to methodically sort out our feelings, assess their importance, and decide if we should act on our belief."[18]

In addition to using the values clarification approach to consider personal and professional issues and concerns, HCPs can use similar strategies to help patients deal with the dizzying array of emotions they feel when faced with disruptions of normal coping patterns caused by illness, surgery, hospital stays, and so on. Values clarification can help patients to (1) identify conflict areas; (2) isolate, evaluate, and choose alternatives; (3) set goals; and (4) act.

Helping patients in these ways is important because HCPs, generally speaking, have been unsuccessful in influencing patients' health behavior so that a higher degree of wellness can be achieved and maintained. Facts are taught, concepts developed, and sophisticated instructional material used in an attempt to change patient health behavior. But more often than not, such efforts have fallen short of the mark. Why does patient health behavior remain unchanged? Diane B. Uustal, registered nurse thinks: "It may be that we have not taken the most important first step, that of finding out what the *patient* values and wants to know first and then teaching those concepts and that information first."[19] Her point is well taken. Often there is a gap between what patients see as important to learn and what HCPs view as important to impart. Progress in changing patient health behavior, then, begins with examining and identifying our own values and recognizing how they color the teaching and information shared with patients.

Equally important, HCPs need to identify those concepts that patients re-

17. D. B. Uustal, "Values Clarification in Nursing: Application to Practice," American Journal of Nursing, *December 1978, p. 2061.*

18. S. Smith Coletta, "Values Clarification in Nursing: Why?" American Journal of Nursing, *December 1978, p. 2057.*

19. Uustal, p. 2062.

gard as important and to teach these in addition to the ones that HCPs think are important. Unless HCPs do this, they risk imposing their own values on others, rather than working within patients' value systems. In short, patient compliance with medical regimens ultimately depends on two things: (1) patient awareness of the importance of the regimen to them and (2) patients' valuing the change the regimen is likely to produce. For example, a blizzard of statistics about the hazards of smoking will not get cigarette smokers to stop unless they value the changes in health that occur when one stops smoking and give these higher priority than the value they see in smoking.

How can HCPs help people to value high-level wellness? There is no easy answer to this question, but it is fair to surmise that HCPs who are comfortable with the valuing process will stand a better chance of answering this question than those who are not familiar with it. HCPs who use the valuing process are in a position to help patients examine alternatives and the consequences of each and to help select a path of value behavior that is consistent with identified values. "As a result, positive behaviors are more likely to occur because the patient is in charge of the decision-making process and has chosen values based on options presented by health team members."[20]

But simple awareness of the valuing process is insufficient to clarify values, give them priority, and suggest courses of action. What is needed are practical ways for incorporating the seven valuing subprocesses into our lives and for helping others to do the same. Those working in the area of values clarification have developed some strategies to attain these ends. The values clarification approach does not aim to instill any particular set of values. Rather, the goal of values clarification is to help individuals use the seven subprocesses of valuing in their personal and professional lives.

In his book, Simon provides seventy-nine specific, practical strategies to help us build the valuing process into our lives. These strategies have several features in common. First, they try to help individuals identify beliefs and behaviors that they prize and would be willing to stand up for, in and out of the classroom. Second, they encourage individuals to consider alternative modes of thinking and acting and to weigh the pros and cons and the consequences of various alternatives. Third, they attempt to help individuals consider whether their actions match their stated beliefs and, if not, how to harmonize the two.

Regarding these strategies, Simon has written:

> The small amount of empirical research that has been done of the values-clarification approach, and the large amount of practical experience with this approach by thousands of teachers indicates that students who have been exposed to this approach have become less apathetic, less flighty, less conforming, as well as less over-dissenting. They are more zestful and energetic, more critical in their thinking, and more likely to follow through on their decisions. In the case of under-achievers, values clarification has led to better success in school.[21]

---

20. *Uustal, p. 2063.*

21. *Simon, pp. 20–21.*

Here is a simple example of a strategy exercise that Uustal has adapted from Simon's collection of strategies. It encourages you to think about yourself, on the premise that we rarely take time to ask introspective questions about our directions, characteristics, goals, and values.

Name Tag

Take a piece of paper and write your name in the middle of it. In each of the four corners write your responses to these four questions:

What two things would you like your colleagues to say about you?

What single most important thing do (or would) you do to make your patient relationships positive ones?

What do you do on a daily basis that indicates you value your health?

What are three values you believe in most strongly?

In the space around your name write at least six adjectives that you feel best describe who you are. Take a closer look at your responses to the questions and to the ways in which you described yourself. What have you discovered about yourself? Were any of the questions especially difficult to answer? What values are reflected in the answers you gave? What additional questions would you ask?[22]

A Critique of Values Clarification

Before leaving this discussion, we should consider some cautions about the values clarification approach. These comments are in no way intended to undercut the preceding exposition. Indeed, it is a basic assumption of this book that the values clarification approach has considerable merit. This view is reinforced by the experiences of numerous HCPs who have attempted to implement it with varying degrees of success. At the same time, values clarification alone is inadequate to help us determine what to value and how to behave. It does not provide a basis for value selection or for moral choice.

To illustrate, suppose that in answer to the question "What are the three values you believe in most strongly?" in the preceding Name Tag exercise, you had answered, "autonomy, honesty, and loyalty." Suppose further that someone asks you "Why is loyalty to be preferred?" or "Why should a person value autonomy?" or "What makes loyalty worthwhile?" There is nothing in values clarification to help you respond. The reason is that values clarification only *describes* the value process. In other words, it is nonnormative: It does not provide any standards, principles, or norms on the basis of which you can make value judgments. Values clarification is descriptive of the process involved in valuing. As such, it can be very useful in illuminating what we actually do value and, thus, can help to initiate a critical analysis of those values. However, it does not offer any normative theory of values that may be used as a basis for value selection and priority.

---

22. *Uustal, p. 2063.*

Nor does values clarification offer any normative theory of obligation, which would provide some standard, principle, or norm as a basis for deciding what to do. To see the inadequacy of values clarification in this regard, reconsider the case described at the start of this chapter. A woman, Rita Drury, refuses to give permission to have an autopsy conducted on her husband. The staff continues to pressure her for it. Were you one of the health team members, what if anything would you do? (1) Do nothing; (2) intervene on Rita's behalf; (3) support the health care team; or (4) do nothing at the time but report the entire incident later? Values clarification might help you determine what you probably *would* do, but it cannot help you determine what you *should* or *ought* to do. It is precisely this difference between *would* and *should* that accounts for the difference between nonnormative ethics, such as values clarification, and normative ethics. The former describes; the latter prescribes. Thus questions like "What *should* I value?" and "How *ought* I behave?" require a full-fledged investigation into the theory of moral value and obligation.

Discovering what in fact we do value is important as the beginning of a searching examination of values. Therefore, values clarification exercises are used throughout this book. But the discovery of our values needs to be supplemented with criteria that must come out of, or be justified by, moral theory. Otherwise, the study of health care ethics remains a ship without a rudder. This is why a good deal of our study must also include normative theory.

Summary

This chapter began by noting the profound changes occurring in health care, in the midst of which HCPs are often uncertain about the values they hold and the priority to give them. Value was defined as an assessment of worth and axiology as the study of the general theory of value—including the origin, nature, classification, and place of values in the world. A number of concerns recur in the study of values: (1) the distinction between values and facts; (2) subjectivity versus objectivity in values; and (3) value selection.

Although there is little philosophical agreement about the nature of values, their relative importance, or their relationship to one another, most philosophers agree that individuals should (1) prefer what is of intrinsic value to what is of extrinsic value; (2) prefer values that are productive and lasting to ones that are not; (3) choose their own values according to their own goals and ideals; and (4) always prefer the greater of two values. Personal experience, culture, science, and religion are primary sources of values and influence not only value content but the very approach we take to determining what is of value.

Values can be transmitted through various attitudes: moralizing, modeling, and laissez-faire. Another way that has received considerable attention in the health sciences is values clarification, which attempts to help individuals sort through, analyze, and give priority to their values. Values clarification is concerned with the process of valuing; according to values clarification theorists, this process consists of choosing, prizing, and acting. The values clarification approach is useful in helping individuals discover their values and in initiating a critical examination of these values, but it is strictly nonnormative. It describes but does not prescribe. Values clarification tells us what in fact we do value, but it does not tell us what we should value or should do. Of course, technically speaking, there is a kind of metanormative view implicit in values clarification. If one values X highly, presumably one *should* choose, prize, and act on X over other alternatives all or most of the time. But values clarification is still not normative. Rather, it is part of the analytical meaning of valuing X. Thus, in the final analysis, values clarification needs to be supplemented with criteria that must come out of, or be justified by, normative theory.

Exercise 1

The following values clarification strategy can be termed the Preference Strategy. It allows you to state your opinion on many current issues in health care. This strategy is useful because it allows you to make a choice and then to examine the feelings that influenced your choice. You can enhance the worth of this exercise by sharing your responses with others, especially with those of a dissimilar moral orientation. In discussions, maintain an open mind; work to create a relaxed climate in which others feel free to express their values.[23]

To the left of each statement, place the number which *best* explains your present position on the statement:

1	2	3	4	5
I mostly agree	I somewhat agree	I'm ambivalent or neutral	I somewhat disagree	I mostly disagree

_____ 1. Infants with severe handicaps ought to be left to die.
_____ 2. Extraordinary medical treatment is always indicated.
_____ 3. My role as [an HCP] is to always give resuscitation to clients who could benefit from it, no matter what has been decided previously.
_____ 4. I must follow physicians' orders without question.
_____ 5. Older patients ought to be allowed to die with dignity.
_____ 6. Medical technology has advanced the quality of life.
_____ 7. Children ought not to be involved in giving consent for treatments.
_____ 8. Families or significant others ought to make the decisions about life or death situations without involving the client.
_____ 9. Children ought to participate in human experimentation that is not harmful, even if it has no benefit to them.
_____ 10. Prisoners ought to participate in scientific experiments to repay society for their wrongdoings.

23. *This exercise has been taken from S. M. Steele and V. M. Harmon,* Values Clarification in Nursing *(New York: Appleton-Century-Crofts, 1979), pp. 70–74. Reprinted by permission. Brackets indicate adapted material.*

_____ 11. Vasectomy is the safest and best type of sterilization.

_____ 12. Adults with developmental lag ought to be sterilized.

_____ 13. Women ought to seek medical supervision from female physicians to avoid potential discriminatory practices.

_____ 14. Children whose parents refuse to have them receive medical care ought to be removed from their families through court action.

_____ 15. Research using fetuses ought to be vigorously pursued.

_____ 16. Abortion is the right of the woman and should be decided by collaboration between her and her physician.

_____ 17. Life support systems ought to be discontinued after several days of a flat electroencephalogram.

_____ 18. Health professionals are a scarce resource in many parts of the country.

_____ 19. Nursing is a subservient profession, especially to the medical profession.

_____ 20. As a [member of a health care team], I must relinquish my personal philosophy to support the philosophies of others.

_____ 21. All clients, regardless of differences, ought to be treated in a humanistic way.

_____ 22. I ought to give a mouth-to-mouth resuscitation to a derelict if he needs it.

_____ 23. A child who is disabled has value.

_____ 24. All forms of human life have value.

_____ 25. I ought to be involved in decision-making regarding ethical issues in practice.

_____ 26. Committees should decide who receives scarce resources, such as kidneys.

_____ 27. Clients' individual rights ought to be more important than the rights of society-at-large.

_____ 28. A person has the right to make a Living Will.[24]

_____ 29. Women of childbearing age ought to be sterilized after two pregnancies to maintain zero population growth.

_____ 30. Underdeveloped countries ought to be given health and financial support from developed countries.

_____ 31. I should support all the positions on ethical questions taken by my professional organizations.

_____ 32. I should aggressively support my own values when they conflict with values of others.

_____ 33. Consideration of the cultural values of clients is a waste of time.

_____ 34. The *care* component of nursing practice is not as important as the *cure* component of medical practice.

---

24. *A living will is a document or directive by which individuals inform others of the kind of treatment they wish to receive if and when they become so seriously ill that they can't communicate to others how they wish to be treated. In some cases, a living will allows individuals to request that they not be kept alive by artificial means or "heroic measures."*

_____ 35. The [HCP's] primary role in decision-making on ethical issues is to implement the selected alternative.

_____ 36. I feel afraid when caring for a client who is dying.

_____ 37. Children who have disabilities ought to be institutionalized.

_____ 38. Clients in mental health institutions and prisons ought to be given behavior modification therapy to make them conform to society.

_____ 39. Personal possessions of clients ought to be removed to guarantee safekeeping during hospitalization.

_____ 40. Clients ought to have access to their own health information.

_____ 41. Withholding health information fosters the client's recovery.

_____ 42. A client with kidney failure is always able to get kidney dialysis when needed.

_____ 43. Society ought to bear the cost of extraordinary medical interventions.

_____ 44. Confidentiality is an important part of the [HCP's] role.

_____ 45. As [an HCP], I ought to value responsibility.

_____ 46. Homosexuality ought to be discouraged.

_____ 47. [HCPs] have a right to withhold information to facilitate nursing research on human subjects.

_____ 48. The client who refuses treatment ought to be dropped from the health supervision of an agency or a professional.

_____ 49. Sexually active adolescents ought to be encouraged to use contraceptives.

_____ 50. Transplantations ought to be done whenever needed.

After completing all the statements, add up the number of 1s, 2s, 3s, 4s and 5s that you have. How many statements do you have clear ideas about? _____

Do these outweigh the number of ambivalent or neutral   Yes   No
statements you have?

Do the statements you agree with (include "mostly" and "somewhat") outweigh the statements you disagree with (include "mostly" and "somewhat")?

Look at the questions you "mostly disagree" with. Do you see any relationship between the statements which influenced your responses (e.g., age of client, severity of condition, etc.)?

Look at the questions you mostly agree with. Do you see any relationship between these statements which influenced your responses?

Now go back and look at the way you rated the particular clusters of statements identified below. Do you see any consistency in the way you rated these statements due to such variables as age, sex, etc.?

Try to think why you might be consistent or inconsistent in the way you rate the statements. Statements 5, 8, 16, 17, 28, and 36 relate to issues pertaining to *death*. Do you see any consistency in the way you rated these statements? What variable(s) influenced your decisions?

Statements 11, 12, 16, 29, 46, and 49 relate to *human sexuality and reproductive issues*. Do you see any consistency in the way you rated these statements? What variable(s) influenced your decision?

Statements 3, 4, 19, 20, 25, 31, 35, 44, and 45 relate to [*HCPs and the health care profession*]. Do you see any consistency in the way you rated these statements? What variable(s) influenced your decision?

Statements 2, 6, 15, 17, 24, 42, 43, and 50 relate to the issues raised by *advanced medical technology*. Do you see any consistency in the way you rated these statements? What variable(s) influenced your decision?

Statements 1, 7, 9, 14, 23, 37, and 49 relate to *children*. Do you see any consistency in the way you rated these statements? What variable(s) influenced your decision?

Statements 9, 10, 15, and 47 relate to *human experimentation*. Do you see any consistency in the way you rated these statements? What variable(s) influenced your decision?

Statements 3, 7, 8, 13, 14, 21, 22, 27, 28, 33, 39, 40, 41, 44, and 48 relate to *rights of clients*. Do you see any consistency in the way you rated these statements? What variable(s) influenced your decision?

Statements 9, 10, 27, 29, 30, 32, and 43 relate to the *rights of society*. Do you see any consistency in the way you rated these statements? What variable(s) influenced your decision?

Statements 18, 26, 40, and 42 relate to the issue of *scarce resources*. Do you see any consistency in the way you rated these statements? What variable(s) influenced your decision?

Statements 3, 4, 20, 21, 22, 25, 26, 31, 32, 35, 39, and 45 relate to your perception of what you feel are *obligations* in certain circumstances. Do you see any consistency in the way you rated these statements? What variable(s) influenced your decision?

What have you learned about yourself from completing this exercise? Was it easy to stick to your decision after discussing the choices with others?

Exercise 2

Reconsider your answers to the Preference Strategy questionnaire in the light of the various sources of values. Can you see connections between your choices and your experience, culture, science, and religion?

Exercise 3

After completing the Preference Strategy exercise, suppose that someone remarked, "Doing this exercise was useful in helping me see what I value and in getting me to think about those values, but it didn't help me discover what, if anything, I ought to value or what, if anything, I ought to do in a given situation." (1) Explain what the person means. (2) Do you agree? (3) How would you advise the person?

For Further Reading

Fish, S., Fish, S., and Allen, J. *Spiritual Care: The Nurse's Role.* Downers Grove, Ill.: Inter Varsity Press, 1978.

Fried, C. *An Anatomy of Values.* Cambridge, Mass.: Harvard University Press, 1970.

Fromm, E. *Man For Himself: An Inquiry into the Psychology of Ethics.* New York: Holt, Rinehart & Winston, 1977.

Herman, S. *Becoming Assertive: A Guide for Nurses.* New York: Van Nostrand, 1978.

Jourard, S. M. *The Transparent Self.* New York: Van Nostrand, 1964.

Kirschenbaum, H. *Advanced Values Clarification.* La Jolla, Calif.: University Associates, 1977.

Kluckhorn, F., and Strodtbeck, F. *Variations in Value Orientation.* Elmhurst, Ill.: Row, Peterson, 1961.

Lair, J. *I Ain't Well—But I Sure Am Better.* Greenwich, Conn.: Fawcett Crest, 1975.

Maslow, A. *Toward a Psychology of Being.* New York: Van Nostrand Reinhold, 1968.

Maslow, A., ed. *New Knowledge in Human Values.* Chicago: Henry Regnery, 1970.

May, R. *Man's Search for Himself.* New York: W. W. Norton, 1953.

Moustakas, C. E. *Turning Points.* Englewood Cliffs, N.J.: Prentice-Hall, 1977.

Read, D., Simon, S. B., and Goodman, J. *Health Education: The Search for Values.* Englewood Cliffs, N.J.: Prentice-Hall, 1977.

Read, D., Simon, S. B., and Goodman, J. *Looking In: Exploring One's Personal Health Values.* Englewood Cliffs, N.J.: Prentice-Hall, 1977.

Rogers, C. *Freedom to Learn.* Columbus, Ohio: Charles E. Merrill, 1969.

Rokeach, M. *The Nature of Human Values.* New York: The Free Press, 1973.

Simon, S., and Clark, J. *More Values Clarification.* San Diego: Pennant Press, 1975.

Steele, S. M., and Harmon, V. M. *Values Clarification in Nursing.* New York: Appleton-Century-Crofts, 1979.

Tiselius, A., and Nilsson, S., eds. *The Place of Value in a World of Facts.* New York: Wiley, 1970.

3
LAW, CODES, AND ETHICAL THEORIES

Harry and Vilma Armand are both twenty-three, married, and have two children, one a twenty-two-month-old boy, the other a girl of six months. The Armands live in a two-bedroom walkup in a depressed area of the city. They hope to own their own home some day—when Harry has a decent job. Lacking a skill and any higher education, Harry has little chance of landing something "real good." In fact, he spends most of his time looking for work. Vilma is employed parttime in the classified advertising department of the local newspaper.

Vilma does not particularly enjoy being a mother. "I think I'm just too young for this sort of thing," she has told public health nurse Rhona Steadman more than once. Rhona has made followup visits since the birth of the Armands' second child.

Vilma admits not liking "some outsider snooping around her place," but she has no choice. The hospital required such visits before the baby was permitted to go home. Actually, now that Vilma has come to know Rhona, she is more inclined to open up with her, to talk about the hostility she feels as a mother.

One day during a visit, she says to Rhona, "I don't know how to put this exactly . . . I mean I'm not even sure or anything . . . but I think Harry's beat up on Robbie."

Robbie is the Armands' baby boy. Rhona did notice some bruises on him during her last visit, but Harry—the parent home that time—told her that the boy had fallen, and Rhona did not press the issue.

"Don't get me wrong," Vilma tells Rhona. "Harry's a good father. He loves his kids. It's just that he's having . . . well . . . a hard time right now, and when the kid cries—I mean he cries an awful lot—well, I guess Harry just sort of takes it out on him. I mean he isn't any childbeater, is what I'm trying to say."

Rhona wonders if she should report this suspected case of child abuse. If she does, she is certain she will lose Vilma's confidence and trust and will undo all the progress she has made. On the other hand, if she does not, who knows what might happen to little Robbie?

The decision Rhona faces is a moral one. Whatever she decides will reflect a value that she holds. Reporting the case would reflect a greater commitment to noninjury. Failing to report would reflect an overriding commitment to patient confidentiality. The priority that Rhona gives to these or other values will figure prominently in her decisions. Furthermore, her personal experience, culture, religion, and other value sources probably will influence the relative importance she gives to each value.

These reflections describe the process by which Rhona will arrive at her decision and thus may help answer the question "What will Rhona do?" But they do not address the question of what she ought to do. They do not prescribe what her duty or responsibility is or what she must do to be considered moral—and avoid doing lest she be considered immoral.

There are many ways to resolve moral dilemmas. For HCPs, two of the most popular are reliance on law or on professional codes of conduct. Taking the legal approach, Rhona might reason this way: "There's a state law that requires me to report suspected cases of child abuse. I must abide by the law. So the right thing for me to do is to report this case." In this instance, Rhona has equated what is legally right with what is morally right.

As an alternative, Rhona might rely more on her professional code of conduct for direction. It might obligate her to consider not only law but also client confidentiality and physical jeopardy to clients. Using this standard, her duty probably is not so clear-cut. She must weigh the relevant principles of her code and decide which takes priority.

Is the law or a professional code a satisfactory guide for moral conduct? Is either sufficient by itself to determine what is the morally right thing to do in a health care situation? This chapter looks at this question and suggests that the answer is no. Although both are useful, neither law nor professional code is adequate by itself as an ethical norm. Moral principles of an ethical theory are needed. This chapter examines what are considered the most important and influential ethical theories in the history of Western philosophy. Before beginning to study these theories, however, try responding to the following questions, which are intended to start you thinking about your values and priorities as they apply to ethical theory.

Values and Priorities

1. If I were Rhona Steadman, I probably *would* _____.

2. If I were Rhona, I probably *should* _____.

3. The thing Rhona ought to think about most in deciding what to do is _____.

4. The thing I think about most in deciding what to do is _____.

5. What I do sometimes doesn't correspond with what I think I should do: agree/disagree. For example, _____.

6. If I were Rhona, I'd probably go along with what my supervisor told me to do: agree/disagree.

7. The main reason that I don't always do what I ought to do is _____.

8. I usually/rarely/never know what I *ought* to do.

9. When I don't know what I ought to do, it's usually because _____.

10. I think that people who don't do what they believe is right are

_____.

11. I think Rhona should do what she thinks is in her own best interests: agree/disagree.

12. The law is always a reliable moral guide for HCPs to follow: agree/disagree.

13. I don't think it's ever right for an HCP willfully to break the law: agree/disagree.

14. I doubt that an atheist can ever act morally.

15. Sometimes lying is right; other times it isn't: agree/disagree.

16. Some things are always wrong: agree/disagree. For example,

_____.

17. If HCPs took their professional codes and oaths seriously, they would have little difficulty determining what they ought to do: agree/disagree.

18. Things are right or wrong according to the situation: agree/disagree.

19. I don't think the right thing to do is always the thing that promises the best consequences: agree/disagree. For example, _____.

20. One principle that I always try to keep is _____.

Law

As you saw in the story about Rhona Steadman, law sometimes serves as an ethical standard and a source of values for moral decisions in health care situations. Before assessing the relationship between legality and morality, it might be useful to distinguish the various kinds of law. Basically there are four kinds of law: statutes, regulations, common law, and constitutional law.

Statutes are laws enacted by legislative bodies. The law that obligates Rhona to report a suspected case of child abuse is a statute. So is the law that prohibits touching other people without their consent (battery). Statutes can be enacted by Congress or state legislatures. (Laws enacted by local governing bodies, such as city councils, generally are termed ordinances.) Without doubt, statutes constitute a substantial part of the law and are what most people have in mind when they speak of laws. But there are other important sources of law, too.

Administrative regulations are a second source. Given limitations on their expertise, legislative bodies often establish boards or agencies, one of whose functions is to issue detailed regulations of certain kinds of conduct. For example, state legislatures set up physician and nurse licensure boards. These boards promulgate regulations for the licensing and professional conduct of physicians and nurses. As long as these regulations do not exceed the board's statutory powers and do not conflict with other kinds of law, they are legally binding on those whose qualifications and conduct they are intended to regulate.

Common law refers to law applied in the English-speaking world before there were any statutes. Courts frequently wrote opinions explaining the basis of their decision in specific cases, including the legal principle they deemed appropriate. Each of these opinions became a precedent for subsequent decisions in similar cases. Over the years, a massive body of legal principles accumulated that is collectively referred to as common law. Like administrative regulations, common law is valid insofar as it is consistent with statutory law and with still another source, constitutional law.

Constitutional law refers to court rulings on any law. The courts are empowered by the U.S. Constitution to declare any law unconstitutional. So, although courts cannot make laws, they have far-reaching powers to rule on a law's constitutionality and thereby to declare it invalid. Invested with the greatest judiciary power, of course, is the U.S. Supreme Court, which frequently rules on cases that apply directly to health care. A good example is the Court's legalization of abortion in the *Roe* v. *Wade* decision of 1973.

Is the law, whatever its sources, always a reliable standard for determining moral behavior? It seems that it is if, and only if, what is legal is necessarily moral and what is not prohibited by law is always moral. Both these propositions are questionable.

Is a legal standard necessarily a moral standard? Consider the case of a four-month-old baby with diarrhea and fever. The family physician prescribes medication by telephone on the second day of the illness and sees the child during office hours on the third day. On the fourth day the child's condition worsens. Knowing that the doctor is not in that day, the parents whisk the infant to the emergency room of a nearby hospital, where they are told that hospital policy forbids treating anyone already under a doctor's care without first contacting that doctor. Unable to contact the doctor, the parents take the baby home. Later that day the child dies of what turns out to be bronchial pneumonia.[1]

There was a time when hospitals had a legal right to accept for emergency treatment only those whom they chose to accept. In this case the hospital was exercising this legal right. But did that make its action moral? As it turned out, the case went to court and set the precedent of repudiating the traditional discretionary powers afforded a hospital in running its emergency facility. But even if the court had upheld the institution's right, profound moral questions of injury and fairness would still remain.

What about the second proposition, that what is not prohibited by law is always moral? To examine this claim, suppose that you, an HCP, are driving to work one day and see an accident victim on the side of the road, blood oozing from her leg. The injured person is clearly in need of immediate medical attention. Should you stop?

From the strictly legal viewpoint, you have no obligation to stop and render aid. Indeed, under common law the prudent thing for you to do would be to drive on, since by stopping you would assume a duty to the victim to use reasonable care and, consequently, would incur legal liability if your failure to use such care were to cause damage to the victim. Even in states where Good

---

1.   *Wilmington General Hospital v. Manlove, 54 Delaware 15,174 A. 2nd 135 (1961).*

Samaritan laws have been enacted to protect or give immunity from damages to those rendering emergency aid (except for gross negligence or serious misconduct), one is under no legal obligation to render such aid. And yet, virtually everyone within or outside the health care field would agree that such self-defensive behavior raises moral questions.

Again, health care providers are not legally prohibited from charging whatever they choose; from using drugs to make the institutionalized elderly and mentally ill more submissive; from denying the terminally ill who are in excruciating pain powerful drugs on request. Yet, surely it cannot be argued that the absence of legal prohibitions forecloses serious debate about the moral propriety of such behavior or that it morally sanctions such actions. After all, embedded in these issues, regardless of their legal standing, are profound questions of justice, injury, and autonomy.

In theory and practice, then, the law functions to codify the customs, ideals, beliefs, and moral values of a society. Therefore, it undoubtedly reflects changes in a society's way of thinking, in what it views as right and wrong at a particular time. It is a mistake, however, to look on the law as establishing a society's or profession's standard of morality. The law simply cannot cover the wide range of human conduct that any society or profession exhibits. True, the law prohibits the most outrageous violations of what a society considers ethical standards. But what about the countless cases that do not involve a wanton breach of those standards? For these reasons, then, the law alone, although useful in alerting us to moral issues and in informing HCPs of their rights and responsibilities, cannot be taken as an adequate standard for moral conduct.

Professional Oaths and Codes

There is no question that professional oaths and codes have historically played a formative role in focusing HCPs on the highest values and ideals of their professions. They continue to do that. There are, however, underlying weaknesses in professional oaths and codes that preclude their qualifying as adequate standards of moral conduct in the health fields. The best way to demonstrate this point is to take a close look at the codes themselves. It is not possible for us to consider all the various codes that govern conduct in the health care professions, but those regulating the conduct of physicians and nurses are typical.

Over 2000 years ago, in the fifth century BC, the Greek physician Hippocrates formulated an oath that has subsequently been sworn by physicians throughout the Western world (see Appendix A). Although it is not legally binding, the Hippocratic oath sets forth the highest principles of the medical profession. Those who swear it make a solemn promise to uphold its tenents. It reminds them of their duties to their patients—to provide the best possible care, to hold the physician-patient relationship in trust, to honor the patient's privacy, to moderate their appetite for monetary gain, to respect the profession and their colleagues. Noble ideals, to be sure.

However, the oath has its weaknesses. For one thing, having been formulated in a pretechnological age, it does not address bioethical problems that have resulted from the advances of medical science. For another, it is concerned exclusively with the cure aspects of medicine. Nowhere is there mention of the care component or of preventive medicine, both of which figure prominently in modern conceptions of the role of the physician. What is more, some of the principles expressed undoubtedly conflict with the moral beliefs of many contemporary physicians. For example, the Hippocratic oath forbids euthanasia and abortion. Yet numerous physicians believe that both actions are at least sometimes morally permissible. Finally, the oath requires physicians to keep their knowledge secret from the general public. How can this be reconciled with our current belief in the patient's right to information?

The traditional concerns of the physician have subsequently been reaffirmed in other documents, such as the International Code of Medical Ethics and the Declaration of Geneva, adopted in 1948. In 1971 the American Medical Association (AMA) also formulated a set of ethical principles to guide the conduct of physicians (see Appendix B).

Like the Hippocratic oath, the AMA's Code of Ethics expresses unimpeachable values and ideals. But it, too, has shortcomings. One is an ailment common to most such documents: generality. For example, one section calls for the rendering of service "with full respect for the dignity of man." How should dignity of man be interpreted with respect to abortion, euthanasia, experimentation on fetuses, and genetic engineering? Another section speaks of conditions that cause a deterioration of "the quality of medical care." But what is quality medical care to begin with? Still another section discourages physicians from dispensing their services under conditions that detract from "the free and complete exercise of their medical judgment and skill." Should this be viewed as a prohibition of any form of national health insurance or socialized medicine? Or does it mean that the physician must always be the ultimate, unimpeachable authority, even when functioning as a member of a health team?

Such codes as these are not restricted to physicians. Many other health care professions have similar codes. Nursing is a good example. Convinced that the person and not the disease must be treated, that prevention is better than cure, and that nurses should aspire to the highest ideals, Florence Nightingale (1820–1910), the founder of modern nursing, formulated this pledge, subsequently taken by thousands of nurses:

> I solemnly pledge myself before God and in presence of this assembly;
> To pass my life in purity and to practice my profession faithfully.
> I will abstain from whatever is deleterious and mischievous and will not take or knowingly administer any harmful drug.
> I will do all in my power to maintain and elevate the standard of my profession and will hold in confidence all personal matters committed to my keeping and family affairs coming to my knowledge in the practice of my calling.
> With loyalty will I endeavor to aid the physician in his work, and devote myself to the welfare of those committed to my care.

This pledge suffers from the same weaknesses as the Hippocratic oath and the AMA Code of Ethics. First, it is so general as to be of little practical use.

Injunctions to abstain from whatever is "deleterious and mischievous," to "elevate the standard of my profession," and "to aid the physician in his work" are open to wide interpretation. Second, like the Hippocratic oath, it fails to address the tough bioethical issues of our day, which ensnare nurses as well as physicians and other HCPs. Third, and most important, it does not anticipate the dynamic changes in the practice and organizational patterns of modern nursing, many of which raise moral problems for nurses.

Primarily in an attempt to update its code of ethics, the International Council of Nurses in Geneva formulated the International Code of Nursing Ethics in 1973. This, too, is a most edifying document. Indeed, those formulating this code should be commended for recognizing the fourfold responsibilities of the nurse and for targeting the various parties that have legitimate claims on the nurse. Unfortunately, the code requires the nurse to respect the "beliefs, values and customs of the individual" but does not acknowledge the nurse's own value system and does not indicate whether, in cases of conflict, the nurse's values should be subordinated to the patient's. And although nurses should sustain a "cooperative relationship with co-workers," they are also required to "safeguard the individual when his care is endangered by a co-worker." Surely these two injunctions are not always compatible. How is the nurse to resolve conflicts? In addition, one could ask whether the requirement to help establish and maintain equitable economic working conditions includes the right, even the obligation to strike.

An even more recent code is the American Nurses' Association Code for Nurses with Interpretive Statements. This document is far more detailed than others but still leaves many questions unanswered.

Without doubt, such codes and oaths continue to serve as useful and needed reminders to HCPs about the nature of their calling; the duties that they have assumed; and the ideals, values, and goals that they have embraced. They are testimony to the health care professions' deep and abiding concern to inculcate, transmit, and maintain standards governing the conduct of their members. Indeed, given the plethora of oaths and codes that circulate, one may rightly ask whether they are not enough. Are these codes and oaths not sufficient for providing moral guidance for HCPs?

As vital as these documents are, even our cursory look at them has turned up several weaknesses: generality, lack of topicality, failure to give priority to values. These weaknesses alone make it clear that principles grounded in ethical theory are needed. But there are other difficulties with codes and oaths.

First, given the number and variety of these documents, it is natural to ask which one, if any, HCPs ought to accept. Some might say, "Let nurses adopt the nursing codes, physicians the physician's code, and so on through the professions." The problem with this is that the various codes of the health care professions are not always in agreement. For example, the Hippocratic oath explicitly forbids euthanasia and abortion. But the AMA's Principles of Medical Ethics does not. Which of the two is the physician supposed to endorse? Similarly, the codes of different professions sometimes conflict. The AMA Principles of Medical Ethics is silent on the issue of abortion, but the Directive for Catholic Hospitals expressly forbids it.

Second, there will always be cases in which the values and moral beliefs of individual HCPs clash with the principles of their professions. Some individuals genuinely believe that euthanasia, suicide, abortion, experimenting without informed consent, experimentation on fetuses, and administering drugs even at the risk of addiction are at least sometimes morally justifiable. How are they to reconcile their own deep and sincere moral convictions with a professional oath or code that may proscribe these acts? In order to resolve such dilemmas, one must appeal to some standard outside the codes and oaths themselves, to some fundamental ethical principles that bear on dilemmas like these.

Third, and most important, codes and oaths in the health professions cannot begin to cover the multitude of moral dilemmas that HCPs encounter. For example, none of these codes directly addresses the moral problems raised by institutional understaffing, scarce medical resources, or the delivery of health services generally.

The inadequacies of law and professional codes for setting moral standards suggest the need for moral principles that are grounded in ethical theory. Thus the remainder of this chapter considers six important ethical theories: egoism, utilitarianism, the categorical imperative, prima facie duties, the maximin principle, and natural law ethics. For convenience, these can be grouped under two main headings: *consequential theories* (egoism and utilitarianism) and *nonconsequential theories* (categorical imperative, prima facie duties, maximin principle, natural law ethics).* Each offers a theory of value and obligation and possesses strengths and weaknesses in a health care context.

Consequential Theories

Traditionally, many theorists have contended that the moral rightness of an action can be determined by looking at its consequences. If the consequences are good, the act is right; if bad, the act is wrong. Moralists who argue this way are called consequentialists, or teleologists. *Consequential (teleological) theories, therefore, are ones that measure the morality of actions on the basis of their nonmoral consequences.* There are two prominent consequential theories: egoism and utilitarianism.

Egoism

Egoism contends that an act is moral when it promotes the individual's best long-term interests. In determining the morality of an action, then, egoists consider their best long-term advantage. If an action produces, will probably produce, or is intended to produce a greater ratio of good to evil for the individual in the long run than any other alternative, then that action is the right one to perform. Indeed, the individual is obligated to take that course.

Ethicists distinguish between two kinds of egoism: personal and impersonal. Personal egoists claim that they should pursue their own best long-term interests, but they do not say what others should do. Impersonal egoists, in

---

See Jacques Thiroux. Ethics Encino, California, Glencoe Press, 1980, p. 34.

contrast, claim that everyone should follow his or her own best long-term interests.

A number of misconceptions haunt egoism. One is that egoists do what they want, that they are of the "eat, drink, and be merry" school. Not so. Undergoing unpleasant, even painful, experiences is compatible with egoism, providing such temporary sacrifice is consistent with advancing long-term well-being. Another misconception is that egoists necessarily eschew virtues like honesty, generosity, and self-sacrifice. Again, this is not always so. Whatever is compatible with one's long-term best interests—including self-giving acts—is compatible with egoism. A final misconception is that all egoists are exponents of hedonism, the view that only pleasure is good in itself and worth seeking. Although it is true that some egoists are hedonistic, as was the ancient Greek philosopher Epicurus (341–270 BC), other egoists identify the good with knowledge, power, or rational self-interest or with what some modern psychologists term self-actualization. In reality, ethical egoists may hold many different theories of what is good and what is bad.

EGOISM IN HEALTH CARE.   Two features make ethical egoism an attractive moral theory in a health care context.

First, egoism provides a basis for formulating and testing policies. Regulations usually take the form of directives, guidelines, explicit policies, and codes of conduct. The purpose of such controls is clear: to ease individual decision making by discouraging certain kinds of actions and encouraging others. But on what grounds are actions to be evaluated?

Egoism provides the answer. A policy is legitimate when and if it promotes the best long-term interests of the institution, profession, or entity making it. To test existing policies, then, one need only determine whether they advance or retard best long-term self-interest. For instance, how might the AMA determine whether to throw its weight behind any of the proposed national health insurance programs? If it acted egoistically, it would decide by appeal to its own best long-term interests.

Second, egoism provides individuals in health care with a similarly convenient basis for determining what they ought to do. Rhona Steadman is a good example. In determining what she ought to do, egoist Rhona would determine what is in her best long-term interests. The legality of what she does, or its impact on the Armands, her relationship with them, the child, and the profession at large, are morally significant only insofar as they affect her own long-term interests.

Despite these strengths, many would say that egoism has decided disadvantages as an adequate guide to moral behavior in a health care setting.

WEAKNESSES OF ETHICAL EGOISM.   *First, personal egoism is inconsistent.* The first serious objection to egoism is directed against personal egoism, the position that contends that each individual should act in terms of self-interest but does not say how others should act. Let us suppose that Rhona was a personal egoist and that she decided to report the case because not to do so would imperil her position—that is, would probably produce less pleasure or more personal pain for her than the alternative. If asked whether others faced with a comparable decision ought to use a pleasure-pain calculus for deciding what to do, she could not answer. Why not? If acting in accordance with long-term self-interest is proper for one person, why should it not be proper for all? If personal happiness is a valid ethical standard for the individual, it should be so for everyone. By refusing to allow everyone else the same pursuit of happiness that they afford themselves, personal egoists are logically inconsistent. Consequently, personal egoism is not a very popular position.

Many more egoists today are impersonal egoists—those who maintain that everyone should be egoistic. Impersonal egoism may be more consistent than the personal version, but it too has problems.

Second, whatever its variety, the egoistic standard cannot be verified. Antiegoists frequently object that ethical egoism cannot be proved, that there is no way of verifying that long-term self-interest should determine human behavior. How can impersonal egoists be sure that individuals ought to consider only their own interests? Egoists reply that individuals will be happier that way. But this is a questionable claim, premised on unknown facts about the future. Although acting in what she believed was her best long-term interests may have produced happiness for Rhona in the past, she cannot really be sure that it will continue to do so in this instance. The man with an ulcer who consumes a spicy pizza, a beer, and a hot fudge sundae may not have felt a twinge of pain the last time he so indulged. But the next time, ouch! So it seems that all the egoist can say is that acting out of self-interest will *probably* produce more happiness than not. But even this is debatable.

The more complex society becomes and the more interdependent individ-duals become, the less, it seems, we can act with only our best interests in mind or the less knowledgeable we are about what constitutes our best long-term interests. Perhaps if all acted in their own best interests, then collectively there would be more happiness, as impersonal egoists maintain. But in maintaining this viewpoint, egoists are no longer egoists; they are no longer acting out of self-interest but out of a universal happiness impulse, which moves them closer to utilitarianism and results in another evident inconsistency.

But objections to egoism need not be restricted to personal or impersonal egoism. Egoism is open to attack on more general grounds.

Third, egoism does not provide means for settling conflicts of interest. Let us suppose that it is in Rhona's best long-term interests to report the incident but not in Mr. Armand's. Presumably, ethical egoism would have each do whatever was necessary to promote his or her best long-term self-interest. In that case, what Rhona should do is incompatible with what Harry should do. Apparently there are two opposing ethical obligations here. Can both be right? Egoists insist that egoism is not intended to arbitrate ethical disputes like these. But critics insist that a moral code must at least do that. They claim that, by definition, a moral code must resolve such conflicts. If it cannot, how useful is it? What good is a moral code that does not help resolve conflicts of interests between patients and HCPs, HCP and HCP, or HCPs and institutions?

Fourth, egoism introduces inconsistency into moral counsel. Suppose that Rhona goes to her supervisor for advice. The supervisor happens to be an impersonal egoist. Assuming that it is in Rhona's best interests to report the case, the supervisor advises her to do everything she can to do so. On learning that he is about to be reported for child abuse, Harry Armand calls Rhona's supervisor.

Assuming that it is in Harry's best interests to muzzle Rhona, would have to advise him to do everything he can to stop Rhor seems to be a simple, consistent attitude. True, the impersonal maintain that the supervisor is simply saying that both Rhona an try to pursue their own best interests and that the supervisor, h opinion beyond that, hopes both will win. But the objection supervisor is recommending conflicting courses of action.

Fifth, egoism undermines the moral point of view. Many ethicists moral point of view is a necessary part of moral decision making point of view, ethicists mean the attitude of seeing or attempting of an issue without being committed to the interests of a particular individual or group. Thus the moral point of view demands disinterest and impartiality. If we accept the moral point of view, then we must look for it in any proposed ethical standard, including ethical egoism.

But ethical egoists cannot take the moral point of view, for they are always influenced by what is in their own best interests, regardless of the issue, principles, or circumstances involved. Consider in the preceding example the implications of the supervisor's own egoism. Since the supervisor is an egoist, in theory she must advise Rhona and Harry in *her own* best long-term interests, not theirs.

Sixth, egoism ignores blatant wrongs. This may be the most common objection to egoism, indeed to any consequential ethic. By reducing everything to the standard of the best long-term self-interest, egoism takes no stand on principle against seemingly outrageous acts: murder, unfair discrimination, deliberately false advertising, child abuse, and so on. All such actions are morally neutral until the litmus test of self-interest is applied.

Seventh, egoism is incompatible with the nature of health care. Everyone agrees that the health care professions exist for the service of others, for those needing and wanting various kinds of health care. An ethical standard that focuses exclusively on self-interest undercuts the very reasons that the caring professions and institutions exist.

Utilitarianism

Whereas egoism maintains that the promotion of one's own best long-term interests should be the standard of morality, utilitarianism insists that the promotion of the best long-term interests of everyone concerned should be the moral standard. Stated briefly, *the utilitarian doctrine asserts that we should always act so as to produce the greatest possible ratio of good to evil for everyone concerned.* Again, as in all consequential positions, *good* and *evil* are taken to mean nonmoral good and evil.

As developed by English philosophers Jeremy Bentham (1748–1832) and John Stuart Mill (1806–1873), utilitarianism maintained that what is intrinsically good is pleasure or happiness. This was unequivocally stated in the opening chapter of Bentham's *Introduction to the Principles of Morals and Legislation,* a portion of which follows. Notice in the excerpt how Bentham moves from the pleasure and pain experienced by an individual to that experienced by the group. In so doing, he lays the basis for the utilitarian moral principle that actions are right to the extent that they promote happiness and pleasure for all concerned, wrong to the degree that they tend to produce pain and the absence of pleasure.

I. Nature has placed mankind under the governance of two sovereign masters, *pain* and *pleasure*. It is for them alone to point out what we ought to do, as well as to determine what we shall do. On the one hand the standard of right and wrong, on the other the chain of causes and effects, are fastened to their throne. They govern us in all we do, in all we say, in all we think: every effort we can make to throw off our subjection, will serve but to demonstrate and confirm it. In words a man may pretend to abjure their empire: but in reality he will remain subject to it all the while. The *principle of utility* recognizes this subjection, and assumes it for the foundation of that system, the object of which is to rear the fabric of felicity by the hands of reason and of law. Systems which attempt to question it, deal in sounds instead of sense, in caprice instead of reason, in darkness instead of light.

But enough of metaphor and declamation: it is not by such means that moral science is to be improved.

II. The principle of utility is the foundation of the present work: it will be proper therefore at the outset to give an explicit and determinate account of what is meant by it. By the principle of utility is meant that principle which approves or disapproves of every action whatsoever, according to the tendency which it appears to have to augment or diminish the happiness of the party whose interest is in question: or, what is the same thing in other words, to promote or to oppose that happiness, I say of every action whatsoever; and therefore not only of every action of a private individual, but of every measure of government.

III. By utility is meant the property in any object, whereby it tends to produce the benefit, advantage, pleasure, good, or happiness (all this in the present comes to the same thing) to prevent the happening of mischief, pain, evil, or unhappiness to the party whose interest is considered; if that party be the community in general, then the happiness of the community: if a particular individual, then the happiness of that individual.

IV. The interest of the community is one of the most general expressions that can occur in the phraseology of morals: no wonder that the meaning of it is often lost. When it has a meaning, it is this. The community is a fictitious *body*, composed of the individual persons who are considered as constituting as it were its *members*. The interest of the community then is, what?—the sum of the interests of the several members who compose it.

V. It is in vain to talk of the interest of the community, without understanding what is the interest of the individual. A thing is said to promote the interest, or to be *for* the interest, of an individual, when it tends to add to the sum total of his pleasures: or, what comes to the same thing, to diminish the sum total of his pains.

VI. An action then may be said to be conformable to the principle of utility, or, for shortness sake, to utility, (meaning with respect to the community at large) when the tendency it has to augment the happiness of the community is greater than any it has to diminish it.

VII. A measure of government (which is but a particular kind of action, performed by a particular person or persons), may be said to be conformable to or dictated by the principle of utility, when in like manner the tendency which it has to augment the happiness of the community is greater than any which it has to diminish it.[2]

In contrast with Bentham's original formulation, many modern utilitarians would view things other than happiness or pleasure as having intrinsic worth as well—for example, power, knowledge, beauty, or moral qualities. These views are often termed *ideal utilitarianism*, and they have attracted many philosophers. But since we are considering primarily classical utilitarianism, *good* will mean "pleasure." Statements about classical utilitarianism, however, apply equally to pluralistic positions, if the phrase *intrinsic good* is substituted for *pleasure*.

Before evaluating the utilitarian doctrine, let us consider some points about utilitarianism that frequently lead to its misapplication. First, in speaking of right and wrong acts, utilitarians mean those actions that we can control, those that are voluntary. This does not mean, however, that our actions must be premeditated. For example, suppose that, as you are walking through a parking lot, you observe a child playing behind a car that is about to back up. Since the driver cannot see the toddler, the child surely will be struck. Without deliberation, you snatch the child from the path of the car. Although you did not premeditate this action, you could have acted otherwise; you could have chosen not to save the child. Therefore, this is a voluntary action.

Second, in referring to the greatest possible ratio of good to evil, utilitarians do not indicate a preference for either immediate or remote good. The emphasis is on *greatest*. If the long-term good will be greater than the short-term, we should prefer the long-term—and vice versa. Frequently, however, the long-term good is less certain than the immediate good. In such cases we should prefer the immediate good.

Third, in determining the greatest possible ratio of good to evil for everyone, we must consider unhappiness or pain as well as happiness. For example, if it were possible to calculate pleasure and pain, then we should subtract the total unhappiness that our action would produce from the total happiness antici-pated. The result, in theory, would be an accurate measure of the action's worth. Thus, if an action produces eight units of happiness and four units of unhappi-ness, its net worth is four units. If another action produces ten units of happi-ness and seven units of unhappiness, its net worth is three units. In such a case we should choose the first action over the second. In the event that both acts lead not to happiness but to unhappiness, we should choose the one that leads to the fewest units of unhappiness.

Fourth, when choosing between two actions, one of which we prefer, we should choose the one that produces the greatest net happiness overall. Obvi-ously we should not disregard our own preferences, but they should not carry added weight: Count yourself as just one vote among the many.

---

2. J. Bentham, Introduction to the Principles of Morals and Legislation *(Oxford, England: Oxford University Press, 1823), chap. 1. Originally published in 1789.*

With this understanding as a base, we can now examine the forms that utilitarianism can take: act or rule. Both are influential in health care ethics. *Act utilitarianism maintains that the right act is the one that produces the greatest ratio of good to evil for all concerned.* Thus, in performing an action we must ask ourselves what the consequences of this particular act in this particular situation will be for all concerned. If the consequences produce more general good than those of any other alternative, then that action is right and is the one we should perform.

Were Rhona Steadman an act utilitarian, then, she would consider the consequences of the alternatives facing her not only for herself but for the child, Mrs. Armand, Mr. Armand, her profession, and society at large. She would let an evaluation of all the probable consequences to all the parties suggest the course of her action. Under other circumstances she might take another course. In short, each situation is considered unique, its circumstances calling for a fresh evaluation each time.

Act utilitarianism deals with a number of the objections raised against egoism. For one thing, it provides an objective way of resolving conflicts of self-interest. By proposing a standard outside the self, act utilitarianism greatly minimizes, and may actually eliminate, conflicts of self-interest. Rather than considering only their own interests, which may conflict, parties look to a uniform standard: the general good. In addition, act utilitarianism has the logical consistency that personal egoism lacks and is able to take the moral point of view. But it, too, has weaknesses. One of its chief weaknesses is that, at least in theory, it can allow what appear to be morally objectionable acts.

For example, suppose that a derelict suffering from liver disease is picked up on skid row and brought to a hospital. The man, who appears to be in his sixties, cannot be identified and appears to have no family or friends. Researchers at the hospital, eager to test a new liver drug, contemplate using the derelict as a test subject. The drug may benefit the man, but it carries significant risks, which the researchers minimize in extracting the derelict's "informed consent." Although ordinarily they would respect an experimental subject's right to complete information, the researchers agree that in this case far more common good stands to be gained by "fudging" the facts than by disclosing them forthrightly to the derelict. The research, after all, promises great benefit to humanity as a whole; subjects are needed; and this man obviously is of little or no worth to anybody.

Although act utilitarians might agree that, under such circumstances, the researchers' conduct is justified, not all ethicists would. In fact, not all utilitarians would. Some utilitarians might argue that, even if the researchers' action was calculated to produce the most total good, they acted wrongly. In this reasoning the key factor is not the act but rather the rule under which the act falls.

Rule utilitarianism asserts that we should not consider the consequences of a particular action but the consequences of the rule under which the action falls. Thus the rule utilitarian might argue that a rule obliging HCPs to give subjects complete, accurate information about the benefits and risks of participation in an experiment is a good one because it generally produces a greater ratio of good to evil than does not informing the subject. Reasoning thus, they would call the researchers' action immoral, even though in this particular case the greatest good might have been served.

Rule utilitarians, then, ask us to determine the worth of the rule under which any action falls. If keeping the rule produces more total good than breaking it, we should keep it, regardless of the consequences in any particular situation. Again, like act utilitarians, rulists would evaluate the nonmoral consequences of the rule. Specifically, they would compare the nonmoral consequences of keeping the rule with those of breaking it. Think of all the undesirable consequences of breaking the rule of informed consent: abuse of those who are most vulnerable, distrust of the whole medical research system, contempt for individual autonomy, and so on. It is hard to imagine how such things could add up to more total good than would result from obeying the rule. True, in some situations they might. But for rule utilitarians, situational nuances are irrelevant.

One strong argument for rule utilitarianism was offered by English philosopher George Berkeley (1685–1783). Berkeley reasoned that, if on each moral decision-making occasion a person had to evaluate the consequences of a proposed action, enormous difficulties would arise because of ignorance, prejudice, carelessness, lack of time, and indifference. The result would hardly be in the best interests of the general good. On the other hand, rules that everyone is aware of and attempts to follow simplify such problems and therefore advance the common good.

UTILITARIANISM IN HEALTH CARE. There are several features that make utilitarianism appealing as a standard for moral decisions in health care.

First, like egoism, utilitarianism provides a basis for formulating and testing policies. By utilitarian standards, health care policies, decisions, and actions are good if they promote the general welfare more than any other alternative. To show that a policy is wrong (or needs modification), all that is required is to show that it does not promote the principle of utility as well as would some alternative. For example, consider the distribution of medical benefits. Should we distribute medical benefits in a different way? If some alternative is likely to produce more total good, yes; if such an alternative does not exist, no. We can see then that utilitarians, of whatever persuasion, do not ask us to accept rules, policies, or principles blindly. On the contrary, they require us to test the worth of these concepts and even to indicate the testing standards.

Second, utilitarianism provides an objective way of resolving conflicts of self-interest. This feature, as suggested earlier, contrasts sharply with egoism, which seems incapable of resolving conflicts of self-interest between HCP and patient, HCP and HCP, HCP and bureaucracy, and the health care system and society. By proposing a standard outside self-interest, utilitarianism greatly minimizes and may actually eliminate such disputes. By this approach, all individuals could make moral decisions and evaluate their actions by appealing to a common denominator: the general good. Therefore, issues from adequate health care to informed consent, from abortion to patient autonomy, from euthanasia to the allocation of scarce medical resources would be resolved by determining what is in society's best interest.

Third, like egoism, utilitarianism provides the flexibility in moral decision making that health care seems to need. By refusing to recognize actions of a general kind as inherently good or bad, both schools of utilitarianism allow HCPs and health care institutions to make professional decisions and tailor policies to suit the complexities of the situation. This is as true of rule utilitarianism as it is of the act philosophy. After all, there is nothing static about a rule; it is always open to reevaluation, as time and circumstance require. For example, at one time, abortion for any reasons other than therapeutic might be undesirable, by a rule utilitarian calculus. At a different time, under different circumstances, a more liberalized rule might better advance the common good.

So, utilitarianism seemingly has much to recommend it to the health fields as a standard for moral conduct. At the same time, critics have pointed out serious shortcomings in both act and rule utilitarianism.

WEAKNESSES OF UTILITARIANISM. *First, both act and rule utilitarianism make no provision for actions that appear to be blatant wrongs.* We have already seen how act utilitarianism might be used to approve of actions that, in themselves, seem questionable. Thus, for the greater common good, a man may be denied the information he needs to decide the use to be made of his body. Thus the same criticism raised initially against egoism also applies to act and rule utilitarianism. Remember, like egoism, act and rule utilitarianism focus on the ends of an action, not the means employed to achieve an end. In ethics, *ends* refers to the consequences or results of an action; *means* refers to the action itself, to its nature and characteristics. In effect, for all consequentialists, the end justifies the means. Thus no action or rule, or characteristic of either, is in itself objectionable. It is only objectionable insofar as it results in less good than evil.

Some object to the very flexibility of rule utilitarianism, which was just portrayed as a strength. They point out that, although today a rule ensuring that information be given to experimental subjects might produce the most common good, tomorrow it might not. What then? Similarly, today a rule ensuring care of the mentally ill might produce the most total good. But what if at some point that rule no longer is so productive of the common good as one that sanctions the purging of the mentally ill? Under rule utilitarianism, presumably one would then be obliged to dispose of the mentally ill in whatever way seemed most beneficial to society.

Second, the principle of utility may conflict with that of justice. Critics of utilitarianism point out that a mere increase in total happiness is not of itself good. The distribution of happiness is a further, most important question. Surely it is not unreasonable to argue for a state of affairs in which fewer people—the most deserving—are happy instead of a condition in which more people—including the undeserving—are happy. In fact, this apparent weakness in utilitarianism relates to the preceding one; in both cases utilitarians seem to associate justice with efficiency rather than with fair play. In effect, what is just is determined by a calculation of total benefit rather than by considering need, effort, or merit. Thus, out of concern for institutional efficiency, one would abandon the bedside of a patient precisely when she is about to open up and provide insight into her health care needs. Is this fair or right? If efficiency is taken as the standard of what is just, then it is; but if justice is defined in terms of some other standard, such as meeting individual needs, then it probably is not.

Third, it is very difficult to formulate satisfactory rules. In evaluating rule utilitarianism, consider the area of confidential information and professional secrecy. What should be the operative rule that governs HCPs' disposition of information given them by patients in confidence? The glib reply would be "Never reveal what is told in confidence." But is this the *best* rule that can govern confidential information? Is it the one productive of the most common good? Perhaps it could be argued that, for the welfare of the community, the physician must reveal that the driver of a school bus has cataracts or that the janitor in the school building is a homosexual or that the pilot of a jet plane is an alcoholic or that the nurse on the surgical floor is a drug addict.[3] Maybe the rule "Never reveal what is told in confidence" needs to be qualified to exempt some or all circumstances like these.

The point is that rules are not always easy to formulate. Frequently we make rules so general that they are unrealistic. Other times, they may be so specific as to have little application outside a situational category. And of course, deciding which one among competing rules is likely to produce the greatest happiness can present a problem.

Nonconsequential Theories

Nonconsequentialism (deontology) is the ethical doctrine that denies that an action's or rule's consequences are the only criteria for determining the morality of an action. Thus nonconsequentialists insist that some things are right or wrong, not because they produce a certain ratio of good to evil but because of the nature of the action itself. Nonconsequentialists, then, might argue that lying, breaking promises, or killing is wrong in itself, not only because it may produce undesirable consequences. The grounds for a nonconsequentialist's determining the morality of an act may be arbitrarily decided, divinely revealed, or deducible from metaphysics. So, whereas consequential normative positions argue that we should consider only consequences in evaluating an action, *nonconsequential theories maintain that we should consider other factors as well.* In some cases nonconsequentialists even contend that we should not consider consequences at all.

Like utilitarianism, nonconsequentialism generally takes two forms—act and rule. *Act nonconsequential theories usually maintain that there are no rules or guidelines to govern ethical behavior, that we must evaluate each act as it comes along to determine its rightness or wrongness.* Thus in one context lying may be bad, but in another it may be good; it all depends on the situation. Act nonconsequentialists insist that blanket statements about acts are impossible.

In its extreme form, act nonconsequentialism contends that we can never formulate any rules or guidelines at all, that each situation is fundamentally different from any other; therefore, each situation must be evaluated as a unique ethical dilemma. Thus extreme act nonconsequentialism would not introduce an action's consequences at all in determining its morality.

---

3. H. A. Davidson, "Professional Secrecy," *in* Ethical Issues in Medicine, *ed. E. F. Torrey (Boston: Little, Brown, 1968), p. 107.

A more moderate act nonconsequential position maintains that general principles or rules of thumb can develop from particular situations and can then be used as a basis for making subsequent decisions. This position presumes, of course, that recurring situations are sufficiently similar to a prior one to allow application of a rule. For example, say that you have determined that the last time you lied to a patient you did the wrong thing. The next time a sufficiently similar situation arises, you then have grounds to suspect that it is wrong to lie again. In no case, however, do even moderate act nonconsequentialists claim that a rule would or should override a particular judgment regarding what a person ought to do. In other words, even a moderate act nonconsequentialist would not say "Never lie." Although there may be features to recommend them in health care, act nonconsequential positions have glaring weaknesses.

First, act nonconsequentialism provides little, if any, moral direction. What is Rhona Steadman to do? Act nonconsequentialism seems of no help. After all, Steadman must consider her situation as unique and resolve it accordingly. Extreme act nonconsequentialism cannot even provide general principles, such as "Never endanger someone's safety" or "Always preserve the confidence of your patients" or "Always obey the law." It tells us to rely on our intuition, feelings, and inclinations in order to determine the right thing to do. Thus, extreme act nonconsequentialism leaves us in a state of ethical limbo. Act nonconsequentialism extremists argue that the situation should guide the ethical choice. But this appears to be begging the question, for the situation can only provide data and facts that crystallize the choices. The problem remains: Which option to choose?

Second, act nonconsequentialism makes feelings the guide to moral decisions. Unfortunately, even modified act nonconsequentialism does not seem to provide Rhona much more help. Granting that her case is not unique, there are still enough variables involved to make rule-of-thumb formulation difficult. Even if Rhona has some general rules to apply, according to act nonconsequentialists she arrived at them on the basis of intuition or decisions that she had made in the past. But is the way one feels about a certain situation sufficient reason for claiming that what one does is moral? (Implied in this objection, of course, is criticism of intuition as a sufficient and legitimate source of moral knowledge, a point that will be pursued later.)

Third, act nonconsequentialism precludes the formulation of necessary general ethical rules. The act nonconsequentialist claims that no two situations are the same and that it is therefore impossible to generalize about moral behavior. But this seems a straw man, for although it is true that no two situations are completely identical, the issue is really whether two situations can ever be sufficiently similar to allow moral generalization.

For example, suppose that Rhona asks a colleague what to do and the colleague tells her to report the case. Five minutes later, Rhona asks the same colleague the same question, but this time is told not to report it. Assuming that nothing more than the ticking of a clock has occurred in the interim, should the colleague not respond the same as the first time, since the situation itself has not changed? When Rhona is told that the decision bears no connection to the one made five minutes earlier, she might rightly consider her colleague's logic and ethics strange—certainly inconsistent.

When act nonconsequentialism chooses to ignore similarities between past and present, it relinquishes a potent tool in inductive reasoning—the analogy. We reason analogically when we make judgments about present and future events based on similar past events. When someone reasons, "I won't go to Dr. Jones again because he always treats me like a child," that person reasons analogically. If there are enough similarities and few significant differences, then the analogy is sound. Analogical reasoning is one of the most useful methods we have for maintaining some measure of control over our lives. In ethics, when we attribute goodness or badness to an action, surely we should attribute it to another action that is similar in all relevant respects.

These objections to act nonconsequentialism have led many who eschew consequentialism to adopt rule nonconsequential positions. Indeed, within the category of rule nonconsequentialism fall some of the most influential ethical theories in Western thought. *According to rule nonconsequentialism there are one or more rules to serve as moral standards.* For example, some might hold that you should always tell the truth. No matter how good the consequences might be that result from lying, some rule nonconsequentialists might say that to violate the rule of truth telling is always wrong. Unlike moderate act nonconsequentialism, however, rule nonconsequentialism would maintain that these rules are not derived from particular examples or actions but from the nature of the actions themselves. The following sections consider four rule nonconsequential positions: the categorical imperative, prima facie duties, the maximin principle, and natural law ethics.

Kant's Categorical Imperative

One of the most influential thinkers in the history of philosophy was the eighteenth-century German philosopher Immanuel Kant (1724–1804). Kant's ethics stands as the foremost illustration of a purely nonconsequential (deontological) theory, one that attempts to exclude a consideration of consequences in moral decision making. To understand Kant's thought as expressed in his *Foundations of the Metaphysics of Morals,* it is essential to grasp his idea of intrinsic value.[4]

Kant believed that nothing is good in itself except a "good will." Intelligence, judgment, and all other facets of the human personality are perhaps good and desirable, but only if the will that makes use of them is good. *By "will" Kant meant the uniquely human capacity to act according to the concept of law—that is, principles.* It is this emphasis on will or intention that sets Kant apart decisively from consequential thinkers. For Kant, these laws or principles operate in nature. Thus, a good will is one that acts in accordance with nature's law.

In estimating the total worth of our actions, Kant believed that a good will takes precedence over all else. Contained in a good will is the concept of duty.

---

4. *I. Kant,* Foundations of the Metaphysics of Morals, *trans. L. W. Beck (New York: Bobbs-Merrill, 1959).*

Only when we act from duty does our action have moral worth. When we act merely out of feeling, inclination, or with a view to intended results, our actions—although they might be otherwise identical to ones that spring from a sense of duty—have no true moral worth.

To illustrate, HCPs have a duty not to misinform patients about their conditions. But simply because HCPs do not misinform patients does not necessarily mean that they are acting from a good will. They may be acting from an inclination to promote their practices or to avoid legal entanglement. Thus they would be acting in *accordance with* duty but not *from* duty. In other words, their apparently virtuous behavior happens to coincide with duty, but they have not willed the action from a sense of duty to be fair and honest. Therefore, their action would not have true moral worth. For Kant, actions have true moral worth only when they spring from a recognition of duty and a choice to discharge it.

Kant distinguishes two kinds of duties: perfect and imperfect. A perfect duty is one that we must always observe; an imperfect duty is one that we must observe only on some occasions. Furthermore, *perfect duty* refers to negative obligations—that is, things we must always refrain from doing. *Imperfect duty* refers to positive obligations, things that we must do only on some occasions. For example, we would always have the perfect duty not to injure another person; but we have only an imperfect duty to show love and compassion. There are times when we must indeed show love and compassion, but when, to whom, and how much remain unclear. One other thing about these distinctions: The nature of one's duties determines what others can legitimately claim as a right. Since I am always obligated to refrain from injuring you, you have a right to demand of me that I not injure you. But since I need not always show love and compassion, you have no right to demand that I do so.

Still, we are left wondering what duties we have and how we can know them. Suppose that Rhona Steadman wishes to act from duty. What is her duty? How can she discover it?

Kant believed that through reason alone we can arrive at a moral law that is not based on empirical evidence relating to similar situations or to consequences. Just as we know, seemingly through reason alone, that a triangle has three sides and that no triangle is a circle, so by the same kind of reasoning Kant believed that we can arrive at absolute moral truths.

For Kant, an absolute moral truth had to be logically consistent, free from internal contradiction. To say, for example, that a triangle has four sides or that a square is a circle is to state a contradiction. Kant attempted to construct an absolute moral law free from such contradictions. If he could formulate such a rule, he contended, it would oblige everyone without exception to follow it. Kant believed that he formulated such a logically consistent rule in his categorical imperative.

In Kant's view there was only one command or imperative that was categorical, that presented an action as necessary of itself, regardless of any other considerations. He argued that, from this one categorical imperative, this universalizable command, we could derive all commands of duty. *Simply stated,*

Kant's categorical imperative says that we should act in such a way to will the maxim of our action to become a universal law.

By *maxim*, Kant means the subjective principle of an action, the principle that people in effect formulate in determining their conduct. For example, suppose that despite her promise to Vilma Armand to keep the family's secret, Rhona decides to break that promise if it suits her purposes. Her maxim can be expressed as "I'll make promises that I'll keep until doing so no longer suits my purposes." This is the subjective principle, the maxim, that directs her action. Kant insisted that the morality of any maxim depends on whether we can wish it to become a universal law. Would Rhona's maxim qualify?

That depends on whether the maxim as law would involve a logical contradiction. Obviously, the maxim "I'll make promises that I'll keep until doing so no longer suits my purposes" could not universally be acted on, because it involves a contradiction of will. On the one hand, Rhona is willing that it be possible to make promises and have them honored. On the other, if everyone intended to break promises when they thought it best, then the concept of promises would have no meaning. In other words, it is part of the nature of promises that they be believed. A law that allowed promise breaking would contradict the very nature of a promise. This is true even if desirable consequences might result from breaking the promise.

So crucial is the categorical imperative to Kant's ethics that we should examine Kant's own development of this concept:

But what sort of law can that be, the conception of which must determine the will, even without paying any regard to the effect expected from it, in order that this will may be called good absolutely and without qualification? As I have deprived the will of every impulse which could arise to it from obedience to any law, there remains nothing but the universal conformity of its actions to law in general, which alone is to serve the will as a principle, i.e. I am never to act otherwise than so *that I could also will that my maxim should become a universal law.* Here, now, it is the simple conformity to law in general, without assuming any particular law applicable to certain actions, that serves the will as its principle, and must so serve it, if duty is not to be a vain delusion and a chimerical notion. The common reason of men in its practical judgments perfectly coincides with this, and always has in view the principle here suggested. Let the question be, for example: May I when in distress make a promise with the intention not to keep it? I readily distinguish here between the two significations which the question may have: whether it is prudent, or whether it is right, to make a false promise? The former may undoubtedly often be the case. I see clearly indeed that it is not enough to extricate myself from a present difficulty by means of this subterfuge, but it must be well considered whether there may not hereafter spring from this lie much greater inconvenience than that from which I now free myelf, and as, with all my supposed *cunning,* the consequences cannot be so easily foreseen but that credit once lost may be much more injurious to me than any mischief which I seek to avoid at present, it should be considered whether it would not be more *prudent* to act herein according to a universal maxim, and to make it a habit to promise nothing except with the intention of keeping it. But it is soon clear to me that such a maxim will still only be based on the fear of consequences. Now it is a wholly different thing to be truthful from duty, and to be so from apprehension of injurious consequences. In the first case, the very notion of the action already implies a law for me; in the second case, I must first look about elsewhere to see what results may be combined with it which would affect myself. For to deviate from the principle of duty is beyond all doubt wicked; but to be unfaithful to my maxim of prudence may often be very advantageous to me, although to abide by it is certainly safer. The shortest way, however, and an unerring one, to discover the answer to this question whether a lying promise is consistent with duty, is to ask myself, Should I be content that my maxim (to extricate myself from difficulty by a false promise) should hold good as a universal law, for myself as well as for others? and should I be able to say to myself, "Every one may make a deceitful promise when he finds himself in a difficulty from which he cannot otherwise extricate himself"? Then I presently become aware that while I can will the lie, I can by no means will that lying should be a universal law. For with such a law there would be no promises at all, since it would be in vain to allege my intention in regard to my future actions to those who would not believe this allegation, or if they over-hastily did so, would pay me back in my own coin. Hence my maxim, as soon as it should be made a universal law, would necessarily destroy itself.

I do not, therefore, need any far-reaching penetration to discern what I have to do in order that my will may be morally good. Inexperienced in the course of the world, incapable of being prepared for all its contingencies, I only ask myself: Canst thou also will that thy maxim should be a universal law? If not, then it must be rejected, and that not because of a disadvantage accruing from it to myself or even to others, but because it cannot enter as a principle into a possible universal legislation, and reason extorts from me immediate respect for such legislation. I need not as yet *discern* on what this respect is *based* (this the philosopher may inquire), but at least I understand this, that it is an estimation of the worth which far outweighs all worth of what is recommended by inclination, and that the necessity of acting from *pure* respect for the practical law is what constitutes duty, to which every other motive must give place, because it is the condition of a will being good *in itself,* and the worth of such a will is above everything.

Thus, then, without quitting the moral knowledge of common human reason, we have arrived at its principle. And although, no doubt, common men do not conceive it in such an abstract and universal form, yet they always have it really before their eyes, and use it as the standard of their decision.[5]

Note that Kant is not a consequentialist. He is not arguing that the consequences of a universal law condoning promise breaking would be bad and therefore that the rule is bad. Instead, he is claiming that the rule is self-contradictory, that it is self-defeating, and that the institution of promise making would necessarily dissolve on universalizing such a maxim. A closer look at the promise keeping example actually reveals three formulations of the categorical imperative.

First, to be a moral rule, the rule of conduct must be consistently universalizable, because Kant's moral rule prescribes categorically, not hypothetically.

---

5.   I. Kant, Fundamental Principles of the Metaphysics of Morals, *6th ed., trans. T. K. Abbott (London: Longmans Green, 1909), sec. 1, pp. 9–10.*

A hypothetical prescription tells us what we should do if we desire cert
consequences: "If we want people to think well of us, we should keep prom
ises." In hypothetical prescriptions the goodness of promise keeping depend
on consequences. But Kant's imperative is categorical; that is, it commands
unconditionally, and so the command holds regardless of consequences. A
categorical imperative takes the form "Do this" or "Don't do that,"—no ifs,
ands, or buts. Such a command must be universalizable; if it were not, then its
worth would be determined on empirical grounds—that is, on hypothetical
ones. Put another way, if a person did not follow a moral rule purely on
the grounds that it was a universal law of moral conduct, that person would
allow empirical conditions to determine whether to follow a rule. The rule
would thereby lose its inherent necessity and universality—and thus, its moral
character.

Second, Kant, also believed that in order for a rule of conduct to be a moral
rule, humans should treat each other as ends in themselves when following it,
not just as means to ends. Kant believed that humans, as rational beings, would
be inconsistent if they did not treat everyone else the way that they wanted to be
treated. And since, according to Kant, rational beings recognize their own inner
worth, they would never wish to be used as entities possessing relative worth,
as means to an end.

Third, Kant argues that for a rule of conduct to be a moral rule it must allow
persons who are universally legislating it to impose it on themselves. For
example, in the case of sexually or racially discriminatory employment policies,
where the regulation makers do not bind themselves by the rules that they set
up, the rules have no moral import.

KANT'S CATEGORICAL IMPERATIVE IN HEALTH CARE. Kant's ethics contains
several elements that are applicable to health care. *First, Kant's ethics takes much of
the guesswork out of moral decision making in health care.* HCPs rightly complain that
professional codes of conduct are often so vague as to be useless. Worse, these
codes leave much to individual interpretation, thus making morality into a kind
of guessing game. By contrast, Kant's ethics removes much of this uncertainty
and subjectivity. To act morally is to act on principle. No matter what the
consequences or situational nuances, some actions are always wrong. Thus,
HCPs must never deliberately misrepresent to patients their condition or cause
them needless suffering, for such treatment would not be consistent with what
rational, autonomous creatures deserve.

Second, Kant's ethics introduces a needed humanistic dimension into health care. We
saw that one of the principal objections to egoism and utilitarianism is that they
can treat humans exclusively as means to ends. Kant's ethics expressly forbids
this. As a result, it brings a humanistic tone to moral decisions in health care.
This is especially needed today, when the importance of the individual can so
easily be lost amidst ever-increasing and sophisticated technology. Kant's ethics
establishes the centrality of the individual person in moral decisions. Thus
medical researchers, no matter how lofty their goals or limited their resources,
may never use human beings as research subjects without first obtaining their
truly informed consent. Otherwise, the researchers will be using people as
means to an end, which is never permissible.

Third, Kant's concept of duty implies the moral obligation to act from a respect for rights and a recognition of responsibilities. According to Kant, we act morally when we behave dutifully—that is, according to a concept of law. But the formulation of laws involves the process of demarcating—between individuals and other individuals, between individuals and groups, between groups and other groups. The need for this process exists in health care relationships just as it does in social ones. Indeed, it seems particularly needed in health care, because part of health care's moral quandary, as we saw, is an inability to define and assign rights and responsibilities, which is basic to drawing lines of demarcation. Subscribing to Kant's ethics would necessitate defining and specifying rights and responsibilities clearly and then following the moral imperative in acting with respect to them. For example, a patient can exercise certain rights stemming from the perfect duties that an HCP has. One of these would be the right to information; another would be the right to truth; a third would be the right to adequate health care. Likewise, an HCP could exercise certain rights stemming from the perfect duties of the patient. The HCP is as much entitled to information and truth as is the patient. Specifically, the HCP has a right to expect the patient to reveal anything that will materially affect the HCP's capacity to deliver adequate health care to that patient and to expect that the patient will refrain from doing anything that will undercut the quality of the care.

Even from as sketchy an account as the foregoing, one can begin to see the value of Kant's ethics for health care. Nevertheless, as was true of the other theories, Kant's has its shortcomings.

WEAKNESSES OF KANT'S CATEGORICAL IMPERATIVE. *First, Kant's principles provide no clear way to resolve conflicts of duties.* Problems arise with the categorical imperative when rules conflict. For example, suppose that by not lying to a patient an HCP would cause the person emotional pain. Presumably, one has a perfect duty to refrain from lying and from causing others pain. Which of these obligations takes priority when they conflict?

Second, Kant's application of duties to rational beings can raise problems in a health care context. Kant is quite clear that we have duties to beings who are rational and who have an "autonomous self-regulating will." Generally speaking, this imperative does not raise any difficulties. But in health care, questions often arise about who qualifies as a rational, autonomous being. Do children? The mentally deranged? The senile? The unborn? Are fetuses to be considered persons? What about infants with serious birth defects? Numerous moral questions beset HCPs who treat individuals in these categories. Inevitably, the questions turn on some concept of the status of these individuals, which will determine their rights. Unfortunately, Kant is ambiguous on these matters.

Third, there is no compelling reason why the prohibition against certain actions should hold without exception. Critics of Kant frequently question, even reject, his principle of universalizability. They wonder why the prohibition against actions like lying, promise breaking, killing, and suicide must function without exception. They contend that Kant failed to distinguish between saying that a person should make no exceptions to a rule and that the rule itself has no exceptions. The statement that a person should make no exceptions to a rule simply means that one should never exclude oneself from a rule's application. If lying is wrong, it is wrong for me as well as for you. To say that "Lying is wrong, except when I do it" would not be universalizable, for then lying would be right for all to do. But just because no one may make himself or herself an exception to a rule, it does not necessarily follow that the rule itself has no exceptions.

Suppose, for example, that an HCP decides that lying is sometimes right, perhaps to save a life. Thus the rule becomes "Never lie except when telling the truth will jeopardize the life of a patient." Perhaps this rule could be refined, but the point remains: This rule is just as universalizable as "Don't lie." The clause "except when . . ." can be viewed not as an exception to the rule but as a qualification of it. Critics of Kant ask why a qualified rule is not so good as an unqualified one. If in fact it is equally good, then we no longer need to state rules in the simple, direct, unqualified manner that Kant did.

So the apparent problem with Kant's system is the rigidity of the rules. There are just no exceptions to them once they have qualified under the categorical imperative. A more realistic approach, it seems, would be to replace absolute rules (ones that allow no exceptions) with a series of rules, the duty to which depends on their relative importance in relation to other relevant rules. This, of course, would entail a rule nonconsequentialism based on more than a single rule. Just such an approach has been taken in the twentieth century by English philosopher Wiliam David Ross.

Prima Facie Duties

A half century ago, Ross published *The Right and the Good*, in which he presented an ethical theory that can be viewed as an attempt to join aspects of utilitarianism with those of Kantianism.[6] Ross began his attempt to provide methodology for resolving conflicts of duties by dismissing the consequentialist belief that what makes an act right is solely whether it produces the most good. Frequently, Ross noted, consequences of conflicting behaviors counterbalance each other. Instead of a consequentialism, Ross argued that, when deciding among ethical alternatives, we must weigh the options to determine what duties we would fulfill by performing or refraining from performing each option. Then we must decide which duty among the alternatives is the most obligatory. The various duties spring from the many values that Ross views as intrinsic, such as justice, noninjury, and self-improvement.

At least two characteristics of his thinking place Ross in the nonconsequential camp. First, judgment of an action's morality is not dependent solely on its consequences. Second, there are duties or obligations that bind us morally. These duties, presumably, can be stated in rules. But how can we know these rules?

Ross claimed that we come to a knowledge of these moral rules or principles by intuition, which he defined as a source of knowledge that does not rely immediately on the senses or reason but on direct awareness. He held that the

---

6. *W. D. Ross,* The Right and the Good *(Oxford, England: Clarendon Press, 1930).*

most fundamental part of our moral knowledge is the intuition of certain moral principles. If we reflect on these moral principles, we stand a better chance of doing what is right. These intuited moral principles are of the form that all acts of a certain kind are wrong. For example, "All acts of promise keeping are right," and "All acts of injury are wrong." Such principles, according to Ross, are known intuitively; if we understand what the principles mean, we recognize their truth. We need no additional evidence to know that the principles are true. Once we have recognized the truth of a moral principle, we can use it to determine what we ought to do in a particular case.

As Ross explained, an act may fall under a number of rules at once. For example, the rule to keep a promise may in a given circumstance conflict with the rule not to do anyone harm. In part this is the issue in Rhona Steadman's case. To keep her promise to Vilma, Rhona must apparently jeopardize the child; to prevent additional injury to the child, she must evidently break her promise.

In such cases each act is accompanied by a number of motivational reasons. Each reason in turn appeals to a moral duty—to keep a promise, to prevent injury, to be fair, to help people, and so on. Each of these moral duties provides grounds for doing a thing, and yet no single one provides sufficient grounds. Simply because Rhona has a duty to keep her promise is not grounds enough for not reporting the case.

Ross was particularly sensitive to cases that involve conflicts of duties. To understand how he handled them, we must consider his concept of *prima facie duties*. By *prima facie* duty Ross is referring to the characteristic that an act possesses because it is a certain kind of act—for example, an act of promise keeping or an act of noninjury. Ross cautioned that the term *prima facie* (Latin for "on first appearance") is unfortunate, because it suggests that "one is speaking only of an appearance which a moral situation presents at first sight, and which may be illusory." What Ross was really speaking of is, rather, "an objective fact involved in the nature of the situation or more strictly in an element of its nature. . . ."[7] He insisted that there is nothing arbitrary about prima facie duties; they can be known to be true by intuition.

There are two important characteristics that Ross ascribed to prima facie duties. One is that they are intuited. The second is that they are conditional. By conditional, he meant that they can be overridden, although they still retain their character as duties. Thus, by intuition, Rhona may recognize a prima facie duty to keep a promise. There may be times, however, when she feels justified, even obligated, to break a promise—for example, in order to prevent injury or pain. In breaking her promise under such conditions, she does not for a minute cease to recognize the prima facie duty to keep a promise. Indeed, if the case were such that a single prima facie duty, such as promise keeping, was involved, she would be obligated to meet that duty. But Ross gave little attention to such cases, because they are not problematic. They pose no real concern to persons who recognize intuitively, for example, that they have a duty to keep a promise and that the particular case involves, without any conflict, an instance of promise keeping.

---

7. *Ross, p. 19.*

Therefore, were promise keeping the only prima facie duty involved in Rhona's case, then she ought to honor the promise. But should she *actually* keep it? Not necessarily, because other duties are involved as well. Thus Ross distinguished prima facie duties from actual duties.

By *actual duty* Ross meant what we actually ought to do in a situation. If there is only one prima facie duty involved, then that is what we actually ought to do. If there are conflicting prima facie duties involved, our actual duty is what we are obliged to do *after we have weighed and considered all the prima facie duties*. Unlike prima facie duties, actual duties are *not* intuited. In fact, actual duties are not knowledge at all, in Ross's view. Rather they are closer to an "educated guess" or a "lucky accident."

Ross cited six categories of prima facie duties:

1. *Duties of fidelity are those that rest on prior acts of our own.* Included under duties of fidelity are the duty not to lie (which Ross viewed as implied in the act of conversation), the duty to remain faithful to contracts, the duty to keep promises, the duty to repair wrongful acts. Rhona has a duty of fidelity to honor the terms of her employment, which would include honoring any oaths she has sworn or professional codes that regulate her actions.

2. *Duties of gratitude are those that rest on other people's acts toward the agent.* Ross argued that we are bound by obligations arising from relationships, such as those between friends or between relatives. For example, suppose that annually Rhona contributes $100 to her favorite charity. The week before she is about to make this contribution, a good friend whom she has not seen in years telephones her. Rhona is delighted to hear her friend's voice and fondly recalls the many times that her friend stood by in times of crisis. Now her friend is in need of $100. In this case a duty of gratitude would probably oblige Rhona to give the money to her friend rather than to the charity.

3. *Duties of justice are those that rest on the fact or possibility of a distribution of pleasure or happiness that is not in accordance with the merits of the people concerned; in such cases there is a duty to upset or prevent such a distribution.* Imagine the case of an unqualified, incompetent person getting a promotion simply on the basis of having "connections." A duty of justice would oblige us to prevent such an occurrence.

4. *Duties of beneficence are those that rely on the fact that there are other people in the world whose virtue, intelligence, or happiness we can improve.* For example, HCPs are in a position to provide care for those who, for a variety of reasons, cannot obtain it. They can also help educate the public and serve as advocates for patient rights.

5. *Duties of self-improvement are those that rest on the fact that we can improve our own condition of virtue, intelligence, or happiness.* Suppose, for example, that a young woman is exceptionally talented. She plays several musical instruments, paints very well, seems to have a natural language-learning ability, and exhibits high scientific acumen. While she is in the midst of determining her life's goals, a rich relative dies, leaving her so much money that she need not work another day in her life. If she were to decide to use this inheritance

for long-term self-indulgence and make no attempt to cultivate her talents, then she would be violating a duty to improve herself.

6. *Duties of nonmaleficence are duties of not injuring others.* Ross included this obligation to contrast with duties of beneficence. Although not injuring others incidentally means doing them good, Ross interpreted the avoidance of injuring others as a more pressing duty than beneficence. Rhona might do the Armands more good by keeping quiet. But she risks injuring the child and this would seem to impose an obligation that overrides the beneficence she owes them.

Ross did not claim that his six categories of prima facie duties represent a complete list, but he did claim that these are duties we acknowledge and willingly accept without argument. In the case of two conflicting prima facie duties, Ross was in favor of following the more obligatory duty. Where more than two duties are involved, our actual duty is the one with the greatest amount of prima facie rightness over wrongness.

ROSS'S PRIMA FACIE DUTIES IN HEALTH CARE.   There are two principal features about Ross's theory that make it useful in health care. *First, Ross's list of duties can play an important role in the moral education of people within and outside the field.* His list exhorts each person to ponder the prima facie duties to primary claimants and to set duties aside only when others take priority. Thus, we can speak of HCPs as having duties of fidelity and noninjury that bind them to provide proper health care to patients, duties of self-improvement to keep abreast of developments in their fields, and duties of beneficence to act as patient advocates. Much the same can be said of the duties of the health care professions toward individual HCPs. We can see, then, that Ross's categories provide a useful framework for ordering individual and organizational responsibilities to claimants.

Second, Ross's ethics brings to health care the utilitarian sensitivity to consequences without ignoring duties of undeniable moral force. As we have seen, flexibility is one of utilitarianism's strengths. Utilitarianism does not lock moral agents into actions that are in themselves considered right or wrong. Everything depends on the circumstances and the probable outcome of the action or rule. At the same time, utilitarianism seems to overlook things that in themselves seem repugnant, such as experimenting on subjects without their informed consent or distributing medical resources inequitably. Ross recognized this problem. Thus, his ethics can suffer evident wrongs but only when more pressing duties are present. Even then we must acknowledge the duty flouted through some kind of reparation.

Despite these strengths, Ross's ethics remains controversial. Specifically, it suffers from two chief weaknesses.

WEAKNESSES OF ROSS'S PRIMA FACIE DUTIES.   *First, people disagree about moral principles.* How do we know what prima facie duties we have to begin with? How can we be sure that we have an obligation to tell the truth, to improve ourselves, or not to injure others? Ross claimed that these are self-evident truths to anyone of sufficient mental maturity who has given them enough attention. But what if people disagree with Ross's list of duties? Surely it is possible to disagree about the relative weight they deserve. Suppose some HCPs do not think that they have an obligation to keep a promise or honor a contract or serve as patient advocates? Ross would say that such individuals lack "sufficient mental maturity" or that they did not give the proposition "sufficient attention." But a question arises as to the nature of sufficient mental maturity. We must also wonder when we have given sufficient attention to a moral statement. In Ross's view, people have attended sufficiently to a moral proposition when they recognize his list of duties. Obviously, this response involves circular reasoning.

Second, it is difficult, if not impossible, to determine the relative weight and merit of conflicting duties. When faced with a situation that presents one or more conflicting prima facie duties, how do we determine what our actual duty is? In the case of two conflicting prima facie duties, one's duty is to perform the act that is in accord with the more stringent prima facie obligation. In cases of more than two conflicting prima facie duties, one's duty is to perform the act that has the greatest balance of prima facie rightness over prima facie wrongness. But without assigning weights to duties, how can we determine the "most stringent" obligation or "the greatest balance of prima facie rightness over prima facie wrongness"? This is not to belittle or minimize Ross's theory. In numerous cases HCPs face no particular difficulties about resolving conflicts of duties. Most people would probably agree that, when faced with having to tell a harmless lie in order to save someone from serious injury, HCPs have a more pressing obligation to prevent injury than to stick to the truth. And often in cases where the choice is not so clear, thoughtful reflection underscores the efficacy of Ross's approach. But what about other cases, when the actual duties are in much greater conflict? Ultimately, it seems that Ross must be judged inconclusive at best about the relative priorities of duties.

Third, it is not certain that anyone possesses an intuitive faculty. This objection speaks to the faculty of intuition that Ross uses as his source for recognizing prima facie duties. The disagreement that exists about the prima facie duties themselves certainly suggests that not everyone has this faculty—for if they did, there would be no disagreement. Beyond this, what evidence is there that anyone does have an intuitive faculty, as distinct from the other sources of value discussed in Chapter 2—experience, religion, science, culture? Critics point out that this intuitive faculty, like extrasensory perception, is highly speculative and cannot be corroborated scientifically.

The Maximin Principle of Justice

We have seen that consequentialists and nonconsequentialists are fundamentally at odds. Specifically, utilitarianism suggests the principle of greatest happiness as the standard of morality, which in theory seems to allow injustices. Kant and Ross, in contrast, made intention and characteristics of acts the fundamental moral concerns and emphasized the intrinsic worth of human beings. In 1971 Harvard professor of philosophy John Rawls published *A Theory of Justice*, in which he tried to use the strengths of consequential and nonconse-

tial ethics while avoiding their pitfalls. At the same time, Rawls hoped to
a workable method for solving problems of social morality. Because his
attempts to maximize the lot of those who are minimally advantaged, it is
the *maximin principle*.[8]

Central to Rawls's theory is the question of establishing principles of justice.
Insofar as Rawl's thinking concerns social justice, it is fair to say that he erected
his theory of obligation on the assumption that justice is of intrinsic value,
although not necessarily the only intrinsic value. So the question becomes this:
What principles can serve as a basis for justice in society? In answering, Rawls
asked us to imagine a "natural state," a hypothetical state of nature in which all
persons are ignorant of their talents and socioeconomic conditions. He called
this the "original position." Rawls claimed that in the original position people
share certain characteristics. They are mutually self-interested; rational, that is,
they know their own best interest more or less accurately; and similar in their
needs, interests, and capacities. Given these assumptions, Rawls then asked
what people in the original position would be likely to formulate if asked to
choose a fundamental principle of justice. He surmised that they would choose
two principles to ensure justice: the liberty principle and the difference principle.

By the *liberty principle*, Rawls meant that people in the original position
would expect each person participating in a practice or affected by it to have an
equal right to the greatest amount of liberty that is compatible with a like liberty
for all. And by the *difference principle*, Rawls meant that people in the original
position would allow inequality only insofar as it serves each person's advantage
and arises under conditions of equal opportunity.

Before examining these concepts further, we should understand that by
person Rawls meant not only particular human beings but also collective agencies.
And by *practice* he meant any form of activity that a system of rules specifies,
such as offices, roles, rights, and duties. Now let us examine the heart of Rawls's
theory: the principles of equal liberty and difference.

Equality means that all persons are to be treated the same. Specifically, by
equality Rawls meant the impartial and equitable application of rules that define a
practice. The equal liberty principle expresses this concept.

When Rawls termed equality "impartial," he was referring to the spirit of
disinterestedness that should characterize the distribution of goods and evils, to
the fact that no person should receive preferred consideration. When he spoke
of "equitable administration," Rawls seemed to mean that the distribution must
be fair and just to begin with. Let's apply these two characteristics of equality to
a concrete situation.

Suppose that a hospital draws up job specifications for the position of head
nurse. The equal liberty principle decrees that everybody be judged by these
criteria. The rule would not be administered impartially if the personnel department made an exception for an applicant. Nor would it be administered
equitably if it excluded candidates from consideration on criteria that are not
related to the job, perhaps on grounds of race, sex, or age. Remember that Rawls

---

8. *J. Rawls*, A Theory of Justice (*Cambridge, Mass.: Harvard University Press, 1971*).

was defining liberty with reference to the pattern of rights, duties, powers, and liabilities established by a practice. In this case, anyone applying for the job has a right to expect equal treatment insofar as the same job specifications will be applied and has a right to expect that the distribution be fair and equal to begin with.

But Rawls's equal liberty principle expresses the idea of equality in another, more important way. An essential part of all regulations is that they cannot help but infringe on personal liberty. Suppose, for example, that the job specifications for head nurse call for graduate work in nursing. Obviously, those lacking this formal education cannot compete equally for the job. In his equal liberty principle, Rawls recognized this inherent characteristic of all laws and other practices: By nature they encroach on the equal liberty of those subject to them. But as long as graduate work is required of everyone applying for the job, then such a requirement is just.

Rawls appears more philosophically perceptive on this point than his opponents. He noted that, if a more extensive liberty were possible for all without loss, damage, or conflict, then it would be irrational to settle for a lesser liberty. For example, in this case an argument could be made for a more extensive liberty by allowing on-the-job experience to substitute for graduate work.

Crucial to any theory of social justice is the determination of when inequality is permissible. After all, a just society is not one in which all are equal but one in which inequalities are justifiable. Rawls addressed this problem with his difference principle, which defines what kinds of inequalities are permissible. It specifies under what conditions the equal liberty principle may be violated.

For Rawls, equality is not contingent. It does not depend on something else for its justification, such as on the greatest happiness for the most people. Rawls believed that equality is fundamental and self-justifying. This does not mean that equality can never be violated. It means that inequality works to the advantage of every individual affected—or at least to the advantage of the least well-off. In other words, Rawls would arrange inequalities so that, ideally, they would benefit all affected or at least those most in need. Suppose, for example, that the hospital limits the number of consecutive hours that staff is allowed to work. Although this rule works against those who want to work longer shifts, the standard can be justified by showing how such a rule works to the advantage of all concerned—those who do not want to work overtime, the patients, even those wanting to work more hours. (After all, they will not be able to jeopardize their own welfare or to have it jeopardized by others.)

Although such a rationale may appear to be a case of the principle of greatest happiness, it is not. The difference principle, in fact, does not allow inequalities generally justified on utilitarian grounds—that is, on the grounds that the disadvantages of persons in one position are outweighed by the advantages of those in another. On the contrary, the difference principle ideally allows inequality only when it works to *everyone's* advantage. Thus the key point in the hospital example: Even those limited by the inequality really benefit from it.

The fundamental difference, then, between utilitarianism and Rawls's theory of justice is in their concepts of justice. In utilitarianism, the concept of justice rests on some notion of efficiency. In contrast, Rawls's view is best described by his own word *reciprocity*. Reciprocity is the principle requiring a practice to be such that all members who fall under it could and would accept it and be bound by it. It requires the possibility of mutual acknowledgment of principles by free people, having no authority over one another, to make the idea of reciprocity fundamental to justice and fairness. Without this acknowledgment, Rawls claimed, there can be no basis for a true community. Thus a fair institutional work policy would be one in which all affected can and will accept it and will then be bound by it.

In effect, then, where there is conflict between his two principles, Rawls relied on his first principle of justice, the equal liberty principle. He insisted that it is logically prior to the difference principle. In contrast, the utilitarian sees liberty as contingent on social productivity. In other words, liberty is desirable only insofar as it produces the most happiness for the most people. Obviously, the utilitarian position does not allow the loss of liberty if the greatest number are not served. But Rawls would argue that any position that even allows the possibility of the loss of equal liberty is unacceptable. As Rawls put it, "Each person possesses an inviolability founded on justice that even the welfare of society as a whole cannot override. . . . Therefore . . . the rights secured by justice are not subject to political bargaining or to the calculus of social interests."[9]

Besides agreeing on these basic principles of justice, people in the original position also would recognize "natural duties" that generate one's obligations to another. The duties that Rawls listed are identical with those noted in discussing Ross's prima facie duties. Moreover, Rawls acknowledged the need to rank these as being of higher and lower obligation, although he did not undertake this task, presumably because his primary concern was with justice in social institutions.

Finally, Rawls recognized the legitimacy of paternalism. By *paternalism* he meant that occasions arise when people are not in a position to make decisions for themselves, when others must make the decisions for them. In health care decisions, for example, patients can exercise virtually no influence over the planning, manufacturing, and marketing of medical goods and services. Rawls's recognition of paternalism requires that those in a decision-making capacity take into consideration the concerns, interests, and values of those who will be affected by the decisions but who do not have any substantive way to make their feelings and preferences known.

THE MAXIMIN PRINCIPLE IN HEALTH CARE. *First, Rawls's principle of justice requires of HCPs an inherent respect for individual persons.* Just as Kant's and Ross's ethical theories focus on the primacy of the individual, so does Rawls's. According to Rawls, it would never be right to exploit an individual or a group for the benefit of others. Thus it would always be wrong to misinform patients about a research project in order to get their consent to participate as subjects. Similarly, keeping patients in the dark about the negative side effects of a treatment or

---

9. *Rawls, p. 4.*

medication would never be justified. In both instances, Rawls would argue that people have an inherent right to decide for themselves what risks they are willing or unwilling to take. Therefore, HCPs have an obligation to respect this right.

Second, Rawls outlined areas of broad social responsibilities for health care professions and institutions. We have seen how Ross's categories of duties can be applied in specific health care situations. Rawls's concept of natural duties can be used in the same way.

Third, Rawls's concept of paternalism requires HCPs and health care institutions to consider the preferences of individuals and groups who cannot act for themselves. Although it is certainly not the only area that suggests paternalistic considerations, the health care of children stands out as an obvious concern. Thus, by Rawls's paternalism, HCPs and hospitals would be justified in intervening on behalf of a child with a life-threatening illness that goes uncorrected only because the child's parents refuse to consent to the necessary treatment.

WEAKNESSES OF RAWLS'S MAXIMIN PRINCIPLE.   Since its formulation in 1971, Rawls's maximin principle has been the center of a highly technical philosophical debate. At this point it is impossible to isolate objections that all critics would share. But two criticisms recur often enough to bear mentioning.

First, Rawls's theory is based on several questionable assumptions. According to Rawls, when we don the veil of ignorance we are mutually self-interested; that is, we normally establish practices on the basis of self-advantage. But Rawls assumed that humans would be rational. When he described people as rational, he implied a great many things that seem questionable: (1) that we more or less know our own best interests, (2) that we can trace the likely consequences of one course over another and can resist the temptation of short-term personal gain, and (3) that we are not greatly bothered by the perceptible difference between our own condition and someone else's. Rawls also assumed that people have similar enough needs, interests, and capacities so that fruitful cooperation is possible. At the same time, he would conceal from people in the original position any knowledge of their interests, talents, purposes, plans, and conceptions of the good. How can people so ignorant of these important aspects of their identities and personalities agree on principles to regulate their lives? But even conceding these assumptions, we can detect other weaknesses in Rawls's theory.

Second, Rawls's principle brings a measure of whim to moral decision making. Although Rawls is clear about the primacy of the equal liberty principle, differences in emphasis and interpretation might elevate the difference principle to the position of supremacy and thus lead to a contrasting moral decision. Take, for example, a program that guarantees everyone adequate health care. Under ideal conditions—those that Rawls described as part of his definition of rationality—no one would probably object to such a program, for fear of being left without adequate health care. But this seems certain only if we assume a kind of germ-free, vacuumlike context for decision making, a context in which no one has any more or less access to health care than anyone else.

But what would occur if you asked those same rational, self-interested, uncoerced people how they would redress existing inequities in the accessibility and distribution of health care? They might emphasize the need to redress this injustice as quickly as possible through an appropriate national health care program, which would be funded by taxing people according to their incomes. In so doing, they might employ Rawls's difference principle by arguing that the worst-off will benefit. Also, they might argue that everyone stands to benefit, assuming that the availability of medical care will probably produce healthier people who would, in turn, be more productive.

But would it be altogether irrational for someone to object that there are bound to be individuals who will find themselves worse off under such a system than under the present one? For example, some would have less of their own money to spend the way they desire. But more to the point, let us suppose—as might well happen—that those who would be taxed the most, the rich, would also be excluded from the program, in whole or in part, because of their high income. In other words, the affluent would end up paying for health care services from which they themselves would be excluded. It would not seem to be stretching a point for these people to invoke Rawls's equal liberty principle and claim that such a program violates their rights, from the point of view of logic if nothing else.

Of course, Rawls might reply that his theory has been warped. By definition, people in the original position are not aware of any preferred status that they might have over anyone else. But rational people seemingly would be aware of the *possibility* of their having preferred financial status; and having such a program explained to them, which needs to be done before they can decide, it would seem that they would recognize that they could be victimized by it. Unless we have badly misconstrued Rawls, then, it seems that whether one emphasizes the equal liberty or the difference principle depends largely on a calculation of where one is likely to end up in the scramble: in the class of the sponsors or the class of the recipients.

Natural-Law Ethics: The Roman Catholic Interpretation

Natural-law theory refers to the general view that moral principles are objective truths that can be discovered in the nature of things by reason alone. Such a view has been extremely influential in the development of political and moral theories. Evidence of it can be seen in Kant, Ross, and Rawls. Indeed, natural-law theory has been a most visible framework within which to view biomedical issues.

At the outset, natural law should be distinguished from scientific law, or laws of nature. Laws of nature are empirical generalizations; they are statements about the physical world that are based on sense experience. Statements about gravitation and inertia are examples of laws of nature. In contrast, *natural laws* delimit the behavior that is morally appropriate for human beings; they express what conduct is proper for a human being as a human being. Proper conduct, in turn, is determined by reference to the kind of creature that humans are—that is, by reference to a theory of human nature. Presumably, we can discover what constitutes human nature by reason.

Two characteristics of natural-law theories emerge. One is that moral values and duties depend on the basic nature of human beings. The other is that we can discover what humans are by applying the processes of rational and scientific reflection, just as we do in other areas of human knowledge. Any ethical theory that makes these two basic assumptions can be called a natural-law theory.[10]

A prime example of how natural-law theory operates in ethics can be seen in the moral theology of Roman Catholicism. Over the years Roman Catholicism has given special attention to morality in medicine. Its positions, rooted in natural law, have influenced our laws, institutions, and social policies. So, regardless of one's own religious views, it is important to be aware of how the moral theology of Roman Catholicism applies to issues in medicine. For these reasons, further discussion of natural-law ethics will be limited to the doctrines developed in the moral theology of Roman Catholicism.

The natural-law theory of Roman Catholicism was fully formulated in the thirteenth century by Saint Thomas Aquinas (1225–1274). Aquinas's view of human nature, and his thought in general, was strongly influenced by the texts of the classical Greek philosopher Aristotle (384–322 BC). A basic idea that Aquinas borrowed from Aristotle is the view of the universe as teleological, which means that the universe is governed by purposes or ends. For example, the end—or goal—of an acorn is to become an oak. Under the right conditions, the acorn, following "the law of its nature," will in fact become an oak. The same movement toward purpose or end characterizes everything in the universe, including humans and their institutions. For example, just as organs have goals (for example, reproductive), so do families. The purpose of the family is the propagation and nurture of offspring under stable and wholesome cir-cumstances. In this view, then, the child beating by Harry Armand is a violation of natural-law precepts. For Aquinas, the goals or purposes that permeate the universe are ordained by God.

According to Aquinas, God has given humans a unique trait: reason. Al-though it is true that, like everything else, humans have a material nature governed by laws in its growth and development, they also possess the capacity to reason, which sets them apart from all other creatures. What is more, in order to develop human potentialities to the fullest, humans must follow the direction of reason. The ultimate goal of all human behavior, the only thing of value in itself, is a state of eternal bliss—union with the Creator, or what Aquinas terms "contemplation of God."

What, then, is the connection between reason and morality? As Aquinas conceived it, morality is not an arbitrary set of rules for behavior. Rather, morality is part of the human's teleological nature. In other words, God has built the basis of moral obligation into the very nature of the human in the form of various inclinations, such as the preservation of life, the propagation of the species, and the search for truth and a peaceful society. The moral law, then, is

---

10. G. J. Hughes, "Natural Law," Journal of Medical Ethics, vol. 2, no. 1 (March 1976).

founded on these natural inclinations and the ability of reason to discern the right course of conduct.

To summarize, in Aquinas's view natural law consists of rules of conduct corresponding to inclinations that God has built into the very nature of the human being. The basic precepts of natural law are the preservation of life, the propagation and education of offspring, and the pursuit of truth and a peaceful society. These precepts reflect God's intentions for the human and, most important, can be discovered and understood by reason.

Although the precepts themselves do not vary, their enforcement does. Since different societies are influenced by different topographies, climates, cultures, and social customs, Aquinas believed that different codes of justice are needed. He called these specific codes of justice human law. The function of rulers is to formulate human law by informing themselves of the specific needs of their communities and then passing appropriate decrees. Whether human law is just depends on its conformity with natural law, which is the expression of the will of God applied to human situations.

Natural law as developed in the writings of Saint Thomas Aquinas has provided the basis for specific Roman Catholic doctrines with implications for medicine. Two of the most pertinent principles are the principle of double effect and the principle of totality.[11]

Often in moral decision making, we are faced with the choice of an action that will produce both good and bad effects. Considering only the good effect, we are inclined to perform the action; considering only the bad effect, we are inclined to avoid the act. The classic example is a pregnant woman with a cancerous uterus. The only way to save the woman's life is by removing the uterus. But if that is done, the fetus will die. Is it right to remove the uterus? The principle of double effect is intended to resolve conflicts like these. *Essentially, the principle of double effect holds that an action should be performed only if the intention is to bring about the good effect and only if the bad effect will be an unintended or indirect consequence.* Specifically, four conditions must be satisfied to justify the action:

1. The action itself must be morally indifferent or good.

2. The bad effect must not be the means by which the good effect is achieved.

3. The intention must be to achieve the good effect only.

4. The good effect must be at least equivalent in importance to the bad effect.

In the preceding case, it seems that these four conditions are met in deciding to remove the uterus. First, removing the uterus is a medical procedure that in itself is neither good nor bad; it is morally neutral. Second, the mother's life is not preserved by the death of the fetus but by removal of the malignant uterus. Third, the intention of the surgeon is to preserve the mother's life and not to kill the fetus. Fourth, the mother's life is certainly equivalent in importance to the life of the fetus.

---

11. *R. Munson,* Intervention and Reflection: Basic Issues in Medical Ethics *(Belmont, Calif.: Wadsworth, 1979), pp. 33–34.*

The principle of totality holds that individuals have the right to dispose of their organs or destroy their capacity to function only to the extent that the general well-being of the whole body demands it. The basis for this principle is the recognition that God has designed each of our organs with a goal or end in mind; each is intended to play a role in maintaining the functional integrity of our total bodies. Since we are custodians, not owners, of our bodies, we must hold our organs in trust. When a diseased organ threatens our lives, then we are justified, indeed obliged, to dispose of it. To do otherwise would be to violate our natural obligation to preserve our lives. Thus we are justified in having a diseased lung or a ruptured appendix removed. But when the organ does not threaten our lives, our obligation is to refrain from interfering with it. Therefore, by Catholic moral theology, vasectomies and tubal ligations intended for contraceptive purposes would always be wrong.

NATURAL-LAW ETHICS IN HEALTH CARE.   Since this discussion has been confined to natural-law ethics in Roman Catholicism, the following remarks will focus on that version. There are several aspects of the Roman Catholic version to recommend it for health care ethics.

First, natural-law ethics synthesizes the elements involved in any moral decision. We have already seen how specific consequential and nonconsequential theories can function in health care. A significant problem with each is that it takes a limited view of moral decision making. For example, utilitarianism focuses on consequences, Kant on the goodness of the will, Ross on duties. Each of these elements is undoubtedly an important consideration for a moral decision in health care. It seems to follow, then, that a synthesis of these emphases would provide the strengths of each without any of their apparent constraints. By addressing intention, means, and ends, natural-law ethics as expressed in Roman Catholic moral theology provides HCPs and health institutions with a comprehensive ethical theory that incorporates the three elements of any moral decision. It apprises them of the ennobling ethical principle that the action regarded as right or good without qualification is the one that springs from a good intention, is implemented by good or at least neutral moral means, and produces good results.

Second, natural-law ethics, although always condemning some actions, nevertheless provides health care ethics with needed flexibility. One of the advantages of consequential theories is that, by not looking into concepts of intrinsic wrong, they provide health care with a flexibility to match the complex nature of health care decisions. At the same time, this feature can also be used to justify what appear to be blatant wrongs. Roman Catholic natural-law ethics appears to provide a large amount of decision-making latitude without this liability, because it does not rely exclusively on consequences to determine the morality of an action. The principle of double effect allows ends to justify means providing neither the means nor the end involves a willful violation of the will of God. In effect, the principle of double effect incorporates the utility principle within the context of a nonconsequential ethic.

Third, the principles of double effect and totality have numerous applications in health care. We have already touched on a couple of areas in which the principle of double effect can be applied to health care decisions: abortion and medical operations. But there are many more: the maintenance of life through extraordinary means, scientific experimentation on human subjects, organ transplants, and so on. The chapters ahead examine these applications more closely.

WEAKNESSES OF ROMAN CATHOLIC NATURAL-LAW ETHICS. *The chief weakness of Roman Catholic natural-law ethics is that it is based on a number of questionable assumptions.* One big assumption is the existence and nature of God. Suffice it here to say that this assumption continues to be vigorously attacked in numerous philosophical quarters. Another assumption is that everything in the universe has a goal or purpose. But does it? Many disagree. They argue that the changes we observe in the universe can be explained by the laws of mechanics and by natural selection. Also, critics point out that the growth and development of organisms can be accounted for by the presence of genetic information that controls the processes. In short, contemporary science does not support the teleological view of Aristotle and Aquinas. What is more, reason alone is inadequate to discover or substantiate such a teleology. Only if we assume that the apparent purpose evident in the operation of the universe reflects a divine plan can we endorse the teleological view advanced by Aquinas. But if all the apparent teleology can be explained in nonteleological ways—as many suggest it can—then there is no reason to assume a divine Creator and Creation. Lacking a teleological foundation, natural-law ethics as presented by Aquinas collapses. This, of course, does not mean that some other version of natural-law ethics might not be defensible logically.

Applying the Theories

Chapters 6 through 14 apply these normative theories to various moral issues in health care. Partially in anticipation, but more specifically in order to complete the discussion of ethical theory, this chapter will briefly apply these views to the choice facing Rhona Steadman: Should Rhona report the child abuse at the Armands?

Were Rhona a consistent egoist, she would determine what to do on the basis of her best long-term self-interest. By this standard the overriding consideration might well be the legal one. Since, by law, Rhona is required to report cases of child abuse, not to do so would place her in grave legal and professional jeopardy. Most important, as an egoist, she would consider the potential injury to the baby and the breach of confidentiality only insofar as they affected her own long-term interests. At the same time, it is important to recognize that in theory egoism could also be used to justify Rhona's remaining silent. If more self-benefit would be likely to result from not reporting the incident, then Rhona the egoist would be justified, even obliged, to remain silent.

Were Rhona a consistent act utilitarian, she would base her decision not exclusively on self-interest but on the interests of all who would be affected by her decision. Which of the two options—to report or remain silent—would likely produce more total good would be her standard. By this calculation,

Rhona would consider the interests of the baby, the Armands, society in general, and of course herself. If she reports the case, not only will she stop the beatings, but she will also probably get Harry the help he needs. At the same time Rhona will have benefited society by helping to control an especially cruel and prevalent social problem. Moreover, she will have met her legal obligations, thereby precluding any subsequent problems for herself. Given this analysis, Rhona probably would report the child abuse. At the same time, and most important, if she were a consistent act utilitarian, Rhona would admit that she might handle the next case of child abuse differently. Should conditions be such that not reporting an incident of child abuse would be likely to yield the greatest social benefit, then she would not report it.

As a rule utilitarian, Rhona would again be concerned with consequences. But Rhona the rulist would evaluate the consequences of implementing the rule under which the proposed action fell. The rule here might be "Public health care professionals should report cases of child abuse." Would following this rule be likely to produce more total good than not following it? Most probably it would. First, following this rule would help protect innocent children from harm. Second, it would help child abusers get needed psychological treatment. Third, it would help society maintain social stability. These consequential concerns alone seem to argue for the worth of the rule. Accordingly, were she a rule utilitarian, Rhona should report the case, even if in this particular instance she believed that more total good would result from remaining silent.

Turning to the nonconsequential theories, let us first apply Kant's ethics. If Rhona subscribed to Kant, she would be concerned with a number of things. First, she would realize that she was in a position to do the Armand baby great good or harm. Should she report the incident, presumably the beatings would stop; should she remain silent, the beatings might go on. Recognizing a duty to prevent injury to the baby, Rhona would feel obliged to report the case. Although the duty might be an imperfect one, it would seemingly be urgent all the same because of the potential injury to the baby, which Rhona would have the opportunity to prevent in the future. In addition, Rhona the Kantian would recognize a duty to honor the law, which requires her to report such cases. Flouting this law, she would undercut the protection that all citizens are entitled to and the social stability that a community needs to survive. In effect, Rhona would be acting in a way that is inconsistent with the nature of society and with the legitimate function of this particular law. Rhona might raise other Kantian concerns, but these are enough to indicate the direction of the analysis, which points to her duty to report the case.

Were Rhona a prima facie theorist, she would be especially mindful of the duties of noninjury, beneficence, justice, and self-improvement. First, should she report the case, she could greatly reduce, if not eliminate, the likelihood of further abuse to the baby. By contrast, her silence might invite further injury. By this account, duties of noninjury and beneficence run parallel. Another aspect of her duty of beneficence applies to Harry Armand: Rhona is in a position to get Harry the psychological help he evidently needs, which he seems more likely to get if she reports the case than if she does not. Similarly, Rhona is in a position to improve both herself and her profession morally, as well as to meet a justice duty, by vigorously championing the rights of one of the most defenseless groups in society—children. By this analysis, if she were a prima facie theorist, Rhona would consider it her actual duty to report the case.

If Rhona were a Rawlsian, she would apply Rawls's natural duties much as she did Ross's prima facie duties. The moral course seems the same: Report the incident. Rhona could also argue that in the original position, behind a veil of ignorance, rational creatures with a sense of their own interests would agree to bind themselves under a rule that required the reporting of child abuse. After all, these individuals would recognize that without such a rule they themselves could suffer, either as victims, abusers in need of help, or citizens in a highly unstable social environment. They would also realize that such a rule would have the effect of producing the greatest amount of freedom that is compatible with a like freedom for all. By this reasoning, Rhona the Rawlsian should report the incident.

Finally, if Rhona endorsed Roman Catholicism's version of natural-law ethics, she would also report the case. First, she would recognize that the child beating by Harry Armand was a violation of natural-law precepts, which was far more serious than any breach of confidentiality would be. Should she decide to remain silent, she would have to justify her action in the light of likely good that is at least equal to the further injury she is risking for the baby and to her violation of the law. Given the facts, no reasonable case could be made along these lines. Thus, Rhona would clearly be obliged to report the case.

If the preceding analyses are correct, then any of the theories could be called on to justify reporting this case of child abuse. But notice that the underlying reasons for charting this moral course vary. With consequential theories, the appeal is always to effects, ends, or results. Where nonconsequential theories are involved, the appeal generally is to some principle of duty, social justice, or natural law.

Selecting a Theory

A question that inevitably arises is which theory, if any, should one endorse? There is no clear-cut principle that we can apply in choosing from the smorgasbord of ethical standards. So perhaps the whole enterprise of trying to formulate and justify moral principles is futile and should be abandoned. Although this reaction is understandable—especially from those just beginning a study of ethics in health care—there are good reasons for not acting on such an impulse of despair. After all, to say that an ethical theory is imperfect is not to say that it is empty or that the search for a satisfactory theory is futile. Human relationships present a tangled web of frequently subtle, ill-defined problems. It is little wonder that we do not have a single theory that wins everyone's acceptance.

Each ethical theory discussed in this chapter has an impressive range of application. In criticizing each theory, philosophers inevitably focus on its weaknesses. This approach is consistent with the philosophical enterprise of pursuing truth and certainty. Philosophers cite cases that tend to break down the theory, that test its strength at the most fundamental levels. Failing to grasp this apsect of the philosophical endeavor, one might well conclude that there is little of worth in these ethical theories. This judgment would be a gross over-statement with which not even the theories' harshest critics would agree. It would be as indefensible as scrapping the theories of biological evolution or of quantum mechanics because these are incomplete or unsatisfactory in important ways. As in the world of science, so in the realm of human relationships with which ethics deals: We face complex realities that by nature seem to preclude absolute answers.

Granting that these theories, although imperfect, are extremely useful, is not the ultimate selection of one theory arbitrary? Since none can be proved totally correct, does it really matter which one we choose? Again, such a reaction is understandable, but it is not the answer. Before deciding that the choice of an ethical theory is arbitrary, we should reflect on the meaning of that word.

It probably would be an arbitrary decision to conclude that, since all ethical codes generally agree on basic ideals, it does not matter which code is followed. But this position would be incorrect, because each code commits us to different principles. The thoughtful person recognizes these differences and thereby chooses. Such a choice, based on a consideration of the alternatives, cannot be arbitrary. Instead, it is grounded on the best available evidence, on a careful weighing of all considerations that it could possibly be founded on. In the words of one philosopher:

> To describe such ultimate decisions as arbitrary . . . would be like saying that a complete description of the universe was utterly unfounded, because no further fact could be called upon in corroboration of it. This is not how we use the words "arbitrary" and "unfounded." Far from being arbitrary, such a decision would be the most well-founded of decisions, because it would be based upon a consideration of everything upon which it could possibly be founded.[12]

The selection of an ethical theory for health care seems only arbitrary for the frivolous and unthinking, those who have never undertaken a serious investiga-tion into ethical theory. For those persons who have, the choice, far from being arbitrary, is often slow, methodical, and agonizing. The process suggests not the impudence of those who say it does not matter but the courage of those who say it does. It is an individual choice, not an arbitrary one.

Finally, the fundamental value of studying and understanding ethical thought is not that we thereby have definitive guides to moral conduct. Rather, the value lies in becoming aware of the moral options available to us, of the general paradigm within which moral inquiry can take place as concrete human beings grapple with real-life issues. Individual moral choices are frequently not between obvious right and wrong, good and bad, but between actions and values that contain elements of both. The challenge, then, is not so much find-ing an ethical standard to use but applying a defensible standard in specific instances.

---

12.   R. M. Hare, The Language of Morals *(Oxford, England: Clarendon Press, 1952), p. 69.*

The chapters that follow are designed to help meet this challenge. They not only raise specific issues, but they also show how these theories might be applied to the issues. We will see that all theories are of some use but that some prove more useful than others in particular situations. It is hoped that such an exposure will prove valuable in helping you recognize the various kinds of ethical systems that may be employed in given situations and in determining what values override others in a conflict.

Summary

In this overview of the place of law and professional codes in health care ethics we have seen that, although both are useful, neither is adequate for setting moral standards. Moral principles grounded in ethical theory are of far more value. The ethical theories presented in this chapter have vast application in health care, as we shall discover in the pages ahead.

CASE PRESENTATION
Outrage

Psychologist Lawrence Kohlberg has done extensive research in the area of moral development. The following is an adaptation of a story he often uses for illustrative purposes. [13]

In Europe a woman was suffering and near death from a rare, painful form of cancer. One drug possibly could save her, but it cost more than the woman's husband, Heinz, could afford or raise. Frantic with worry, Heinz begged the druggist to sell him the medicine cheaper or to let him pay in installments. The druggist refused, claiming that, although the price was far higher than its production costs, he had the right to charge whatever he wanted, especially since he had discovered the drug.

In his desperation, Heinz asked a friend for advice. The friend reminded Heinz that stealing was against the law and refused to get involved any further. Meanwhile, Heinz's wife was pleading with her husband to steal, even murder if necessary, to get the medicine.

That night Heinz broke into the pharmacy and stole the drug. In a fitful rage he also destroyed the remaining medicine and the formula as well.

Questions for Analysis

1. *Do you think that the principals acted immorally? Explain.*

2. *Who do you think acted the most reprehensibly? Explain.*

---

13. *L. Kohlberg, Moral Stages and the Idea of Justice (San Francisco: Harper & Row, 1981).*

3. *What values can you identify as operating in this episode, from the viewpoints of the principals?*

4. *In your view, what is the main value guiding each of the principals? Explain.*

5. *Apply each of the ethical theories presented in this chapter to the following actions, and explain how each action's morality would probably be evaluated in terms of each theory.*
    a. *The druggist's charging what he did*
    b. *The druggist's refusal of Heinz's request*
    c. *The wife's plea*
    d. *The friend's decision to stay uninvolved*
    e. *Heinz's breaking into the pharmacy and stealing the drug*
    f. *Heinz's destroying the remaining medicine and formula*

Exercise

Reconsider your responses to the questions in the Values and Priorities section. Would you answer any of them differently in the light of what you've read in this chapter? Explain.

For Further Reading

Broad C. D. *Five Types of Ethical Theory.* New York: Harcourt, Brace, 1930.

Copleston, F. C. *Aquinas.* Baltimore: Penguin Books, 1965.

Frankena, W. *Ethics.* 2nd ed. Englewood Cliffs, N.J.: Prentice-Hall, 1973.

Fried, C. *An Anatomy of Values: Problems of Personal and Social Choice.* Cambridge, Mass.: Harvard University Press, 1970.

Kant, I. *Foundations of the Metaphysics of Morals.* Translated by L. W. Beck. New York: Bobbs-Merrill, 1959.

Ladd, J. *Ethical Relativism.* Belmont, Calif.: Wadsworth, 1973.

Mill, J. S. *Utilitarianism.* New York: Bobbs-Merrill, 1957.

Muller, H. J. *The Children of Frankenstein.* Bloomington: Indiana University Press, 1970.

Nietzsche, F. *Genealogy of Morals.* Translated by F. Golffing. Garden City, N.Y.: Doubleday, 1956.

O'Conner, D. J. *Aquinas and the Natural Law.* New York: St. Martin's Press, 1969.

Oraison, M. *Morality for Moderns.* Translated by J. F. Bermard. New York: Doubleday, 1972.

Plato. *The Republic.* Translated by Benjamin Jowett. New York: Random House, 1957.

Ramsey, P. *Basic Christian Ethics*. New York: Scribner's, 1950.

Rawls, J. *A Theory of Justice*. Cambridge, Mass.: Harvard University Press, 1971.

Ross, W. D. *Foundations of Ethics*. New York: Oxford University Press, 1954.

Stace, W. T. *The Concept of Morals*. New York: Macmillan, 1965.

Warnock, M. *Existential Ethics*. New York: St. Martin's Press, 1968.

Part II

The Health Care Professional within the Organization

4

THE HEALTH CARE SYSTEM

Between 1973 and 1976, 1,978 patients died in California state mental hospitals. Of these deaths 120 occurred under questionable circumstances. In 1977 the state attempted to get to the bottom of these deaths by stationing an investigator at each of the hospitals. But the authority of the special investigators was immediately resisted in the legislature by the medical lobby, thus igniting a wave of resentment among the investigators. In the spring of 1980, a publication of California's medical association denounced the investigation as "shades of 1984" and a "symbol of the totalitarian spectre." A month after this article appeared, a state senator introduced a resolution calling for an end to review of state hospital patient deaths by special investigators. Instead, the resolution said, the state medical staff should conduct the investigations of its members through a peer review system. The state association drafted the resolution, and the senator introduced it virtually as it was dictated to him by the medical association. Within a month it passed through the senate's health and welfare committee and passed the full senate 39–0.[1]

This case is interesting for several reasons. For one thing, it raises the issue of adequate health care for the mentally ill. For another, it raises the issue of patient confidentiality; the inquiry was stymied in part because investigators were denied access to patient records. But what interests us most about this case is what it tells us about the health care system—about how it is organized and how it sometimes operates. These are the concerns of this chapter. Although it is not possible to cover every aspect of the health care system in this one chapter, we will discuss the system's components and their roles, with emphasis on the role of organized medicine and its values, and will consider how the health care system functions as a mechanism for social control.

There are at least three reasons for considering the health care system in a book about morality and health care. First, many of the moral problems in health care have systemic origins: They have their roots in the conceptual and operational aspects of the system. So to understand the nature of these problems and

---

1.   J. Lewis, *"Records Law Blocked Probe of Hospital Deaths,"* The Sacramento Bee, *December 28, 1980, p. A3;* J. Lewis, *"Doctors Battle Hospital Investigations,"* The Sacramento Bee, *December 29, 1980, p. A1.*

decide how to deal with them intelligently, one must have an understanding of the system that breeds them. Second, HCPs must operate within the system. As we saw in Chapter 2, a good measure of the moral conflict that HCPs feel results from a clash between personal and professional values. But professional values largely emanate from the system. So, again, to fully understand the conflict and to deal with it intelligently, one must have some grasp of the organization and operations of the health care system. A third reason for studying the system is to understand the rights of patients and concomitant responsibilities of HCPs. Subsequent chapters will look at these. Suffice it here to say that many of the moral contraints that HCPs face in meeting their obligations to patients stem in part from the structure of the system and from the interplay among its various parts.

Values and Priorities

Before beginning, consider the following questions, which again are intended to ferret out your values and priorities.

1. Do you think that the California state medical staff should conduct the investigation of its members under a peer review system?

2. Do you think that peer review is always the best way to ensure responsible health care?

3. How do you feel about outsiders evaluating your professional performance?

4. Suppose that you were an HCP in a California state hospital and you knew that some of these deaths occurred as a result of patient neglect or abuse. Would you report what you knew, even if several of your colleagues might be disciplined as a result?

5. What is the difference between being responsible for patient care and being held accountable for patient care?

6. Do you think that HCPs should be held accountable for patient care? If so, by whom?

7. Which of the following do you think should have the most influence over the health system: physicians, other HCPs, government, or patient-consumers?

8. How do you feel about a system that bases health care on a person's ability to pay?

9. Name three rights that you believe you have as an HCP. Which is the most important to you?

10. What do you think the functions of a professional association, such as the American Medical Association or the American Nursing Association, should be?

11. How do you feel about the fact that the health professions are physician-dominated?

12. Do you believe that HCPs should always subscribe to the official positions of their professional associations?

13. How do you feel about unions in nursing and other health care professions?

14. Do you think that patient-consumers have about as much influence over the health care system as they should, should have more influence, or should have less influence?

15. Do you think that physicians overprescribe, prescribe just about right, or underprescribe?

16. How do you feel about the increasing role of technology in the health sciences?

Components of the Health Care System

Before considering the structure of the health care system, a word is in order about systems in general. In essence, a system is a relatively stable whole composed of parts that are integrated and coordinated in their activities by some kind of communications network.[2] Systems can be natural or human-made. For example, a human being could be regarded as a natural system within a hierarchy of other natural systems, which range from the bisophere to subatomic particles. Falling within this hierarchy of natural systems are culture and society, which consist of numerous human-made systems. Included among these are educational, legal, and health care systems.

Like all complex systems, the health care system is made up of interrelated parts, each a system in its own right as well as a part of the greater system. Both the parts and the whole are goal-directed, the primary goal being the delivery of health care. In addition, each of the components can itself be broken down into smaller and simpler systems, thus forming a hierarchy of systems ranging from the complex to the simple. *The health care system may be defined, then, as a distinct, separable, and permanent part of our society that consists of an elaborate set of arrangements and interrelated parts, whereby the technology of the health sciences are made available to us.*[3] The bewildering variety of its parts and their interrelationships make it difficult to provide a complete sketch of the health system, but being aware of its main components will serve as a useful point of departure for understanding the system as a whole.

In general, any health care system can be visualized as consisting of three major elements: people and equipment, the organization of these two, and the financial arrangements between suppliers and consumers.[4] Specifically, the

---

2. *C. W. Aakster. "Psychosocial Stress and Health Disturbances,"* Social Science in Medicine, *February 1974, p. 77.*

3. *D. B. Smith and A. D. Kaluzny,* The White Labyrinth: Understanding the Organization of Health Care *(Berkeley, Calif.: McCutchan Publishing, 1975), p. 5.*

4. *D. F. Stroman,* The Medical Establishment and Social Responsibility *(Port Washington, N.Y.: Kennikat Press, 1976), p. 12.*

health system consists chiefly of secondary and primary providers, financing mechanisms, and patient-consumers.

Secondary providers are those who, although not providing direct patient care, shape the system in which health services are delivered and organized. They supply the human resources, the technology, and the overall coordination of health services. Among the most important secondary providers are medical and nursing schools, as well as other professional training programs; professional associations; pharmaceutical companies; equipment suppliers; and various planning agencies. The influence of these secondary providers may not be apparent to outsiders, but they dominate the health care system in the United States. We cannot understand the system without understanding the influential role played by secondary providers. Understanding this role provides insight into why the system operates as it does, why HCPs face the moral problems they do, and why and how their capacity to manage these problems is often constrained.

Primary providers, the second component, are those who provide direct patient care in an office, health center, institution, or clinic. They can conveniently be distinguished on the basis of the kind of services rendered: institutional, ambulatory, and community. Institutional providers treat patients within the confines of an institution. Hospitals, nursing homes, and extended and home care programs are among the chief institutional providers. In contrast, ambulatory care provides health services within the context of a physician's office, a hospital outpatient clinic or emergency room, or one of the many group practice facilities that are emerging. Community or public health care focuses on the treatment of the community rather than the individual. Local health departments and public health agencies carry out most of the activities in this area.

The third key component in the health system is the elaborate set of mechanisms for financing primary providers. Among these are private and public charities, Blue Cross– Blue Shield, commercial insurance companies, Medicaid and Medicare, and other federal subsidies. Traditionally, primary providers, especially physicians, have viewed the emergence of these "third parties" with considerable suspicion and distrust, even with hostility. Presumably, part of their fear is that these financing mechanisms will wrest financial control away from primary providers. As a result, the third-party mechanisms that have developed have been controlled largely by primary providers.

Consumers represent the final component in the health system. By and large they exercise very little influence over the other parts, and thus their influence on the system is negligible. In effect, consumers of health services have no choice but to accept what is offered and to use what is available. Most of their costs are subsidized by Social Security and payroll deductions. Rarely do consumers understand precisely how these various subsidies apply to health care, nor do they exercise any control over how the contributions that are withheld are to be used.

The health care system, then, can be viewed as an interaction among providers, financing mechanisms, and consumers. The inputs of each are combined in various services, such as treatment or therapy. These services produce specific results—health, death, profit, loss, high morale, staff turnover, and so on. These results interact in turn with the four major components of the system to produce new kinds of inputs, such as more or fewer financial and human resources, government policies, and malpractice suits, among others. The system must then process and adapt to these new inputs, in such a way that a dynamic equilibrium is maintained and the basic character of the system is preserved. All the rules, regulations, standards, policies, and procedures that HCPs labor under are presumably part of the adaptive mechanism by which the stability and nature of the system are established and maintained.

Although no single entity is solely responsible for managing the health care system, secondary providers have traditionally exercised most control. Today their influence is being challenged on several fronts, but they continue to play a formidable part in shaping health care in the United States. For this reason, we need to inspect the instrumental role of secondary providers in the system.

Key Secondary Providers

As we have seen, secondary providers comprise a large number of individuals and entities. It is impossible in this brief sketch to discuss the roles of all of them, but there are three that exercise enormous power in shaping the system: medical schools, professional organizations, and drug companies. This section will be restricted to these key secondary providers.

Medical Schools

"Medical schools" in this context are institutions that train physicians. So defined, medical schools help shape the health system in several important ways. First, they produce the essential humanpower for the system. Second, they shape the technology of the system. Third, they determine the technology that is appropriate to the provision of health care. Fourth, they define the standards of health care. Fifth, they serve as models for the education of nonphysician HCPs. Sixth, they help shape nonphysician HCPs by influencing their licensing and accreditation procedures.

But the training of both physicians and nonphysician HCPs goes beyond teaching skills and imparting knowledge necessary for practice. Indeed, it is crucial to our study of morality in health care to recognize that such training attempts to impart a system of values, a sense of professional loyalty, and a profound commitment to independence and self-determination. This attempt to transmit a distinctive group identity and a homogeneous set of values, which underlies the lengthy training period of medical students (and to a lesser extent the training of other HCPs and paraprofessionals), has far-reaching implications for how HCPs will subsequently fit into the system and interact with patients. Commenting on this phenomenon, two authors in the field of public health express its significance this way:

> This helps to justify and at the same time increase the autonomy and
> insulation of the professional groups from outside pressures and control. From

the point of view of the student, the process comes close to that experienced by all those processed by "total institutions." His experience has certain similarities with that of a prisoner in a concentration camp, a new recruit in an army boot camp, and, ironically enough, a patient in a large hospital. He is placed in a tightly scheduled, isolated environment, under careful surveillance. He is stripped of his previous identity, often goaded and humiliated, and given restricted opportunities for contact with those outside the institutional framework. He is given a new identity, one that he shares with his peers, all of whom are treated the same way. His dependence on the institution and the profession is emphasized at each stage of indoctrination. Such a process helps induct the individual into a tight, cohesive, and insulated social world. He learns to function well in such a world and to adhere to norms that go far beyond the limited confines of technical proficiency.[5]

So, clearly, medical schools exert influence that extends far beyond transmitting a body of necessary knowledge and skills. They impart values, group identity, and a sense of autonomy. They also determine the technology that shall be used in health care and define the nature of health care itself. Finally, they help shape the various healing professions, as well as determine who will be admitted to them.

Professional Associations

Other key secondary providers are the various professional associations, which function in large part to advance the interests and standing of their members. Again, as in the education of all HCPs and paraprofessionals, the medical profession makes its formidable presence felt, for the American Medical Association (AMA) serves as the prototype on which all other professional health associations have been modeled. It also exercises immense influence over secondary providers generally.

Although the AMA claims the membership of only about half of today's physicians and its membership figures continue to decline, by numerous indices it wields power far beyond its numbers. In organizational stability, membership size, percentage of total practitioners, stability of membership, fidelity of members, status, wealth, and numbers and circulation of association periodicals, the AMA remains the dominant medical organization among all the health care professions. As one author puts it, "while the medical establishment dominates health care, the AMA dominates the medical establishment."[6] For this reason, the term *organized medicine* generally refers collectively to the AMA, its constituent state and territorial societies, and its component county and district medical societies.

Organized medicine, as embodied in the AMA, has numerous professional and public functions. Professionally, as we saw in Chapter 3, it provides ethical standards and guidance for its members. It also provides resources for the

---

5. *Smith and Kaluzny, pp. 27–28. For a discussion of "total institutions," see E. Goffman,* Asylums *(Garden City, N.Y.: Doubleday, Anchor Books, 1961).*

6. *Stroman, p. 100.*

maintenance of professional skills and working conditions, speaks to and for the profession, and protects physician autonomy and financial and social interests. On the public side, organized medicine helps ensure high standards of medical care and provides people with information on health matters. But the professional and public functions of organized medicine do not always run parallel. For example, zealously guarding physician autonomy and financial interests is not always compatible with providing health care to all who need it. Again, urging unconditional professional loyalty among its membership is not always consistent with public service. We need only remember that organized medicine, operating through the AMA, has opposed most major health bills for over fifty years. Indeed, the only major pieces of legislation it has supported have been for federal funds for hospital construction and medical research.[7]

One of organized medicine's important functions that is most relevant to all nonphysician HCPs involves its effort to coordinate various health groups, their training programs, and their activities. The groups include specialists and their boards, nurses' associations, the American Hospital Association, and a large number of allied health organizations. For example, the AMA's Council of Medical Education is responsible for approving curriculum and training programs and for certifying a whole range of occupational titles, including assistant to the primary care physician, electroencephalographic technician, laboratory assistant, medical assistant, nuclear medicine technician, occupational therapist, physical therapist, radiologic technologist, and specialist in blood bank technology, to name just a handful.

The AMA also exercises considerable indirect influence over the operations of health care. For example, through its Joint Committee on Accreditation, it can pressure for withdrawal of accreditation from hospitals engaged in the training of "unacceptable occupations."[8] The mere threat of such withdrawal can undercut hospital-initiated programs, as for example with the nurse-midwifery program, whose uncertain fate can in part be attributed to opposition from obstetricians and resulting reluctance on the part of hospitals and medical schools to establish such a program. In short, organized medicine (the AMA) influences the way that work is organized and the existing division of labor in most primary provider settings.[9]

By exercising control over various health care professions and institutions, organized medicine maintains physician dominance over health care policies. One can only imagine the complex organizational problems that enshroud this enterprise, which in effect involves one voluntary organization, the AMA, attempting to coordinate other voluntary groups, some of whose vested interests inevitably clash with the AMA's. As a consequence, the creation of specialty boards seems largely the result of a political process rather than a rational or scientific one. In effect, this means that, although these boards may have

---

7. Stroman, p. 193.

8. C. A. Brown, "Division of Laborers: Allied Health Professions," International Journal of Health Services 3 (Summer 1973): 438.

9. Smith and Kaluzny, p. 74.

quasi-legal certifying powers, for the most part they are unaccountable to the public.[10]

A number of moral concerns arise out of these realities, all of which involve the issue of raw power. First, given the enormous influence of the AMA, one must wonder about the public interest and whether it is best served by such a concentration of power. A second concern relates to professional and personal autonomy. Given this physician-dominated organization, how much autonomy do other health care professions and HCPs retain? Apparently very little. How much should they have? This is a question that would seem to need debate urgently, within and outside the professions, and there is some evidence that it is beginning to receive attention. Third, a similar question can be asked about health care consumers. Given the virtually complete control that organized medicine has over who will practice health care, what sorts of services will be permitted, and what kinds of institutions, treatments, and therapies will be licensed, how much autonomy in health matters do consumers have? How much should they have? Again, these questions deserve an airing, because at the very least they raise serious social and moral issues about where the line should be drawn between paternalism and individual autonomy.

Drug Suppliers

The third group of secondary providers that interacts with enormous day-to-day influence on the provision of health care consists of drug suppliers. To get some idea of the extent of their influence, consider that in 1971, 1.5 billion prescriptions were filled in the United States—an average of twenty per family and an increase of 150 percent in prescriptions per capita in a ten-year period. In 1972, $11 billion out of a total of $70 billion spent for health care was for drugs, medical supplies, and equipment.[11] Today the figures are even greater. But that is only part of the picture.

Drug companies must market prescription drugs. In order to do so, they spend about $1.5 billion per year on advertising and promotion, a figure that represents roughly one out of every four dollars they make wholesale on their products and nearly four times what they spend annually on research and development.[12] This advertising is not aimed at the ultimate consumer but primarily at the instrumental consumers, the physician and pharmacist. The marketing effort includes thousands of persons who visit physicians, pharmacists, and institutional purchasers to sell their company's products. Those contacts and the advertising that appears in professional journals are the primary ways that physicians become aware of new pharmaceutical products and new applications for old ones.

To get some idea of the money involved, consider that one pharmaceutical company, Hoffman–La Roche, spent $200 million in ten years and engaged

10. R. Stevens, American Medicine and the Public Interest (New Haven, Conn.: Yale University Press, 1971).

11. J. L. Goddard, "The Medical Business," The Scientific American, September 1973, p. 161.

12. Goddard, p. 162.

about 200 physicians per year to produce scientific articles extolling the benefits of Valium.[13] Similarly, in 1973 the drug industry spent about $4500 per physician for advertising and promotion, which was roughly the equivalent of the cost of a year in medical school. In the same year, the industry contributed less than 3 percent of the budget of American medical schools.[14] What is more, industry estimates peg Hoffman–La Roche's profits from minor tranquilizers such as Valium and Librium, and hypnotics such as Dalmane at about $100 million in 1978.[15]

Unquestionably, then, the pharmaceutical industry has a huge financial stake in the health care system. Naturally enough, it will seek to protect this investment by influencing the nature of health care delivery. In itself, there is nothing objectionable about this. Nevertheless, the impact of drug suppliers on the health system raises serious moral issues. For one thing, it could be argued that the availability of a galaxy of drugs helps foster a reliance on drug technology to solve health problems. For another thing, the whopping commercial interests that are involved raise important questions about the proper distribution of society's resources relative to health and about whether the primary concern is with people or profit. But there are additional moral considerations. Assuming that physicians' drug selections are based largely on the saturation advertising in medical journals and the sales pitches of drug company representatives, one must wonder about the validity and efficiency of at least some prescriptions. Indeed, given that the physician's or pharmacist's knowledge about a drug comes primarily from data generated by pharmaceutical interests and the physician-controlled AMA publicity department, physicians can end up prescribing useless, even dangerous drugs and even promoting drug abuse. A case in point can be found in the history of two drugs: Chloromycetin and Darvon.

During the 1960s, chloramphenicol was packaged as Chloromycetin by Parke-Davis and constituted about one-third of the company's overall profits. Even then it was known that people who take chloramphenicol stand a chance of dying of aplastic anemia, an incurable blood disease, and that typhoid is almost the only disease that warrants such a risk. Nonetheless, through the 1950s and early 1960s, Parke-Davis spent large sums of money to promote the drug. Evidently the promotion worked, for physicians in the United States prescribed chloramphenicol to almost 4 million people per year to treat acne, sore throat, and the common cold. Given that typhoid is rare in the United States, it can safely be assumed that only a small fraction of patients actually needed the drug.[16]

---

13. I. Illich, Medical Nemesis: The Expropriation of Health (New York: Bantam Books, 1976), p. 66.

14. Illich, p. 66.

15. R. Hughes and R. Breuin, The Tranquilizing of America (New York: Harcourt, Brace, Jovanovich, Warner Books, 1979), p. 21.

16. U.S. Senate, Select Committee on Small Business, Subcommittee on Monopoly, "Competitive Problems in the Drug Industry," 1967–1968, part 2, p. 565.

A footnote to the chloramphenicol tale underscores the power that the drug industry wields, not only in health care but in the area of government control of dangerous drugs. At the same time it brings out another moral dimension of the issue of pharmaceutical influence. Shortly after the dangers of chloramphenicol were exposed by a Congressional hearing, its use dwindled, and Parke-Davis was forced to insert strict warnings about its hazards. However, this new restriction did not extend to exports. As a result, the drug continued to be used indiscriminately in Mexico, thus apparently helping to breed a drug-resistant strain of typhoid bacillus that today not only infects Central America but is spreading to the rest of the world.[17]

A similar story can be told of Darvon, which kills not only pain but apparently people as well. According to estimates by the National Institute of Drug Abuse, Darvon kills about 1,100 people a year.[18] Eli Lilly and Company introduced Darvon in 1957, promoting it as an effective painkiller without the potential for abuse or addiction. But many researchers disagreed. In their view Darvon, which shares some of the chemical properties of methadone, was addictive and should have been classified as a narcotic. By 1976 the Justice Department had determined that, Lilly's claims notwithstanding, Darvon was indeed a narcotic. It also found that the drug was being widely abused and was being sold in the streets for anywhere from 25¢ to $1.50 a capsule. Subsequent action by the Department of Health, Education, and Welfare, the Attorney General, and the U.S. Senate has led to Darvon's reclassification as a dangerous drug. Yet incontrovertible evidence to the contrary, Lilly (whose Darvon sales produced $140 million in 1977 alone) insisted that the drug was perfectly safe.[19]

The reasons for the widespread use of drugs in the United States are complex. In part, it seems that this is the result of a health care system that is tightly organized around what some authors have termed

> an unbreakable drug-dispensing circle. . . . This circle consists of a profit-motivated drug manufacturer that must spend heavily to develop new drugs and spend equally heavily to promote them; an often harried physician who has read drug ads and believes in them; and a patient who is seeking and has come to believe it is right to receive instant relief for whatever problem, mental or physical, plagues him. All play equal roles in this circle, and until the circle is broken, the use of psychoactive drugs—and the problems associated with them—will continue to grow.[20]

Each of the preceding is a key component among secondary providers, but organized medicine is the most important. For that reason, we should take a close look at the values of organized medicine, how they help shape the entire system, and what moral questions they raise.

---

17. Illich, p. 62.
18. Hughes and Breuin, p. 21.
19. Hughes and Breuin, p. 32.
20. Hughes and Breuin, p. 31.

The Values of Organized Medicine

Although this discussion concerns the values of organized medicine, organized medicine has no monopoly on these values. In fact, the values are so woven into the fabric of our socioeconomic system that any institution—medical, legal, educational—will reflect them. Indeed, any individual will also do so, to a greater or lesser degree. In considering the values of organized medicine, then, we are simply considering societal values that are emphasized in the conceptualization and practice of health care in the United States. It is important for HCPs to be aware of these values, because they work within the system that embraces them and because their own personal values may conflict at times with the system's values, thus giving rise to moral dilemmas. In addition, many of the problems that the health system faces today have their roots in these systemic value assumptions, especially in the conflicts among them.

To begin with, even a cursory look at the many professional codes and oaths taken by HCPs, the literature in the field, and the selfless actions of health providers testifies to the value of humanitarianism in the health system. Among other things, this humanitarian value commits HCPs to provide services to a patient regardless of social or economic status, personal attributes, or the nature of the health problem; to safeguard the patient's right to privacy; to protect the patient and the public when health care and safety are affected by the incompetent, unethical, or illegal practices of any person; to protect the public from misinformation and misrepresentation; to maintain the integrity of health care; and to collaborate with members of the healing professions and other citizens in promoting community and national efforts to meet the health needs of the public. In brief, a profound orientation to the worth, dignity, and welfare of each individual human being is an inherent value of organized medicine and the health system.

At the same time, other values operating in the health system—and in society at large—often war against this noble humanitarian impulse. Indeed, the tension between humanitarianism and these other values can be great enough to leave HCPs impaled on the horns of numerous moral dilemmas. Two such values are free-enterprise medicine and technology.

Free-Enterprise Medicine

The value of free-enterprise medicine should be understood in terms of economics, for the practice of free enterprise is a defining characteristic of the economic system known as capitalism. For purposes of simplicity, capitalism can be defined ideally, as an economic system in which the major portion of production and distribution is in private hands, operating under what is termed a profit or market system. Notice that this definition contains two features that are generally accepted as defining a capitalistic system: (1) the means of production are privately owned, that is, individuals own society's capital equipment and have the right to use what they own for private gain; and (2) reliance on the market system to determine the distribution of resources and wealth. Both of these features of capitalism underpin the operations of the health care system in the United States today. To see how this is so requires an understanding of the market system and of how self-interest functions in it.

MARKET SYSTEM.   The market system is a complex mode of organizing society, in which an essentially unfettered and spontaneous pursuit of private interest is thought to affect social order and efficiency. In other words, under pure capitalism (which does not exist), the economic factors operating in society are free to interplay with one another and to establish an equilibrium that will provide the best use of resources in satisfying human material wants. Thus, in the United States no control agency or commission dictates the nature of what will be produced, its quantity, or its means of production (with the exception of government expenditures for health, education, welfare, military equipment, and highways). The "laws" of supply and demand supposedly determine the use of resources; what is produced is determined by consumer wants. In an economy like ours, then, suppliers of goods and services keep adjusting supplies and charging prices in response to variations in consumer demand. The ultimate goal is a balance between supply and demand. In effecting this goal, the market system allocates its resources among society's various interests, thereby establishing income levels for different social classes.

SELF-INTEREST.   The belief that individuals are basically creatures of self-interest plays a formative role in the development of classical capitalism. Classical capitalism also believes that individuals should pursue self-interest because, by so doing, economic progress and a constantly increasing social dividend will be ensured. Generally speaking, then, capitalism holds that the surest way to undermine social well-being is to interfere with the pursuit of economic self-interest as practiced in a market economy.

To get the flavor of this capitalistic stress on self-interest and subsequently to understand its impact on the health system, one should at least place it in the context of the laissez-faire individualism formulated in the economic and social philosophy of the eighteenth and nineteenth centuries. The doctrine of laissez-faire individualism maintained that business and commerce should be free from government control, thus enabling the entrepreneur to pursue free enterprise.

For example, Adam Smith (1723–1790), author of *Wealth of Nations* and the leading exponent of laissez-faire economics, insisted that government interference in private enterprise must be reduced, free competition encouraged, and enlightened self-interest made the rule of the day. Similarly, John Stuart Mill, mentioned earlier in connection with utilitarianism, feared government interference in the economy. Mill believed that government interference was justifiable only when society itself could not find solutions to problems. Such problems should then be resolved according to the principle of utility: what produces the greatest happiness for the greatest number of people. But under no circumstances, Mill argued, should the government unnecessarily restrict individual freedom, including the individual's right to realize as much pleasure and progress for self as possible.

This emphasis on self-interest and noninterference took concrete form in the capitalistic claim that individuals have the right to own and manage property and to accumulate profits. Classical capitalists thought that in this way property and profit would serve as incentives for individuals to pursue self-interest, which would produce the greatest social benefit. In a word, economic and social progress was thought to be ensured when individuals pursued self-interest.

We can isolate, then, three beliefs that accompanied the emphasis on self-interest and individual freedom as it appeared in the eighteenth and nineteenth centuries:

1. Individuals should be free to pursue their own interests without interference, providing that they do not impinge on the rights and interests of others.

2. Individuals should be allowed to earn as much money as they can and to spend it however they choose.

3. Individuals should not expect government to aid or inhibit their economic growth, for such interference destroys individual incentive and creates indolence.[21]

Understanding even this bare-bones sketch of capitalism is necessary, because the market system and self-interest are major building blocks in the edifice of organized medicine. Furthermore, without even this rudimentary awareness of our economic system, it is easy and tempting to indict organized medicine as some kind of grotesque anomaly on the landscape of contemporary society. In fact, it is in part a natural outgrowth of an economic system that we have taken pains to cultivate. Therefore, any indictment of organized medicine is ill advised and really unfair outside a broader critique of our whole economic system, or at least apart from an examination of why organized medicine ought not incorporate the enabling assumptions of capitalism.

MARKET SYSTEM, SELF-INTEREST, AND HEALTH CARE. The market system traditionally has been supported by organized medicine in the United States, and this has effectively blocked outside interference with the management of the health care system. Historically, the thinking of organized medicine has taken root squarely in the soil of capitalistic economics, arguing as it has for the free choice of individual consumers and for the private relationships between them and health care providers. It is such a free play of the market that is thought to provide the optimal coordination of health care resources. Viewed this way, health care is simply another commodity that the consumer may choose or refuse. Outside interference is neither necessary nor desirable.

Beyond this, one need not look far in AMA literature to find the themes of free enterprise, the market system, and self-interest. Implied in its editorial philosophy is that medicine, practiced by private physicians, is a small-business enterprise. *Medical Economics,* a bimonthly for physicians, and *Prism,* another of the AMA's many journals, consistently entertain themes and articles devoted to good office management, ways to increase productivity, gains in earning by specialty, elimination of time-consuming procedures, and ways to attract pa-

---

21. *V. E. Barry, Moral Issues in Business (Belmont, Calif.: Wadsworth, 1979), pp. 361–362.*

tients. Most important, fee for service, a cornerstone in the health system structure, is viewed as parallel to the business practice of setting prices for goods and services and as the ideal way to organize a medical practice.

Fee for service refers to the system of paying physicians and other HCPs for each service rendered. Nearly all current physician charges outside the military are calculated this way. There are several reasons solidly rooted in free enterprise for organized medicine's support of the fee-for-service method. First, this system allows physicians to set prices on their services. Since, by professional ethics, physicians are not allowed to be openly competitive and since they have a monopoly on health services, they can charge largely on the basis of what they think patients will pay or what they think is a fair rate for their services. Second, fee for service supposedly makes health services available to all; it ensures that no one will be denied medical care. The argument here is that, if some additional party were involved in controlling costs, then patients might be denied services that they needed. But under the fee-for-service system, physicians alone determine what services are needed. Third, fee for service is said to create an incentive for physicians to get their patients well and to provide services as they are needed.[22] Whether or not these arguments for the fee-for-service system are compelling or logical, one thing is certain: Under this sytem health care providers, especially physicians, retain a sizable measure of control. Thus their own self-interests are generously served.

Technology

Organized medicine places great importance on technology. But as with its commitment to free-enterprise medicine, the value it attaches to technology must be understood in a broader historical context.

Since the Industrial Revolution, our society has carried on a love affair with technology—both with the invention of tools and with the subsequent creation of new skills, information, and theories. In the late twentieth century, this relationship has reached a fever pitch. And little wonder.

The advances of modern science and technology in our own times have led to unprecedented progress. Largely through technology, we move inexorably closer to unraveling the secrets of nature, achieving an economy of abundance, and conquering disease. The average American today is better off in terms of food, shelter, and medical care than anyone else in human history. The back-breaking labor we once did has been replaced by electronic gadgets that liberate us to engage in more creative pursuits. Whereas once the pleasures of the arts, travel, sports, and learning were available chiefly to the upper class, they are now accessible essentially to all of us. Likewise, through literature, films, television, cable television, satellite broadcasts, and cassettes, we come in contact with experiences that formerly could have been fantasized only in the richest of imaginations. Indeed, learning experiences that were once the prerogatives of the lordly are today the claim of anyone with access to the media.

---

22. *Stroman, p. 21.*

Turning to health care, we can see that advances in medicine have for the most part made us the healthiest people of all time, have increased our life expectancy, and have made us eager for and expectant of a bountiful future. The technology and technical achievements of our age are literally mind-boggling, so much so that it is difficult to begin to catalog them all.

Physician and professor of social medicine Victor Sidel conveniently identifies groups of technological and biomedical innnovations that continue to exert enormous influence in four different areas of health care.[23] First are technologies to maintain health and prevent disease, including vaccines for elimination of contagious diseases such as smallpox, diphtheria, poliomyelitis, and measles. Also within this group of technologies are contraceptives, which are used not only to control population growth but also, in connection with genetic counseling, to reduce the number of disabled and defective children. A second group of technologies is involved in diagnosing disease. Many new skills and techniques have appeared here, including epidemiological case studies, automated multiphasic health tests, x-rays, and ultrasound. Such diagnostic technologies offer an impressive array of health benefits. A third set of technologies is used in treating diseases and disability. In this category falls a large number of innovations: drugs, placebos, artificial organs, hyperbaric chambers, organ-freezing techniques, heart-lung machines, organ transplants, coronary bypass surgery, joint replacements, and advances in the treatment of certain forms of cancer. The fourth set of technologies deals with the organization of medical practice. Examples are data processing skills and equipment and the team approach to medicine.

All told, modern medicine's commitment to technology has paid rich dividends for health care consumers. It has also made America's health care system the most technically advanced in the world. In short, the value placed on technology has helped produce a system that, although far from perfect, is in many ways the envy of every nation on earth.

At the same time, the medical organization's emphasis on technology has led to a number of problems of moral concern. The same can be said of its adaptation of the free-enterprise ethos. HCPs need to be aware of these problems in order to grasp the systemic origins of some of the ethical dilemmas that they face and to understand the nature of the criticism that the health system and its providers are receiving from many quarters.

A Critique of the Values

For simplicity, this section reconsiders each of the basic values of organized medicine in terms of its conceptual and operational difficulties. Although it will not say all that can and should be said, it initiates a critical assessment that should help HCPs begin to look more searchingly and honestly at the system they operate in.

---

23. V. Sidel, "New Technologies and the Practice of Medicine," in Human Aspects of Biomedical Innovation, ed. Everett Mendelsohn, et al. (Cambridge, Mass.: Harvard University Press, 1971).

At the outset, we can note a fundamental tension between the values of humanitarianism and of free-enterprise medicine and technology. The humanitarian impulse might commit HCPs to providing services to patients regardless of economic status, but the free-enterprise value defends, even encourages, viewing the patient as a source of revenue. Again, by a humanitarian standard, the needs of the individual patient always come first; but with the emphasis on free enterprise, individual needs can and perhaps should be sacrificed to efficiency. In short, humanitarianism is directed toward benefit for others, whereas the practice of free enterprise is directed toward benefit for self. A clash is inevitable.

A similar clash is evident in the application of technology. Technology can enhance the humanitarian urge by reducing suffering. But when it is used indiscriminately and followed blindly, it can refocus attention from human to machine and can ultimately lead to a net increase in human suffering.

Free-Enterprise Medicine Reconsidered

The basic free-enterprise assumptions that underlie the modern practice of health care are open to the same philosophical criticism as capitalism. Those criticisms will not be detailed here. Suffice it to say that, obviously, since the founding of classical capitalism much has occurred that makes its key assumptions at least questionable. Exorbitant costs, complex technology, increasing demands, and intense competition (although not in health care) have all conspired against individual productiveness; specialization in certain industries has significantly undermined job satisfaction; relatively few corporate giants dominate our contemporary economy. The point is that the conceptual weaknesses in capitalism as it is currently practiced in the United States are in a general sense also applicable to the practice of free enterprise in health care. But beyond this, there are specific operational features about how these assumptions work in health care that raise moral concern.

For one thing, in health care the market system does not operate as it is supposed to in pure capitalism. In the market system consumer demand determines what is produced; thus supplies and prices are controlled by the consumer. Ideally, then, consumers have tremendous power in a pure capitalistic economy. But in health care, the consumer has virtually no influence over the kinds of services supplied, since the provider determines what services the patient needs. In addition, a good portion of the demand is generated by the provider. This is so because consumers of medical services generally are far more ignorant of their needs and of the comparative merits of the goods and services available than, say, consumers of automobiles or education. Indeed, ordinarily they buy, use, and do precisely what they are told to buy, use, and do by their health care providers—who also happen to be the individuals who stand to profit financially from the transaction. It is not surprising, therefore, that the fee-for-service system does not always work out in practice as it is supposed to. Some patients who cannot afford services are ignored and so financially intimidated that they forsake needed health care. Although there may be considerable merit in letting physicians alone determine patients' needs, physicians are free to create their own demand for services. Similarly, although fee for service may function as an incentive to physicians, the incentive may not always benefit the patient. In fact, fee for service may create incentives to provide a variety of unnecessary or more expensive services, as well as hurried care, and unnecessary return visits—for patients are not in a position to know whether the services rendered were either necessary or of high quality.[24]

None of this is meant to imply that HCPs unscrupulously exploit their economic power over consumers. Rather, the point is that the market system in health care operates differently from its idealized conceptualization in the doctrine of capitalism. Recognizing this can alert HCPs to some of the problems and tensions they will face in practice. It can also help them to understand the barrage of criticism that often comes their way and, most important, can help them to start dealing with it constructively rather than reacting defensively.

Technology Reconsidered

Without doubt, ongoing technological innovations help the health care system adapt and maintain dynamic equilibrium. But at the same time, they can introduce a stress that is felt by both providers and consumers. For example, sophisticated technology can be vital in diagnosing and treating disease, but it can also so mesmerize those operating it that they lose sight of their primary humanitarian concern: the individual patient, whose brush with modern technology can produce more anxiety than marvel.

Similarly, it is easy and tempting for professionals to confuse sophisticated techniques with quality care. Thus computerized scheduling and multiphasic screening may streamline operations, but it is still doubtful whether they can alter the social-psychological environments of clinics, the resources that flow into them, or the discrepancies in the care that patients receive in them.

Then there is the question of how well the technology actually does what it is supposed to do. For example, intensive cardiac units, which are quite expensive to operate, are generally assumed to provide the best possible, even the only medically appropriate, care for the heart attack victim. Yet a British study done in 1971 raises serious doubts about these assumptions.[25] In this study, 343 men with episodes of acute myocardial infarction were assigned randomly either to hospital treatment in an intensive care unit or to home care by a family doctor. The results indicated no difference in the mortality rates of the two groups. Indeed, those with a hypotensive history seemed to do better at home.

The questionable efficacy of technological innovations is even more pronounced when the technology involves chemicals and the chemicals are drugs. Take, for example, just one kind: psychotropics. Psychotropic drugs are those that act on the brain or the central nervous system. They include minor tran-

---

24. *Stroman, p. 21.*

25. *H. G. Mather, N. G. Pearson, and K. L. Read, "Acute Myocardial Infarction: Home and Hospital Treatment," British Medical Journal 3 (1971): 334–338.*

quilizers like Librium, Equanil, Valium, and Miltown; sedative-hypnotics like Seconal and Dalmane; antidepressants like Elavil and Quaalude; painkillers like Darvon; and major tranquilizers like Thorazine. Hundreds of millions of prescriptions for these drugs are written annually in the United States. Thus, as a group, Americans consume many billion doses of psychotropics a year. These drugs represent a welcome innovation in the management of stress, but they also take a high toll in human suffering.

Consider, for example, the case of the tranquilizer Valium. According to the National Institute of Drug Abuse and its Drug Alert Warning Network (DAWN), 54,400 people sought emergency treatment for the use, overuse, or abuse of Valium during the twelve-month period from May 1976 through April 1977. In the same period, Valium was the largest-selling prescription drug in the country: 57.1 million prescriptions containing an estimated 3.2 billion pills. The DAWN report also indicated that in the same year at least 900 deaths were attributable to Valium use, another 200 to its chemical predecessor Librium, and still another 100 to Dalmane, a drug with similar properties."[26]

The personal suffering associated with the use of such drugs is rather easy to see, but the social consequences are not. Yet they are just as serious. Numerous researchers and scholars have suggested links between drug use and escape from urgent social problems, such as the changing relations between people and classes, economic uncertainty, intense competition at school and work, a collapse of traditional values, and a loss of faith and confidence in social institutions. Instead, the escape to drugs simply produces another social problem: reliance on chemicals to help solve whatever problems beset us. Thus, what may essentially be social and economic problems are being treated exclusively as medical problems.

Related to this on the conceptual level, technology has had the effect of intensifying the Cartesian approach to health care, as opposed to a configuration approach.[27] The Cartesian approach is named after the seventeenth-century French philosopher René Descartes. It entails reducing a problem to the smallest possible dimension, then attempting to solve it at that level. In patient care this has led to oversimplification of the total organism. The patient is in effect first divided into body and mind, then into organ systems, enzyme systems, cells, and even smaller parts of cells. In contrast, the configuration approach involves viewing problems in the context of total patterns.[28] Applied to health care, this means that all of a patient's problems must be evaluated with the immediate complaint before a decision is made.

In the Cartesian approach, disease is the *sine qua non*. In the absence of a defined organic problem, the patient is assumed healthy. By contrast, the configuration approach recognizes that a patient can have both a disease and symptoms but that the symptoms may be unrelated to the disease. It also

---

26. *Hughes and Breuin, p. 19.*

27. *B. F. Fuller and F. Fuller,* Physician or Magician? The Myths and Realities of Patient Care *(New York: McGraw-Hill, 1978), p. 42.*

28. *Peter Drucker,* Age of Discontinuity *(New York: Harper & Row, 1969).*

acknowledges that a patient who is free of disease may nevertheless have disabling symptoms that need to be treated. In short, in the configuration approach a total pattern must be defined before appropriate health care can be given.

No doubt, technology has reinforced the Cartesian approach to patient care by intensifying a concept of disease that focuses on a single, specific cause for each illness, as opposed to the interplay of many causative factors. In large part it has done this by providing a sophisticated basis for diagnosis and treatment, by means of numerous eye-popping advances of the sort previously cataloged. At the same time, a strict adherence to the Cartesian approach can lead to many disquieting operational problems. It also promotes a health system that does not take social components into account and has probably contributed to diffused, complicated, and unnecessarily expensive care by encouraging HCPs to deal exclusively with tissues and organs rather than with the complex interplay of social, physical, and emotional factors that must be taken into account in disease prevention.

Although the health system's emphases on free enterprise and technology have yielded an impressive litany of personal and social benefits, they are not lacking in their own particular conceptual and operational disadvantages. The overriding problem is that these values clash with the system's humanitarian purpose. This clash produces considerable inner conflict for the sensitive, conscientious HCP. It is also at the root of what is already shaping up as a major social debate of our time: the capacity of our health system to provide adequate health care to all citizens.

Another issue that is beginning to garner attention is the social function of the health care system as a maintenance or control mechanism. Because this issue raises important concerns about individual autonomy, we should consider it before concluding our brief look at the health care system.

The Social Function of the System

The health care system serves many purposes. It provides health care; it employs huge numbers of professionals and paraprofessionals; it acts as a source of information and disseminates knowledge about health matters. But one of its most important functions, and one with heavy moral overtones, is its social role. Specifically, the health care system functions in part as a mechanism of social control and maintenance. Indeed, in a classic essay, sociologist Irving Kenneth Zola has argued that medicine has become an institution of social control.[29] Whether or not the system is primarily a social control mechanism might be disputed, but there can be no doubt that the health system exercises enormous control in both the health and nonhealth spheres, and this controlling function raises moral considerations.

---

29. I. K. Zola, "Medicine as an Institution of Social Control," in The Cultural Crisis of Modern Medicine, ed. John Ehrenreich (New York: Month Review Press, 1978).

Health Management

Undoubtedly, the health care system, like the educational or legal system, is part of the larger societal maintenance or control system. One way that the health system performs a control function is by reducing health threats to society. For example, health clinics attempt to control such diseases as venereal disease and tuberculosis; physicians give opinions at mental competence hearings; various health associations educate the public to life-threatening and disease-producing lifestyles. Such control activities seem harmless, even socially beneficial, but the issue of social control always raises the important moral issue of personal freedom or autonomy. After all, by definition any social control mechanism encroaches on the personal liberties of those constrained by it. So, in functioning as a control institution, the health system is bound to confront philosophical and moral questions concerning the proper line of demarcation between individual and society. This is always a controversial issue, and drawing this line becomes even more problematic in the light of the particular operational and conceptual difficulties that beset the health system as an institution of social control.

For example, conflicts often arise among various control institutions in society about areas of responsibility. Today it is rather common for prisons to incorporate educational and health components. Similarly, hospitals frequently devote resources to patient education. Again, in some old-age institutions, treatment is punitive rather than supportive. As society's various control institutions begin to share traditional responsbilities, a serious question emerges as to who is responsible for what.

Pinpointing responsibility may seem little more than an abstract task, but the problem can become real and immediate in the day-to-day operations of HCPs. For example, health clinic workers, who daily see the connection between disease and ignorance, misinformation and neglect, might well puzzle over what part patient education should play in their duties as health providers. Likewise, HCPs in contact with drug addicts might understandably reflect about what part, if any, law enforcement should play in their health care responsibilities. Clearly, the more society charges both individual and system with control functions, the heavier such considerations weigh on both.

Health care facilities, too, can find themselves engaged in activities that traditionally have fallen to other control institutions. A case in point is New York City's Bellevue Psychiatric Hospital, which employs a battery of lawyers to look after the civil rights of patients. The result is that under one roof two powerful professions—the legal and medical—compete to assert their own professionally determined definitions of deviance. What hangs in the balance is the appropriateness of involuntary commitment to a mental hospital as opposed to possible imprisonment. The dilemma looms even larger when one realizes that by some estimates 90 percent of all patients in state mental hospitals are involuntary commitments.[30]

---

30.   T. Szasz, Law, Liberty, and Psychiatry: An Inquiry into the Social Uses of Mental Processes (New York: Collier Books, 1963), p. 46.

But pinpointing responsibility for social control is only one of the conceptual and operational problems that currently beset the health system. Another concerns the ultimate social impact of collaboration between various institutions that perform the control functions. Take, for instance, the collaboration between medicine and law. At the very least, such an alliance poses a most formidable mechanism for social control. To illustrate, recently the National Institute of Mental Health and the Legal Assistance Administration of the Department of Justice financed a study on the use of lobotomies to deal with behavior control. A similar collaboration between medicine and law can be seen in the practice of reducing the criminal sentences of heroin addicts, providing that they agree to participate in methadone maintenance programs. But the health system's collaboration need not be restricted to the legal establishment. When physicians in a state refuse to perform deliveries on welfare mothers who will not consent to sterilization as part of the birth, by this coercion they are in effect acting as a punitive arm of the welfare system. Such examples place in high relief the roles of the health system and HCPs in social control; the important moral questions that implementing those roles raise; and the immediate impact that functioning as agents of control can have on HCPs, patients, the system itself, and society in general.

Undoubtedly, then, defining and assigning responsibility for control, defining and processing "deviants," and the collaboration between control institutions raise troublesome issues for the health system. Such matters lead not only to speculation about the proper role of the system in society but, more specifically, to a question about the proper limitations of its power and influence. This question of limits to the social control function of the health care system becomes crucial when the discussion turns to the system's management of nonhealth matters.

Nonhealth Management

As our lives fall more and more under the influence of the health system—that is, as we become "medicalized"—the system acquires increasing power to shape and influence us. In some instances, HCPs who are supposedly morally neutral and scientifically objective make judgments with profound implications that have little or nothing to do with health and illness. This can be especially true of the physician whose status as expert has been extended into nonmedical areas of our lives.

Take, as an example, the case of obstetrician-gynecologists. Presumably legitimate experts on the female reproductive system, these medical specialists have on many occasions succeeded in extending their sphere of influence to include the female sex role, psychology, and sexuality. Yet, as a group, they appear to have no particular expertise in these areas; indeed it could be argued that, given their training, they probably bring a biased viewpoint to these subjects. Nevertheless, they often do offer their views and, because of their medical status, find an audience for them. A look at gynecology textbooks illustrates the point. Here, for example, is some advice offered budding physicians by a Harvard gynecologist in a 1971 textbook. The questionable value assumptions should need no comment.

> If the sexual inadequacy on the part of the wife stems from a fundamental immaturity and inability or failure to assume the normal adult role in the marital relationship, [the gynecologist] may be able to help by gradually imparting to her the nature of what her role should be—as Sturgis has described it so well, the fact that although the instinctive sexual drive of the male, who carries the primary responsibility for biologic survival of the race, is greater than hers, it is nevertheless of fundamental importance for the woman, his wife, particularly in a monogamous society, to make herself available for the fulfillment of this drive, and perfectly natural and normal that she do it willingly and derive satisfaction and pleasure from the union. Herein lies her power and purpose—to preserve the family unit as a happy, secure place for both man and wife and for the rearing of their children. Only by understanding and assuming this role can a woman throw off childhood inhibitions and taboos and attain the feminine maturity essential to a happy, successful marital adjustment.[31]

What is happening here is essentially that women are being categorized by role. Categorization is always dangerous, and it is particularly questionable when done by those whose training has not equipped them to do it. Yet this is precisely the nature of the extension of medical influence into nonhealth areas. In a medicalized society, the influence of HCPs, especially physicians, extends not only to medical matters but to the categories to which people are assigned. Social critic Ivan Illich thinks the tendency is pervasive. He writes:

> Medical bureaucrats subdivide people into those who may drive a car, those who may stay away from work, those who must be locked up, those who may become soldiers, those who may cross borders, cook, or practice prostitution, those who may not run for the vice-presidency of the United States, those who are dead, those who are competent to commit a crime, and those who are liable to commit one. . . . Each kind of certificate provides the holder with a special status based on medical rather than civic opinion. Used outside the therapeutic process, this medicalized status does two obvious things: (1) it exempts the holder from work, prison, military service, or the marriage bond, and (2) it gives others the right to encroach upon the holder's freedom by putting him into an institution or denying him work.[32]

There are several dangers associated with such widespread categorization. First, any categorization by definition focuses on similarities among group members while ignoring differences. When applied to persons, categories can have the effect of overlooking, even denying, the uniqueness of each individual. Second, in overlooking differences among like members of a group, categorization can gloss over individual needs, thus leaving them unmet. We will have many occasions in the pages ahead to see how this can undercut adequate health

---

31. *T. H. Green*, Gynecology: Essentials of Clinical Practice *(Boston: Little, Brown, 1971), p. 436. Quoted in D. Scully* Men Who Control Women's Health: The Miseducation of Obstetrician-Gynecologists *(Boston: Houghton-Mifflin, 1980), p. 21.*

32. *Illich, p. 71.*

care. Third, and most important, extensive categorization in health care inevitably threatens individual autonomy. When individuals are subjected to lifelong medical supervision, when "society is so organized that medicine can transform people into patients because they are unborn, newborn, menopausal, or at some other age of risk,"[33] then individuals surely lose some of their self-determination to their healers.

The health care system therefore exercises social control not only over health matters but over concerns that relate only remotely, if at all, to health. Social control always raises moral questions about the proper limits of societal intervention in the lives of its citizens, but control by the health system in nonhealth areas raises even more serious ones. Chief among them is what justification, if any, the health care system and its members have to encroach on individual autonomy in nonhealth matters. This question must—and should—be of utmost moral concern to the health care system, its members, and to society in general. Its urgency promises to increase if the influence of the health system continues to expand into nonhealth areas.

Summary

This chapter defined the health care system as a distinct, separable, and permanent part of our society consisting of an elaborate set of arrangements and interrelated parts whereby the technology of the health sciences are made available to us. The chief components of the system are secondary and primary providers, financial mechanisms, and patients. Most of the system's power and influence is concentrated in the hands of secondary providers, especially organized medicine. Traditionally, organized medicine has espoused the values of humanitarianism. At the same time it has also been committed to free-enterprise medicine and technology. Each of these values has yielded an impressive array of personal, professional, and social dividends. At the same time, organized medicine's humanitarian orientation frequently clashes with its emphases on free enterprise and technology. Finally, the health care system performs important social control functions in and outside health management. Each of these topics—the interplay of various components, the domination of the system by physicians, the values of organized medicine—raises serious moral and social questions that deserve airing within and outside the health care professions.

CASE PRESENTATION
Making Flowers and Getting Educated

Dr. Penzias knew his friend and colleague well enough to recognize all the signs of a gathering storm. The mottled neck, the clenched fist, the pursed lips—there

---

33. *Illich, p. 72.*

was no doubt about it. Dr. Wyeth was showing all the signs of a volcano about to erupt. But why?

Dr. Penzias couldn't figure it out. Was it something he had said? He did have a way of getting rather unruly at these CME (Continuing Medical Education) meetings. No, he decided, Fay Wyeth knew him too well not to take any of his tomfoolery to heart. Just as he was about to explore other possibilities, Dr. Wyeth shoved the journal she was reading under Dr. Penzias's nose and barked, "Here, read this and weep."

"Do I have to?" Dr. Penzias asked meekly.

"If you want a ride home you do."

What Dr. Penzias read was an advertisement for a powerful tranquilizer, which was being promoted as a means of controlling disruptive behavior in nursing home patients with minimal risk of sedation and hypotension. The ad ran three pages, although Dr. Penzias never got beyond the first, which pictured a smiling, alert and wrinkled woman holding a red paper flower under the caption "I made a flower today." On the facing page, the ad boasted of the drug's potential for "keeping nursing home patients under control."

"Sounds pretty good," Dr. Penzias muttered. "It should do the trick."

"*Do the trick?*" Incredulity fired the blood in Wyeth's face. "Are you a doctor or a magician?"

"Huh?"

"We're supposed to care and to cure, Jerry, not to turn people into zombies!"

"Now hold it just a minute, Fay. Don't you think you're exaggerating?" Dr. Penzias held up the picture of the endearing old lady with the red paper flower. "Really, now—is this a zombie or an elderly woman made functional with the help of a little medication?"

Dr. Wyeth smiled—the tight little smile of someone about to deliver the *coup de grace.* "May I bring to your attention, Dr. Penzias, the fine print in the corner of the page?" The fine print indicated that the woman actually had made the flower as part of a vocational group therapy program and was *not* a patient receiving the advertised tranquilizer.

"Oh, well," Dr. Penzias grinned sheepishly, dismissing the disclaimer with a wave of his hand. "All's fair in love and war—and advertising, I guess."

"It's not just the ad, Jerry—although I think this sort of Madison Avenue hucksterism is unprofessional and downright dangerous. In this case it reinforces the notion that our professional relationship to the elderly is one of custodian to inmate. Maybe some of these people are difficult to manage because of appalling institutional conditions. But rather than facing up to the conditions, what do we do? We give them a pill. And it's not just the elderly. The whole nation is turning into a tranquilized society."

Before she could go on, Dr. Penzias glanced at his watch and reminded her that they had only a few minutes to get to the next seminar panel, which happened to be on anxiety.

"Very informative, very informative," Dr. Penzias judged the seminar at its conclusion.

"Really?" Dr. Wyeth said skeptically. "How many panelists advocated an alternative to drugs as a maintenance approach to anxiety?" Out of nine panelists, only one had. "And did you notice the repeated endorsements of particular drugs?"

"Well, sure," said Dr. Penzias, "but that's only because. . . ."

"Because why, Jerry?"

"I know what you're going to say, Fay. Because they're the drugs of the company sponsoring this CME program."

"Precisely."

"So what's wrong with that? After all, the company is underwriting this program. They're making it possible for you and me to keep up with advances in medicine, to update our knowledge and maintain our licenses. What's so horrible about that?"

"How deliriously altruistic of the pharmaceuticals!" Dr. Wyeth laughed. "Did it ever occur to you, my dear Dr. Penzias, that all these seminars that the drug companies so generously sponsor for us are simply another device to sell their drugs? Instead of just besieging us with representatives, journal ads, and direct mail, they're sponsoring so-called educational programs that we slavishly attend for additional brainwashing."

"That's pretty harsh, Fay."

"Is it?" And with that Dr. Wyeth produced, seemingly from thin air, an article from the 1978 health issue of *Advertising Age* titled "Education: New Drug Promoter." She proceeded to quote at length to Dr. Penzias. The article reported that the outlook for educational programs as promotional tools was excellent, according to those involved in producing the programs. It went on to point out that such educational programs probably would continue to account for a greater percentage of the total promotional budget. One reason for the growth in this area was cited by the vice president of a company specializing in educational programs for the medical profession. In his view, there were lots of problems in getting the physician's attention. Coming to physicians in a pseudoeducational setting with credible information and with advertising unopposed by competitors in that setting was a most prudent way to promote drugs.

When Dr. Wyeth finished, she said, "Conflict of interest—clear and simple. And we, you and I, Jerry, are unwitting accomplices."

Dr. Penzias sat silent, unaware that his colleague had finished. Somewhere along the way his mind had begun to wander. Other more interesting concerns had crowded out Dr. Wyeth's diatribe. Indeed, Dr. Penzias now was beset by one overriding concern: Should the seminar run much longer, he would miss his weekly racquetball match.

Questions for Analysis

1. *Identify all the parties who have an interest in drug-sponsored educational programs. Make a list of what is of value to each—that is, what each has at stake. Indicate where values parallel and clash. A conflict of interest may simply be defined as a situation in which a party in a transaction stands to benefit*

financially from some official action. Do you think that the drug companies who sponsor CME programs, like the one in this case, are involved in a conflict of interest? Do you think such sponsorship is immoral? Explain.

2. *What do you think of the sort of drug advertisement cited in this case? Dr. Wyeth has raised one moral concern about it. Do you agree? Are there other moral concerns that such ads raise?*

3. *What, if anything, is objectionable about the fact that instrumental consumers (physicians and pharmacists) receive virtually all of their information about drugs from drug company representatives, journal ads, direct mail, and now company-sponsored educational programs?*

4. *At one point, Dr. Wyeth refers to herself and Dr. Penzias as accomplices. What does she mean? Does the charge have merit? What responsibilities do HCPs have in this area?*

5. *"Drug companies, like the medical professions, are caught between humanitarian values on the one hand and free-enterprise and technological values on the other." Do you agree with this statement? Explain.*

Exercise 1

At the urging of its largest state chapter, the American Psychiatric Association (APA) creates a committee to look into the possibility of abolishing advertisements in its professional journals (*American Journal of Psychiatry* and *Psychiatric News*). The committee discovers that the net loss to the association will be $100,000 a year if all drug support is eliminated, including the drug ads and the $70,000 a year the APA gets from drug companies in the form of direct grants and contributions to annual meetings. Some committee members want the ads banned because they think they influence doctors' attitudes and thus endanger their objectivity in deciding what's best for patients. Others think the drug companies have a right to advertise their products where the ads will do the most good. Also, they say that particularly offensive ads can be settled on their own merits. With which aspects of this issue would the various ethical theories be particularly concerned? What course of action would they suggest?[34]

Exercise 2

Reconsider your responses to the Values and Priorities questions. Are there any you would now change or enlarge on?

---

34. *In 1976 the California chapter of the APA made precisely the demand cited in this case. California psychiatrists subsequently banned drug ads in their publications. Hughes and Breuin, p. 243.*

Collective Responsibility and the Nursing Profession

James L. Muyskens

In the following essay, professor of philosophy James L. Muyskens oberves that the nurse is the linchpin of the health care delivery system. If the system is substandard, nurses are not merely victims but also accomplices. Nurses must share some responsibility for the system's deficiencies, but at the same time they are often so constrained in their choices that they cannot rightly be held personally responsible. The paradoxical plight of nurses is that they are powerful and powerless, responsible yet not responsible.

To deal with this paradox, Muyskens introduces the idea of collective responsibility, responsibility of the group, profession, or system rather than the individual alone. In Muyskens' view, the health care system is constituted and functions so as to preclude ascribing responsibility to the individual nurse. Who or what, then, is to blame for substandard health care? The system.

But do not misunderstand the author. Although collective responsibility may exonerate the individual nurse for substandard practices or action, it may also show that group's behavior is substandard. When this is the case, the nurse has a responsibility to help change the group's conduct. The ultimate value of the author's concept of collective responsibility is that it can be used by HCPs working to upgrade the nursing profession and the health care system generally.

Members of the nursing profession, for a variety of reasons including the nature of the profession but also economic exploitation and sexism,[1] have been "caught in the middle." On the one hand, for example, the nurse is hired to carry out the directives of the physician and to support the policy of the hospital administration. The system cannot function as presently constituted without such cooperation and support in carrying out the decisions and policies of those higher up in the hierarchy. Yet, on the other hand, the nurse is legally and morally accountable for her or his judgments exercised and actions taken. "Neither physician's prescriptions nor the employing agency's policies relieve the nurse of ethical or legal accountability for actions taken and judgments made."[2]

A common predicament of nurses is expressed in the April issue of *Nursing 78* by a nurse at a West Coast university hospital. She says:

> Our biggest problem right now is that our nursing leadership at the administrative level is completely impotent. They have no voting rights on any committee that has direct control over the hospital and/or nursing. Worse, the acting director and her associate have no idea of taking any power into their own hands, where it rightfully belongs. They ask permission to improve staffing ratios, by increasing or closing beds, and when they're turned down, say to us "Sorry girls! Work doubles." . . .[3]

The overwork and understaffing not only make working conditions less than desirable for the nurse, they clearly endanger patients. When, for example, one registered nurse and an aide must try to care for thirty to thirty-six patients who have just undergone surgery, the situation is very dangerous and health care cannot be delivered in accordance with acceptable standards.

We can all sympathize with the nurse who wrote the following:

> I am supposed to be responsible for the control and safety of techniques used in the operating theatre. I have spent many hours teaching the technicians and the aides the routines necessary for maintaining aseptic conditions during surgery. They have learned to prepare materials and to maintain an

adequate supply for all needs. They have learned to handle supplies with good technique.

I find it is extremely difficult to have these appropriate routines carried out constantly by employees with little theoretical background or understanding. The surgeons are frequently breaking techniques and respond in a belligerent manner when breaks in technique are brought to their attention. I find a reminder of techniques often brings a determined response to ignore the reminder and proceed with surgery. For a male surgeon to be questioned by a female nurse is a serious breach of respect to them.

One day a surgeon wore the same gown for two successive operations even though there were other gowns available. I quietly called this to his attention, but I had no authority which really allowed me to control his behavior for the good of the patient. In this situation even the hospital administrator was of no help to me.[4]

This nurse is responsible for the control and safety of techniques used in the operating rooms. The conditions over which she is responsible have fallen below acceptable standards. Although she has done her best, the assigned task has not been accomplished. The patients who have a right to expect, and have paid for, a safe and aseptic operating room have been let down.

Nursing is the largest group of health-care professionals within the vast health-care delivery system—a system that, despite some dramatic achievements, is increasingly under attack as dehumanizing, exploitative, and cost-ineffective. Despite the seeming powerlessness of any individual nurse, taken collectively nursing more than any other health-care profession is a necessary component in the emergence of the present health-care delivery system. The present system could not have developed without nursing. If all nurses were to walk out tomorrow, the system would collapse. This cannot be said for any other group of health-care professionals, including physicians. Hence, if the health-care delivery system is substandard (as I believe it is), the nurse is not merely a victim of the system (along with the rest of us), but she or he is also an accomplice. As an accomplice she shares responsibility for the system's deficiences. The nurse's plight is by no

means unique. The paradoxical plight of the nurse of being both powerless and powerful, responsible yet not responsible, is a plight in which we almost all find ourselves in some aspects of our lives.

One way to try to make sense of these paradoxical situations—the way to be explored in this essay—is to introduce the notion of collective responsibility. Two dramatic and widely discussed illustrations of this are the prosecution's case against certain middle-level Nazis after World War II and the defense's case for First Lieutenant William Calley charged with murder at My-Lai.

In the prosecution case, blame for the actions of certain individual members of the collective is ascribed to all members. Karl Jaspers expressed this view when he said: "Every German is made to share the blame for the crimes committed in the name of the Reich . . . inasmuch as we let such a regime arise among us."[5] In condemning every German, Jaspers is not merely blaming each German for his or her active or passive tolerance of the Nazis. He is saying that "the world of German ideas," "German thought," and "national tradition" are to blame. Collective responsibility is used as a net from which no member of the collective can escape.

In the defense case, the individual whose behavior has fallen below the acceptable standard is shielded from the full weight of blame, because the weight is shifted to the collective. It is the collective, the system, that must bear the brunt of the burden rather than the individual. In the Calley case it was claimed that Americans as a group failed to perform as they could have been expected.

In a recent survey of nurses' attitudes[6] this defense strategy was tacitly used. It was reported that, although nurses saw themselves as performing well given the work conditions, they "felt they ought somehow to deliver even when the system won't let them." The writers of the report indicate that this blame is misplaced ("not deserved"). Although performing below the acceptable standard, they were not to be blamed because as individuals each was doing the best possible for her in the situation. The system itself was to be blamed.

If the blame appropriately ascribed in a situation is no greater than the sum of all the ascriptions of blame to the individuals, we do not have a case of collective responsibility except in a weak (distributive) sense. By collective responsibility in the strong (nondistributive) sense—as the term is to be used in this essay—we mean that the responsibility of the group is not equivalent to that of the individuals. That is, the whole is not equal to the sum of its parts.

It is incontrovertible that we do ascribe responsibility to collectives in this strong sense. To use an example of D. E. Cooper,[7] if we say that the local tennis club is responsible for its closure, we don't necessarily or usually mean that the officers of the club or any particular members are responsible for its closure. If you were to question the speaker, he or she may be unwilling to blame any particular individuals or the officers of the club. It is not that any person failed to do what was expected of him or her. Yet something was missing. "It was just a bad club as a whole."[8] From the claim that the local tennis club is responsible for its closure no statements about particular individuals follow. "This is so," as Cooper says, "because the existence of a collective is compatible with a varying membership. No determinate set of individuals is necessary for the existence of a collective."[9]

As R. S. Downie has argued, ". . . to provide an adequate description of the actions, purposes, and responsibilities of a certain range of collectives, such as governments, armies, colleges, incorporated business firms, etc., we must make use of concepts which logically cannot be analyzed in individualistic terms."[10] The reductionists who deny this have the principle of parsimony on their side, but little else. Although the reductionist says the ascriptions of collective responsibility could be reduced to statements about individuals, he or she does not do it. These reductionistic attempts suffer from the same problems and deserve the same fate as the discredited reductionist programs in theory of knowledge and philosophy of science.

The question to ask then is what set of conditions must obtain in order properly to ascribe nondistributive, collective blame or responsibility. The conditions advanced by Cooper in his essay "Responsibility and the 'System' " are sufficiently accurate and refined for purposes of this essay. These conditions are:

1. Members of a group perform undesirable acts.

2. Their performing these acts is partly explained by their acting in accordance with the "way of

life" of the group (i.e., the rules, mores, customs, etc., of the group).

3. These characteristics of the group's "way of life" are below standards we might reasonably expect the group to meet.

4. It is not necessarily the case that members of the group, in performing the acts, are falling below standards we can reasonably expect individuals to meet.[11]

A few comments about these conditions are in order. Clearly we do not *hold* an individual or a group responsible—that is, following its etymology: having liability to answer to a *charge*—if undesirable acts have not been performed. When no undesirable acts occur, the question of blame or responsibility in the sense of liability does not arise. Hence we see the need for Condition 1.

The second condition is not strictly necessary. It does seem, as Virginia Held has argued,[12] that when special conditions obtain, even a random collection of individuals can be held responsible (a claim denied by Condition 2). However, for present purposes—consideration of collective responsibility of members of a profession—this stronger claim need not be defended. The most plausible cases for ascribing collective responsibility are those cases in which the group has distinctive characteristics, has a sense of solidarity and cohesion (for example, feels "vicarious pride and shame"[13]), members identify themselves as members of the group (for example, "Who are you?" "I am a nurse."), and some of these group feelings or characteristics are appealed to in explaining the acts in question. For example, if the citizens of Syldavia can be characterized as being rather hostile and distrustful of foreigners and their customs, laws, and policies reflect this, then, when (say) some border guards—in overzealously carrying out the Syldavian policy—kill some visiting dignitaries, we blame not only the border guards but the Syldavians. In contrast, if these border guards steal from the visiting dignitaries but in accounting for this behavior we would not be inclined to appeal to any larger group feelings or characteristics, we definitely would not wish to ascribe collective blame.

We have seen above in the variety of cases discussed that it is when a collective fails to live up to what can reasonably be expeced of it—i.e., it falls below an acceptable standard—that it can incur collective blame. Hence we see the need for Condition 3.

Condition 4 is necessary because the standards applied to groups may be different from those applied to individuals. For example, we may feel that the nurse (in the case cited above) who was charged with responsibility for the control and safety of techniques used in the operating rooms adequately met her obligations. She did not fall below standards we can reasonably expect an individual to meet. After all, as Joel Feinberg has argued, "no individual person can be blamed for not being a hero or a saint." Yet, as Feinberg goes on to say, "a whole people can be blamed for not producing a hero when the times require it, especially when the failure can be charged to some discernible element in the group's 'way of life' that militates against heroism."[14] Although Feinberg was not talking about this case or collective responsibility of the nursing profession (he was talking about a Jesse James train robbery case), his remarks are especially apt for this case and many other situations within the nursing profession.

One can readily see that conditions outlined for properly ascribing nondistributive collective responsibility obtain in many situations within professions. Professions more than most other collectives are bound together by common aspirations, values, methodologies, and training. In too many cases, they also have similar socio-economic backgrounds and are of the same sex and ethnic group. As we have seen, the more cohesive the group, the less problematic the ascription of collective responsibility. The fact that professions such as nursing promulgate codes of ethics or standards of behavior toward which they expect members to strive, provides a clear criterion for judging whether the actual practices of the profession fall below standards to which we can reasonably hold the group.

In addition to meeting these formal criteria for ascribing collective responsibility, there are several other reasons unique to professions for ascribing collective responsibility in certain situations.

A. There are several ways by which one becomes responsible. One can be *saddled* with it by circumstances, one can have responsibility *assigned* to one, or one can deliberately *assume* responsibility.[15] Typically a profession is chosen. In choosing the profession, one *assumes* the responsibility concomitant with being a professional. One chooses to adopt the values, methodology, and "way of life" of the profession. Such choice is much less prominent with most other basic group affiliations. One does not choose family membership, region of birth, usually not citizenship, and often not military service. Once in the profession, of course, as people go about their jobs, they will also sometimes be saddled with responsibility by circumstances and be assigned responsibility. But these assignments are all within the context of choice to assume professional responsibility. This choice to assume professional responsibility provides the backdrop for all one's professional activities. Hence, as a professional, more than most other group affiliations, one sees oneself as a member of the group and has—with eyes open—chosen the identification.

B. Nurses (as is, of course, also the case in several other professions) have been vested by the state with the power to regulate and control nursing practice. This collective power or right—given exclusively to the profession—has concomitant with it a collective responsibility or duty to see to it that acceptable standards are maintained. Since it is possible that each individual nurse, including officers of the American Nursing Association, is meeting acceptable standards in her or his own assignments and yet the group's "way of life" must be characterized as below an acceptable standard, appeal to collective responsibility is one of the tools the public has at its disposal to try to insure adequate nursing and general health care. Obviously in these cases (when no individual has failed to meet her or his legal obligation) the public does not have recourse to lawsuits against individuals.

C. Supposedly as a means to protect the public, the licensing statutes of the states allow only those who have passed certain requirements set down by the state to practice nursing. One result of this is that the profession which is by law also self-regulatory becomes a protected monopoly. If a person is going to receive nursing care, this care must be provided by a member of the profession. If nursing care is to be upgraded, it must be from within with at most prodding from without. Quite clearly one of the most effective tools for such prodding is that of demonstrating collective responsibility, a responsibility that goes beyond the sum of each individual's responsibility.

From the discussion thus far, it is evident that the appeal to collective responsibility when some substandard behavior or undesirable acts have occurred is a two-edged sword. It can be used to show that, despite undesirable performances or actions or conditions within a collective, a particular member of the collective is not individually responsible. However, it can also be used to show that, despite the fact that the behavior of individuals does not fall below standards we can reasonably require individuals to meet (given that we cannot *demand* that an individual be a hero), the group's conduct is below standards we can reasonably expect the group to meet. One of the reasons the weapon of collective responsibility looks suspect in the widely discussed World War II prosecution and Vietnam conflict defense cases is that only one edge of the sword is used while the other edge is conveniently ignored.

If conditions for properly ascribing collective responsibility are satisfied, to the extent that the individual is exonerated, the group is indicted. To the degree the individual *qua* individual is indicted, the group is exonerated. Either way the individual group member bears responsibility. For any member of a collective but especially (for reasons cited above) a professional, it is not enough just to know that one has done all that could be expected of him or her strictly as an individual. The arm of responsibility for a professional has a longer reach than that of the individual.

Specific situations within the nursing profession illustrate the two edges of the sword of collective responsibility. These situations should be seen within the context of the rapid evolution of the nursing profession in recent years. In recent years there has been considerable effort both within and outside the profession (e.g., the medical profession) to upgrade the requirements for licensure. These efforts have borne results. The scope of the professional nurse has expanded greatly as exemplified by medical-assistant programs and the use of nurses as paramedical practitioners to relieve the shortage of medical doctors in certain areas. The history of the struggle first to adopt a code of ethics for American nurses and then to revise it reflects this evolution. Tentative codes were presented in the twenties, thirties, and forties. These efforts were met by opposition from those who feared the professionalization of nursing. A striking instance of this is the advice given by a physician to one of the earliest advocates of a code of ethics for American nurses: "Be good women but do not have a code of ethics."[16] It was not until 1950 that a code of ethics was adopted.

The code has been changed several times since then, the most recent being in 1976. Two of the most interesting changes from our vantage point have been the following: Earlier versions stated that the nurse had an obligation to carry out physician's orders. The 1968 and 1976 versions of the code stress instead the nurse's obligation to the patient (called client in the 1976 version). The physician just mentioned who advised against having a code may have foreseen this development! Whereas earlier versions of the code point to an obligation to sustain confidence in associates, this has been replaced by the obligation to protect the patient from incompetent, unethical, or illegal practice from any quarter.[17]

With this background one can see why it is especially interesting to look at the nursing profession when speaking of collective responsibility in the professions. The fundamental issue in the ongoing struggle to upgrade the profession—reflected in the code changes—has been that of accountability, the willingness to make decisions and accept responsibility for these decisions. The crucial question in the attempt to upgrade the profession is that of the interface of individual and collective responsibility.

The author of an article in the *Quarterly Record of the Massachusetts General Hospital Nurses Alumnae Association* wrote about "blame avoidance" behavior in nurses. As explained, blame avoidance behavior is exhibited when the nurse says such things as "I did this because the supervisor told me to do it," or "the doctor ordered it," or "the hospital rules demanded it." The author maintains that accountability requires that the nurse can say, "I did this because in my best judgment it is what the patient needed."[18] Notwithstanding the many good qualities common to nurses, blame avoidance behavior does seem to be one of the more prevalent, endemic faults of the nursing profession. As we have seen, a concerted effort by many within the profession has made inroads on this "way of life" of the profession.

These efforts have been made without explicit appeal to the concept of collective responsibility. As a result, judgment in cases of blame avoidance and other unacceptable or undesirable behavior has tended either to be too harsh or too lenient. That is, either (a) one judges that the individual nurse caught in the middle and in difficult circumstances has done all one can reasonably expect her or him to do. After all, we can not expect or demand that such a person be a hero or a saint. Hence, the nurse is exonerated. Yet the unacceptable practice or condition continues unabated. Or (b) one focuses on professional responsibility and the fact that, if some individuals do not stand up against substandard practices—no matter what the odds of thereby improving the situation and no matter at what price to the individual—these practices likely will not be stopped. From this perspective the individual nurse who fails to do all within her or his power—including actions that will likely jeopardize the nurse's position—to insure the best care possible for patients in the nurse's care, is judged to be a moral coward.

For example, in the case of the nurse charged with responsibility for maintaining a safe and aseptic operating room, without appeal to the concept of collective responsibility we are likely to say either: (1) that she has done all we can require of her—(She has asked the surgeon to comply. She does not have the authority or status to demand compliance to proper procedures. The lack of compliance quite properly was followed by a report to the hospital administration) or (2) that she had not done all we can require of her—(She cannot allow dangerous violations of operating room aseptic standards to take place. In doing so, she is failing to carry out her assignment and is allowing the patient's life to be placed in jeopardy. She should not be cowed by the surgeon's arrogance and sexism. Even at the risk of losing her job, she cannot allow the operation to take place in these conditions).

The problem is that (1) is too lenient a judgment and (2) is too harsh. We cannot require the

nurse *qua* individual to do more than she has done. But the nurse *qua* nurse shares blame with her colleagues in such cases despite the much greater blame which must be placed on the surgeon violating reasonable requirements. The lack of aggressive advocacy for the patient's welfare, the willingness to be dominated by the (usually male) physician or surgeon—unfortunate even if understandable "ways of life" of the nursing profession—which partially explain this nurse's behavior are below the standard we can rightfully expect the group authorized to provide nursing services to meet. Appeal to collective responsibility yields a judgment neither too harsh nor too lenient.

This judgment conforms to the moral intuitions of the nurses surveyed who were mentioned earlier. Despite a feeling that as individuals they were doing all that could reasonably be required of them in their circumstances, they still felt dissatisfied with their performance. As nurses they felt blame for falling short of the mark set for the profession.

This dissatisfaction, when seen in the light of collective responsibility, can be turned to positive use. The nurse who has done all she is required to do as an individual need not suffer debilitating guilt. Guilt, in such cases, is misplaced. Her individual actions do not warrant guilt. And, in contrast to nondistributive collective responsibility, there is no nondistributive collective guilt. "Guilt," as Feinberg has said, "consists in the intentional transgression of a prohibition. . . . there can be no such thing as vicarious guilt."[19] However, although rightfully free of guilt, she cannot be complacent. She is a member of a group that stands judged (i.e., is liable) and must, with her colleagues, take appropriate steps to alleviate the undesirable conditions. It is not enough for a professional to do all that is required of her or him as an individual. With the nurse having freely accepted the privileges and benefits of the profession, her or his responsibilities in the areas of professional competence are greater than would be those of an equally skilled and knowledgeable individual who was not a member of the profession.

In order to meet this larger responsibility, as the American Nursing Association has recognized, "there should be an established mechanism for the reporting and handling of incompetent, unethical, or illegal practice within the employment setting so that such reporting can go through official channels and be done without fear of reprisal. The nurse should be knowledgeable about the mechanism and be prepared to utilize it if necessary."[20]

Paradoxically if such machinery which collective responsibility requires were put in place, individual accountability would increase and the need to appeal to collective responsibility would decrease. If reporting incompetent, unethical, or illegal conduct could be done effectively through official channels and done without fear of reprisal, such reporting—which under more dangerous and less effective circumstances is not required—would be morally required of the individual. Hence, it may be that a profession should strive to organize itself and regulate itself to such a degree that the conditions for proper ascription of collective responsibility do not arise. But this is not the situation within the nursing profession at the present. Therefore, I conclude that the notion of collective responsibility is a timely weapon of considerable force for those who are working toward upgrading the nursing profession and the health-care delivery system.

Notes

1. See Jo Ann Ashley, *Hospitals, Paternalism, and the Role of the Nurse* (New York: Teachers, 1976), for a discussion of economic exploitation and sexism, which have plagued the nursing profession.

2. "Code for Nurses with Interpretive Statements" (Kansas City, Mo.: American Nurses' Association, 1976), p. 10.

3. Marjorie A. Godfrey, "Job Satisfaction—Or Should That Be Dissatisfaction? How Nurses Feel About Nursing," Part I, *Nursing 78* (April 1978), pp. 101–102.

4. Barbara L. Tate, ed., *The Nurse's Dilemma* (New York: American Journal of Nursing Company, 1977), pp. 47–48.

5. Quoted by D. E. Cooper, "Responsibility and the 'System,' " in Peter French, ed., *Individual and Collective Responsibility* (Cambridge, Mass.: Schenkman, 1972), p. 86.

6. Godfrey, ibid., Part II, *Nursing 78* (May 1978), p. 110.

7. D. E. Cooper, "Collective Responsibility," *Philosophy*, vol. XLIII, no. 165 (July 1968), pp. 260–262.

8. Ibid., p. 262.

9. Ibid., p. 260.

10. R. S. Downie, "Responsibility and Social Roles," in French, op. cit., p. 69.

11. "Responsibility and the 'System,' " in French, op. cit., pp. 90–91.

12. Virginia Held, "Can a Random Collection of Individuals Be Morally Responsible?" *The Journal of Philosophy*, vol. LXVII, no. 14 (July 23, 1970), pp. 471–481.

13. Joel Feinberg, "Collective Responsibility," *The Journal of Philosophy*, vol. LXV, no. 21 (Nov. 7, 1968), p. 677.

14. Ibid., p. 687.

15. Kurt Baier, "Guilt and Responsibility," in French, op. cit., p. 52.

16. Lavinia L. Dock, *A History of Nursing*, vol. III (New York: Putnam, 1912), p. 129.

17. See Kathleen M. Sward, "An Historical Perspective," in *Perspectives on the Code for Nurses* (Kansas City, Mo.: American Nurses' Association, 1978), for a discussion of these and other changes in the versions of the code.

18. Quoted by Barbara Durand, "A Nursing Practice Perspective," in *Perspectives on the Code for Nurses* (Kansas City, Mo.: American Nurses' Association, 1978), p. 19.

19. Feinberg, ibid., p. 676.

20. Sward, ibid., p. 8.

For Further Reading

Alford, R. R. *Health Care Politics: Ideological Interest Group Barriers to Reform.* Chicago: University of Chicago Press, 1975.

Augenstein, L. *Come Let Us Play God.* New York: Harper & Row, 1969.

Edwards, M. H. *Hazardous to Your Health: A New Look at the Health Care Crisis in America.* New York: Arlington House, 1972.

Ehrenreich, B., and Ehrenreich, J. *The American Health Empire: Power, Profits, and Politics.* New York: Random House, 1970.

Ginzborg, E., and Ostow, M. *Men, Money, and Medicine.* New York: Columbia University Press, 1967.

Greenberg, S. *The Quality of Mercy.* New York: Atheneum, 1971.

Knowles, J. H., ed. *Doing Better and Feeling Worse; Health Care in the U.S.* New York: Norton, 1977.

Kunz, R. M., and Fehr, H., eds. *The Challenge of Life: Biomedical Progress and Human Values.* Basel, Switzerland: Birkhauser Verlag, 1972.

Ostheimer, N., and Ostheimer, J., eds. *Life or Death—Who Controls?* New York: Springer, 1976.

Stevens, R. *American Medicine and the Public Interest.* New Haven, Conn.: Yale University Press, 1971.

Veatch, R., and Branson, R., eds. *Ethics and Health Policy.* Cambridge, Mass.: Ballinger, 1976.

5
THE HEALTH CARE SETTING

To understand why head nurse Ellen McGregor is marching to the clinical director's office with both resolve and resignation, one must understand a little about Hillsboro. Hillsboro is a private nonprofit mental health center that has been providing superb care for a long time. In recent years, however, a real problem has developed in the inpatient component of Hillsboro's service.

As is the custom with such facilities, Hillsboro's inpatient services are used not only by the center's psychiatrists but by those practicing in the community as well. In other words, psychiatrists in private practice have admitting privileges. This has caused a serious conflict of authority between private physicians and hospital staff and rules.

An example of the conflict can be seen in how group therapy sessions and activities like assertiveness training and art therapy are handled. Staff finds that patients who attend these group sessions and engage in the exercises benefit most. In order to incorporate group sessions into each patient's treatment plan, Hillsboro has been trying to implement a team approach that includes daily staff meetings with psychiatrist, nursing staff, mental health worker, social worker, and whoever else may be a part of the patient's health team. Out of these meetings evolves a treatment plan tailor-made for each particular patient. Naturally, attendance by the patient's outside psychiatrist is crucial at these meetings if treatment is to be coordinated. The problem is that only about half of the private psychiatrists ever attend.

Psychiatrists who do not attend these meetings offer various excuses. Some say they are too busy; others voice skepticism about the whole approach. All resent having their authority diluted. One psychiatrist puts it this way: "I'll decide which institutional programs, if any, my patient will participate in." This psychiatrist and others have effectively undercut the hospital's authority by telling patients who object to the institutional sessions that they need not attend them. Obviously, this attitude encourages patients to play the staff against their psychiatrists, and it is making other nonprivate patients uncooperative, because they feel that they do not have so much freedom of choice as some of the others.

The upshot of all this is that relations between nurses and private psychiatrists have deteriorated to the point that patient care is suffering. Indeed, the nurses have told Ellen McGregor that something must be done. Thus Ellen is on her way to the clinical director's office.

The clinical director of inpatient service at Hillsboro is Dr. Weyland, a psychiatrist. Ellen does not expect much from this doctor, but in the hierarchy of control he is the one she must see first. The reason for her pessimism is that Dr. Weyland has consistently expressed resentment of what he terms the "deprofessionalization" of psychiatry. In fact, just one day before, Dr. Weyland had harangued the staff about how any "quack" could now do his job. He was referring to a recent law that allows a psychologist or someone with a master's degree in social work (M.S.W.) to serve as clinical director of inpatient service. Several of the staff objected to Weyland's characterization and told him so. In any event, it is clear that he perceives his profession as being under attack. So Ellen is sure that Weyland's loyalties in this matter will lie with the profession, not the institution, and for this reason he will probably do nothing to make the private psychiatrists more responsive to the center's rules and regulations. She is proved right. Rather than responding to her complaint, Weyland waxes eloquent again about the "deprofessionalization" of medicine generally and of psychiatry in particular.

Undaunted, Ellen proceeds to the office of the hospital administrator, Pat Berman. Berman, like most people with this title, is a nonphysician with a master's degree in business administration (M.B.A.). Berman tells Ellen to try to work things out. Ellen reminds him that she has tried and has failed, that relations between nurses and private physicians have soured, and that the inpatient therapy program is unraveling. "We'll have to force the issue," she argues. "We must inform the outside physicians that they will play by our rules or not play at all."

Berman is sympathetic, but he suggests that Ellen is not looking at the "big picture." "You must realize that one-quarter of our inpatients have uncooperative physicians. We simply cannot afford to lose these patients without jeopardizing the whole service."

"What you're saying, then," Ellen replies, "is that we cannot afford to offend these outside doctors."

"What I'm saying is that we should be circumspect in what we do." Then Berman adds, "I suggest you have the nurses keep accurate and detailed records of these incidents. Yes, that's what we'll do—generate some data to find out how serious a problem this is."

Ellen McGregor leaves the office, crestfallen. Walking slowly down the narrow, neat-as-a-pin hallway, her mind turns to thoughts of cannibalism. I wonder, she thinks, how long it will take the nurses to eat me alive when they hear this?*

HCPs with any institutional experience will recognize the preceding situation as not at all uncommon. The tension between bureaucracy and professionalism, the divided loyalties, the struggle for autonomy, the clash between administrative and professional tasks, all characterize the occupational lives of health providers. Indeed, the tugs and pulls that HCPs feel in working within a health setting can be as wrenching as any other occupational experience and in many cases more difficult to resolve.

It is fitting, then, that this chapter consider the health care setting. Since many of the problems arise because of the setting's organizational system, this chapter examines two models of organization that interact in health care settings: bureaucracy and professionalism. After explaining what these involve, the chapter will show how they operate in health care and why they are fundamentally at odds. Most important, it will consider the moral problems that organizational tension raises for patients and providers within the system.

---

*Thanks is due Jeannine Cox, M.S.W., for suggesting this case.

Values and Priorities

Before beginning this chapter, try to answer the following questions, which are intended to stimulate introspection about health care values:

1. Should Ellen McGregor try to get her nurses to carry out Berman's directive or encourage them to lodge further complaints?

2. Do you think that the psychiatrists are right in permitting their patients to forego group therapy sessions?

3. What do you think of Dr. Weyland's loyalties?

4. Do you agree with Pat Berman's priorities?

5. What would you say are the chief values of any bureaucracy?

6. Do you consider health care a unique *service*? If so, in what sense does it differ from other services?

7. Should health care be primarily a basic right or a service available to those who can afford it?

8. Should HCPs try to preserve order in an institution at almost any cost, or should they sometimes subordinate order to other values?

9. In cases of conflict, should HCPs put the welfare of the institution before personal opportunities for upward mobility?

10. How would you feel about taking orders from someone outside the health fields, such as a hospital administrator?

11. Should the duties of HCPs working in institutions be clearly specified, or should they be highly flexible and offer a large measure of autonomy?

12. Should HCPs be hired and promoted according to whether they will accomplish bureaucratic goals or whether they embody professional values and likely will achieve the goals of their profession?

13. Do you think that only people in the health fields (that is, doctors and nurses) should run hospitals?

14. Would you agree that under no circumstances a nurse should disobey a physician's order?

15. Do physicians always know what's best for their patients?

The Bureaucratic Model

An *organization* may be defined as a social unit consisting of a network of relationships that orient and regulate behavior among people in the pursuit of special goals.[1] In health care settings the predominant organizational system is the bureaucracy. So to understand what does and does not occur in a health service setting, one must have some understanding of the nature and problems of all bureaucracies.

---

1. M. Kramer, Reality Shock (St. Louis, Mo.: C. V. Mosby, 1974), p. 11.

Although the bureaucracy as a system of work organization is very old, its modern formulation came about in the aftermath of the industrial revolution— and for good reason. Inasmuch as the bureaucratic system was, and continues to be, based on precepts related to the production of goods and services, it lends itself nicely to industrialization. Some of these precepts involve the replacement and interchangeability of parts; others the standardization of components. One extremely important precept concerns the division of labor.

Division of labor means dividing the production process and labor into areas of specialization, which is thought to increase capital and strengthen economic productivity. Implied in the division of labor is a clearly segmented way to organize a task or a piece of work. Individual workers are rarely expected to do the whole task but rather to accomplish a part of it. As a result, they need master only the requisite skills, which workers usually can learn on the job in a way that ensures that they will perform them the same way each time. In theory, such repetition should make workers quite proficient at doing their particular tasks. Another economic advantage of such labor fragmentation is that, since workers learn on the job, the hiring standards can be kept quite low, which has the effect of maintaining a plentiful source of relatively cheap labor. Furthermore, worker performance can easily be measured in terms of productivity. Of course, the supervisors who evaluate performance must have knowledge of the entire task or operation, which entails a clearly defined hierarchy of control and authority.

Even from this sketchy account, a profile of the bureaucratic system emerges that has at least four traits. First, in any bureaucracy there is a fragmentation of roles and tasks. Individuals are expected to master and perform only a part or segment of the total task. Second, all work is directed toward specific quantifiable goals. Third, worker performance is evaluated according to whether or not these goals are achieved. Fourth, all positions are organized into a hierarchy of authority and control. These four characteristics imply a set of values that underlie any bureaucracy, foremost among them the values of efficiency, predictability, impersonality, and speed.

With this common understanding as a base, we can now turn to an important question: How appropriate is a bureaucracy for organizing work in a health service setting? The answer depends largely on how well the bureaucracy delivers health care to patients. To the degree that the bureaucracy is incompatible with health care settings, it will probably not create a work atmosphere that is conducive to patient care.

Bureaucracy in the Health Setting

A number of aspects about bureaucracies pose problems in a health service setting. These can conveniently be grouped into conceptual and operational difficulties.

Conceptual Problems

One conceptual problem concerns the bureaucratic view of health care as another economic good. In this view, health care is subject to the same economic laws as other goods and requires the same kind of economically efficient market controls to assure the best production for consumers.[2] The fee-for-service system, discussed in Chapter 4, is solidly embedded in this concept.

But is health care another economic good? Perhaps not. For one thing, health care is a service, not a good. Granted, that fact alone does not disprove the bureaucratic claim; there is no reason that services cannot be arranged economically, the way goods are. Services such as hairdressing, pest control, pool maintenance, gardening—to mention a few—make this point. The issue, then, is whether and how a service differs from a good and, specifically, whether health care is a unique service. Are there features about health care that give pause to the bureaucratic concept of it as just another economic good? In fact, there seem to be a number of features about the nature of health care that belie this conceptual assumption.

To begin with, health care is a service that all of us at some time probably will need—not just want—and will not be able to provide for ourselves. Thus we must seek it from a health care provider. In seeking the care, we rarely have any specific idea of what treatment or therapy we need. Furthermore, the highly technical array of care possibilities, treatments, and therapies makes it very difficult for us to educate ourselves enough to make intelligent service judgments. Even when we are adequately informed, we often are legally denied the opportunity for self-administration and thus must seek care from someone licensed to provide it. Lacking the ability and opportunity to minister to self, patients who seek professional care must trust their HCPs, specifically physicians, to provide the needed service. But physicians, and health providers who work with them, are in the unusual market position of not only providing services but of determining them and even setting their price. Of course, patients ultimately can refuse prescriptions or reject proposed services, but generally they do not. They do not because, being ill, they are vulnerable and desirous of relief; and being unknowledgeable, they are in no position to question medical authority. Furthermore, physicians always have the power to withhold services from the uncooperative, which further discourages consumer challenges to their authority.

Clearly, then, health care differs from economic goods in being a service and, more importantly, a service that involves the satisfaction of basic maintenance needs. Generally speaking, people are in no position to satisfy these needs on their own or else are prevented by the providers from obtaining them independently. To treat health care like an economic good, then, seemingly invites a philosophy of care that views it as a prerogative of the wealthy. In short, money gives one a claim to care, as it does to any other economic good. At the very least, this raises a moral question of social justice. It could be argued vigorously that human need, and not just financial capacity, must be the determining factor in the distribution of health services.

The view of health care as an economic good is only one conceptual problem with the bureaucratic model. Another concerns the basic values that underlie

---

2. *D. B. Smith and A. D. Kaluzny,* The White Labyrinth: Understanding the Organization of Health Care *(Berkeley, Calif.: McCutchan Publishing, 1975), p. 124.*

bureaucracies: efficiency, predictability, impersonality, and speed. These values are basically ideals of industry. Without doubt, they have served many industries marvelously well in terms of productivity. For example, an automobile, which decades ago took the better part of a work day to assemble, can today be put together in a fraction of that time. But is that which is suitable for the production of goods appropriate for the production of services, specifically health services? Will that which works on things work as effectively on human beings?

Some would argue that it is at least possible, perhaps likely, that when the ideal of bureaucratic efficiency is brought into the health service setting, individuals can suffer. HCPs can be overloaded with work, forced to choose between economy and quality of health care, or thrown into conflict with various elements of the bureaucratic structure. Even worse, patient care can suffer, because worker efficiency and consumer need frequently do not correspond. In fact, for the good of the bureaucratic institution, the needs of the individual patient may go begging. All these problems are evident in the Hillsboro case.

It could also be argued that the industrial value of predictability may be misplaced in a health service setting. After all, individual patients, for all their similarities, remain unique, not just in a philosophical sense but in a therapeutic one. How many times have HCPs been upended by patients' reactions to drugs and medical procedures? Indeed, how many times have HCPs been taken aback by their own responses to work situations, by their own on-the-job behavior? The point is that, while they share with science the ideal of predictability, the health sciences must deal with the human dimension of both patient and provider under stressful circumstances and therefore must be prepared to incorporate a large measure of variability.

Again, although impersonality is understandable in the production of goods, its value in a health setting is questionable. Where impersonality means lack of bias and prejudice and functions to ensure equitable treatment for all, it has everything to recommend it. But when impersonality takes a form so clinically detached that the patient's or provider's very human needs for emotional support go unmet, one has to wonder. And what about the impersonality that allows HCPs to deny their own feelings of grief, inadequacy, defeat, or, as in the case of Hillsboro, frustration and outrage?

Much the same can be said of the industrial value of speed. Although it is true that the speed with which a worker can assemble a car partly determines productivity and is a measure of proficiency, this equation cannot be applied so easily in providing a health service. Again, in health care it is not a good that is being produced but a service that is being performed—a service necessarily complicated by the complexity of human personalities and innumerable other variables.

Thus there appear to be a number of conceptual difficulties in applying the bureaucratic system of work organization to the health care setting. One involves the bureaucratic concept of health care; others concern the bureaucracy's industrial ideals. These raise significant moral problems about patient care and treatment of health providers. Apart from these, there are a number of operational problems, which also raise moral issues.

Operational Problems

Many problems beset bureaucracies in their day-to-day operations in a health setting. Among them are problems relating to (1) routine versus non-routine tasks, (2) the internal dynamics of the institution, (3) conflicting sets of values, (4) separation of patients from their social context, (5) information control, (6) communications breakdown, (7) dilution of authority, (8) displacement of responsibility, and (9) development of subgroups.[3]

1. ROUTINE VERSUS NONROUTINE TASKS.   One operational problem of the bureaucratic system in a health setting stems from the bureaucracy's division of labor. Any system that is based on a part-task operation is designed to deal primarily, perhaps exclusively, with routine or inert tasks. Routine tasks are those in which the resistance to be overcome remains constant and therefore predictable. In contrast, nonroutine or active tasks are ones in which the resistance is variable across performances and therefore less predictable.[4] A simple example will illustrate.

Assembling a car can be viewed as a routine task. The degree of resistance to be overcome (which is always a part of task activities) for each task in the process is predictable. The amount of inertia that must be overcome to lift a fender and place it on a car or the energy required to tighten the wheel lugs is predictable and standardized and remains constant over repeated performances. Given that the resistance is known and standardized and that workers have been trained to overcome it, it is likely that the job will be done accurately and inexpensively; although even here actual outcome often belies desired outcome, as the owner of any new car will testify. In any event, the point is that bureaucracies lend themselves to routine tasks.

Take as an example of a nonroutine task the apparently simple task of helping a postoperative patient out of bed.[5] Although in common parlance this may seem an altogether "routine" task, it is anything but that in terms of the two types of tasks under consideration. Any HCP or layperson who has ever tried to get a postoperative patient out of bed will agree that the task has a large measure of variable resistance, depending, for example, on whether the patient is a child, adult, or octogenarian; whether the patient is recovering from an appendectomy, mastectomy, or colostomy; whether the task is being performed very soon after surgery or after a period of recovery. In short, getting a postoperative patient out of bed is a chore with a high degree of variability; it is a nonroutine task.

The fact is that services rendered in a health setting, especially those performed by HCPs, involve many nonroutine tasks. But a bureaucracy, geared as it is to manage routine tasks, is not structurally equipped to perform such tasks accurately. Thus, if the task of getting a patient out of bed is done strictly on a

---

3.   Kramer, pp. 11–16; E. A. Krause, Power and Illness: The Political Sociology of Health and Medical Care (New York: Elsevier, 1977), pp. 111–117.

4.   W. R. Scott, "Some Implications of Organization Theory for Research on Health Services," Millbank Memorial Fund Quarterly 44 (1966).

5.   Kramer, p. 15.

bureaucratic principle, then presumably someone (probably a nurse's aide) will be assigned the task. But since, as we have seen, this task involves considerable variability, whether or not it will be done competently is in doubt. Of course, in theory the task could be further fragmented, and a specialist could be employed to meet the unique needs of each postoperative patient. But this solution would be prohibitively expensive.

Although the bureaucratic system is designed chiefly to deal with routine tasks, many of those done in a health setting by HCPs are nonroutine. Thus a moral question arises as to how well the bureaucratic structure can meet patient needs in a health setting, as well as the fairness of holding health providers, such as nurse's aides, fully responsible for performing nonroutine tasks when their training has equipped them largely to deal with routine tasks.

2. **INTERNAL DYNAMICS.** In any bureaucracy where highly skilled people work, there is a kind of delicate balancing act of attitudes, positions, and prejudices to keep the system functioning. This process makes up what is termed the "internal dynamics" of the system.[6] Negotiations are conducted, positions negotiated, bargains struck. Obviously, informal arrangements are necessary to keep the health setting functioning. As a result, necessary, basic changes can go unmade, for strong ideological positions often meet firm resistance from administrative authorities, who would prefer to tolerate some inefficiencies rather than risk disrupting the system. The Hillsboro case is a perfect example. Faced with Ellen McGregor's demand that outside psychiatrists be made to abide by hospital rules, the hospital administrator in effect hedges rather than "rock the boat." Again, faced with charges of sexism in an institution, chances are the authorities will move to correct it only if their action will not disrupt operations. A glaring example of this tendency to resist change can be seen in the staunch opposition that has met efforts to unionize health care groups, other than physicians. In brief, the internal dynamics of bureaucracy like a health care setting requires the kind of careful accommodation of attitudes that often has the effect of sacrificing basic change in favor of order and predictability. This attitude also forces workers into almost reflexive opposition to any radically different way of doing things, for the sake of preserving order and the status quo.

3. **CONFLICTING VALUES.** The array of personalities in a health setting inevitably produces conflict between individuals and bureaucratic values. Chapter 1 noted that this phenomenon is one reason to study health care ethics. A good example of such a conflict can be seen in old-age institutions or mental hospitals, which regularly employ a large corps of nonprofessional workers to perform the daily tasks of caring for patients. Numerous studies reveal that, more often than not, these workers bring to their jobs the prejudices about the elderly and mentally ill that prevail in their socioeconomic class, as well as a strict custodial conception of their jobs. As a result, the behavior of some orderlies in these institutions often resembles that of a prison guard toward inmates more than that of a caring provider of health care to those in need.

---

6. *H. Cohen,* Demonics of Bureaucracy *(Ames: Iowa State University), 1965.*

But value conflicts arise just as frequently between professionals and the bureaucracy. For example, the bureaucratic system generally rewards loyalty to the institution. Thus professionals who follow rules, who are faithful to the demands of the system's hierarchy of authority and control, and who build up longevity on the job usually reap the rewards. But HCPs rarely subscribe wholeheartedly to these bureaucratic reward prerequisites. One reason they do not is that HCPs generally are socialized to value opportunities for upward mobility and thus do not necessarily feel anything more than a short-term commitment to the organization. Once they have learned all they can, if they have the opportunity to move on to greener employment pastures, they do so with little or no compunction. Another reason that helps account for the rift between bureaucratic and professional expectations is that, as skilled workers, HCPs do not mesh with the bureaucracy's highly structured machinery of authority and control so smoothly as nonskilled workers do—especially not when they are expected to take orders from someone outside the health field, such as a hospital administrator. We will hear more about this conflict between bureaucracy and professionalism shortly, in a discussion of the professional model. Suffice it here to note that such conflicts interest not only the medical sociologist but the ethicist as well; how bureaucrat and professional interact in a health setting in part determines the nature and quality of patient care.

4. **SEPARATION OF PATIENTS FROM THEIR SOCIAL CONTEXT.** A fourth problem besetting the service bureaucracy in a health setting is that, in the name of efficiency, patients often are separated from their social contexts. It is true that the separation is rarely one of total isolation or confinement; nonetheless, in many cases bureaucratic rules and regulations effectively cut patients off from their normal social environment. Having assigned the patient a place in the complex operation of the health setting, the typical service bureaucracy approach will then consider the time spent in accommodating the patient's social network as an "extra" to be weighed against time spent on direct patient care and operating the facility. Specifically, institutions generally orient new patients to institutional rules and expectations. Since such indoctrination is best done apart from family and friends, most health care settings with "total institution status" (those housing patients twenty-four hours a day) try to interrupt the patient's relations with outside contacts during this orientation period. Just as important, they limit and control these contacts afterward.[7]

Are such restrictions of social continuity and invasions of privacy always necessary and justifiable? If efficiency is the measuring rod, then they probably are. But efficiency that operates primarily to serve the ends of the institution and not the patient raises a fundamental question about the institution's purpose and function and with it a matter of legitimate moral debate. This issue takes on even greater moral significance when the patients happen to be especially vulnerable, as are children, the elderly, and the dying. Since cases involving these special patients are considered elsewhere, their unusual circumstances will

---

7. *Krause, p. 115.*

not be addressed further here. But it is important to understand the point that a casual denial of a patient's social contacts raises serious moral concerns about proper patient care.

5. INFORMATION CONTROL.   Because of their highly structured compartmentalization of work activities, bureaucracies naturally breed a sort of rivalry among various compartments, out of which the various specialities establish and maintain their identities and loyalties and firmly embed themselves in the bureaucratic organization. For example, studies indicate that in bureaucratic health settings low-level groups often try to increase their independence by keeping those above them ignorant of their activities, keeping their own skills private as much as possible, and not sharing insights with others.[8] Furthermore, observability of a group in a health setting is apparently a delicate and volatile issue. Saddled with too-close supervision, workers may rebel; left with too little, they may not do their jobs or may even sabotage operations. Either way, the patient stands to lose, and this possibility is what gives this facet of information control moral significance.

But there is another aspect of information control in the bureaucratic setting that raises moral concerns. Information control often functions to deny patients the information they have a rightful claim to and need in order to make judgments about their health care and to give informed consent. In part, the tendency to control information is bred outside the bureaucracy in the professional training of HCPs, which inculcates elitism and paternalism. At the same time, the bureaucratic line of authority and specific organizational policies encourage and reinforce that tendency to control information. Thus, by professional fiat and bureaucratic enforcement, HCPs with the most day-to-day contact with patients—nurses—have no authority to communicate with them in any substantive way about their conditions. Presumably, nurses are expected to evade or humor patients who ask reasonable questions about their health and welfare. The potential harm to patient and nurse is considerable.

An example can be found in a study of information seeking by tuberculosis patients in a sanatorium.[9] Unable to get a true picture of their conditions from physicians, the patients resorted to comparing their progress with one another, threatening to leave the institution against medical advice, concealing relevant information from HCPs, and bargaining not only over the information but over the timetable of treatment itself. The upshot of such institutional combat is often political compromises that preclude ideal patient treatment.

6. COMMUNICATIONS BREAKDOWN.   Bureaucratic stratification can undercut communications. Because of the highly structured bureaucratic hierarchy of authority and control, there are numerous points at which communications between echelons can be disrupted or even break down. This is particularly true when communication flows from lower to higher levels. Since information must

---

8.   M. Crozier, The Bureaucratic Phenomenon *(Chicago: University of Chicago Press, 1964).*

9.   J. A. Roth, Timetables: Structuring the Passage of Time in Hospitals and Other Careers *(Indianapolis, Ind.: Bobbs-Merrill, 1963).*

be relayed from one level to the next without bypassing any stage along the way, it must pass a long chain, and the original message can get quite distorted or even lost altogether by the time it reaches the end of the line.

7. **DILUTION OF AUTHORITY.** The bureaucracy's highly structured hierarchy of control has the effect of diluting authority, short-circuiting responsibility, and blunting individual initiative. For example, the charge nurse on a unit may be having difficulty in getting the laundry room to deliver needed supplies. When a patient's physician asks the nurse why the patient has not been receiving proper care, the nurse explains the problem. The physician calls for the supplies and receives them. Although this problem should have been resolved at a lower level, it was not; the nurse was dependent on the physician to secure the supplies.[10]

8. **DISPLACEMENT OF RESPONSIBILITY.** The bureaucratic structure can allow "buck passing" and thereby undermine motivation. Given the bureaucracy's stratification and its overlapping spheres of responsibility, it is easy for workers to deny that a task is part of their job or to shift blame to the next higher or lower person. In addition, individuals at the lowest bureaucratic levels often feel that they and their work are unimportant and unappreciated. Consequently, they perform at a minimal level.

9. **DEVELOPMENT OF SUBGROUPS.** The bureaucracy's stratification of personnel breeds isolated subgroups, because it inevitably establishes a pecking order within which various subgroups tend to align themselves for mutual protection in the social milieu and for social intercourse. The upshot can be professional clannishness. Commenting on this phenomenon within the health setting, one author has written,

> So simple a thing as a coffee break can demonstrate this. Looking over the people in the coffee shop, one can see operating room nurses together, medical unit nurses together, clinical laboratory personnel together, and so on, for as many different services as may be represented. Each tends to stay within its own narrow boundaries, refusing the stimulus of other points of view. Group cohesiveness shuts out the "interloper" unless he makes a conscious effort to introduce himself.[11]

What are the effects of such institutional cliques? The same author writes, "It is well known that groups impose discipline upon those who wish to remain 'in.' The restrictions can as easily extend to work norms as to social norms. Frequently, a group member may be ostracized by his peers if he is more productive than their norms require, or even if he is more cheerful about carrying out his assignments."[12]

The bureaucratic organization, then, although offering many advantages, has a number of features that make it of questionable appropriateness in a health setting, raise serious moral questions, and help explain the moral constraints

---

10.   *E. L. Brown,* Newer Dimensions of Patient Care: Improving Staff Motivation and Competence in the General Hospital *(New York: Russell Sage Foundation, 1962).*

11.   *R. M. French,* The Dynamics of Health Care *(New York: McGraw-Hill, 1974), p. 49.*

12.   *French, p. 50.*

that HCPs must work within. These conceptual and operational problems are further compounded by the presence of another organizational feature that is prevalent in health settings: professionalism.

The Professional Model

Many of those employed in health settings are or consider themselves to be professionals or members of professions. The term *professional* refers to a special kind of person, whereas *profession* refers to a special kind of occupation. Both are influenced by the model of professionalization, which has accumulated certain indicators—among them education, code of ethics, licensing, association, and peer control.[13]

Education is taken to indicate a body of knowledge within the occupation. The level of education is viewed as relating to skills, level of work competence, or a specific body of information. Thus, professionals are thought to have higher-than-average levels of education.

Professions also have codes of conduct or ethics. These, as noted earlier, express the noblest ideals of the various health professions.

In addition, professions have a licensing process with the purpose of demonstrating to society that certain individuals have met specific requirements for skills and knowledge, have had their competence measured, and have been deemed fit to serve the community. This credibility effect is strengthened because the whole licensing procedure is legitimized by the state, which empowers a control agency to oversee it.

The essence of a profession also includes a professional association. As we saw in Chapter 4, professional associations such as the AMA and ANA work to maintain and improve standards, update skills, control the conduct of members, standardize licensing, and in general serve the public and its own members.

Finally, a profession generally has some mechanism of peer control. Whether formally part of the association or informally implemented at the local level, such a mechanism gives the appearance of professional self-control and orientation to service.

These five characteristics, then, are often regarded as the primary indicators of a profession. Understandably, a model of professionalization has arisen around these indicators. They are in many cases the route to actual status as a profession, and thus many occupations are striving to score high on them in order to achieve professional status. But more important for our purpose is the impact on individual HCPs of passage through this professionalization process.

In general, HCPs absorb their profession's values, which are embedded in the indicators. For example, the value of autonomy is particularly apparent in licensing, associations, and peer control, all of which a profession can use to bargain for increased control over health management. A second value is collegiality, which refers to the collective responsibility shown by each of one's

---

13. J. A. Denton, Medical Sociology *(Boston: Houghton Mifflin, 1978), pp. 182–186.*

colleagues. An integral part of collegiality is the management of one's own affairs from within the occupation. A third value, service to others, is especially implied in codes of conduct and ethics. A fourth value is skill, which is evident in the licensing process. Autonomy, collegiality, service to others, skill—these are some of the values that underlie the concept of a profession and, most important, that individuals absorb in the process of professionalization.

Not only is the whole concept of profession erected on such values, but members of professions are greatly influenced by them. So much so that HCPs find themselves in recurring conflict because the health setting is organized primarily according to a bureaucratic model, not a professional one. Neither the core values of a profession nor its ways of organizing work necessarily synchronize with the bureaucratic setup. When clashes result, the health setting and patient care can be disrupted.

Professionalism in the Health Setting

To be sure, the professional model has adherents who are quick to proclaim its perceived strengths in a health setting. For example, some claim that it almost always enhances the morale of a highly skilled work force. Treated with respect and given independence, these workers respond in kind—with a truly professional performance. Others suggest that professionalism generally means higher quality, improved efficiency, and greater patient responsiveness. Still others argue that the professional model seems better able to deal with nonroutine or active tasks, because the professionalization process aims to produce workers who possess total knowledge and skills. Recall that a bureaucratic setup seems an efficient and effective way of organizing tasks aimed at constant resistance. But tasks aimed at overcoming variable resistance seem better organized in a professional manner, since skilled nonroutine responses are better for meeting unpredictable resistance with a minimum of errors.[14] Whether or not these and other claims for professionalization can be fully substantiated is questionable, but most would agree that there are at least some things about the professional model to recommend it as a way of organizing work in a health setting.

At the same time, like the bureaucratic perspective, professionalism seems to have inherent traits that can cause problems in a health setting. If work groups in a health setting were truly professionalized, then presumably they would succeed in dividing up the organizational territory in such a way that each of them controlled a piece of the turf. Indeed, their associations would vigorously try to maintain and perhaps expand the group's territorial holdings and would certainly buck any internal or external interference. The upshot of this triumph of professionalism could easily be what some term "quasi-feudal stagnation,"[15] in which each group is pitted against the other in an attempt to control its own territory.

---

14. *Kramer, p. 16.*
15. *Smith and Kaluzny, p. 337.*

Such a development would not bode well for institutional operations or for patient care. For one thing, outside attempts to make services more efficient and responsive to patients, personnel, and changing technology would meet resistance. For another, a state of tension, even destructive competition, would be likely to exist among groups, thus making coordination of activities very difficult. Furthermore, each group, probably through its association, would establish hierarchies of control that would not only regulate the activities of its members but would also influence the health setting in which members work. As for any outside regulatory agency, it could play no objective role in managing the health setting; being made up primarily of the group's members, it would be little more than an extension of the professional association itself. In fact, such an agency might actually undercut the public interest by giving the misimpression that it is an objective, impartial regulatory agency concerned only with the common good. Assuming this, the public would rest easy in the mistaken notion that its interests were being safeguarded. Finally, a fully professionalized work group would rely exclusively on peer control to ensure that its members maintained the highest standards of the profession. This means that the conduct of HCPs would be evaluated by other HCPs within the particular professions. But would such a system breed objectivity? Or would the profession's own interests bias the review of any one of its members? Could HCPs truly be said to be accountable to society when not society but they themselves, through just their associations, developed their own evaluation technology and then conducted the evaluations?

This picture of the fully professionalized work force does not seem far-fetched. In fact, some would say it approximates the existing scene in health care. Whether or not this is so is of less importance here than that the professional model is capable of developing such a picture, which seemingly holds out little promise for enhanced patient care.

So, like the bureaucratic model, professionalism has much to recommend it but also many grounds for rejecting it as appropriate for the health setting. In reality, of course, both models operate in most health settings, in what is often less than peaceful coexistence. Indeed, the fundamental tension between the bureaucracy and professionalism is of major moral concern because it poses a threat to worker and patient alike.

Professional-Bureaucratic Tensions

To understand the nature of the tension between the bureaucratic and professional views in a health setting, one must have some knowledge of how a health setting is organized. Generally speaking, any health setting has two lines of authority, the administrative and the medical. The administrative is organized primarily on the bureaucratic model; the medical on the professional model.

The administrative subsystem generally consists of a governing board, an institutional administrator, sometimes administrative assistants, various departmental supervisors, and subsection supervisors. The administrative organization of a health setting functions primarily to secure enough capital to establish, maintain, operate, and expand its facility; to secure acceptance in the form of basic legitimization of activity; to marshal needed skills; and to coordinate the activities of its members, interorganizational relations, and interactions with clients and patients.[16]

The medical organization consists of a select group of physicians and dentists who, by their membership, are granted practicing privileges; they can use the facilities for their patients. The medical staff has its own pattern of organization, which in some respects parallels the administrative setup. Its organizational line of authority usually consists of an organized medical staff, a chief of staff, various committees, and chiefs of the different services. This line of authority is not subordinate to the administrative organization but constitutes a second structure of authority within the institutional organization. Balancing these two separate structures of equal authority is not always easy. Indeed, sometimes it presents problems that reverberate throughout the whole system, even at the lowest levels, and that threaten patient care.[17] Before turning to specific problems with moral overtones, let us take a closer look at the root cause of these problems.

The main reason for the problems is the incompatibility between the bureaucratic model, which shapes administrative functions, and the professional model, which underlies medical functions. The incompatibility, which leads to institutional conflicts, is apparent in contrasting the bureaucratic and professional views on five basic issues.[18]

First, we saw earlier that the bureaucratic perspective views health care as an economic good, subject to the same economic laws as other goods and requiring the same kind of market controls to ensure the most economical product. In contrast, the professional view considers health care "a service, which is provided by an elite with sole command of a socially valued body of knowledge and with both a social and moral obligation to provide the highest quality technical service to all consumers."[19]

Second, in the bureaucracy the duties of individual workers are clearly specified, and a sharp division of labor is prescribed based on efficiency data. But professionalism entails a far more informal specification of the division of labor, as well as a large degree of flexibility.

Third, whereas the bureaucracy has a structure of authority and control based on institutional functions, professionalism bases its authority on technical expertise and collegial decision making. Similarly, the bureaucracy provides a clearly defined career ladder within the organization, but professionalism allows for the pursuit of professional goals largely outside the organization's structure, through recognition in the larger professional community.

Fourth, in the bureaucracy, recruitment and promotion are based largely on

16. C. Perrow, "The Analysis of Goals in Complex Organizations," American Sociological Review 26 (December 1961): 854–866.

17. See Brown; also R. P. Sloan, Today's Hospital (New York: Harper & Row, 1966).

18. Smith and Kaluzny, p. 125.

19. Smith and Kaluzny, p. 125.

an objective evaluation of the candidate's ability to achieve bureaucratic goals. However, in the professional view, recruitment and promotion are based on the candidate's embodiment of professional values and ability to achieve the goals of the profession.

Fifth, in the bureaucracy the personal, subjective values of officials ideally play no part in their decisions. But the same cannot be said of professionalism, in which recognition is paid to the uniqueness of individual situations and the role of professional values in decisions and actions.

Given these fundamental differences between the bureaucratic and professional perspectives, it is little wonder that problems can arise when the administrative (bureaucratic) and medical (professional) lines of authority cross in a health setting.

When Lines of Authority Cross

The crossing of administrative and medical lines of authority can lead to numerous conflicts for HCPs. Among the chief ones are subordination of professionals to the supervision of nonprofessionals, interrole conflicts, intrarole conflicts, and incompatible role expectations.

Subordination of Professionals to Nonprofessional Supervision

Perhaps the most apparent conflict develops when medical professionals are subordinated to the rank authority of nonprofessional superiors. As suggested earlier, when Ellen McGregor reported to the nurses about her conversation with hospital administrator Pat Berman, they became incensed. Not only did they object to Berman's impassivity, but they resented their professional judgment being overridden by a "business mentality." Indeed, this is a recurring problem at Hillsboro, especially now that state and federal funds are drying up. Numerous programs, such as the center's day care program for the chronically ill, are in jeopardy. The staff has pushed for their continuance and has even claimed that the integrity of the whole program hangs in the balance, but the administration has intimated that the cutbacks will probably be necessary. Furthermore, a reduction of staff also appears imminent. For its part, staff has argued that any such reductions will result in a work overload that cannot help but hurt patient care.

According to Pat Berman, one way out of this predicament is to reclassify some of the jobs, so that less highly skilled workers can hold certain positions. Indeed, through their association, hospital administrators were instrumental in lobbying for the new regulation that allows a psychologist or an M.S.W. to hold the position of clinical director of inpatient services. However, as Dr. Weyland's reaction shows, this regulation has caused intrainstitutional problems, not only at Hillsboro but also statewide. The professional medical staff feels that the administration's authority is far too great over what staff sees as strictly medical or professional matters. Underlying this problem is a clash between the bureaucratic goal of efficiency and the professional goal of autonomy; this, in turn, cannot be separated from the bureaucratic view of health care as another economic good and the professional view of health care as a service.

Conflicts over the subordination of professionals to the rank authority of nonprofessionals are less likely in settings where physicians carry formal administrative responsibilities, so that they are ultimately responsible for work performance. The same goes for facilities where bureaucratic and professional patterns overlap. For example, in many teaching and psychiatric hospitals, bureaucratic functions are commonly transferred to professionals. A physician "medical administrator" might have parttime or fulltime administrative functions in the medical-clinical area or on a hospitalwide basis. At Hillsboro, Dr. Weyland performs not only clinical functions but also administrative tasks within the component of inpatient services. This is precisely why Ellen McGregor went to see him. But note that Dr. Weyland could not or would not help McGregor because of his own conflict between loyalty to the institution and to his profession, a recurring problem faced by HCPs performing administrative tasks. Even when the conflict is not this sort, the medical administrative staff still faces problems: Its executive functions are rarely defined sufficiently, and it is difficult to demarcate bureaucratic from professional responsibility. The result often is interrole conflict.

Interrole Conflict

Interrole conflict refers to incompatibility between the bureaucratic and administrative functions that a worker must perform. For example, as head nurse, Ellen McGregor performs both clinical and administrative tasks: The two lines of authority, professional and bureaucratic, converge in her position. What sent her to Dr. Weyland and Mr. Berman were professional concerns. In response to persistent nursing complaints, supported by her own observations that patient care was suffering, she brought the matter to administrative attention. At the same time, as head nurse she is expected to carry back institutional administrative policies to the staff. How should she deal with the nurses' outrage and frustration about Berman's position? Should she follow her professional instincts and support, even encourage, them? Or should she be a good "company" person and ensure that the nurses carry out the directive? The fact is that Ellen, like all head nurses, coordinates not only care and cure structures but also specialized functions and demands within each of these two subsystems. Thus, in the words of one author, the nurse is

> First, by virtue of her nursing function, the person who performs those bedside tasks which are nursing care rather than medical therapy, and which therefore are part of the hospital's direct responsibility and her profession's prerogative. Secondly, by virtue of the absence of an administrative hierarchy, the nurse has become the administrator of the territorial unit, the patient care area. . . . Thirdly, by virtue of the growing specialization and diversification of the tasks that are involved in patient care, and by virtue of the continuity of time, she has become the mediator and coordinator of the various functionaries who,

with but episodal responsibility for their specialties, come and go on the nursing care unit, without assuming any responsibility for overall coordination and continuity.[20]

The organizational role of graduate professional nurses in hospitals, then, has at least two distinct facets. On the one hand, they must implement the therapeutic-clinical aspects of patient care as indicated by physicians. On the other, they must implement the administrative and organizational aspects of patient care. Thus, they exercise coordinating functions as administrators, supervisors, and lower-level representatives of the nonmedical administration; and they function as "line" workers responsible for the implementation of medical directives and the provision of "independent" nursing care.[21] We can see, then, that the range of the professional nurse's functions extends across services. This range is increased by the nurse's relative scarcity in relation to auxiliary assistance, a factor that increases the number and kind of nurses' interpersonal contacts on the job and, most important, the chances for error.[22]

These interrole conflicts carry several specific moral implications. First, questions about HCPs' primary loyalties obscure HCP responsibilities, thereby making decisions and evaluations difficult. In addition, the wide range of duties stemming from multiple role responsibilities can result in worker overload, in which case patient care may suffer. But beyond interrole conflicts and those stemming from the subordination of professionals to the rank authority of nonprofessionals, still another kind of conflict can arise when the two lines of institutional authority cross: intrarole conflict.

Intrarole Conflict

Intrarole conflict refers to the fact that the expectations of the individual performing a role may have built-in contradictions.[23] For example, a patient's first contact at Hillsboro is often the admissions clerk, Beth Broheimer. Broheimer is expected to be considerate of patients, but she is also expected to get pertinent financial information from them. In fact, treatment at Hillsboro cannot begin until Broheimer obtains this information. As a result, she has more than once complained about how torn she feels between what she calls the compassionate and mercenary aspects of her job. Indeed, on occasion she has misrepresented records in a patient's interests.

For instance, not too long ago an adolescent male turned up at Hillsboro seeking psychiatric help. Although the young man was holding down a parttime job at a hamburger stand, he was living at home and being supported largely by his parents. Therefore, he did not qualify for financial aid. At the same time, the

20. H. O. Mauksch, "Nursing Dilemmas in the Organization of Patient Care," Nursing Outlook 5 (1957): 31–33.

21. W. V. Heydebrand, Hospital Bureaucracy (New York: Dunellen Publishing, 1973), p. 204.

22. R. W. Habenstein and E. A. Christ, Professionalizer, Traditionalizer, and Utilizer, 2nd ed. (Columbia: University of Missouri, 1963), p. 155.

23. Smith and Kaluzny, p. 112.

youth himself could not pay for the services he needed, and he didn't want his parents to know about his treatment. So Broheimer indicated on the record that the young man was self-supporting. Since his low income allowed him to qualify for aid, she was then able to initiate a treatment program for him.

A similar case involved a young divorced mother of two, whose income from waitressing and child support put her just beyond financial assistance. Since the woman could not afford to pay for the services, Broheimer again "fudged" on her application by omitting the child support. Thus the woman's income was reduced to a level that qualified for assistance. Incidents like these have caused Beth Broheimer profound moral unrest. On the one hand, she feels an obligation to help such patients qualify for the health services they need and worries about their fate should they be turned away. On the other hand, she cannot help thinking that she is defrauding the county and state, or at the very least is violating her job responsibilities. In addition, Broheimer feels hypocritical when she insists that other workers adhere strictly to regulations.

Similar intrarole conflicts can be seen among the nurses at Hillsboro, whose contacts with patients involve both physical tasks and emotional support and reassurance. The same can be said of the physician's contacts. On the one hand, nurses and physicians are expected to be warm and supportive—but not to the point of sacrificing scientific objectivity. On the other hand, they are expected to be clinically detached—but not to the point where they are cold and unfeeling. Exactly where appropriately warm, cordial patient relations end and inappropriately close, personal relations begin is not easy to say. In this ambiguity lies the tension for HCPs. A comparable tension can arise when one entity in the system does not view a second entity's role as the second sees itself. This can lead to yet another kind of conflict.

Incompatible Role Expectations

A conflict arises when one group or individual in a health setting does not share with another group common expectations of that other's role in the organization. For example, medical record librarians, nurses, and other HCPs often find themselves divided between administrative (bureaucratic) regulations and physicians' demands. For instance, a physician may expect immediate access to a patient's records, but the medical records department may not permit such access without proper authorization. Again, private psychiatrists feel that Hillsboro's function is primarily custodial and expect full control over their patients at Hillsboro, but the staff insists that all patients follow institutional policy. Similarly, Hillsboro's governing board views the local health services planning organization as a source of support and additional revenue for services wanted by the community. But this organization, and all agencies empowered to evaluate the health care needs of a community, are evaluated at the state and federal levels in terms of their ability to control costs. Thus they are expected to limit the expansion of new services. In each of these cases, some receiver—medical librarian, hospital administrator, planning agency—receives at least two incompatible messages. Hence, the conflict.

Such conflicts are particularly evident in health settings because of the multiple lines of authority. An even greater problem is that there is little agreement among individuals about the nature and scope of their own authority and that of others. For example, one study has revealed that each group (nurse, physician, administrator, board member, and so on) sees itself as having more influence than it is perceived to have by other groups.[24] As might be expected, the most serious conflict is between physicians and administrators, since both see themselves as the ultimate authority. This problem is of moral significance, because the most acute physician-administrator clashes involve hospital policies regulating the general treatment of patients.[25]

Compounding this problem are the conflicting bureaucratic and professional work orientations mentioned earlier. Administrators establish regulations largely with routine situations in mind. But physicians have to deal with non-routine situations. Administrators expect all situations to be handled through established routines, whereas physicians expect to handle nonroutine situations in appropriately unorthodox ways. (Actually, administrators recognize that standard operating procedures cannot be maintained in nonroutine situations, which creates interrole conflict for them.) This tension between bureaucratic and professional expectations can give rise to situations in which medical personnel feel obliged to circumvent institutional regulations in order to meet their own professional expectations. Administrators, in turn, feel compelled to take appropriate corrective measures to ensure that bureaucratic expectations are met. The upshot can be short-term chaos, inefficiency, ill will, and inadequate patient care.

A good example can again be seen at Hillsboro, whose administrative sector has defined a routine for admitting patients of private, outside physicians. The private physician fills out a form that the patient is to present at the center as a request for admission. Admissions clerk Beth Broheimer and other admissions personnel study the form and decide, on the basis of institutional rules, whether to admit the patient. Reasons for refusing admission include an unpaid institutional bill or a bad credit reputation. In such cases, patients are returned to their physicians. Faced with a bureaucratic impasse but convinced of the necessity of hospital admission for their patients all the same, some outside physicians have taken to writing *emergency* across the form and having the patient take it back to the center. This is the doctors' way of indicating that the admitting routine should be overridden, that an exception should be made. The patient may or may not be a true emergency case; at any rate, the physician is merely using the technique to gain a concession from the administrative bureaucracy. But Broheimer and her staff have begun to look askance at such tactics, aware that some physicians routinely abuse the emergency override. For their part, the physicians simply increase their use of it. Ultimately, Broheimer will probably take the problem to hospital administrator Pat Berman, who will try to reconcile

24. F. Bates and R. White, "Differential Perceptions of Authority in Hospitals," Journal of Health and Human Behavior, Winter 1961, pp. 262–267.

25. Smith and Kaluzny, p. 264.

the conflict. For a while, the problem will probably abate, but eventually it will start up again—and with it the game of bouncing some patients back and forth between organizational courts.[26]

A similar conflict involves physician and nurse. Because nurses and physicians are largely responsible for the primary care of patients in health settings and because there are so many flash points in their relationship, this conflict deserves special consideration.

Physician-Nurse Relationships

Like physicians and administrators, physicians and nurses appear to harbor different perceptions of the legitimate authority of each other's occupation. To be specific, nurses generally perceive themselves as having more decision-making authority than physicians perceive them as having.[27] Therefore, although physicians may actually make a particular decision, nurses believe that they should have the authority to do so; similarly, a nurse may in reality make decisions and take action that physicians believe they have no right to take. These different perceptions of authority really clash when the physician happens to be an intern. In that case, an inexperienced doctor dons the physician's mantle of authority over nurses who, in fact, have far more working knowledge. Given the frequency of nurse-intern interactions and the overlap in their responsibilities, incompatibility of their role conceptions may result in all sorts of acute complaints, frustrations, and difficulties. One author characterizes the problem this way:

> The frustrations that the nurse experiences in the nurse-doctor relationship breed mutual resentment. Interns complain that when "you have to deal with nurses, they're bitchy and administrative" or that "sometimes one doubts whether nurses know how to read." But the intern's relationship with the head nurse on a ward is a very temporary one and his main concern is with his future career. The nurse, on the other hand, is tied to the hospital in which she will continue to deal with a succession of such "youngsters," as she calls them.[28]

A number of factors help account for the problems between nurses and physicians and for their divergent perceptions of decision-making authority.[29] First, nurses and physicians differ in their educational backgrounds, physicians generally receiving far more formal training. Since professional status is in part measured by the amount and intensity of educational training, physicians gain a

---

26. C. Taylor, In Horizontal Orbit: Hospitals and the Cult of Efficiency (New York: Holt, Rinehart & Winston, 1970), pp. 23–26.

27. M. K. Davis, "Intrarole Conflict and Job Satisfaction on Psychiatric Units," Nursing Research 23 (November-December 1974): 482–488; L. P. Christman, "Nurse-Physician Communications in the Hospital," Journal of the American Medical Association 194 (November 1965): 541; F. L. Bates and R. F. White, "Differential Perceptions of Authority in Hospitals," Journal of Health and Human Behavior, Winter 1961, p. 264.

28. R. Laub Coser, Life in the Ward (East Lansing: Michigan State University Press, 1962), p. 25.

29. Christman, p. 543.

professional edge over nurses, which translates into perceived and real power, authority, and influence.

Second, major differences in career patterns divide physicians and nurses. Physicians are primarily self-employed; nurses are not. Again, this gives the physician advantages in autonomy, authority, and control.

Third, semantic differences sometimes stand between the two groups, and common words often carry different meanings. For example, studies indicate that when physicians describe a nurse as cooperative, they mean that she obeys orders. But when nurses describe their interactions with physicians as cooperative, they mean equal.[30]

Fourth, there are major class differences separating nurses and physicians. Physicians make more money, have more prestige, and interact with those higher in authority both within and outside the health setting. Just as such differences in social class tend to divide groups in general, so do they in health care. Furthermore, they divide them in such a way that the physician holds the dominant or "superior" position.

Fifth, differences of authority divide physicians and nurses. Physicians have special status in hospitals—unrestricted access to facilities and services and power over personnel. Indeed, hospitals are frequently viewed as an extension and instrument of medical practitioners, particularly since they are the only members of the medical profession who may legally "practice medicine." All of this gives physicians tremendous authority in their interactions with other HCPs, including nurses.

At the same time, it should be noted that the physician's role in a health setting is currently undergoing reevaluation and is even being circumscribed by the institution, which is assuming a more active role in patient care. This means that nurses are acquiring more intraorganizational power by the fact that they are expected to limit the physician's use of hospital resources.[31] Ultimately, this change may establish an atmosphere of shared authority, but in many instances it is causing even more friction between physician and nurse, as the Hillsboro case again illustrates.

Sixth, sex differences divide physician and nurse. Although more women are becoming physicians than ever before, the medical profession continues to be dominated by men and the nursing profession by women. This division of occupations by gender prevails even in such recently established occupations as the female-dominated nurse associate and the male-dominated physician assistant. Many physicians in private practice appear to prefer the nurse associate as a helper, because of the combination of sexual and occupational deference that is still imparted in nursing schools. Nevertheless, nurse associates are paid significantly less than physician assistants, even though in most cases nurse associates have as much or more training and experience.[32]

---

30. E. D. Pelligrino, "What's Wrong with the Nurse-Physician Relationship in Today's Hospitals: A Physician's View," *Hospitals* 40 (1967): 70.

31. Taylor, p. 116; R. S. Duff and A. B. Hellingshead, *Sickness and Society* (New York: Holt, Rinehart & Winston, 1968), p. 71.

32. Krause, p. 62.

The issues facing nurses today resemble those facing all women. Indeed, for nurses they may be more pronounced, since in health care these issues receive widespread institutional legitimization. The issues distill to the struggle to break free of the traditional subservient role of handmaiden. Given the many sexist obstacles to the development of nursing as a viable and autonomous profession, it is surprising that nurses have been so slow to engage in activities calculated to improve the standing of women in society in general. Such apparent impassivity seems especially puzzling since some of nursing's greatest leaders were involved in some way with the movement for women's rights, among them Florence Nightingale, Lavinia Dock, Lillian Ward, and Margaret Sanger.[33] At the same time, of course, the subordinate position of the nurse goes back at least to Nightingale's original definition of the modern professional nurse's role. Commenting on this formulation, one author has observed that

> Nightingale . . . refused to allow any of her nurses to undertake to give any service at all on their own initiative. Her nurses' services were to be granted only when specifically requested by the doctors. No nurse could give food to any patient without the doctor's written order. No nurse could soothe or clean a patient without the doctor's order. Nuns were forbidden to engage in religious visiting. Nightingale thus required that what the nurse did for the patient was a function of what the doctor felt was required for the care of the patient. . . . *Nursing may thus be defined as a subordinate part of the technical division of labor surrounding medicine.*[34]

Although more men are entering nursing, they still make up only a small percentage of the profession. This fact can probably be attributed in part to a stereotype that associates the central concepts of nursing—nurturing and support—with women. In modern times, it has become fashionable to attribute such alleged gender qualities exclusively to social influences. Ultimately we may find this to be true. However, recent studies by biomedical researchers, such as the University of Chicago's Jerre Levy, suggest that men and women do have profoundly different abilities, which are traceable to the different organizations of their brains.[35] Some of these differences may indicate that *on the whole* women are better equipped to perform the nurturing, supporting functions of nursing. Even if these early findings prove accurate, they indicate only *statistical* differences—trends within a whole population of men and women. They say nothing about *individual* differences. In other words, it would be bad science to conclude on the basis of such research that a particular man would not make an effective nurse because he is a man or that a particular woman would not make an effective doctor because she is a woman.

The seventh reason for conflicts between nurses and physicians is that they do not share the same orientation to patients. Both perform care and cure

---

33. *K. Creason Sorensen and J. Luckman,* Basic Nursing: A Psychophysiologic Approach *(Philadelphia: W. B. Saunders, 1979), p. 61.*

34. *E. Freidson,* Profession of Medicine: A Study in the Sociology of Applied Knowledge *(New York: Dodd, Mead, 1970), p. 61.*

35. *J. Durden-Smith, "Male and Female—Why?"* Quest, *October 1980, p. 15.*

functions, but the nurse's primary function is to care, the physician's to cure. Nurses are oriented to support and nurture patients and to provide for their overall well-being. In contrast, physicians are oriented to attend to patients' immediate medical needs and comfort.

In addition, nurses in health settings care directly for the daily needs of patients and are charged with the supervision of those needs, whereas physicians spend limited time with the institutionalized patient. As a result nurses have close and continuous interactions with the patient, and physicians have remote, fragmented, and fleeting ones. Thus, nurses often feel that they are in a better position than physicians to know what is best for the patient. A nurse may recognize that a patient who has consented to undergo surgery is having second thoughts. The nurse may recommend that surgery be delayed for a day or so in order to give the patient time to resolve the anxiety; otherwise, the patient may not have a full and speedy recovery. But the surgeon, determined to proceed with the operation as scheduled, is likely to forge ahead. Or again, a nurse may suspect from conversations with a terminally ill patient that this person genuinely wants to know his prognosis. If the doctor has been reluctant to inform the man, he is unlikely to take the nurse's advice. Examples like these raise important moral questions about the decision-making authority of HCPs as it relates to patient care.

Given these differences between nurses and physicians, it is understandable that conflicts arise. The moral significance of these clashes is that they pose a potential threat to the health team as a team. Fundamental to the meaning and success of the team is shared authority—intensive and equal interaction of all HCPs in order to resolve the patient's problem. Conflicts between physicians and nurse, and between physicians and other HCPs, can easily preclude such equal sharing and thus undercut the health team. Even where health teams are not involved, intrainstitutional tensions of the kind sketched here are bound to pit one work group or individual against another in a power struggle that promises no winners and threatens the welfare of the patient who gets caught in the middle. Obviously, then, incompatible role expectations are of fundamental moral concern because they pose real and direct threats to patient care.

Summary

In this chapter we saw that two lines of authority—bureaucratic and professional—operate in most health care settings. These two organizational models are fundamentally at odds on a number of concepts: health care, the division of labor, the structure of authority, evaluation procedures, and the nature of decision making. Because the lines of authority disagree on basic values, conflicts can arise when they cross. Among the most common clashes are the subordination of professionals to the rank authority of nonprofessionals, interrole conflicts, intrarole conflicts, and incompatible role expectations. The moral significance of these conflicts is that (1) they can undercut patient care and (2) they can constrain the HCP's capacity to make moral decisions.

CASE PRESENTATION
Caught in the Middle

It didn't seem fair, but fair or not, Barbara Martin was about to lose her job. Barbara was still an aide at the convalescent hospital introduced in Chapter 1, which sought to provide a more homelike atmosphere by assigning aides to the same group of patients for an entire month. Barbara was the aide who befriended seventy-eight-year-old Rachel, who had suffered a stroke and was nearly deaf, fiercely independent, and suspicious of strangers. Indeed, among staff Rachel was considered "a real bear."

It took about a week for Rachel and Barbara to get used to each other. In the interim, Barbara discovered that Rachel was being neglected. She reported some areas of broken skin and a large sore on Rachel's ear that was caused by leaving her hearing aid in at night. Barbara's report made her somewhat unpopular with the evening shift, but the head nurse promised to investigate.

By the third week, Barbara had established such rapport with Rachel that the old lady decided she didn't want to be touched by any other aide or nurse. Told that such an arrangement was impossible, Rachel took to giving everyone but Barbara an especially bad time. This further undermined Barbara's relationships with her colleagues. To compound matters, during the fourth week Rachel seemed to be coming down with a case of the flu—which made Barbara feel even more sorry for the old lady.

You'll recall that at this point Rachel's daughter-in-law cornered Barbara and demanded some frank answers. Evidently the doctor had painted a bright picture of Rachel's prognosis, totally out of phase with the realities as the daughter-in-law observed them. Barbara knew that, when the charge nurse called the doctor for medication and treatment orders, he just instructed her to make Rachel as comfortable as possible, "because she isn't going to last very long anyway." To make matters worse, the doctor hadn't been to see the old lady for two months. When the daughter-in-law asked Barbara point-blank what was going on and whether she thought the physician was doing an adequate job, Barbara's immediate response was that she should ask the charge nurse. The daughter-in-law said she had done that, but that the nurse had merely mumbled something about professional ethics and refused to discuss the matter further.

When she heard this, Barbara told the woman that she ought to contact the doctor. "I'd go straight to the source," Barbara told the daughter-in-law. "Then decide for yourself."

The woman did just that. How well she did it! Just two days later Barbara was summoned by the charge nurse, who reported that the physician was furious. The charge nurse told Barbara that she had exceeded her authority inexcusably by advising the daughter-in-law at all. "We have a chain of command here," the nurse told her. "As far as you're concerned, that means you bring any and all problems to me. I'll take it from there." Barbara pointed out that she advised the daughter-in-law in what she thought was Rachel's best interests. The charge nurse told her that she was not qualified to make decisions about the interests of patients, that her only responsibility was to conform to the rules and regulations of the institution as interpreted by her supervisors. The nurse furthermore informed Barbara that her employment would be reevaluated in light of Barbara's "gross insubordination."

Questions for Analysis

1. *What values underlie the decisions and actions of the people in this case?*

2. *Do you see any points at which bureaucratic and professional values clash?*

3. *Do you agree with the charge nurse's interpretation of Barbara's responsibility?*

4. *Do you think that professional jealousy of the sort described here can develop among HCPs? If so, what impact might it have on patient care?*

5. *When Rachel refused to be nursed by anyone other than Barbara, did Barbara have an obligation to intercede on behalf of the institution and to attempt to reason with the patient?*

6. *If you were part of the committee called to look into this matter, how would you evaluate Barbara's behavior?*

7. *Apply the ethical theories discussed in Chapter 3 to the decision Barbara faced. Indicate what aspects of the situation each theory would emphasize and show what moral course each might chart.*

Exercise 1

Commenting on the physicians who were chosen to staff medical schools, Alexander Rush, associate professor of clinical medicine at the University of Pennsylvania School of Medicine, has written:

> To these new leaders fell the responsibility of training a new generation of young physicians. . . .
> The direction of their efforts was toward the production of graduates steeped in the basic sciences and nurtured to become faculty and hospital-based specialists.
> Emphasis became centered upon disease and what modern scientific technology might be able to do about it, not upon the unfortunate being afflicted with the disease.
> Recent graduates in medicine possess an awesome knowledge of the basic pathophysiology of disease coupled with supreme confidence in the ability of medical technology to solve all problems. What they lack is a compassionate approach to the patient.[36]

36. *A. Rush,* The Magazine of Rush—Presbyterian—St. Luke's Medical Center *9 (Winter 1976– 1977): 42–43.*

Given the accuracy of this description, how might this medical orientation clash with the orientation of other HCPs, especially nurses? In what areas of patient care does this "cure" orientation become totally irrelevant and the "care" function become essential?

Exercise 2

Reconsider your responses to the Values and Priorities section. Would you wish to change or amend any of your answers in light of what you have read in this chapter?

Ethical Dilemmas for Nurses: Physicians' Orders versus Patients' Rights

E. Joy Kroeger Mappes

Chapter 6 examines four basic patient rights and indicates how meeting them can raise moral problems for HCPs. Two of these rights, the right to adequate health care and the right to self-determination, concern professor of philosophy and registered nurse E. Joy Kroeger Mappes in the following essay.

Mappes focuses on two sets of ethical problems that hospital nurses face in honoring these basic patient rights. Both involve physician orders. What are nurses to do when following physician orders will violate the patient's right to adequate care? What are they to do when following orders threatens the patient's right of self-determination? In the author's view, hospital nurses are often caught on the horns of a dilemma: If they respect patient rights, they violate physician orders; if they obey physician orders, they violate patient rights. This is a classic dilemma that nurses and other HCPs in subordinate positions within a health care setting must inevitably face.

For Mappes, the nurse's overriding obligation is to the patient. But she concedes that classist and sexist forces sometimes make it difficult for nurses to meet their responsibilities. Accordingly, she argues for changes in the workplace that will allow nurses to meet their obligations to patients.

The American Hospital Association, in a widely promulgated statement entitled "A Patient's Bill of Rights," makes explicit a number of the generally recognized rights of hospitalized patients.[1] Among the rights expressly articulated in the AHA statement is a cluster of rights closely associated with a more general right, the right of self-determination. The "self-determination cluster" includes: (1) the right to information concerning diagnosis, treatment, and prognosis; (2) the right to information necessary to give informed consent; and (3) the right to refuse treatment. The AHA statement duly recognizes several other important patient rights but, importantly, fails to explicitly recognize the patient's right to adequate medical care.[2] Surely, if the purpose of a statement of patients' rights is to catalogue patients' rights, we ought not to overlook this one. After all, the patient has agreed to enter the hospital setting precisely for the purpose of obtaining medical treatment. To the extent that adequate medical care is not forthcoming, the patient has been done an injustice. That is, the patient's right to adequate medical care has been violated.

This paper explores two types of ethical dilemmas related to patients' rights that arise for the hospital nurse.[3] (1) The first set of dilemmas is related to the patient's basic right to adequate medical care. (2) The second set of dilemmas is related to the cluster of rights closely associated with the patient's right of self-determination. Dilemmas arise for a nurse if adequate medical care for a patient would be jeopardized by following the expressed or understood orders of a physician. Dilemmas also arise for a nurse if the patient's right to self-determination would be violated by following the expressed or understood orders of a physician. In each case, the logic of the dilemma is similar. The dilemma arises because the nurse's apparent obligation to follow the physician's order conflicts with his or her obligation to act in the interest of the patient. To carry out the physician's order would be to act against the interest of the patient. To act in the interest of the patient would be to disobey the physician's order.[4] I will argue that when this conflict arises the nurse's obligation to the patient is overriding and that nurses must act and be allowed and encouraged to act to protect the rights of the patient.

I. Nursing Dilemmas and the Patient's Right to Adequate Medical Care

In a hospital the primary responsibility for a patient's care rests with a physician. Physicians determine the medical diagnosis, treatment, and prognosis of patients' illnesses and write orders to arrive at and effect these determinations. In general, physicians' orders govern what a patient is to do and what is to be done for a patient, i.e., the degree of activity, diet, medication, diagnostic and treatment procedures to be performed. Nurses carry out physicians' orders themselves, delegate tasks to others, or make the orders known to those responsible for carrying them out. They are not generally allowed by law to diagnose or prescribe.[5] Although this is a greatly oversimplified picture of what goes on, as anyone familiar at all with the functioning of a hospital will realize, at least some of the complexities involved in the interaction among physicians, nurses, and patients in a hospital setting will emerge as we proceed.

The complexity of the ethical dilemmas arising for nurses regarding the patient's right to adequate medical care can best be understood by examining various examples. The following are suggested as being not atypical of situations arising in hospitals:

1. A patient who has had emphysema for a number of years is admitted to a cardiac unit for observation with a tentative diagnosis of myocardial infarction. Oxygen is ordered in a concentration commonly given for patients with this diagnosis. The nurse, knowing that oxygen is contraindicated for patients with emphysema, must decide whether to carry out or question the order through appropriate channels.

2. A patient admitted to the hospital for a diagnostic work-up has been on a special and fairly extensive drug therapy regimen. This regimen is common to patients of a particular private physician, seemingly regardless of their diagnosis. The private physician orders the drug therapy program continued after admission. However, accepted medical practice would ordinarily call for ceasing as many drugs as is safely possible, thus avoiding unnecessary variables in arriving at an accurate diagnosis. In general the private physician is viewed by other physicians as incompetent. The nurse is aware that the orders do not reflect good medical practice, but also realizes that she[6] will be dealing with this physician as long as she works at that hospital. The nurse must decide whether to follow the orders or refuse to carry out the orders, attempting through channels to have the orders changed.

3. A frail patient recovering from recent surgery has been receiving intra-muscular antibiotic injections four times a day. The injection sites are very tender, and though the patient now is able to eat without problems, the intern refuses to change the order to an oral route of administration of the antibiotic because the absorption of the medication would be slightly diminished. The nurse must decide whether to follow the order as it stands or continue through channels to try to have the order changed.

4. A nurse on the midnight shift of a large medical center is closely monitoring a patient's vital signs (blood pressure, pulse rate, respiratory rate). The physicians have been unable to di-

agnose the patient's illness. In reviewing the patient's record, the nurse thinks of a possible diagnosis. The patient's condition begins to worsen and the nurse phones the intern-on-call to notify him of the patient's condition. The nurse mentions that the record indicates that diagnosis X is possible. The intern dismisses the nurse's suggested diagnosis and instructs the nurse to follow existing orders. Concerned that the patient's condition will continue to deteriorate, the nurse contacts her supervisor, who concurs with the intern. The patient's blood pressure gradually but steadily falls and the pulse increases. The nurse has contacted the intern twice since the initial call but the orders remain unchanged. The nurse must decide whether to pursue the matter further, e.g., calling the resident-on-call and/or the patient's private physician.[7]

What are the obligations of nurses in such cases? Under what circumstances are nurses obligated to rely on their judgment and to question the physician's order? To what extent must nurses pursue the questioning when, in their view, the patient's right to adequate medical care is being violated? It is often taken for granted that when the medical assessments of physician and nurse differ, "the physician knows best." In order to see both why this is thought, perhaps correctly so, to be generally true and yet why it is surely not always true, it is necessary to consider some of the factors that account for the difference in physician and nurse assessments.

A nurse's assessment of what constitutes adequate medical treatment may differ from a physician's assessment for at least three reasons. (a) There is a difference in the amount and the content of their formal training. Physicians generally have a number of years more formal training than nurses, though that difference is not as great as it once was. More nurses now continue formal training in various ways, i.e., by pursuing graduate work and/or by becoming nurse practitioners, nurse clinicians, or nurse anesthetists. In addition, proportionately more nurses than ever before are college graduates. However, a physician's formal training is more extensive and detailed. Moreover, and perhaps most importantly, physicians are explicitly trained in the diagnosis and treatment of

illness, with the emphasis of the training placed on the hard sciences. Nurses are trained to be knowledgeable about illness in general, the symptoms and treatment of illness, and the complications and side effects of various forms of therapy. While this formal training includes both the hard sciences and the social (primarily behavioral) sciences, there is an emphasis on the behavioral sciences. Nurses are trained to concentrate on the overall well-being of the patient. (b) There may be a difference in the length or concentration of their experience. For example, nurses who have worked in special care units (in medical and surgical cardiac units, burn units, renal units, intensive care units) for a number of years acquire a great deal of knowledge which may not be possessed by interns, and perhaps even residents and nonspecialty private physicians. Nurses who have worked for years in small community hospitals may well be more knowledgeable in some areas than some physicians. (c) There may be a difference in their knowledge of the patient. Nurses often have more detailed knowledge about patients than do physicians, who often see a patient only once a day. Nurses who are "at the bedside" are thus in a position to recognize small changes as they happen. Because of the possibility of more detailed knowledge, nurse assessments may be more accurate than physician assessments. Where physician and nurse assessments differ then, it is not necessarily the case that the physician's assessment is the correct one simply because of the amount and content of the physician's formal training. Physicians do make mistakes and, when they do, nurses must be in a position to protect the patient.[8]

Ethical dilemmas of the kind typified in the above four examples arise when to follow physician's orders would be to act against the medical interest of the patient. Given the fact that the *basic* obligation of both the physician and the nurse is to act in the medical interest of the patient, it is rather striking that anyone should suppose that the nurse's obligation to follow the physician's orders should ever take precedence. What, after all, is the foundation of the nurse's obligation to follow the physician's orders? Presumably, the nurse's obligation to follow the physician's orders is grounded on the nurse's obligation to act in the medical interest of the patient. The point is that the nurse has an obligation to follow physicians' orders because, ordinarily, patient welfare (interest) thereby is ensured. Thus when a nurse's obligation to follow a physician's order comes into *direct* conflict with the nurse's obligation to act in the medical interest of the patient, it would seem to follow that the patient's interests should always take precedence.

For instance, Example 1 provides a clear case of a medically unsound order. In fact, it is such a clear case that a nurse not questioning the order would be judged incompetent. The medically unsound order may be the result of a medical mistake or of medical incompetence. If the order is the result of an oversight, the physician is likely to be grateful when (if) a nurse questions the order. If the order is a result of incompetence, the physician is not likely to be grateful. Whatever the reason for the medically unsound order, the nurse is obligated to question an order if it is clearly medically unsound. The nurse must refuse to carry out the order if it is not changed, and to press the matter through channels in order to protect the medical interest of the patient. Example 2 is similar to Example 1 in that it involves a medical practice that is clearly unsound. If the orders are questioned, the physician here again is not likely to be grateful. Indeed, since the physician's practice may ultimately be at stake, the pressures brought to bear on a nurse may be overwhelming, particularly if the physician's colleagues choose to defend him or her. It is undeniable, however, that medical incompetence is not in the best medical interest of patients, and thus that a nurse's obligation is to question the order. It is of course true that a nurse acting on behalf of the patient in this situation may pay a heavy price, perhaps his or her job, for protecting the medical interests of patients. However, the nurse's moral obligation is no less real on this account.

With Example 3 the murkiness begins, since it does not provide a clear case of unsound medical practice, though perhaps it presents us with a case in which the physician is operating with a too-narrow view of good medical practice. As I mentioned earlier, nurses often are more concerned with the overall well-being of the patient. Physicians often are concerned only with identifying the illness, treating it, and determining how responsive the illness is to the treatment. To the extent that Example 3 resembles Example 1, i.e., the

lumps are very bad and the difference in the absorption of the medication in the two routes of administration is small, the nurse has an obligation to question the order. Example 4 is like Example 3 in that it does not provide a clear case of a medically unsound order. However, in Example 4, much more is at stake. Since life itself is involved, any decision must be considered very carefully. The problem here is not a problem of weighing or balancing but a problem of being either right or wrong. Both Examples 3 and 4 force a nurse to assess this question: "How strong are my grounds for thinking that the orders are not in tune with the patient's best medical interest?" The murkiness comes in knowing exactly when the physician's order is in direct conflict with the patient's medical interest. As the last two examples illustrate, it can be very difficult to know exactly what is in the best medical interest of the patient. To the extent that the nurse has carefully considered the situation and is sure that his or her view is in accord with good medical care, the order should be questioned. The less sure one is, the less clear it becomes whether the order should be questioned and the matter pressed through channels.

In arguing the above, I am not advocating uncritical questioning. Clearly, questioning at some point must cease. Otherwise, hospitals could not function efficiently and as a result the medical interest of all patients would suffer. But if there is little or no opportunity for nurses and other health professionals to contribute their knowledge to the care of the patient, or if they are directly or indirectly discouraged from contributing, it would seem that they will find it difficult, if not impossible, to fulfill their obligations with respect to the patient's right to adequate medical care.

II. Nursing Dilemmas and the Patient's Right of Self-determination

The complexity of the ethical dilemmas arising for nurses regarding the patient's right of self-determination can best be understood by examining various cases. Again, the following are suggested as being not atypical of situations arising in hospitals:

1. A patient is scheduled for prostate surgery (prostatectomy) early in the morning. Because

he is generally unaware of what is happening to him due to senility, he is judged incompetent to give informed consent for surgery. The patient's sister visits him in the evenings, but the physicians have not been available during those times for her to give consent. The physicians have asked the nurses to obtain consent from the patient's sister. When she arrives the evening before surgery is scheduled, the nurses explain to her that her brother is to have surgery and what it would entail. The sister had not been told that surgery was being considered and questions its necessity for her brother, who has not experienced any real problems due to an enlarged prostate. She does not feel she can sign the consent form without talking with one of his physicians. Should the nurses encourage her to sign the consent form, as the physicians have requested, or call one of the physicians to speak with the patient's sister?

2. A patient in the cardiac unit who was admitted with a massive myocardial infarction begins to show signs of increased cardiac failure. The patient and the family have clearly expressed their desire to the medical and nursing staff to refrain from "heroics" should complications arise. The patient stops breathing and the intern begins to intubate the patient, requests the nurse's assistance, and orders a respirator. Should the nurse follow orders or attempt to convince the intern to reconsider, calling the resident-on-call should the intern refuse?

3. A patient is hospitalized for a series of diagnostic tests. The tests, history, and physical pretty clearly indicate a certain diagnosis. The physicians only tell the patient that they are not yet sure of the diagnosis, reassuring the patient that he is in good hands. Each day after the physicians leave, the patient asks the nurse, coming in to give medications, what the tests have shown and what his diagnosis is. When the nurse encourages the patient to ask the physicians these questions, he says he feels intimidated by them and that when he does ask questions they simply say that everything will be fine. Should the nurse reinforce what the physicians have said or attempt to convince the physicians that the patient has a right to information about his illness, pressing the matter through channels if they do not agree?

4. A patient suffering from cancer is scheduled for surgery in the morning. While instructing the patient not to eat or drink after a certain hour, the nurse realizes that the patient is unaware of the risks involved in having the surgery and of those involved in not having the surgery. In talking further, the nurse sees clearly that the option of not having surgery was never presented and that the patient has only a vague idea of what the surgery will entail. She also appears to be unaware of her diagnosis. The patient has signed the consent form for surgery. Should the nurse proceed in preparing the patient for surgery, or, proceeding through channels, attempt to provide for a genuinely informed consent for the surgery?

What are the obligations of nurses in examples such as these? To what extent must nurses pursue questioning when, in their view, the patient's right of self-determination is being violated? Unless we are willing to say that a patient upon entering a hospital surrenders the right of self-determination, it seems clear that physicians' orders, explicit and implied, should be questioned in all of the above examples. After all, in each of the above examples, one of the rights expressly outlined by the American Hospital Association is in danger of being abridged. The rights involved are (or should be) known by all to be possessed by all. Here a difference in the formal training and knowledge based on experience is not relevant in any difference between a physician's and a nurse's assessment. No formal training in medicine is necessary to arrive at the conclusion that the patient's right of self-determination is endangered.

The tension that exists for nurses in situations typified in the above four examples is not really that of a moral dilemma, but rather, a tension between doing what is morally right and what is least difficult practically, a tension common in everyday life. The problem is not that the nurse's obligation is unclear, but that in actual situations fulfilling this moral obligation is extremely difficult. What we must consider now in some detail are the social forces that make it so difficult for a nurse to act on behalf of the patient.

III. Nursing Dilemmas and the Importance of a Classist and Sexist Context

I have argued that when following a physician's order would violate the patient's right to adequate medical care or the patient's right of self-determination, the nurse's moral obligation is to question the order. If necessary, the matter should be pressed through channels. It is well and good to say what nurses should do. It is quite another thing, given the forces at work in the everyday world in which nurses must work, to expect nurses to do what they ought to do.

To begin with, we must recognize that there is an important class difference between physicians and nurses, the difference between the upper middle class or upper class of physicians and the lower middle class of nurses.[9] A large proportion of physicians both start out and end their lives in the upper class. Though the economic status of a physician is not as high as that of a high-level corporation executive, the social status of a physician is very high because of the prestige of the profession of medicine in the United States today.[10] Physicians have a high social status in American society and they understand and identify with people who have a similarly high status.[11] "Physicians talked with physicians; nurses talked with nurses," is an observation of one sociological study.[12] Generally physicians do not understand or identify with nurses (or with most patients), in part because of a difference in social status. Correspondingly there is an educational difference between physicians and nurses. As mentioned earlier, the formal educational training necessary for a physician is generally much longer than that necessary for a nurse, and their training differs in content.

The differences in the composition of each profession on the basis of sex is clear. Most physicians are male (93.1 percent) and most nurses are female (97.3 percent).[13] In accordance with traditional sex roles, physicians are encouraged to be decisive and to act with authority. Studies indicate that physicians view themselves as omnipotent.[14] Nurses are encouraged to be tactful, sensitive, and diplomatic. Tact and diplomacy are necessary to make a physician feel in control. Put another way, nurses' recommendations for patient treatment must take a particular form. These recommenda-

tions must appear to be initiated by the physician. Nurses are expected to take the initiative and are responsible for making recommendations, but at the same time must appear passive.[15] Nurses who see their roles, partially at least, as one of consultant must follow certain rules of the "game."[16] If they refuse to follow the rules, they will be made to suffer consequences such as snide remarks, ostracism, harassment, or job termination.

Again, in accordance with traditional sex roles, nurses in hospitals are viewed much the same as are wives and mothers in the family. This is the view of nursing held both by society and by physicians. Nurses as women are expected to be subservient to physicians as men, to provide "tender loving care" to whomever may be in need, and to be responsible and competent in the absence of physicians but to relinquish that responsibility upon request, i.e., when physicians are present.[17]

As in society, women in hospitals (here women nurses) are typically viewed as sex objects, a situation which encourages physicians to discount the input of nurses with regard to patient care. The observation that women are viewed by male physicians as sexual objects was prevalent in a project which studied the discriminatory practices and attitudes against women in forty-one United States medical schools as seen from questionnaires completed by 146 women medical students. As the author notes, "The open expression of the notion that any woman—even if she is a patient—is fair game for lecherous interests of all men (including physicians) is in some ways the most distressing fact of these student observations."[18] Responses showing the prevalent attitude of physicians toward women in general or toward women as patients included: "[I] often hear demeaning remarks, usually toward nurses offered by clinicians. . . ." "[There is] superficial discussion of topics related to women. . . . Basic assumption: women are not worth serious consideration." "[The] most frequent remarks concern female patients—women's illnesses are assumed psychosomatic until proven otherwise."[19] Perhaps the most frequent response of women medical students depicting the attitudes of male medical students, professors, and clinicians centered around the use of slides in class of parts of women's anatomy and slides of nude women from magazine centerfolds. Those slides were introduced by medical-student colleagues or instructors often to bait women medical students or belittle them. One student relates, ''My own experience with [a professor who had included a ''nudie'' slide in his lecture] was an interesting and emotional comment ending on, 'Men need to look down on women, and that's why I show the slide,' ''[20] The response of the male members of the class to the slides was generally one of unmitigated laughter and approval. With such a negative and restricted view of women as persons, nurses, not to mention all women, are at a disadvantage in dealing with most male physicians.

Another aspect of the sexism that permeates the physician-nurse relationship is reflected in divergent standards of mental health for men and women. A study of thirty-three female and forty-six male psychiatrists, psychologists, and social workers showed that they held a different standard of mental health for women and men. The standard agreed upon for mentally healthy men was basically the same as the standard for mentally healthy adults. The standard for mentally healthy women included being more easily influenced, less objective, etc., in general characteristics which are less socially desirable.[21] Women then who are mentally healthy women are mentally unhealthy adults and women who are mentally healthy adults are mentally unhealthy women. This is clearly a ''no win'' situation for all women. Women nurses are no exception.

It is the just described classist and sexist economic and social context of the physician-nurse relationship that often inhibits the nurse from effectively functioning on behalf of the patient. Nurses have a moral responsibility to act on behalf of the patient, but in order to expect them to carry out that responsibility, changes must be made in the workplace. Nurses must be in a position to act to protect the rights of patients. They must be allowed and encouraged to do so. Therefore, those operating and managing hospitals and those responsible for hospital policies must establish policies which make it possible for nurses to protect patients' rights without risking their present and future employment. Those operating and managing hospitals cannot eradicate classism and sexism, but they must be aware of the impact it has on patient care, for again the ultimate goal of everyone connected with hospitals is adequate medical care within the framework of patients'

rights. As potential patients it is important to all of us.[22]

Notes

1. The statement can be found, for example, in *Hospitals*, vol. 4 (Feb. 16, 1973).

2. I am aware of the difficulties in determining what constitutes adequate medical care. For example, is adequate medical care determined solely by reference to past and present medical practices, by the established wisdom of knowledgeable health professionals, or by knowledgeable recipients of medical care? And how is the standard for knowledgeability determined? I am presuming that problems such as these, though difficult, are not insoluble. I am also aware of the related difficulty in distinguishing medical care from health care. And what distinguishes medical care and health care from nursing care? In this paper, ''medical care'' will refer to the diagnosis and treatment of illness. Health professionals then, who aid physicians in the process of diagnosing and treating illness, aid in providing medical care.

3. A large majority of all working nurses work in hospitals.

4. The International Code of Nursing Ethics is ambiguous in addressing such a dilemma. The relevant section (#7) of the code merely states: ''The nurse is under an obligation to carry out the physician's orders intelligently and loyally and to refuse to participate in the unethical procedures.'' What exactly is the nurse supposed to do when to carry out the physician's orders is in effect to participate in unethical procedures? The most recent (1976) version of the *Code for Nurses* (available from the American Nurses' Association) adopted by the American Nurses' Association directly addresses this problem. Section 3 states, ''The nurse acts to safeguard the client and the public when health care and safety are affected by the incompetent, unethical, or illegal practice of any person.'' The interpretive statement of section 3 begins, ''The nurse's primary commitment is to the client's care and safety. Hence, in the role of client advocate, the nurse must be alert to and take appropriate action regarding any instances of incompetent, unethical, or illegal practice(s) by any member of the health-care team or the health-care system itself, or any action on the part of others that is prejudicial to the client's best interests.''

5. The area of practice which is solely that of the nurse and the area of practice which is solely that of the physician is presently in a state of flux. The

submissive role that nursing has held in relation to the physician's practice of medicine is being rejected by the nursing profession. Nurse practice acts, which regulate the practice of nursing, in many states reflect the change toward an expanded and more independent role for nurses. For example, a definition of a nursing diagnosis, as distinct from a medical diagnosis, is a part of some nurse practice acts. Daniel A. Rothman and Nancy Lloyd Rothman, *The Professional Nurse and the Law* (Boston: Little, Brown, 1977), pp. 65–81.

6. The overwhelming majority of nurses are women and the overwhelming majority of physicians are men. Because the examples are intended to reflect the hospital situation as it exists, I will use the feminine pronoun to refer to nurses and the masculine pronoun to refer to physicians.

7. In an actual case of this description, the intern dismissed the nurse's diagnosis by asking if her woman's intuition told her that diagnosis X was the correct one. The nurse's decision was to not pursue the matter and early in the morning the patient sustained a cardiac arrest and was unresponsive to resuscitation efforts by the resuscitation team.

8. Obviously, nurses also make mistakes, but physicians are clearly in a position to protect the patient when they become aware of nurses' mistakes.

9. Vicente Navarro, "Women in Health Care," *New England Journal of Medicine*, vol. 292 (Feb. 20, 1975), p. 400.

10. Barbara Ehrenreich and John Ehrenreich, "Health Care and Social Control," *Social Policy*, vol. 5 (May/June 1974), p. 33.

11. Raymond S. Duff and August Hollingshead, *Sickness and Society* (New York: Harper & Row, 1968), p. 371.

12. Ibid., p. 376.

13. Navarro, p. 400.

14. Robert L. Kane and Rosalie A. Kane, "Physicians' Attitudes of Omnipotence in a University Hospital," *Journal of Medical Education*, vol. 44 (August 1969), pp. 684–690, and Trucia Kushner, "Nursing Profession: Condition Critical," *Ms*, vol. 2 (August 1973), p. 99.

15. Kushner, p. 99.

16. Leonard I. Stein, "The Doctor-Nurse Game," in Edith R. Lewis, ed., *Changing Patterns of Nursing Practice: New Needs, New Roles* (New York: American Journal of Nursing Company, 1971), p. 227.

17. Jo Ann Ashley, *Hospitals, Paternalism, and the Role of the Nurse* (New York: Teachers College, 1976), p. 17.

18. Margaret A. Campbell, *Why Would a Girl Go into Medicine?* (Old Westbury, N.Y.: Feminist Press, 1973), p. 73.

19. Ibid., p. 74.

20. Ibid., p. 26.

21. Inge K. Broverman, Donald M. Broverman, Frank E. Clarkson, Paul S. Rosenkrantz, and Susan R. Vogel, "Sex-Role Stereotypes and Clinical Judgments of Mental Health," *Journal of Consulting and Clinical Psychology*, vol. 34 (February 1970), pp. 1–7.

22. I wish to thank Jorn Bramann, Marilyn Edmunds, Jane Zembaty, and especially Tom Mappes for their helpful comments on earlier versions of this paper.

For Further Reading

Abdellah, F. G., et al. *New Directions in Patient-Centered Nursing.* New York: Macmillan, 1973.

Auld, M. E., and Birum, L. H., eds. *The Challenge of Nursing.* St. Louis, Mo.: C. V. Mosby, 1973.

Brown, E. L. *Nursing Reconsidered: A Study of Change.* Philadelphia: J. B. Lippincott, 1970 (Part I), 1971 (Part II).

Browning, M. H., and Lewis, E. P., eds. *The Expanded Role of the Nurse.* New York: The American Journal of Nursing Company, 1973.

Bullough, B. *The Law and the Expanding Nursing Role.* New York: Appleton-Century-Crofts, 1975.

Bullough, B., and Bullough, V., eds. *Issues in Nursing.* New York: Springer, 1966.

Bullough B., and Bullough, V., eds. *New Directions for Nurses.* New York: Springer, 1971.

Davis, M. Z., et al., eds. *Nurses in Practice: A Perspective of Work Environment.* St. Louis, Mo.: C. V. Mosby, 1975.

Lewis, E. P., ed. *Changing Patterns of Nursing Practice.* New York: The American Journal of Nursing Company, 1971.

Part III

The Health Care Professional and the Patient

6
PATIENT RIGHTS

A patient is brought to the emergency room bleeding from a gunshot wound in his arm. A tourniquet is applied and left on to the point where the man loses his arm.

Another patient suffering from breast cancer has a breast surgically removed. Radiation therapy to the site of the mastectomy and the surrounding area is instituted. The treating radiologist mumbles something about the risks involved. As a result of the treatment, the patient suffers radiation burns and skin breakdown.

Still another patient, this one in a teaching hospital, is visited by nearly two dozen HCPs one day: three pathologists, a diagnostic radiologist and a radiological therapist, a half dozen or more surgeons, several medical specialists, a clutch of nurses, and seven house officers. Without having been consulted, the patient has been made the subject of a medical lesson.

Each of these cases raises a question of patient rights, the concern of this chapter. It is hard to imagine a term more bandied around today than *rights*. In recent years we have heard many arguments, often eloquent ones, for women's rights, gay rights, black rights, criminal rights, prisoner's rights, children's rights, students' rights, even animal rights. We continue to hear about the rights of every American to a job, an education, housing, food, and a healthful environment. With the reinstitution of draft registration, the land resounds with the cries of those who claim that their rights are being violated. When a former U.S. Attorney General traveled to Iran in hopes of gaining the release of American hostages, U.S. government officials, including the President himself, claimed that he had no right to do that; the Attorney General replied that he had every right. There is no question that we are living at a time when the rights of individuals are receiving intensive interest and widespread exposure.

The situation is no different in health care. There we hear of the rights of pregnant women, the rights of the terminally ill, the rights of the dying, the rights of the mentally ill, the rights of the mentally retarded, the rights of the fetus. Furthermore, for some time now, people within and outside the health care professions have vigorously been debating the right of the individual to choose to die, to have an abortion, to choose or have access to unconventional medical treatment. In brief, many of the most pressing health care issues of the day crackle with allusions to, assumptions about, or debates over rights.

What rights do individuals have with respect to health care? There is perhaps no more basic question in medical ethics, for the answer to it not only helps carve out areas of patient entitlement but also helps define HCP responsibility. Indeed, answering this question provides a foundation for the claims of special patients—such as children, the elderly, and the disadvantaged—and of patients in special circumstances—such as those facing an operation, those needing scarce medical resources, or those in excruciating pain that can only be relieved by powerful drugs. Related to the question of rights are other important issues in health care that will also be discussed in this chapter: the concepts of health and paternalism and the role of advocacy in patient care.

Values and Priorities

Before proceeding to study the issue of rights, try to decide whether you agree or disagree with the following statements. Your answers should help you become aware of your values and priorities in this matter.

1. In the gun wound case, the patient did not receive adequate care.

2. In the mastectomy case, the radiologist was wrong in not thoroughly informing the woman of the risks of radiation.

3. The radiologist would not have been wrong if the treatment had not resulted in injury to the patient.

4. The patient with the clutch of medical visitors has no reason to complain, since he or she is getting the services of nearly two dozen HCPs free of charge.

5. Health is best defined as the absence of disease.

6. Patients who request information about their condition are always entitled to the truth.

7. HCPs have an obligation to give patients information about their condition, but only if they request it.

8. It would never be justifiable to misinform patients about their condition.

9. Patients have a right to participate in all decisions about their health care.

10. HCPs are justified in concealing information from patients about their condition when they think it is in the patients' best interests to do so.

11. A patient has no right to refuse life-sustaining medical treatment.

12. A HCP is justified in overriding a patient's refusal to consent to life-sustaining medical treatment.

13. If many lives will probably be saved, it is all right to use an incorrigible prisoner in a medical experiment without first gaining the person's consent.

14. Patients should be allowed access to their medical records.

15. Health is a state of complete physical, mental, and social well-being.

16. HCPs ought to work actively to secure the rights of patients.

17. HCPs sometimes are obliged to take action that is calculated to advance a patient's interests but that goes against the patient's desires or limits the patient's freedom of choice.

18. When it comes to knowing what's best for their health, patients rarely know as much as HCPs, especially physicians.

19. Nurses should be required to take relicensing tests every five years.

20. People should be allowed legal access to drugs like Laetrile.

21. HCPs should never violate patient confidentiality.

Toward a Definition of "a Right"

It is one thing to speak of rights but quite another to say what a right is. But why should we even try to define *a right*? After all, many of us use the word *right*; we claim rights and profess to respect the rights of others. If we can get along without a further definition of *a right*, then why bother to try to find one?

The reason we must attempt to pin down the meaning of *a right* is that without it we can neither formulate nor evaluate intelligently the moral and ethical arguments for patient rights. To illustrate, suppose someone argues that people have a right to equal health care because unequal health care is unjust. Assuming for the moment that unequal health care is unjust, just how does that assumption imply that equal health care is a right? Well, some people define *a right* as "a just claim." If a right is a just claim, and if unequal health care is unjust, then it follows that the claim of equal health care is a just claim and therefore a right. But if *a right* cannot be defined correctly as "a just claim," then it is unclear how the justice or injustice of any practice relates to the existence or nonexistence of a right; it is unclear how the injustice of unequal health care relates to the right to equal health care.

One of the most common arguments against the right to equal health care is that it would be useless, even harmful, to legislate such a right, because society simply could not afford it. Therefore, individuals do not have a right to equal health care. Granted, legislating equal health care might be socially undesirable. But even if it should not be legislated, how does it follow that people have no right to it? At best, the folly of legislating equal health care seems to imply that equal health care should not be enforced by law. But the imprudence of such legislation does not appear to prove that people do not have a right to equal health care. Or does it? Those who argue that it does prove the point assume that saying someone has a right to something implies that society should provide that thing to individuals and should protect them in its possession. In other words, proponents of this argument assume that *a right* must be defined in terms of social sanction and protection. They say that to talk of a right in any other way is to speak gibberish. Thus, since society cannot provide equal health care to everyone, individuals have no right to it.

Each of the arguments for and against equal health care, then, hinges on what constitutes a right. So it is with other rights. Therefore, before examining patient rights, we should try to establish a philosophically helpful definition of *a right.*

The Meanings of "a Right"

Various meanings attach to *a right.* For example, the seventeenth-century English philosopher Thomas Hobbes (1588–1679) associated a right with a liberty. In this view, to say that individuals have a right to do something means that they are free or at liberty to do it, that they are under no obligation to refrain from doing something. Therefore, to assert that patients have a right to discharge themselves from a hospital means that they are not obliged to remain against their will or simply that they are at liberty to leave if they wish.

Another meaning of *a right* connects it with the force of law. In this view, a right is a permission to do something that is secured and protected by law. Thus patients are not only at liberty to leave a hospital if they wish, they have legal sanction to do so. This means that society would not only allow patients to leave hospitals but would restrain others from preventing them from so doing.

Still other definitions associate a right with a duty. A duty is something one is obliged or bound to perform. Philosophers sometimes speak of relative duties, by which they mean a duty to some assignable person other than oneself or the state. For example, HCPs have a relative duty to provide adequate medical care to those who have enlisted their services. They also have a relative duty to keep secret whatever communications between patients and themselves may arise during treatment. In this view, a right is associated with a duty as viewed by the individual to whom the action is due. Thus patients have a right to adequate health care from the HCP they enlist to treat them; patients also have a right to privacy.

Other philosophers associate a right with a conditional duty—that is, a duty that binds only on the condition that the beneficiary of it chooses to have it exercised. By this analysis, a patient's right to privacy means that HCPs have a duty to respect the patient's privacy if the patient wants it respected. If the patient does not, then the HCP is under no obligation to respect it.

There is something to recommend each of these interpretations of the meaning of *a right* for health care. The liberty definition has the advantage of associating a right with freedom, which is often defended by appeal to rights. The legal definition associates a right with a permission and protection, which recognizes that a right is something that a person may or may not choose to exercise. The duty definition recognizes that rights imply obligations that others have, which may bind unequivocally or may bind only if the claimant chooses to exercise the right.

None of these definitions seems complete in itself. For example, if liberty is associated only with the absence of any contrary obligation, then it ignores the obligations that others have not to interfere with a right. If I have the right to discharge myself from a hospital, then presumably others must respect this right; but the liberty definition does not seem to obligate this. Associating a right with legal force recognizes the duty of others to refrain from interfering with my right, but it does not seem to acknowledge any rights the law does not sanction. For example, a pedestrian who has just been struck by a speeding car might be said to have a right to attention from a passing physician. But if this is a right, it is a moral right, not a legal one, because the law does not recognize it. A moral right is one derived from some principles of a moral theory. Thus the injured pedestrian may have a moral right to attention from a passing physician, but the physician is not required by law to provide it.

Relative duties always involve relations between particular individuals. A relative duty is a duty of one or more specific persons to one or more other specific persons. But what about cases where duties cannot be so definitely assigned? If we assume, for example, that people have a moral or political right to health care, to which assignable person or persons does the duty fall to provide it? As for conditional duties, they seem to ignore social obligations. For instance, suppose a victim of a clearly egregious act of medical malpractice chooses to take no action against those who are responsible. Does that mean that society and the medical profession have no obligation to do anything to protect the basic rights of the individual who has been injured?

Given the variety of perspectives on the concept of a right, it is easy to become confused and to despair of formulating a helpful definition. But on closer inspection, these definitions and others suggest three basic characteristics inherent in the concept of a right.

Characteristics of "a Right"

The characteristics of a right have been identified by philosophy professor Carl Wellman, who has done much to clarify the definition of *a right*.[1] According to Wellman, *the first characteristic of a right is that it is permissive of its possessor*. This means that those possessing the right may exercise it if they wish but do not have to. For example, patients have a right to be informed about the benefits and risks of a proposed medical procedure, but they are not required to exercise the right. *Second, a right implies a duty of other individuals*. Although patients may or may not demand information about medical procedures as they wish, HCPs are required to provide the information on request, whether or not they wish to do so. *Third, a right is or ought to be protected by society or secured by law to the individual*. For example, the patient's right to know about the risks and benefits of a medical procedure is protected by law.

In Wellman's view, the challenge to the jurist or legal philosopher is to sum up all three characteristics in a single definition. Wellman attempts to do this by defining *a right* as "a sphere of decision that is or ought to be respected by other individuals and protected by society." By his own admission, this definition does not sum up clearly and precisely the nature of a right. For one thing, *sphere of decision* needs further clarification. For another, the dichotomy between *is* and *ought* implies that there is no single nature shared by all rights. Although legal

---

1. *C. Wellman*, Morals and Ethics *(Glenview, Ill.: Scott, Foresman, 1975).*

rights are secured by society and often respected by individuals, it does not necessarily follow that they carve out "spheres of decision" that should be respected and protected. On the other hand, moral rights, assuming that there are some, should indeed be protected and respected. Frequently, however, they are not, a fact that partially explains the formulation of various patient bills of rights. In addition, the connection between the obligation of individuals to respect spheres of decisions and the obligation of society to protect these—if there is a connection—remains unclear. Finally, Wellman's definition seems to restrict rights to spheres of decisions. It leaves unclear whether a comatose, unconscious, or incompetent patient has any rights, since such patients cannot engage in decision making.

Despite these shortcomings, Wellman's definition is philosophically useful and has much to recommend it for our discussion of patient rights. First, it directly addresses the purpose or function of a right as it applies to health care. In a health care context, rights arise primarily out of situations in which individuals are likely to choose conflicting courses of action. For example, HCPs may choose to deny patients information about less-conventional methods or places of treatment, even though patients may want *all* relevant information; or HCPs may choose to mislead patients about the kind of medication they are receiving (as with placebos), even though patients may want to know exactly what they are taking. In short, the function of defining a right in a health care situation is primarily to prevent conflicts by identifying some areas of decisions within which one person's wishes or choices take priority over those of another person.

Wellman's definition also allows the possibility of rights even where they are not legally secured and enforced. In health care this is particularly important because so many of the questions about rights that arise do not have the force of law to support either side of the conflict. This is especially true of so-called moral and political rights. For instance, assuming that adult patients have the moral right as autonomous beings to be treated in a mature, adult way, then HCPs have an obligation not to treat them like children, not to patronize or humor them, not to deny them dignity and respect. But there is no law to enforce this right.

Finally, close inspection of Wellman's definition can be interpreted as providing possible rights for the comatose, unconscious, and mentally incompetent. It is true that such patients cannot engage in decision making, but this fact does not preclude their having rights that substitute decision makers may and should exercise. Indeed, it could be argued that, precisely because these patients are not capable of making decisions of their own, surrogate decision makers are needed to protect their interests. How far such paternalism should extend, or who should be the decision makers, is, of course, problematic. But it seems that one could validly interpret Wellman's definition of *a right* to include cases of patients who are incapable of making decisions.

Therefore, in all that follows, *a right* will mean a sphere of decision that is or ought to be respected by other individuals and protected by society. This definition is useful, although not perfect, for a consideration of patient rights.

Precisely what rights, then, are patients thought to have? Before attempting to answer this question, it is important for us to understand the concepts of health and paternalism. The former inevitably influences the nature and extent of patient rights. The latter frequently takes the form of an action perceived by the HCP as a professional obligation but that conflicts with a patient right.

Concepts of Health

The main reason that defining health and disease is difficult is that they are not static conditions. They are dynamic, subject to the same changes as humans themselves. In recent years, research in health care has expanded our views of health and disease in at least four major ways: (1) broader and more inclusive definitions of health and disease; (2) awareness of changes in disease patterns; (3) changes in our concepts of disease causation; and (4) changes in our patterns of health care delivery.[2]

Not too long ago, *disease* was defined simply as the absence of health and *health* as the absence of disease. Such circular definitions led nowhere. What is more, definitions that associated health and disease exclusively with states of physical discomfort or well-being totally ignored the emotions and social pressures that affect us all.

Today experts continue to disagree about what constitutes health as opposed to disease. Small wonder, for defining and understanding these complex states involves, among other things, value judgments, abstractions that cannot always be measured in objective terms, and cultural determinants. Nevertheless, individuals and groups continue to work at definitions. We will concentrate on the competing concepts of what constitutes health, since these relate directly to the rights that patients are thought to have.

One view associates health with the lack of evident physical or psychological pathology or deviation from a group average. This perspective derives largely from the medical view that considers physical health exclusively as the absence of disease. Given that, on the physical and psychological levels, the human system tends toward normality and health, health is viewed as a universal phenomenon under favorable conditions, and development is considered healthy when it does not deviate significantly from the norm.[3]

This view is properly termed negative, because it considers the mere absence of disease to be synonymous with health. But is such a negative description satisfactory? Some have pointed out that, for one thing, it is inadequate to deal with personality development. Take, for example, the case of an adolescent who shows no physical abnormalities, gets along well with peers, passes all courses in school, and shows no sign of serious maladjustment. Yet it is apparent to parents, teachers, and friends that the person is using only a smidgen of his or

2. *K. Creason Sorenson and J. Luckmann,* Basic Nursing: A Psychophysiologic Approach *(Philadelphia: W. B. Saunders, 1979), p. 100.*

3. *J. C. Coleman and C. Hammen,* Contemporary Psychology and Effective Behavior *(Glenview, Ill.: Scott, Foresman, 1974), p. 305.*

her capabilities. Granted, such a young person may not be sick. But is he or she truly healthy?

A second view associates health with optimal realization of the individual's potential. Here health is viewed positively, as something that transcends the "normal" or "average." Thus, health and the healthy person are associated with self-realization and the fulfillment of potential. This view is captured in the World Health Organization's (WHO) definition of *health* as "a state of complete physical, mental, and social well-being and not merely the absence of disease or infirmity."

The WHO view may have the advantage over the first of integrating the physical, psychological, and social aspects of health. But is it precise and realistic? Some have pointed out that, life being what it is, we can never aspire to anything more than fleeting interludes of "complete physical, mental, and social well-being." Others see real dangers in the WHO definition. They charge that the word *complete* leads people to expect that life can and should deliver protection and, as a result, that health care must deliver the impossible.[4] Whether or not such criticisms are justified, there is no consensus about what constitutes "optimal development." Any definition will be influenced by personal and cultural values, customs, and beliefs. Thus, at best, optimal development appears to be more a general principle than an operational standard that we can define and measure.

A third view identifies health with the attainment of certain criteria that are thought to be essential for psychological health. Such criteria might be self-acceptance and a realistic appraisal of one's assets and liabilities; a realistic view of oneself and the surrounding world; unity of personality, freedom from disabling inner conflicts, and good stress tolerance; or development of essential physical, intellectual, emotional, and social competencies for coping with life problems. Although such criteria may be useful in determining development, they remain arbitrary. One can always ask why these and not others? They are also ambiguous. Just what constitutes, say, "a realistic appraisal of one's assets and liabilities" or "good stress tolerance"? What is more, these criteria are always subject to change based on human development.

A fourth view of health springs from the conception of the human as a living system. By this account, humans, like all other living systems, have interdependent parts or subsystems that allow the system to function as an integrated unit. The system also has a built-in tendency to maintain this functional integrity. Thus, the human system would be healthy to the degree that it is in a state of harmonious and dynamic equilibrium. Or more simply, it is healthy when all parts in the system are in harmony.

But when is that? Individual persons have much in common with other persons and with all natural systems, but at the same time, individuals are unique: No two of us are identical. The uniqueness makes it difficult, maybe impossible, for anyone outside the individual to know when that person is in "sync," when he or she is in the state of harmonious and dynamic equilibrium defined as health in this view.

---

4. D. Callahan, "The WHO Definition of 'Health,' " *Center Studies* 1 (1973): 77.

The reason for mentioning these different perspectives of health—and, by implication, of disease—is not to endorse one or the other but to point out that one's concept of patient rights is bound to be influenced by one's view of health. We have already seen this in the ongoing debate about whether or not individuals have a right to health care. We will see it again in our examination of patient rights, especially the right to adequate health care. These views also influence the concept of paternalism.

Paternalism

In general, *paternalism refers to an action that is intended to advance a person's interests but goes against the person's desires or limits the person's freedom of choice.* A paternalistic act, then, has two noteworthy features: (1) In intention, it is directed toward the other's best interests; (2) in means, it disregards the other's desires or limits the other's freedom. Usually, those who act paternalistically defend their actions by appeal to their intentions, which are considered to override their means.

Recall the case cited at the beginning of the chapter about the woman who suffered radiation burns and skin breakdown. Suppose that the radiologist had not informed her adequately of the possible dangers of radiation therapy because he felt he would unduly alarm her, and as a result, she might forego the treatment that she so desperately needed. Clearly, in that case the radiologist did not inform her because he believed that in so doing he served her best interests. This case illustrates *personal paternalism,* which refers to an individual's deciding on the basis of his or her own values and principles what is in another's interests and then acting in a way that disregards the other person's wishes or limits that person's freedom of choice. Based on his own values, the radiologist decided not to inform the woman "for her own good."

That HCPs tend to be paternalistic is altogether understandable. The dynamics of the patient-HCP relationship invite it. Patients, after all, are very vulnerable. They are ill, in need of professional knowledge and skill, and desirous of care. In contrast, HCPs are well, knowledgeable and skillful, and capable of providing the care. The patient-HCP relationship, then, is one in which one party, the patient, has little power and the other, the HCP, has vast power. Little wonder that HCPs often act paternalistically.

But how far should paternalism extend? For example, should HCPs and health care institutions intervene to save the life of a child who desperately needs an operation that her parents will not consent to on religious grounds? Should an HCP conceal from a patient a terminal diagnosis when the HCP feels that the person cannot handle the truth? Should an HCP, fearing to influence a young couple's desire to have another child, conceal from them that their infant died shortly after birth because of an unpredictable birth defect?

Instrinsic to all such cases is the problem of drawing a line between paternalism and autonomy. When is paternalism acceptable? When is it not? When does it become an unjustifiable constraint of personal freedom? These are tough questions that should and do receive profound moral debate.

The same questions can be raised about another form of paternalism: state paternalism. *State paternalism* refers to actions taken not by individuals but by government bodies or agencies, over particular kinds of actions or procedures. Usually such controls take the form of laws and regulations. In each instance, the regulating body disregards personal desires or limits personal freedom for "one's own good."

The vast majority of such paternalistic acts in health care are probably uncontroversial. We welcome most instances of licensing and technical specifications in health care, since these presumably minimize risks to patients and threats to the integrity of the professions. At the same time, such regulations inevitably restrict freedom of choice. Although licensing and drug standards may be said to protect patient-consumers, they also limit the variety of medical viewpoints, therapies, and medications that consumers can choose from.

An outstanding example of this constraint feature of state paternalism can be seen in the celebrated case of actor Steve McQueen, who sought treatment in Mexico for a rare form of lung cancer. The treatment was not available in the United States—at least not legally. Similarly, considerable debate surrounds the allegedly anticancer drug Laetrile, currently prohibited by the Food and Drug Administration. Limiting the treatment, therapy, and drugs from which the ill can choose understandably raises a question about the rightful authority of the state. Specifically, to what extent is a government morally justified in restricting the actions and choices of its citizens "for their own good"?

Paternalism certainly ranks among the most important issues in the study of health care ethics. As we will see in this chapter and throughout the book, it emerges under different circumstances and takes various forms. But paternalism invariably raises a moral question about the propriety of the intervention. Whether paternalistic intervention is justified in a particular case entails a close inspection of the patient rights that are involved—a subject we will now examine.

A Patient's Bill of Rights

In an attempt to hammer out a statement of patients' rights, a study was undertaken in 1970 by the American Hospital Association's board of trustees and four consumer representatives. Three years later, in January 1973, the group released "A Patient's Bill of Rights." Spokespersons for the American Hospital Association (AHA) acknowledged that hospitals would not lose accreditation if they failed to adopt the statement, but they expressed the hope that all member hospitals would adopt it. On February 6, 1973 the document was approved by the House of Delegates of the AHA. Most hospitals in the United States have subsequently adopted the statement or some version of it. Here is the exact text of "A Patient's Bill of Rights":

1. The patient has the right to considerate and respectful care.
2. The patient has the right to obtain from his physician complete current information concerning his diagnosis, treatment, and prognosis in terms the patient can be reasonably expected to understand. When it is not medically

advisable to give such information to the patient, the information should be made available to an appropriate person in his behalf. He has the right to know, by name, the physician responsible for his care.

3. The patient has the right to receive from his physician information necessary to give informed consent prior to the start of any procedure and/or treatment. Except in emergencies, such information for informed consent should include but not necessarily be limited to the specific procedure and/or treatment, the medically significant risks involved, and the probable duration of incapacitation. Where medically significant alternatives for care or treatment exist, or when the patient requests information concerning medical alternatives, the patient has the right to such information. The patient also has the right to know the name of the person responsible for the procedures and/or treatment.

4. The patient has the right to refuse treatment to the extent permitted by law and to be informed of the medical consequences of his action.

5. The patient has the right to every consideration of his privacy concerning his own medical care program. Case discussion, consultation, examination, and treatment are confidential and should be conducted discreetly. Those not directly involved in his care must have the permission of the patient to be present.

6. The patient has the right to expect that all communications and records pertaining to his care should be treated as confidential.

7. The patient has the right to expect that within its capacity a hospital must make reasonable response to the request of a patient for services. The hospital must provide evaluation, service, and/or referral as indicated by the urgency of the case. When medically permissible, a patient may be transferred to another facility only after he has received complete information and explanation concerning the needs for and alternatives to such a transfer. The institution to which the patient is to be transferred must first have accepted the patient for transfer.

8. The patient has the right to obtain information as to any relationship of his hospital to other health care and educational institutions insofar as his care is concerned. The patient has the right to obtain information as to the existence of any professional relationships among individuals, by name, who are treating him.

9. The patient has the right to be advised if the hospital proposes to engage in or perform human experimentation affecting his care or treatment. The patient has the right to refuse to participate in such research projects.

10. The patient has the right to expect reasonable continuity of care. He has the right to know in advance what appointment times and physicians are available and where. The patient has the right to expect that the hospital will provide a mechanism whereby he is informed by his physician of the patient's continuing health care requirements following discharge.

11. The patient has the right to examine and receive an explanation of his bill regardless of source of payment.

12. The patient has the right to know what hospital rules and regulations apply to his conduct as a patient.

It is important to point out that patient rights do not emanate, either historically or legally, from the AHA's Patient's Bill of Rights. Well before this document came into existence, the law accorded certain rights to patients in their relationships with HCPs. These rights can be subsumed under four headings:

the right to adequate health care, the right to information, the right to privacy, and the right to self-determination. The AHA's statement addresses these same four general topics in its twelve separate principles. A closer look at these main areas is in order, not only to gain a better understanding but also to see their inadequacies and to grasp the moral questions they pose.

Adequate Health Care

Traditionally, the law provides that patients are entitled to adequate medical care. This concept is addressed in principles 1, 7, and 10 of the Patient's Bill of Rights. The failure to deliver proper medical care is called *malpractice,* a term that has no precise legal definition but that usually refers to professional negligence or to any kind of professional misconduct that gives rise to legal liability. Thus the malpractice law entitles a patient to a certain level of care from an HCP. But precisely what level constitutes "proper" or "adequate" care is unclear, left to be determined on a case-by-case basis by the courts. The innocent language in which the AHA couches this right does nothing to clarify it.

Patients are told that they have "the right to considerate and respectful care" (principle 1). But *considerate* and *respectful* are vague words, inviting myriad interpretations. Again, patients are told that they have the right to expect a hospital to make "reasonable responses to the request of a patient for services," but the question of what constitutes reasonableness is entirely ignored. For example, a pattern of nurse staffing may be reasonable enough when the majority of patients are only moderately ill and ambulatory but altogether inadequate when some critically ill patients in need of highly specialized and intensive care are admitted into the hospital. To what extent, then, must a hospital provide for, and patients pay for, a pattern of overstaffing to meet such periodic demands? We can also ask whether requests of the moderately ill and ambulatory for, say, explanation and reassurance are reasonable when HCPs are busy tending to the more seriously ill. Such questions raise issues not only of adequate health care but of justice and injury as well.

Similar moral concerns can arise in interpreting the terms in "provide evaluation, service and/or referrals as indicated by the urgency of the case." Are health care services to be provided at the request of the patients and family or of the hospital? And just what factors enter into the decision—need, available resources, ability to pay?[5]

Other aspects of the right to adequate health care as stated in the Patient's Bill of Rights raise concerns. For one thing the right is limited—some might say vitiated—by the hospital's "capacity" to respond to the request. For another, patients are told that they have the right to "reasonable continuity of care," but again *reasonable* goes undefined, as does *continuity* and even *care.* The language of principle 10 clearly implies that responsibility for continuity of care is a joint responsibility of physicians and hospitals, but in practice it requires the collaboration of all members of a hospital staff, knowledge of community health

---

5. *E. Bandman and B. Bandman, "There Is Nothing Automatic about Rights,"* American Journal of Nursing, *May 1977, p. 871.*

services, and effective use of resources. Thus, provision for care requires not just the mobilization of physicians but of a wide range of community care providers. So, in emphasizing the responsibility of hospitals and physicians for providing continuity of care, principle 10 acknowledges the needs of the sick that are generally met by these agents but ignores the fact that continuity of care may also encompass the promotion and maintenance of health, which often falls to other HCPs. This is especially true in caring for patients with special needs, such as the poor and the elderly. Finally, some would argue that the main oversight of principle 10 is that it fails to acknowledge patients as active participants in planning their own health care and continuing care. It delegates the responsibility entirely to hospitals and physicians.

Underlying the obscurity that surrounds the right to adequate health care seems to be a failure on the part of HCPs and society in general to decide unequivocally whether health care is a right or a privilege. If it is a right, then precisely what does this right include?

For example, some who have considered the issue believe that patients have a right not just to adequate health care but to the highest quality of health care available. One of them is George J. Annas, director of the Center of Law and Health Services of the Boston University of Law. Annas has argued that a patient has the right to "access to the highest degree of medical care without regard to sources of payment for the treatment and care."[6] Presumably, he views this as a political or moral right.

Those who agree with Annas usually find the source of this right in some ultimate moral principle. For example, some have pointed to the Kantian moral principle that every person is of inherent and equal worth. On the basis of this principle, they claim that every person has an equal right to medical care, simply by virtue of being a person. Others seek to justify equal health care in terms of principles that express the aims and commitments of society. They argue that a society that endorses justice and equality must offer health care to all if it offers it to anyone.

But not all agree that everyone has a right to health care, let alone equal or even adequate health care. There are those, for example, who subscribe to a position commonly termed *medical individualism*. In effect, they argue that recognizing a right to health care in effect violates the rights of HCPs. Medical practitioners, if equal care were required, would be obliged to employ their time, talent, and energy as society dictates. But this arrangement would deprive HCPs of their autonomy.

Others who are willing to elevate health care to the status of a right point out that health care is just one social good among others: jobs, education, housing, defense, and so on. To admit all these as rights would place an economic burden on society that it simply cannot shoulder. Such an argument differs from medical individualism in that it does not deny that health care is a right. Rather, it asks, just what sort of rights do we want to support? Do we want to claim that

---

6. G. J. Annas, "The Patient's Rights Advocate: Can Nurses Effectively Fill the Role?" Supervisor Nurse, 5 (1974):20; and G. J. Annas and F. X. Healy, "The Patient's Rights Advocate: Redefining the Doctor-Patient Relationship in the Hospital Context," Vanderbilt Law Review 27 (1974):243.

everyone has a right to equal health care, as Annas does? If we do, then we must be prepared to allocate a large part of society's economic resources to health services, to give health care high priority in relation to other local, regional, and national concerns. Or on the other hand, do we want to claim that everyone has the right to minimum care but that those who can afford it have a right to buy additional superior care?

The innocuous-sounding claim that individuals have a right to health care is one that raises profound conceptual and moral problems, shot through with economic and political overtones. These questions have far-reaching implications for those seeking health care as well as for those providing it.

Information: Informed Consent

The right of the patient to information is addressed directly in principles 2, 3, 8, 11, and 12 of the Patient's Bill of Rights. Since principle 3 deals with a unique and complex aspect of the information issue, informed consent, we will consider it last.

One problem with principle 2 is that it does not specify what is to be considered "complete" current information regarding diagnosis, treatment, and prognosis. In fact, this interpretation is left entirely to the physician. So is the decision about whether it is "medically advisable to give such information to the patient." Therefore, this principle encourages a paternalism that says, in effect, "You have a right to information, providing I think you should have it."

Furthermore, there is a serious question about the precise meaning of "complete current information." Does this mean that patients are entitled to all the information relevant to their conditions, even information about recent research and experimental drugs? For example, many credible scientific sources attribute preventive and curative value to a variety of vitamins. Noble Prize winner Linus Pauling, for one, has written compellingly about the usefulness of vitamin C in the prevention and treatment of various cancers; others have spoken of the value of vitamin E in preventing and treating heart disease. Such vitamin therapy, however, does not receive the endorsement of the "medical establishment." Do patients have a right to this kind of information, and do HCPs have an obligation to provide it? One can almost hear physicians, nurses, and other HCPs object that giving patients all relevant information would be inordinately time-consuming and, ultimately, do little more than confuse patients. Such a paternalistic reaction is understandable, perhaps even justified. But it points up other inadequacies in principle 2, because this item says nothing about how much time and effort HCPs should give to educating patients about their medical condition and whether HCPs have a right to prejudge a patient's capacity to understand information.

One of the most controversial patient rights ignored by principle 2 and the rest of the Patient's Bill of Rights is the right of patients to receive a copy of their medical record at the conclusion of their treatment or simply for their own use and information. The traditional argument against patient access to medical records is that patients cannot understand what is often a highly technical recording of their health care and consequently might become alarmed. One objection to this paternalistic view is that open records are a way to educate patients about their own health so that they can participate more knowledgeably in decisions affecting their continuing care. Of course, this objection assumes that patient participation is valued.

Annas, for one, does value patient participation, and his version of the Patient's Bill of Rights states that patients have a right to see their records. Similarly, the Department of Health, Education, and Welfare's Report on Medical Malpractice had this recommendation on the subject:

> The Commission finds that patients have a right to the information contained in their medical records and recommends that such information be made more easily accessible to the patients and the Commission further recommends that states enact legislation enabling patients to obtain access to the information contained in their medical records through their legal representative, public or private, without having to file a suit.[7]

In fact, a number of states now allow patients direct access to their records under specified conditions. What is more, several studies indicate greater patient cooperation and less anxiety when patients have access to their health records.[8]

Principle 8 addresses a most important and often violated patient right—the right to information about the nature and quality of the professional services that are available to them through their HCPs and institutions. It is important because often it is in patients' best interests to find out about the credentials of the health care and educational institutions affiliated with the hospitals providing care. By definition, teaching hospitals add training and research to their care and cure functions. These added functions may provide distinct advantages to patients with complex diagnostic and treatment problems. But they may be distinctly disadvantageous to patients facing intractable or terminal illness who seek only loving care and who do not want the constant intrusion of hordes of interns and students.

This right to information about services and institutions is often flouted because the HCP reaction to legitimate patient inquiries is still largely defensive, often resentful. Too often, patients are made to feel that asking means indicting. As a result, being as vulnerable as they are, many patients prefer to say nothing rather than risk reprisals. Changes in HCP attitudes apparently will occur only after HCPs themselves fully understand this right and its value to the patient and learn to view it not as a threat or an indictment of their own standards and services but as a legitimate request about a legitimate concern.

Principle 11 addresses the patient's right to billing information but it says nothing about the kind of explanations that are needed. Medical consumers are neither technicians nor bookkeepers; few of them understand the accountancy jargon that passes for bill explanations. Worse, those at the desk who present the bills are often ill equipped to illuminate patients. Compounding the problem is

7. *U.S. Department of Health, Education, and Welfare,* Report of the Secretary's Commission on Medical Malpractice, *Pub. N. 0573-88 (1973).*

8. *E. Bandman and B. Bandman, "How to Reduce Patients' Anxiety: Show Them Their Hospital Records,"* Medical World News, *January 13, 1975, p. 871.*

the fact that patients' reactions to bills are frequently colored by their total medical and hospital experience. Feelings of gratitude for, or resentment of, the treatment they have received are apt to influence patient reactions to their bills. Some in the field have suggested that the right to receive and examine explanations of bills should be supported by offering the bill in advance of discharge from the hospital.[9]

Information issues like these raise moral concern, but they pale in importance compared to the concerns raised by the question of informed consent. This right, addressed in principle 3, is so crucial as to warrant special consideration. Informed consent is perhaps the most controversial right of all addressed in the Patient's Bill of Rights. It is also the basis for numerous lawsuits. For example, a nineteen-year-old patient who was left paraplegic after exploratory surgery sued on grounds that the surgeon had not fully informed him of the risk of paralysis. Had the surgeon informed him, the patient claimed, he never would have agreed to the procedure. In another case, a woman who developed a vesicovaginal fistula after an elective hysterectomy sued because, she said, she would have refused the surgery had she been informed of the risk. In still another case, a middle-aged schoolteacher, wife, and mother of five sued after being facially disfigured and left without speech following surgery for malignancy of the jawbone. She claimed that she had not been informed of the extent of the surgery, which included removal of the whole lower jaw and two-thirds of the tongue and was followed by x-ray therapy to the vocal cords.[10] Such cases raise serious questions about the nature and limits of informed consent.

The central issue in these examples is not whether patients agreed or consented to certain medical procedures, for they obviously did. The issue is whether the consent was informed, meaning valid or legitimate. The informed part of the consent raises an important moral concern if persons are valued as autonomous agents. If they are, then it follows that they must have the opportunity to decide whether to participate in medical procedures. Of course, if people are not valued as autonomous beings, then the question of the validity of consent does not arise so forcefully. But since the AHA statement emanates from a recognition of the value of individual autonomy, we will take that position as our point of departure.

For consent to be truly informed, at least two things are required: deliberation and free choice. This means that patients must understand what they are consenting to and must then voluntarily choose it. Thus, deliberation requires not only the availability of facts but also understanding of the information as well. Principle 3 alludes to information, but it says little about understanding, although understanding may be implied in the phrase "necessary information." In order to understand their medical situation, patients need not only enough information but the right kind of information, meaning usable information, information that they can readily comprehend. Patients can be provided reams

---

9. Bandman and Bandman, "How to Reduce Patients' Anxiety," p. 872.

10. These three cases are cited in D. C. Rubsamen, "What Every Doctor Needs to Know about Changes in Informed Consent," Medical World News, February 9, 1973, p. 66067.

of information so technical that it obscures and intimidates rather than enlightens. Thus, they may understand so little of it that their consent cannot be termed valid.

Complicating the issue is the highly scientific nature of medicine and medical procedures. Obviously such a field contains a large amount of esoteric material and an extremely technical vocabulary.[11] This is precisely the reason why HCPs often argue that trying to provide patients with relevant information is futile. Rarely can patients understand what is offered, and frequently they are overwhelmed and alarmed by it. Therefore, the argument goes, health care specialists should decide on the basis of what they believe is in the patient's best interests—that is, they should act paternalistically.

But this argument fails to distinguish between the difficulty and the impossibility of communication. The fact is that in many fields—physics, astronomy, economics, chemistry, psychology, even ethics, for that matter—professional communication with the uninitiated looms as a major challenge. But a challenge is not the same thing as an impossibility. A challenge is something that can, and often must, be undertaken and managed. To view challenge to communication as an insurmountable obstacle is to polarize the universe into experts and nonexperts: The experts know, and what they know can never be communicated to nonexperts. Such an elitist view seems particularly destructive to health care, for it results in medical experts assuming a proprietary right over the total health of the nonexpert patient. In addition, it discourages people from actively participating in their own health care. It seems ironic that, at a time when *malpractice* is for many HCPs the dirtiest word in the legal vocabulary, more HCPs do not enthusiastically grasp the opportunity to dissipate the responsibility for medical decisions, which sharing information with patients would do.

There are studies that belie the claim that patients cannot understand the information provided them, will be frightened by it, or really do not want the information. In one study, patients were fully and nontechnically informed of the possible risks arising from various angiographic procedures. The consent form explained the techniques as well as such possible complications as clotting (with the need for surgery), allergic reactions to the dye, paralysis, bleeding, or pseudoaneurysm (with the need for surgery). It also mentioned the statistical probability of these complications compared to the probability of injury in a car accident. Patients were asked for their reactions to the consent form and were given an opportunity for discussion with physicians in making up their minds. Out of the 232 patients involved, 80 percent said they appreciated receiving the information; less than one-third said they were made uncomfortable by the information; less than 15 percent requested further information; and only four of the 232 patients refused the procedure because of the information provided.[12]

In short, if consent is to be truly informed, patients must be allowed to deliberate on the basis of enough usable information—information that the

---

11. R. Munson, Intervention and Reflection (*Belmont, Calif.: Wadsworth, 1979*), p. 222.

12. R. J. Alfidi, "*Informed Consent: A Study of Patient Reaction,*" Journal of the American Medical Association 216 (1971):1325–1329.

layperson can understand. On that basis, they are then in a position to decide. But usable information is not of itself enough to guarantee informed consent. Free choice is also important. In other words, the consent part of informed consent is as significant as the informed part.

Everyone agrees that, for consent to be legitimate, it must be voluntary. Patients must willingly agree to the procedures. This means that a person must be in a position to act voluntarily—that he or she must be competent. But what does it mean to be competent? Principle 3 of the Patient's Bill of Rights does not tell us; in fact, it ignores this point entirely.

The consensus of medical and legal opinion ties competence to rationality. When people are capable of acting rationally, they are considered competent. But what does it mean to be capable of acting rationally? Consider, for example, the case of a person who refuses a needed operation or blood transfusion on religious grounds. Is this rational behavior? Is a person who holds such unusual, perhaps eccentric, religious views capable of acting rationally in such matters? By conventional standards, we would probably say no. What about the confirmed altruist who, for the betterment of humankind, is willing to submit to an experiment that is calculated to advance the interests of others, but that will probably sacrifice the individual? From the viewpoint of self-interest, such a view is irrational, and anyone subscribing to it could be judged incompetent. What about the elderly patient who has recently squandered away a fortune on individuals she hardly knows and on the most frivolous activities? Can such a person be considered rational enough to consent to medical treatment? There is a body of case law addressing such cases, but the fact remains that there are degrees of competence and that incompetence in one area (for example, managing one's financial affairs) does not imply incompetence to decide medical treatment. In short, we can determine only when someone is incompetent, not when someone is competent.

Apart from such theoretical concerns there are practical problems of determining competence in the cases of children, the aged, the mentally retarded, and the mentally disturbed. How do we decide who in these groups, if anyone, is competent to give consent? If we decide that none of them is ever capable of giving consent, are we justified in securing consent from a third party?

In addition, the circumstances under which medical procedures are done, especially research, can also have an impact on the voluntariness of consent. Nursing homes, mental health facilities, and other institutions create social pressures on patients that often elicit behavior based more on patient perception of institutional expectations than on personal preference. The ordinary patient in a hospital is no less constrained. The structure of hospitals and their institutional pressures, together with the psychology of the patient, are such that patients are inclined to act in ways they think will please doctors and nurses, not themselves. Indeed, it is not at all uncommon for HCPs to exert social and psychological pressures on patients that have the effect of compromising the patients' voluntariness of consent. Thus some patients pop sleeping pills not because they need them but because their nurse expects them to take the pills "for their own good" (translation: "so that the nurse can enjoy an uninterrupted evening"). HCPs who ignore the pressures on patients created by health care institutions risk committing violations of the patient's right to make voluntary choices.

We can see, then, that the question of informed consent always requires close examination of deliberation and voluntariness. By treating these issues casually, HCPs can undercut a fundamental patient right.

Privacy

So important is the right to privacy that all statements of patient rights recognize it. In the AHA's Patient's Bill of Rights, principles 5 and 6 deal with this issue. Before looking closely at these principles, it would be helpful to distinguish among three kinds of privacy.

Authors in various fields often talk about three spheres of privacy: psychic, physical, and social.[13] Psychic privacy is the privacy of the inner sphere, the privacy of one's thoughts, ideals, ambitions, and feelings. It represents an intensely personal area of the human personality. So intimate is the union between this sphere and the individual person that it is often impossible to distinguish them. Because of its nature, the psychic sphere of privacy demands special respect.

Sometimes we do glimpse a person's inner life and psychic sphere through gestures, handwriting, facial expressions, and language. Indeed, individuals often reveal themselves profoundly, when with loved ones or with professionals attempting to help them. Still, most of us generally conceal large portions of this inner sphere from the public. In this way we protect ourselves from exploitation and injury, and we frequently provide ourselves rich opportunities for personal growth. Because of the unbreakable bond between our innermost thoughts and self-identity, then, most would agree that we have a right to psychic privacy and sometimes even an obligation to protect it. Furthermore, others have the obligation to respect our psychic privacy.

Psychic privacy is enlarged and actually made possible by physical privacy. Physical privacy is secrecy or seclusion from view or contact with others. All humans need a certain amount of physical privacy to function as human beings. They need time to be alone and opportunities to share their innermost feelings with others, if they so choose. Without such solitude and such contacts, individuals would be deprived of a great human need and of the potential good resulting from intimate social intercourse. Numerous relationships require physical privacy in order to function: the relationship between husband and wife, lawyer and client, doctor and patient, counselor and client. Humans have a right to physical privacy, and others have an obligation to respect it.

Social privacy refers to the various roles people are called on to play in their daily lives—to their total sense of social identity. Patients and HCPs are not just people who need care and people who provide it. They are also spouses, parents, sisters and brothers, children, religious adherents, political activists, employers, employees, and so on. Thus, patients in a hospital may wish to keep

---

13. T. Garrett, *Business Ethics (Englewood Cliffs, N.J.: Prentice-Hall, 1966)*, pp. 64–67.

informed of what is happening at work or in the world; they may want to function in roles other than the patient role alone. Of course, sometimes this just is not possible, because they are too ill. But often it is. Nevertheless, the patient's right to this kind of social privacy is often subordinated to unnecessarily strict bureaucratic routine and efficiency. What is being overlooked when this happens is that patients do not shed their history, interests, values, and concerns on entering a hospital. In fact, for many patients the hospitalization period can be a rich opportunity for self-expression and self-definition, if they are allowed to play the other roles that interest them in this new setting. Social privacy, then, refers to the need to keep insulated and separate the social roles that we must play. Because individuals require social privacy to function as complete, self-governing entities, they have a right to it, and others have the obligation to respect it.

Few would argue that any of these rights is absolute or can never be violated. Indeed, conflicts often arise in health care between HCPs' obligations to medical institutions and to society and their obligations to respect the privacy rights of patients. Just how far HCP and health institution should go to protect the privacy of patients continues to be a much-debated moral question.

Principles 5 and 6 imply recognition of all three spheres of privacy. In acknowledging the need for privacy in "case discussions, consultations, examinations, and treatment," principle 5 recognizes privacy as a necessary component of the patient-HCP relationship. In the absence of such privacy, the relationship simply cannot function. Patients will not be inclined to provide HCPs the often intimate information needed to diagnose and prescribe. Furthermore, the trust that patients feel toward HCPs when they realize that what they reveal will be held in confidence is probably an important psychological factor in medical therapy, certainly in psychiatry. Finally, patients' revelations may be very damaging to them if made public. For these reasons, various codes and documents acknowledge the importance of privacy as a basic patient right. Among them are the Hippocratic oath, the International Code of Nursing, and the statement by the Joint Commission on Accreditation of Hospitals.

The final sentence of principle 5 seems to relate especially to teaching hospitals, where the emphasis on instruction of medical students can be given priority over the patient's right to privacy. For example, probing interviews by a classroom full of student-strangers can easily violate the patient's right to psychic privacy. Repeated and indiscriminate use of the patient's body for student examination can invade physical privacy. And "volunteering" or coercing patients into such an instructional role violates the right to social privacy, the right to choose for oneself whether one wishes to assume the role of a human guinea pig or instructional prop. Of course, such involvement can be legitimate, even heroic, as long as the patient gives informed consent. But lacking that consent, HCP and institution may be violating all three spheres of patient privacy.

Principle 6, while specifying that all communications and records should be kept confidential, does not say anything about *sharing* information—for instance, with family, employer, or some government agency—without the patient's consent. Yet the problem of disseminating information is frequently encountered in health care, which principle 6 implies. A number of states have statutes that protect HCPs, especially physicians, from being compelled to testify about their patients in courts. But such "privileged communication" laws are far from universal. In thirty-four states, patient-physician communications are privileged: Doctors may not be compelled to reveal in court what they learned from a patient in the course of professional attention. In the other sixteen states, physicians must, under penalty of contempt of court, testify about their patients. In New York, dentists as well as physicians are privileged. In Georgia and Tennessee, psychologists are privileged, physicians are not. In Arkansas, nurses are privileged. But whether or not an HCP is privileged, moral questions always arise. For one thing, HCPs have sworn oaths or operate within codes that respect privacy. For another, confidentiality is essential to preserving the relationship between HCP and patient. Revealing confidential information always raises a question of breaking trust. So, privileged communication laws or not, HCPs face moral choices stemming largely from the clash between their professional and social roles, between what they think they should do as members of a curing-caring profession and what they think they ought to do as concerned and responsible members of society.

Specifically, physicians are required by law to report to health departments the names of carriers of such communicable diseases as syphilis. The law also requires that they report gunshot wounds and other injuries that might be associated with criminal actions. Before the legalization of abortion, physicians in some states were required to report all cases of attempted abortion. Inevitably, such requirements create a conflict for physicians between preserving the patient's right to confidentiality and acting in accordance with the law or as a responsible citizen. But it is not only the physician who becomes ensnared in these dilemmas; other members of the health care team get caught up in these conflicts as well. For example, what are the obligations of the nurse who knows that a physician is failing to comply with the law?

To illustrate, suppose a victim of an automobile accident lies severely injured in the emergency room of a hospital. As a doctor administers the Babinski test, a packet of heroin falls out of one of the patient's socks. The doctor faces the dilemma of how to resolve the conflict between the confidentiality of the physician-patient relationship and the duty to preserve the law. If the doctor fails to report the incident to the police, he or she might face a charge of concealing a wrongful act. But if the doctor permits the law to take its course, then he or she will risk a lawsuit for violating the physician-patient relationship. Suppose the doctor in this case ignores the packet. A member of the health team picks it up and turns it over to a nurse. What should the nurse do? When this actually occurred, the nurse called the police, and the patient was subsequently arrested, tried, and convicted.[14]

Is a patient's right to privacy and confidentiality absolute? Is it exceptionless? Most would say no. If we assume that the patient's right to privacy is limited, then these are the central moral questions for HCPs: To what extent am I

---

14.   *State v. Anonymous, 269 New York State (2nd 459).*

obligated to protect patient privacy? Under what conditions am I justified in violating it? These are questions that require, and are getting, considerable attention from those in and outside the health care professions. Subsequent chapters will examine this issue more closely.

Self-Determination

When patients put themselves in the hands of HCPs, they expect to receive consideration and judgment aimed at maximizing their health. Patients have a right to expect adequate health care, and HCPs are obliged to provide it. In order for this to be achieved, patients must surrender a certain amount of their autonomy, and as we saw in the discussion of paternalism, they generally do so. When they enlist the services of an HCP, patients explicitly or implicitly agree to follow professional advice, to do certain things at specified times in prescribed ways. Thus, the patient's relationship to the professional is largely one of voluntary dependence. As a result, the HCP acquires a considerable amount of power, which entails great responsibility. In general, the HCP's overriding responsibility is to do what is in the patient's best interests. But meeting this responsibility raises serious questions about the extent to which HCPs should use their power and the limits of the patient's dependence on them. How much power should an HCP exercise over a patient, and how much autonomy or self-determination can a patient reasonably expect to retain? Principles, 3, 4, and 9 of the Patient's Bill of Rights address these issues.

In essence, principles 4 and 9 recognize the patients' rights to reject violation to their bodies and to refuse to participate in medical experiments. Specifically, principle 4 recognizes people's right to refuse treatment, particularly to refuse extraordinary measures used to preserve their lives. Later chapters say more about this. Principle 9 recognizes the patient's right to refuse to participate in research affecting his or her care and treatment.

But the right to refuse is only part of the picture. The validity of the consent or refusal always depends on the kind and amount of information provided, as we saw in the discussion of informed consent. In research, then, the patient's rights to autonomy and information are inseparable. If patients are to make health care decisions with open eyes, they must have knowledge of all relevant facts. In regard to research, one important fact is that most research offers little if any therapeutic value to the patients who participate in it. Granted, eventually the result may benefit patients with conditions similar to those of the experimental subjects, but usually the research does not significantly benefit the subjects themselves. This is an important piece of information for subjects to have because it can easily affect their decision to participate. After all, a patient might reason, "If I don't stand to benefit directly, why should I subject myself to the imposition of this research?" In any case, such information provides patients an opportunity to clarify and exercise their deepest value commitments. A different patient might reason, "Although I myself don't stand to benefit immediately, I want to do this in the hope of improving the lot of others, of ensuring that they don't have to go through what I've had to go through."

Moreover, the design of experiments can be important for patient-subjects to be aware of if they are to exercise autonomy. For example, where double-blind studies are involved, as in all drug testing, the requirements of scientific objectivity can clash with the patient's right to know. In a double-blind test design, a certain number of subjects are given the drug; the remainder are given placebos. A placebo (Latin for "I shall please") is an inactive substance, commonly termed a "sugar pill." Because neither investigator nor subject knows who received what, the design is termed *double blind*. Obviously, some of the subjects in such experiments receive neither any genuine medication nor the best available treatment for their conditions. Thus, they stand no chance of benefiting from the experiment in which they are participating.

Of course, some might argue that even those receiving placebos stand to benefit from the well-documented psychological phenomenon called the *placebo effect*. Even seriously ill patients will frequently show improvement when given any kind of medication, including placebos. On the other hand, not all placebos are harmless. Some contain chemicals that produce side effects. But even if a subject did experience the placebo effect, the benefit would be ephemeral. What are these subjects to believe during the "honeymoon period" of the placebo effect? That their conditions have improved permanently? This seems a cruel hoax.[15]

Everyone agrees that patients should be informed of the risks they run in being research subjects. Not receiving any genuine medication or the best medical treatment available, and possibly being one of those subjects who stand to gain nothing from the experiment, are decided risks. Therefore it seems clear that patients have a right to know that these risks are part of the package. Indeed, there is no way they can exercise real self-determination without such information.

A patient's right to information and autonomy can be intimately connected, but this is not always the case. Often the issue is primarily one of providing or not providing individuals the opportunity to decide for themselves. For example, whether or not children should have some control over health care decisions that affect them is largely a question of whether children are considered to be self-governing. Similarly, whether or not the poor are capable of exercising free choice in health care matters depends largely on what practical options are available to them. Again, the institutionalized patient's right to choose activities depends largely on what activities institutions make available and on the regulations that govern their use. So, as with the other three categories of rights, the AHA's Patient's Bill of Rights leaves many aspects of the right to self-determination unexplored.

The preceding discussion may appear to derogate the Patient's Bill of Rights, but it is not intended to. On the contrary, those formulating the Patient's Bill of Rights are to be commended for having organized and set down patient claims that heretofore were recognized only informally, if at all. In addition, the AHA's

---

15. *Munson*, pp. 225–226.

statement has helped initiate dialogue on the important subject of patient rights. Indeed, the remarks in this chapter should be viewed as part of this ongoing dialogue.

The preceding discussion, then, should be considered primarily as an attempt to uncover what appear to be inadequacies in the Patient's Bill of Rights. These inadequacies result largely from obscurity and lack of specificity. In the last analysis, these shortcomings may be accounted for by the nature of the beast itself: Such documents are, almost of necessity, highly generalized formulations. But this does not make an airing of the issues any less important. Indeed, it seems to necessitate debate all the more. The remainder of this part of the book will therefore try to relate the four basic patient rights to considerations involving special groups: children, the elderly, the dying, the poor, and the mentally ill. Moreover, Part IV will show how these rights bear on such special bioethical issues as abortion, euthanasia, human experimentation, and the allocation of medical resources.

Advocacy

Before concluding this chapter, we should consider one additional concern—advocacy. Granted that patients have certain basic rights and that these can be hammered out, how will they be enforced? Certainly the patients themselves cannot enforce them. Given their state of dependence and vulnerability, they are in no position to. Nor can the law be counted on to perform that function. For one thing, agencies that regulate the health care system are primarily concerned with individual HCP qualifications and conduct and with the institutional environment. For another, in order for a question of rights to reach the courts, somebody has to appeal or sue. But contrary to popular thinking, most patients avoid lawsuits. In the overwhelming number of instances, they prefer to give their HCPs the benefit of the doubt or simply "grin and bear it." As a result, usually only the most egregious violations of patient rights end up in the courts. Many others occur daily and are suffered in silence.

Given this picture, it seems that the obligation of securing, protecting, and implementing patient rights falls partly to the health care professions, partly to institutions, and partly to individual HCPs. Another way of saying this is that these entities seem to have moral responsibilities in the area of advocacy.

The term *advocacy* is used here in its classical sense of "summoning to one's assistance, defending, calling to one's aid." In health matters, those who advocate patient rights plead or defend the patient's case. Associate professor of nursing Mary F. Kohnke has proposed a useful definition of advocacy in health care: "Advocacy is the act of informing and supporting persons, so that they can make the best decisions possible for themselves."[16]

By this definition, any act of advocacy has two characteristics: to inform and

---

16. *M. F. Kohnke, "The Nurse as Advocate,"* American Journal of Nursing, *November 1980, p. 2038. Presumably "persons" would include substitute decision makers acting in the place of those who cannot act for themselves—for example, the comatose, the unconscious, and the incompetent.*

to support. Informing patients would include not only making them aware of their rights in a particular situation but providing them all the necessary information to make an informed decision. Supporting may involve several actions or nonactions. According to Kohnke, one action is to actively reassure patients that it is their decision and that they have the right to make it. Thus, they do not need to cave in to outside pressures, whether from family, friends, or other HCPs. A second action might be related to this: HCPs should not allow others to undermine their patients' confidence in their own decision-making ability.

Kohnke suggests that nonactions may be even more important than actions, yet harder to accomplish. Nonaction usually means keeping oneself from subtly undercutting the patient's decisions. It may also involve refusing the requests of others—family, friends, other HCPs—to talk to patients and convince them they are wrong. The personal repercussions for HCPs of such nonaction can be serious. For example, physicians could easily report staff nurses to their superiors if they refuse to talk patients into signing consent forms for a medical procedure.

There are further related constraints on practicing advocacy. First, any act of advocacy is very difficult for those who have superior knowledge and whose training predisposes them to act paternalistically. Second, and more important, the major functions of advocacy are essentially at odds with the culture of the hospital system. Kohnke puts it graphically:

> First, you inform. That, in itself, is a great sin because you will be labeled an "informer." By telling patients things that will make them ask questions, you get labeled a troublemaker. We give lip service to patients' rights. That's great until the patient begins demanding his rights. Then, watch out! If you are the instigator of this through information you gave, you are in trouble.
>
> You may survive the first step, but if you then go so far as to support patients in their decisions, you have had it. For example, you have not only provided the information, but then you are told by the head nurse, the physician, or the administrator to go and convince patients that they are wrong. They *must* have the test or the operation or leave the hospital. If you refuse, you are, indeed, in a fix. You have, in effect, disobeyed a direct order from your superiors.[17]

Such constraints can make individual acts of advocacy nothing short of heroic. Which is another way of saying that until basic changes are made in the system, we are likely to remain in the hypocritical position of expecting HCPs to meet the obligations of advocacy in a climate decidedly hostile to meeting those obligations.

Today most would agree that HCPs have responsibilities of advocacy. The precise nature and extent of those responsibilities, however, remain unsolved. One curious development is the appearance of the patient's rights advocate, a person with both legal and medical knowledge whose primary responsibility is to assist patients in learning about, protecting, and asserting their health care rights. Obviously, this position supposes that patients do have rights, that these rights have been clearly spelled out, and that the institution has adopted the

---

17. *Kohnke, p. 2040.*

statement of rights as a policy. These provisions are themselves formidable, but perhaps the biggest obstacle facing the establishment of effective patients' rights advocates will be the systemic constraints mentioned here, together with those catalogued in Chapters 4 and 5.

Summary

This chapter dealt with the rights of patients. After indicating the various meanings of "a right," it endorsed the definition that views a right as a sphere of decision that is or ought to be respected by other individuals and should be protected by society. Before considering specific patient rights, this chapter explored the concepts of health and paternalism and indicated how they relate to discussions of patient rights. It then examined the American Hospital Association's "A Patient's Bill of Rights," which includes twelve principles that can be grouped into four categories of patient rights: adequate health care, truth and information, privacy and confidentiality, and self-determination or autonomy. Although this pioneering document carves out broad areas of patient rights, it seems inadequate in itself for charting decisive moral direction in a great many health care areas. Finally, the concept of advocacy was introduced, along with some of the conflicts HCPs face in their role as patient advocates.

Ethical Theories

Before leaving this chapter, you are encouraged to study how the ethical theories outlined in Chapter 3 might be applied to the subject of patient rights. A section on ethical theories is included in all the remaining chapters. The hope is that, by studying these suggested applications, HCPs will find help in integrating moral theory into their own personal health care transactions.

In this chapter, the ethical theories will be applied to the questions of patient rights and paternalism. Specifically, this section will try to determine which theories provide a theoretical basis for patient rights and paternalism and how they would resolve conflicts between the two. The case of the radiologist who must determine whether to reveal to his patient the risks of radiation therapy will be used for illustration. Assume that the patient is the woman recovering from the mastectomy.

Egoism

Egoism would not acknowledge that patients have any inherent rights. The only obligation that an HCP has is to act in a way that advances his or her own best long-term interests. This means that HCPs are obligated to provide adequate health care, make information available, respect privacy, and allow self-determination, but only insofar as these actions are consistent with the HCP's self-interests. If they are not, then the HCP is justified, even obliged, to deny them to the patient. Were the radiologist an egoist, he would decide whether to inform the woman not on the basis of her best interests but of his own.

He would do this by comparing the probable results of telling her and not telling her.

A major egoistic consideration in this evaluation would be the likelihood of a malpractice suit. If he does not tell the woman of the risks and she is subsequently injured, she may sue him. On the other hand, if he does tell her and she refuses treatment and subsequently suffers a recurrence of the cancer, she still may sue him. (By today's malpractice laws, HCPs are accountable for not fully explaining the possible consequences of *not* undergoing treatment.)

Other considerations besides possible lawsuits would figure into the egoistic calculation, including one's own reputation, profit, and—perhaps most important—self-respect and self-image. The weight that these would carry would depend on the *value* the radiologist places on them. What, then, would the egoistic radiologist do? The answer is unclear. But one thing is certain: If he decides to conceal the information from the patient, he does so because he judges it to be in his own best interests to do so, not necessarily in hers.

It follows that egoists could never act paternalistically, for by definition, paternalism is an act on behalf of someone else. However, egoists could support acts of state paternalism, if those acts were likely to advance the egoists' best long-term interests.

Act Utilitarianism

Act utilitarians would not acknowledge that the patient has any inherent rights like those stipulated in such documents as the AHA's Patient's Bill of Rights. Rather, they would examine each case as it arose and decide what to do in the light of everyone's best interests. Adequate health care, information, privacy, and self-determination would be important considerations, but only insofar as they bore on the total happiness of everyone concerned.

The act utilitarian radiologist, then, would consider the probable effects of the alternatives facing him on everyone concerned. If he does not inform the patient, she will undergo the treatment, incur risks, but also may be cured. On the other hand, if he informs her, she may refuse treatment and thus invite a resurgence of the cancer. In addition, he would calculate other consequences, such as the possibility of suits, and the ripple effects of each alternative—the impact on other HCPs, patients, institutions, professions, and society at large. The point is that, were the radiologist ultimately to inform the woman of the risks, he would do so not out of any respect for her inherent right to information but because he thought the action would be likely to produce the most good for the most people concerned. The same rationale would apply to a decision to withhold information.

Thus, act utilitarianism seems to endorse limited personal paternalism. It would also seem to endorse limited state paternalism. At the same time, we should note how vigorously John Stuart Mill opposed the notion of compelling people to act in certain ways "for their own good." In Mill's view, paternalism was justified only if it could be shown that unregulated choice would cause harm to other people.

Rule Utilitarianism

Rule utilitarianism also would take a strict consequential approach to the claims of patients. But in contrast to act utilitarians, rulists would examine the rule under which the act falls. The rule here might read: Never conceal from patients information that they need to help them decide whether to submit to medical treatment. Will following this rule likely produce more total good than not following it? If so, then the radiologist has an obligation to inform the woman, even if in this case more good might result from withholding the information. In determining the efficacy of the rule, rulists would consider its impact on all the following: patients, HCPs, the patients' families, institutions, professions, and society generally. Certainly a major impact, and therefore a major rulist consideration, would be the distrust and suspicion that might infect HCP-patient relationships if patients thought that their autonomy could be flouted at the discretion of the HCPs in charge. Like act utilitarianism, rule utilitarianism could apparently endorse limited personal and state paternalism.

The Categorical Imperative

Kant's ethics suggests a rationale quite different from the consequential positions. By Kant's account, patients are entitled to certain things simply because they are human beings. Adequate health care, information, privacy, and self-determination would probably fall under these inherent rights. Thus, for Kant there is nothing contingent about these rights, as there is in consequential ethics. Rights do not depend on an evaluation of the consequences. On the contrary, patients have a claim, say, to information about their medical conditions simply because they are rational autonomous creatures and, as such, deserve to be told the facts. To deny them relevant information is to violate the autonomy, dignity, and respect that they deserve as human beings. So, by a Kantian standard, the radiologist may not exercise personal paternalism; he must inform the patient of the risks, regardless of the results.

Prima Facie Duties

Ross's prima facie duties also imply basic human rights that can be extended to patients, such as the rights to information, truth, and autonomy. At the same time, Ross would not view these claims as absolute; they can be overridden.

Applied to the radiologist's case, we must remember that Ross's duty of justice springs in part from the belief that individuals are autonomous beings and deserve to have that autonomy respected. It follows that any case of denying patients information about their medical situation violates this prima facie duty. This does not mean, however, that HCPs are always obliged to provide that information. Other duties may be operating, such as the duty of fidelity, which obliges HCPs to operate in the interests of their patients. There is also the duty of noninjury, which enjoins HCPs from harming their patients. It could very well be that in cases like the radiologist's, a disclosure of information would not be in the patient's best interests; that indeed, it would have the effect of harming her. In that event, the HCP's actual duty might be to withhold the information.

Whether or not this is an accurate description of the radiologist's situation, the fact remains that Ross's theory could allow a violation of the patient's right to information. This fact suggests that it could permit personal paternalism as well as state paternalism. At the same time, since such intrusions always upset the duty of justice, they cannot be engaged in casually and must always be accompanied by some measure of reparation, in whatever form.

The Maximin Principle

Like Ross, Rawls would probably recognize certain basic patient rights. After all, given the original position, people would surely agree to be bound by rules that secured their autonomy or provided them information in a health care situation. At the same time, these people presumably would also agree to a certain amount of paternalism, which they would recognize as necessary to advance and protect their basic rights. Thus, they probably would agree to judicious licensing, drug, and institutional regulations. How far such paternalistic acts should go, however, is problematic.

As for personal paternalism as it might function in the radiologist's case, Rawls's natural duties could allow it, but only when it is consistent with the principle of justice—that is, with the equality and difference principles. Thus, it appears that Rawls could allow the radiologist to withhold information if and only if in so doing he actually enlarged the woman's real freedom. For example, if she were unable to function rationally with the information, if she were beset by intense fear and anxiety, then presumably her ability to choose the needed therapy would thereby be constrained.

Roman Catholicism's Version of Natural-Law Ethics

The Roman Catholic version of natural-law ethics seems to juxtapose a patient's right with a conflicting HCP obligation. Thus it could be argued that, according to natural law, patients are entitled to certain rights, among them full disclosure of the facts surrounding their treatment. This conclusion follows from the Kantian-like claim that, being creatures with inherent worth, patients are entitled to full autonomy, which they can exercise only if given information. On the other hand, natural-law doctrine obliges HCPs to act in the best interests of their patients. This does not mean that HCPs can lie, but it does mean that they can withhold information as long as they have the patient's best interests in mind. In theory, then, the radiologist may be justified in withholding the information from his patient.

Assuming that this analysis is correct, the Roman Catholic version of natural-law ethics clearly could endorse limited personal paternalism. It could also endorse limited state paternalism, when the state acts to bring about such "natural goods" as the health and welfare of its citizens.

The preceding analyses are offered as tentative first steps toward a full-scale airing of the issues of paternalism and patient rights within the context of ethical theories. You are encouraged to embellish upon and disagree with these theoretical applications. Only in that way will you improve your understanding of these conceptual frameworks and refine your ability to apply them.

CASE PRESENTATION
A Question of Consent

On a cold, bitter night in December, Anna Drew was admitted to Memorial Hospital for a cesarean section. She was scared. What woman in her position wouldn't be? Anna, twenty years old and single, was about to have her first child and had no family or friends to comfort her.

Preparations were made. The consent form was signed, and Anna's vital signs were taken. Because she had elevated blood pressure, the fetal monitor was attached, which showed a somewhat rapid heartbeat but no immediate danger to the fetus.

About 2:00 A.M. Anna started to have strong contractions about three minutes apart. The monitor showed slight fetal distress. The doctor ordered x-rays to confirm fetal position, and Anna was prepared for surgery.

Preoperative medication was given to Anna, and an I.V. was started. About twenty minutes later, after viewing the x-rays, the physician telephoned and asked that another procedure be added to the consent form: the excision of a cervical tumor. He told the nurse to fill out a new form and obtain Anna's signature. The nurse objected that the pre-op medication had left the patient very drowsy. The physician explained that the tumor had developed rapidly and appeared to be bleeding, although it was impossible to say for sure. He ordered the nurse to obtain the signature regardless of the patient's mental condition. Concerned about the ethical and legal implications of obtaining consent under such circumstances, the nurse refused. A short time later the physician appeared, filled out the consent form himself, and obtained Anna Drew's signature.

The procedure was a complete success. The tumor proved benign, and Anna subsequently took home a healthy baby boy. The incident was never mentioned again.

Questions for Analysis
1. *Do you think any of Anna Drew's rights were violated?*
2. *Was the nurse morally justified in refusing the physician's order? Describe the values clash between nurse and physician.*
3. *Do you think Anna's consent could rightly be described as fully informed and volitional?*
4. *Is Anna entitled to know what surgical procedure was performed on her?*

5. *Do you think the nurse is under any moral obligation to report this incident to hospital authorities?*

6. *Using each ethical theory as a conceptual frame, evaluate the actions of both nurse and doctor.*

Exercise 1

The Medical Information Bureau (MIB), located in Greenwich, Connecticut, is a centralized computer system that stores medical records. It is run and paid for by 700 insurance companies and contains medical data on 11 to 12 million people. MIB's multimillion-dollar computer supplies information to member insurance companies so that, in theory, they can identify bad risks. However, the files contain not only medical information but also data on people's sex lives, drinking habits, and hobbies. Where does MIB get its information? From member insurance companies, who get it from doctors, hospitals, credit bureaus, employer medical records, government files, police and court records, newspapers, and the individuals themselves when they apply for insurance policies. Individuals also feed the system information when they file claims, for they cannot receive benefits until they release all relevant medical information to their insurance companies. The companies then forward this information to MIB to be filed in its computer for the use of all participating companies. Individuals do not have access to the data about themselves.[18] How do you feel about such a system? What is its potential for violating individual privacy? Explain.

Exercise 2

Reconsider your responses to the questions in the Values and Priorities section. Are there any you would alter in light of the material in this chapter? Are there any that you feel more convinced of?

---

18.   G. J. Annas, The Rights of Hospital Patients *(New York: Avon Books, 1975), pp. 127–128.*

The Nurse-Patient Relation: Some Rights and Duties

Dan W. Brock

What is the basis for the nurse-patient relationship? This question concerns professor of philosophy Dan W. Brock in the following article. Brock concedes that several interpretations of the relationship can be identified but maintains that only one sufficiently explains the various roles that nurses are expected to perform with respect to the patient. He calls this role "the nurse as contracted

---

This paper originally appeared in a longer version as "The Nurse-Patient Relation: Some Rights and Duties," in Nursing: Images and Ideals, *edited by S. Gadow and S. Spicker (New York: Springer Publishing Co., 1980). Reprinted by permission.*

clinician." By this account, a patient contracts to have specific care by a nurse in return for payment by the patient. In so doing, the patient gives the nurse permission to perform actions the nurse would otherwise have no right or duty to perform. For their part, nurses contract to perform these duties, thereby incurring an obligation to do so as well as a right to be paid for doing so.

In Brock's view, a most important reason for insisting on this contractual model is that it crystallizes, better than any other alternative, the patient's fundamental right to self-determination. It recognizes that a patient's treatment and care originates and remains with the patient. Brock says that HCPs miss this basic point and thus feel not only justified but obliged to practice personal paternalism in the care of their patients. But the right of HCPs to act in the patient's interest, Brock believes, is limited by the permission or consent that the patient has given in contracting for services.

There are at least two sorts of moral considerations relevant to a full understanding of the moral relationship between the nurse and patient. First, those general moral considerations, rights and duties, that the nurse and patient would have simply as individuals and apart from their roles as nurse or patient—e.g., on most any moral theory it is prima facie wrong to kill or seriously injure another human being, and this holds for persons generally, not merely for nurses and patients. Second, there are moral considerations which arise only out of the particular relationship that exists between a nurse and her[1] patient, just as there are in other relationships such as parent and child, public official and citizen, and so forth. A complete account of the nurse's moral situation must include both sorts of considerations, and would be far too complex and lengthy to attempt here; I shall emphasize considerations of the second sort, and even then my discussion will not be at all comprehensive. On virtually any account of the nurse-patient relationship, the nurse owes at least some care to her patient and the patient has a right to expect that care; this is not, however, an obligation the nurse has to just anyone, but only to her patients, nor a right a person has towards just anyone, or even towards any nurse, but rather only towards his nurse. How then does each get into a relationship at all, how does he become *her* patient, and she become *his* nurse? If we pose the question in this way, I think it is clear that the common alternative accounts of the nurse-patient relationship are not all plausibly construed as even possible answers to this question, and that more generally, they address two different questions—some speak to the origin of the relationship, how it comes about, while others speak to the nature or content of the relationship. I think we gain a

clearer understanding of the relationship if we separate these two issues, because the account of the origin of the relationship will affect in turn the account of its content. Consider six of the more common accounts of the relationship, of the role of the nurse vis-à-vis the patient:[2]

1. The nurse as parent surrogate.
2. The nurse as physician surrogate.
3. The nurse as healer.
4. The nurse as patient advocate or protector.
5. The nurse as health educator.
6. The nurse as contracted clinician.

I do not want to deny that nurses do not at times, and at times justifiably, fill each of the first five of these roles, though none of them are unique to her. And at least most of these first five roles refer to professional duties a nurse assumes in entering the profession of nursing. But how is it a nurse has any duty to perform in any of these roles towards a particular person (patient), and how is it that a person (patient) has any right to expect a particular nurse to perform in these roles toward him? Only the last model, the nurse as contracted clinician, can explain that—we must be able to make reference to a contract, or better an agreement, between the two to explain this. This point may be obscured somewhat by the fact that what a nurse would do for a patient in any of the first five roles can be generally assumed to be beneficial for the patient, or at least intended to be beneficial. If it is for the patient's good, why must he agree before she is permitted to act? But just imagine someone coming up to you on the street and giving you an injection, even one intended to be, and in fact, beneficial to you. A natural response would be, "You have no right to do that," and underlying that response would likely be some belief that each person has a moral right to determine what is done to his body, however difficult it may be to determine the precise nature, scope, and strength of that right. Or, imagine a strange woman in a white uniform coming up to you and lecturing you about the hazards of your smoking or failing to exercise. Well-intentioned though it might be, a natural response again might be, "What business is it of yours, what right do you have to lecture me about my health habits?" Again, the point would be that it is a person's right to act even in ways detrimental to his health if he chooses to do so and bears the consequences of doing so, a particular right usually derived from some more general and basic right to privacy, liberty, or self-determination (autonomy). Yet both these actions are, of course, of the sort frequently performed by nurses toward their patients. Likewise, any duties of a nurse to provide care to a particular person cannot come simply from duties she assumes from her role as a nurse, nor can any right of a particular patient to care from a specific nurse.

If we think of the nurse-patient relationship as arising from a contract or agreement between the nurse and her patient, then these otherwise problematic rights and duties become readily explicable. The patient contracts to have specified care provided by the nurse, in return for payment by the patient, and the patient in so doing grants permission for the nurse to perform actions (give him injections, perform tests, etc.) that she would otherwise have no right or duty to do. In agreeing to perform these duties, the nurse incurs an obligation to the patient to do so, as well as a right to be paid for doing so.

A natural objection to such an account is that it seems to rest on a fiction, since in the great majority of cases nurses and patients never in fact make any such agreement; rather, the patient finds himself in a physician's office, or in a hospital, where the nurse as a matter of course performs certain tasks, while the nurse if she makes any such agreements at all, makes them with the physician or hospital that employs her. This reflects the fact that the provision of health care is considerably more complex and institutionalized than any simple nurse-patient account would suggest, but it does not, in my view, show the contract or agreement model to be mistaken. The patient makes his agreement generally with the physician or the hospital's representative, and that agreement is to have a complex of services performed by a variety of health care professionals. The nurse is indirectly a party to this agreement, and can become committed by it, by having contracted or agreed with the employing physician or hospital to perform a particular role, carrying out its attendant duties, in the health care context.

A related objection to this account is that at both these intervening agreement points, it is still the case that the contract or agreement often, if not generally, never takes place, certainly not where what is to be done is spelled out in any detail, and so the account still rests on a fiction. But these agreements can and do have implicit terms, terms which can be just as binding on the parties as if they had been explicitly spelled out. These implicit terms are to be found in the generally known and accepted understanding of the nature of such health care relationships, and in the warranted social expectations the parties to them have concerning who will do what in such relationships. The content of such expectations will in large part derive from the nature of the training of various health care professionals, the professional codes and legal requirements governing their conduct, as well as more general public understandings of their roles.

Why insist on a contract or agreement model of the nurse-patient relationship that requires appeal to agreements between intervening parties, as well as to implicit terms that are generally not spelled out? The reason is that such an account makes clearer than does any other alternative the fundamental point that the right to determine what is done to and for the patient, and to control, within broad limits, the course of the patient's treatment and care, originates and generally remains with the patient. One important reason for insisting on this is that it is insufficiently appreciated and respected by health care professionals. Many health care professionals believe that if what they are doing is in the best interests of the patient, that is sufficient justification for doing it. That, however, is in my view a mistake of primary importance, because it does not take adequate account of the patient's right to control the course of his treatment.

An important part of at least one common

understanding of the physician-patient relationship, and in turn of the patient–other-health-care-professional (including, but not limited to, the nurse) relationship, is that the health care professional will, with limited exceptions (e.g., public health problems arising from highly contagious diseases), act so as maximally to promote the interests of his or her patient. Treatment recommendations and decisions are to be made solely according to how they affect the interests of the patient, and ought not be influenced by the interests or convenience of others.[3] The confidence that the health professional will act in this way is especially important because of the extreme vulnerability and apprehension the ill patient often feels, the patient's incapacity to provide for himself the care he needs, and the patient's often very limited capacity to evaluate for himself whether a proposed course of treatment and care is in fact the best course for him. This focus on the patient's interests to the exclusion of others, however, is different from and should not be confused with, the physician or nurse being justified in acting in whatever manner they reasonably believe to be in the patient's best interest.

The right of the physician or nurse to act in the patient's interest is *created* and *limited* by the permission or consent (from the patient-nurse/physician agreement) the patient has given. To take two extremes, a patient might say to his physician or nurse, "I want you to do whatever you think best, and don't bother me with the details," or he might insist that he be fully informed about all factors and alternatives concerning his treatment and that he retain the right to reject any aspect of that treatment at any stage along the way. In my view, should the patient desire it, either of these arrangements can be justifiable, as well, of course, as many modified versions of them. This has the important implication that the various expectations referred to above concerning what the health professional will do, which generally give content to the nurse-patient relationship, only partially determine that relationship, and it is subject to modification determined principally by what the patient desires of the relationship, and how he in turn constructs it. This is the other difficulty, besides their failure to explain how the relationship comes into being between particular persons, of the five alternatives mentioned above

to the contract or agreement model of the nurse-patient relation. One cannot speak generally about the extent to which the nurse ought to act or has a duty to act, for example, as health educator or parent surrogate, because it ought to be the patient's right to determine in large part the extent to which the nurse is to take those roles with him. What the patient wants will often only become clear in the course of treatment, but to put the point in obligation language, the nurse's obligation is in large part to accommodate herself to the patient's desires in these matters.

I have been considering the case of patients who satisfy conditions of competence, that is persons who possess the cognitive and other capacities necessary to being able to form purposes and make plans, to weigh alternative courses of action according to how they fulfill those purposes and plans, and to act on the basis of this deliberative process; such persons are able to form and act on a conception of their own good.[4] Some of the more difficult moral problems in health care generally arise in cases where these conditions of competence are not satisfied, e.g., with infants and young children, with cases of extreme senility, and with some forms of mental illness.[5] However, I think we must first understand the nurse-patient relation in the case of the competent patient, before determining how that relationship may have to be modified when the patient is not competent. It may, then, be useful to consider what the contract or agreement model of the nurse-patient relation, with its emphasis on the patient's rights, might imply for some typical moral problems the nurse encounters with the competent patient. Common to many such problems insofar as they involve only the nurse and her patient is a conflict between what the patient wants and believes is best for him, and what the nurse believes is either in his best interests, in the interests of all persons affected, or morally acceptable. Consider the following cases:

Case 1. Patient A has requested of the nurse that she inform him fully of the nature of his condition and of the course of treatment prescribed for it. However, the treatment called for, and which the nurse believes will be most effective in his case, is such that given her knowledge of the patient, she believes that fully informing the patient will reduce his ability and willingness to cooperate in the treatment and so will significantly reduce the likely effectiveness of the treatment. What should she tell him?

Case 2. Patient B instructs his nurse that if his condition deteriorates beyond a specified point, he considers life no longer worth living and wishes all further treatment withdrawn. The nurse believes that life still has value even in such a deteriorated state, that it would be wrong for the patient to deliberately bring about his own death in this way, and in turn wrong for her to aid him in doing so. Should she follow his instructions?

Case 3. Patient C, after being fully informed of principal alternative treatments for his condition, has insisted on a course of treatment that the nurse has good reason to believe is effective in a substantially smaller proportion of cases than an alternative treatment procedure would be. She considers the additional risk in the rejected treatment, which seems to have affected the patient's choice, completely insignificant. Should she insist on the more effective treatment, for example, even by surreptitiously substituting it, if she is able to do so?

Each of these cases lacks sufficient detail to allow a full discussion of it, and in particular, each artificially ignores the presence and role of other health care practitioners, most notably the physician, who is generally prominent if not paramount in such decision-making. But the cases are instructive even in this oversimplified form. Case 3 is perhaps the least difficult. It would be permissible for any interested party, and a duty of the nurse following from her roles as health educator and healer, to discuss the treatment decision with her patient, and to attempt to convince him that he has made a serious mistake in his choice of treatments. But just as the patient should be free to refuse any treatment for his condition if he is competent and so decides, he is likewise entitled to select and have the treatment that the nurse (or physician, for that matter) would not choose if it were her choice; the point simply is that it is not her choice. She has no moral (or professional) right to insist on a treatment the patient does not want, even if it is clearly the "best" treatment, and it would be still

more seriously wrong to surreptitiously and deceptively substitute the treatment she prefers.

Case 2 can be somewhat more difficult because it may at least involve action in conflict with the nurse's moral views rather than a conflict over what course of action is all things considered medically advisable, as in Case 3. Case 2, of course, raises the controversial issue of euthanasia and the so-called right to die. This is a complex question that I have considered elsewhere, and here I only want to note that Case 2 involves only fully voluntary euthanasia, generally accepted to be the least morally controversial form of euthanasia.[6] I shall suppose here, as I think is the case, that a patient's right to control his treatment, and to refuse treatment he does not want, includes the right to order withdrawal of treatment even when that will have the known and intended consequence of terminating his life. If we interpret the nurse's view that life under the circumstances in question would still be worth living as merely her own view about what she would do in similar circumstances, then her view is relevant only to what she would do if she were the patient and nothing more; it entails nothing about what should be done here where it is another's life and his attitude to it that is in question. She would not waive her right not to be killed in these circumstances, but the patient would and does, and it is his life and so his right that is in question. I would suggest as well that a mere difference over what it is best to do in the circumstances (apart from moral considerations) does not justify the nurse's refusal to honor the patient's expressed wishes. However, her difference with the patient may be a moral one, in particular she may hold as a basic moral principle that she has an inviolable moral duty not deliberately to kill an innocent human being. In that case, to assist in the withdrawal of treatment in order to bring about the patient's death will be on her moral view to commit a serious wrong, to participate in a serious evil. The nurse's professional obligations to provide care should not, in my view, be understood to require her to do such, just as she should not be required to assist in abortions if she holds fetuses to be protected by a duty not deliberately to take innocent human life. While there should be no requirement in general for her to participate in medical procedures that violate important moral principles that she holds, that of course in no way implies that another nurse who does not hold such duty-based views about killing should not assist in the withdrawal of treatment. (Of course, if she holds killing to be wrong because it violates a person's right not to be killed, then she will correctly reason that the patient in Case 2 has waived that right, and so no conflict between her moral views and what the patient wants will arise.)[7]

Finally, consider Case 1. The "Patient's Bill of Rights" proposed by the American Hospital Association specifically allows that "when it is not medically advisable to give . . . information to the patient" concerning his diagnosis, treatment, and prognosis, such information need only "be made available to an appropriate person in his behalf" but need not be given to the patient himself. This would seem clearly to permit the nurse to withhold the information from the patient in Case 1. However, I believe the Patient's Bill of Rights is mistaken on this point. This is a particular instance of a general over-emphasis on and consequent over-enlargement of the area in which health professionals should be permitted to act towards patients on the basis of their own judgment of the "medical advisability" of their action toward the patient. It is perhaps natural that health professionals, trained to provide medical care for patients, and who undertake professional responsibilities to do so, should consider medical advisability a sufficient condition generally for acting contrary to a patient's wishes, and here for withholding relevant information from the patient. But once again, so long as we are dealing with patients who satisfy minimal conditions of competence to make decisions about their treatment, then unless the patient has explicitly granted the nurse the right to withhold information he seeks when she considers it medically advisable to do so, medical advisability is not sufficient justification for doing so. Our moral right to control what is done to our body, and our right in turn not to be denied relevant available information for decisions about the exercise of that right, does not end at the point where others decide, even with good reason, that it is medically advisable for us not to be free to exercise that right. In general, one element of the moral respect owed competent adults is to respect, in the sense of honor, their right to make decisions of this sort even when their doing so may not be deemed medically advisable by others, and even when those others are health professionals gener-

ally in a better position to make an informed decision. When other health professionals are in a better position to make an informed decision, a patient may have good reason to transfer his rights to decide to them, or to allow himself to be strongly influenced by what they think best, but he is not required to do so, and so they have no such rights to decide for him what is in his best interests when he has not done so.

I want to emphasize that my views on moral rights generally, and in particular the rights of the patient relevant in the three cases above, are not absolute in the sense that they are never justifiably overriden by competing moral considerations. But such justifiable overriding requires a special justification, and that human welfare generally, or the welfare of the person whose right is at issue, will be better promoted by violation of the right is *not* such a special justification.[8] Thus, rights need not be absolute in order to have an important place in moral reasoning. And it is not necessary that a nurse never be justified, for example, in withholding information from a patient for the patient's right to control his treatment to limit importantly when she may do so. Cases involving young children and non-competent adults are important instances where specifically paternalistic interference with a person's exercise of their rights can be justified.

Perhaps the point of emphasizing the contract or agreement between the patient and the health care professional is now a bit clearer. That contract model emphasizes the basis for and way in which the right to control the directions one's care will take ought to rest with the patient. It is not, of course, that the nurse is mistaken in taking her role to be a healer, or health educator, for these are important professional services she performs, but rather that her performance in these roles ought to be significantly constrained and circumscribed by the rights of her patients to control what is done to their bodies.

I should like to end by noting that focusing on the nurse-patient relationship, as I have done here, has the effect of ignoring at least one extremely important aspect of most nurses' overall moral situation. Specifically, most nurses now work in hierarchic, institutional settings in which they are in the employ of others—hospitals, physicians, etc. Many of the most important moral uncertainties and conflicts nurses experience concerning their rights, duties, and responsibilities derive from their role in this hierarchical structure, and from questions about their consequent authority to decide and act in particular matters. The patient has a place in such issues, but the issues do not arise when only the nurse and patient are considered. To fill in the picture would require consideration of what might be called the nurse-physician/hospital relation, but that is a complex matter that cannot be pursued here.

Notes

1. I have chosen throughout this paper to use feminine pronouns to refer to nurses and masculine pronouns to refer to physicians rather than to adopt gender-neutral pronouns. Since something on the order of 96% of nurses are women, while the great majority of physicians are men, gender-neutral usage in this context seems to mask a significant social reality and problem. My pronoun usage acknowledges the gender distribution among nurses and physicians, while in no way endorsing it.

2. I have drawn these from the very helpful paper by Sally Gadow, "Humanistic Issues at the Interface of Nursing and the Community," in *Nursing: Images and Ideals*, eds. S. Gadow and S. Spicker (New York, 1979).

3. Such a view, with specific reference to the dying patient, is advocated in, among other places, Leon Kass, "Death as an Event: A Commentary on Robert Morrison," *Science* 173 (August 20, 1971), pp. 698–702. To what extent this account of the physician-patient relationship is defensible, or is in fact adhered to in practice by physicians, is problematic.

4. Plans of life, and their relation to a conception of one's good, are discussed in John Rawls, *A Theory of Justice* (Cambridge, 1971), ch. 7, and Charles Fried, *An Anatomy of Values* (Cambridge, 1970).

5. For philosophical accounts of the principles of paternalism relevant to treatment of the incompetent see, for example, Gerald Dworkin, "Paternalism" in *Morality and the Law*, ed. R. Wasserstrom (Belmont, Calif., 1971) and John Hodson, "The Principle of Paternalism," *American Philosophical Quarterly* 14 (1977), pp. 61–69. I have discussed paternalism with specific reference to the mentally ill in "Involuntary Civil Commitment: The Moral Issues," in *Mental Illness: Law and Public Policy*, eds. Baruch Brody and H. Tristram Engelhardt, Jr. (Dordrecht, Holland, 1978).

6. I have discussed some implications of a rights-based view for euthanasia in my "Moral Rights and Permissible Killing," in *Ethical Issues Relating to Life and Death*, ed. John Ladd (New York, 1979). See also the paper by Michael Tooley, "The Termination of Life: Some Moral Issues," in the same volume.

7. On the general distinction between duty-based and rights-based moral views, see Ronald Dworkin, *Taking Rights Seriously*, ch. 6, (Cambridge, 1977).

8. For an attempt to specify the limits of such special justifications, see R. Dworkin, *op. cit.*, ch. 7.

For Further Reading

Ashley, J. *Hospitals, Paternalism, and the Role of the Nurse.* New York: Teachers College, 1976.

Corea, G. *The Hidden Malpractice: How American Medicine Treats Women as Patients and Health Care Professionals.* New York: Morrow, 1977.

Creighton, H. *Law Every Nurse Should Know.* Philadelphia: W. B. Saunders, 1975.

De Young, L. *The Foundations of Nursing,* 3rd ed. St. Louis, Mo.: C. V. Mosby, 1976.

Health Law Center. *Nursing and the Law,* 2nd ed. Rockville, Md.: Aspen Systems, 1975.

King, J. H. *The Law of Medical Malpractice in a Nutshell.* St. Paul, Mn.: West, 1977.

Spaulding, E. *Professional Nursing: Foundations, Perspectives, and Relationships.* Philadelphia: J. B. Lippincott, 1970.

U.S. Department of Health, Education, and Welfare. *Report of the Secretary's Commission on Medical Malpractice.* Pub. No. OS 73-88 (1973).

Veatch, R. M. *Case Studies in Medical Ethics.* Cambridge, Mass.: Harvard University Press, 1977.

7
THE CHILD PATIENT

A fifteen-year-old boy suffers from von Recklinghausen's disease, a disfiguring familial disease. The boy's mother, a Jehovah's Witness, refuses to consent to the blood transfusion necessary for the operation to remove the multiple neurofibromas at least from his face.

A sixteen-year-old unmarried girl is pregnant. She wants to carry the fetus to term, but her parents object. They insist that she have an abortion and make their feelings known to the girl's physician.

A thirteen-year-old boy, thinking he might have venereal disease, goes to a neighborhood health clinic for an examination. "Before anyone touches me," he tells the staff, "you have to swear not to tell my parents what you find."

A five-year-old boy is admitted to a hospital for the correction of an obstruction in the neck of his bladder and for reimplantation of a ureter. By his questions, the child clearly reveals that he does not understand why he is in the hospital. When he asks the pediatric staff, they fend him off with various evasive answers.

The preceding cases have a number of features in common that concern us in this chapter. First, they all involve children. Second, they raise questions of patient rights as they apply to children. Third, they suggest the variety of moral situations that HCPs face in dealing with children.

In this chapter, the terms *children* and *minors* are used as defined by state statute. In some states, individuals are considered minors until they reach the age of twenty-one; in others, the age of majority is eighteen.

There are several reasons for giving special attention to children in our study of health care ethics. First, the age of majority is relevant in the area of consent to treatment, and so it bears directly on the patient's right to information and autonomy. Second, given that, legally, parents have almost complete control over children, HCPs seem to be responsible primarily to the parents, not to the child. Again, this has implications for the child's rights to information and autonomy, as well as to privacy. Third, children are significantly different from adults, which implies that children may need special care. If so, a question arises as to what constitutes adequate health care for the child.

Although this chapter cannot begin to analyze all the moral issues involved in HCP-child relations, it can examine the more important ones. It can also try to flesh out the four basic patient rights as they apply to children. Thus, it will begin with a look at some preliminary conceptual and operational matters that bear on the child's right to adequate health care. Then it will inspect recurring moral problems that relate to information, consent and autonomy, and privacy. Finally, it will examine some of the special concerns raised by cases of dying children.

Values and Priorities

Before turning to an analysis of these issues, start thinking about your own values in this area by responding to the following questions.

1. In the case of the boy with von Recklinghausen's disease, should the hospital team attempt to override the mother?

2. Which of the following rights do you think is the greater: the parents' right to control the lives of their children or the childrens' right to their own life, health, privacy, and safety?

3. What limits are there, if any, to parents' right to control the lives of their children?

4. In the cases like those at the beginning of the chapter, to which of the following should HCPs give priority: (a) the parents' wishes, (b) the law, (c) the child's welfare, or (d) institutional policy?

5. Which of the following best expresses your views? (a) Staff should attempt to override parental decisions only if they are life-threatening to the child, (b) staff should always attempt to override a parental decision that is not in the best interests of the child, (c) staff should never attempt to override parental decisions.

6. In the case of the pregnant sixteen-year-old, should the physician defer to parental wishes?

7. Should HCPs comply with the demands of the boy with veneral disease?

8. Who is primarily responsible for informing young children about their health status and the medical procedures they may have to undergo?

Do you agree or disagree with the following?

9. The parents' role of parenting ceases with the institutionalization of their children.

10. Parents are rarely of any help to HCPs in providing information that significantly contributes to the health management of their children.

11. Most parents are too biased to present an objective view of their children.

12. Parents have a right to information about their children.

13. Parental "rooming-in" in pediatric hospitals and wards generally causes more harm than good.

14. Parents should be allowed to participate in the care of their institutionalized child.

15. Reluctant parents should be forced if necessary to inform dying children of their imminent death.

16. Children are rarely able to understand the nature of their illness and the treatment that's required.

17. It's probably best never to inform dying children of their impending death.

18. "Adult" status ought to be granted strictly on the basis of age.

19. In a conflict between the two, HCPs ought to honor the child's right to privacy over the right to adequate health care.

Preliminary Considerations: Adequate Health Care

As the cases at the start of this chapter illustrate, a number of specific issues relating to truth and information, privacy, and self-determination arise in the case of children. Before turning to these, we should look in a general way at matters that affect the care the child needs and receives. This section focuses on three issues: (1) the factors that influence children; (2) the relationship among parent, child, and HCP; and (3) the institutional environment.

Factors Influencing Children

If HCPs are to meet their responsibilities to the child, they must develop special skills for dealing with children. Children express their needs, feelings, attitudes, and beliefs differently from adults, according to the factors that influence them. Obviously, numerous factors influence children, including constitutional characteristics (characteristics present at birth), sex differences, the child's physical and mental health, and the interactions of parents and child. The study of child development properly addresses these and other factors, and understanding them is crucial for those who deal with children. Here the concern is to show how an understanding of some of the factors bears on the HCP's obligation to provide the child with adequate health care. In this connection, three factors stand out: sociocultural, family, and developmental circumstances.

SOCIOCULTURAL FACTORS. Sociocultural factors encompass the norms, attitudes, and practices of the parents as a function of their social class and cultural heritage.[1] These norms, attitudes, and practices directly affect children, in the way they perceive themselves and their world. These perceptions have a pro-

---

1. J. C. Coleman and C. L. Hammen, Contemporary Psychology and Effective Behavior (Glenview, Ill.: Scott, Foresman, 1974), p. 308.

found influence on how children react to stress, illness, hospitalization, health care, and HCPs. So it is fair to say that fully meeting one's obligation to provide children with adequate health care requires, possibly begins with, an understanding of the specific sociocultural factors that influence a given child.

One of the strongest influences on people everywhere is their cultural background. Cultural background refers to patterns of perceiving values, attitudes, belief systems, interactions, goals, and the world. A most powerful influence, cultural background affects how children react to their environment and make use of their experiences. In addition, culture is a major influence on language and the thought patterns resulting from language.

Especially noteworthy for us is the fact that culture affects values. Whether children are individual- or group-oriented, perceive themselves as exercising a large or small measure of environmental control, or consider people basically good or evil are all values that cultural background helps to shape. Most important, these value patterns affect the behavior of the ill child and influence the child's and parents' perceptions of illness, hospitalization, health care, and staff. In short, families often react to stress according to their cultural heritage. These familial reactions affect the child, and as a result, they affect interactions between HCPs and child and between HCPs and parents.

Contrast the orientations, for example, of foreign-born working-class Italian-Americans with middle-class Americans. The difference in time orientations alone is significant: Italians generally emphasize the past far more than future-oriented Americans do. Realizing this, HCPs can better understand the concerns of a child's Italian grandparents with the origins of the child's illness and the concerns of the Americanized parents about the impact of the illness on the child's schoolwork and future. If they can anticipate conflict between parents and grandparents or between foreign-born parents and staff, HCPs can be prepared to deal with the conflict, which facilitates the delivery of adequate health care to the child.[2]

Socioeconomic background is another powerful influence on the child. Enough studies have been done to confirm the fact that low socioeconomic status contributes to the incidence and prevalence of illness, as well as to difficulties in school and social interactions. Because of the large numbers of poor children in our health care institutions, as well as poor adults, HCPs must have a thorough knowledge of the interplay between socioeconomic level and the capacity of the institution and profession to deliver adequate health care.

A study of the literature in the field is useful in gaining this knowledge. One could profitably begin with the *Report of the Joint Commission on Mental Health of Children*.[3] In part, this document generalizes about the impact of poverty lifestyles on the emotional health of children. Although one must always guard

2. *M. Petrillo and S. Sanger,* Emotional Care of Hospitalized Children *(Philadelphia: J. B. Lippincott, 1972), p. 35; and J. P. Speigel, "Cultural Strain, Family Role Patterns, and Intrapsychic Conflict," in* Theory and Practice of Family Psychiatry, *ed. J. G. Howells (New York: Brunner-Mazel, 1971).*

3. Report of the Joint Commission on Mental Health of Children: Crisis in Child Mental Health *(New York: Harper & Row, 1969).*

against treating people unfairly through stereotyping, the Report's delineation of poverty lifestyle patterns, if viewed as cautious generalizations, can be of help in understanding the child and determining how to meet the child's right to adequate health care.

For example, the poverty lifestyle is often characterized by a lack of goal commitment and lack of belief in long-range success. The main objective of parent and child is to "keep out of trouble." This attitude can lead to a fatalism that influences the child's attitude toward illness and, more importantly, the capacity to deal with it. What is more, such predispositions may produce feelings of alienation that color the child's perceptions of all institutions, health care institutions included.

Poverty lifestyles generally are also characterized by low parental self-esteem and a sense of defeat. As a result, parents may easily blame themselves for the illness of the child, perhaps even perceive the illness as a personal failure. Thus, a potentially more troublesome problem for HCP-parent and HCP-child interactions is that parents may fear, even believe, that HCPs blame them for the child's illness. These misconceptions can give rise to all sorts of communication problems and can eventually undercut the parental support that the child needs.

Still another characteristic of very poor families is limited verbal communication with family members. As a result, the control that parents exercise over children is largely physical. This can present unique communication problems for the HCP who is attempting to inform poor children of their condition or treatment, allay their anxieties, and restore their confidence and hope. Even more important, children whose home lifestyles do not include open, free verbal communication are likely to keep misconceptions, fears, and anxieties "bottled up," with unfortunate consequences for effective health care.

We need not examine the Report's entire list of poverty lifestyle traits and their effects on children. Suffice it to repeat that the moral obligation to provide adequate health care requires a thorough knowledge of the sociocultural factors influencing the child.

THE FAMILY.   Numerous influences related to the family operate on the child, from the family's size and the child's position in it to how long it has been established and special family experiences. The extent of the child's involvement with the family; the family's emotional climate; the degree of restriction or punishment the family imposes on the child; the amount of social, cognitive, and emotional support the family offers the child—all these influence the child's perceptions and responses to the environment. It is important that an awareness of these influences shape the HCP's interactions with the child.

Equally important is an awareness of the various "family constellations"— that is, profiles of families that seem to occur in all cultures.[4] An awareness of these constellations not only enables HCPs to interact more effectively with children but also helps them anticipate some moral problems common to certain constellations.

For example, take the constellation often termed "The Deceptive Family." The deceptive family generally misinforms the child about an illness. When the

---

4. *Petrillo and Sanger, pp. 38–48.*

child is to go to the hospital, how long the child will probably remain there, what treatment the child can expect, and similar information is withheld or misrepresented. Should the child really press for information, the deceptive family typically portrays the upcoming experience as something akin to a trip to Disneyland. Furthermore, such a family often makes impossible promises to the child about the outcome of the treatment. One child might be told that hospitalization means a change in size, another the winning of more friends, still another the capacity to see or hear or walk better. Imagine the disappointment the child faces after treatment that does not deliver on the promises. When it is sufficiently profound, such disappointment can undercut the child's convalescence. The deceptive family intends no harm, but the potential injury to the child is great: Feelings of anxiety, confusion, and betrayal produce general distrust of adults.

HCPs who must deal with deceptive families face a real moral challenge. On the one hand, HCPs who purposely avoid intervening become accomplices to this harmful behavior. On the other hand, HCPs who tell the truth do so at the risk of turning the child against the family. Morally speaking, the behavioral pattern of the deceptive family inevitably raises serious questions for HCPs, questions of truth and information and of injury. We will consider these problems in more detail later.

Similar observations could be made of other family constellations, which HCPs are urged to study. Knowledge of these family types, when combined with information gleaned from a family diagnostic interview about the size of the family and its development, yields a revealing view of the child's world. It is a view that HCPs must take pains to develop and keep in mind if they are to meet the child's right to adequate health care.

LEVELS OF DEVELOPMENT. Just as important as sociocultural and family factors is the child's level of development. Like the other factors, level of development affects the child's understanding of the hospital experience, the child's ability to express feelings about it, and the way the child copes with the stress. Language development influences whether and how the child asks for help and expresses fears and anxieties. Similarly, the child's level of emotional and social development dictates the kind and amount of interaction and support the child needs during different phases of illness and hospitalization.

The literature on child development is vast. It is not appropriate in this context to go into further detail. Rather, view this discussion as a first step in what should be extensive reading in the field—something the obligation to provide adequate health care seems to entail. Such a study will yield general knowledge that bears directly on the treatment HCPs provide children: It will tell HCPs what to look for and will help them order the information they find. In short, knowledge of the facts and theories in child development serves a diagnostic and integrative function, both of which are vital to the compassionate and effective practice of health care.

It is important to conclude this section on the basic concepts of health care for children with a caution: In applying concepts, there is always the temptation to apply them blindly, thoughtlessly. But in doing so, we risk stereotyping individual children in a way that precludes meeting their special health needs. There is danger, and immorality, in ignorance but equally so in a reckless application of general knowledge. In the last analysis, then, the paramount professional and moral challenge in this area is the need to synthesize a sensitive and caring awareness of the individual child with an understanding of the many factors that influence every child.

In addition to the conceptual matters we have just considered, there are numerous operational matters that bear on the quality of care for children, including the availability of resources, the preparation and quality of the staff, and staffing and other institutional policies. Because of the limited space available, this chapter will focus on just two: the relationship among parent, child, and HCP in the health care setting and the institutional environment. Both raise immediate moral concerns in the context of the child's right to adequate health care.

Relationships among Parent, Child, and Professional

Obviously, some kind of relationship will always exist among HCPs, parents, and children. And there is no question that the relationship will have an impact on children, including how they view their illness, how they react to treatment, how they convalesce, and so on. But there is a question about what constitutes a proper relationship. It is impossible to say precisely what the proper relationship should be, but there seem to be several assumptions that function as a basis for defining the relationship.

First, it seems reasonable to acknowledge that the parents' role of parenting does not cease with the institutionalization of their child. By ignoring or flouting this assumption, HCPs seriously undermine the parents' right to autonomy in the matter of their child's health care and invite intolerable stress into their own relationships with parents. What is more, since parents have natural apprehensions about their role being usurped, it is especially important for HCPs to acknowledge explicitly the continuity of the parenting role—indeed, the new and vital dimensions they expect this role to take on.

A second assumption that seems reasonable and basic to a proper relationship is that parents will harbor a variety of feelings about their child's illness, not the least of which will be guilt. Feelings of failure and inadequacy, of poor parenting, are altogether common among parents of ill and hospitalized children. If these feelings are left unaired and are allowed to fester, they can seriously impair any subsequent relationship between HCP and parent. Since parental involvement is a necessary part of the care of the child, HCPs must deal openly and forthrightly from the start with this problem of parental guilt.

A third assumption is that parents are potentially a valuable source of information about their child. This point may seem obvious, but it is often overlooked or minimized in the rush to treat the sick child rather than the child who is sick. Although parents are not always the most reliable source of information about their child, more often than not they are, and they are in ways that contribute subtly and crucially to treating the child. But whether or not individual parents are in fact reliable information sources, it is essential that HCPs make them feel that their unique knowledge of the child automatically warrants for them an important and useful role in the child's health care.

A fourth assumption is that parents have a right to information about the child. Since the right to truth and information will be discussed shortly, we need not consider it here.

These four assumptions are the cornerstones of a proper relationship among HCP, parent, and child. If they are to be viewed as anything but mere pieties, they must be connected to daily procedures in such a way that HCPs and institutions acknowledge a moral and professional obligation to implement them.

For example, if HCPs accept the assumption that parenting continues during the child's hospitalization, then it seems they must devise ways to provide for family continuity. If they do not, they strike the hypocritical pose of acknowledging the principle but refusing to act on it—or worse, of acting inconsistent with it.

The issue of family continuity is of great moral concern, for the health and welfare of the child are at stake. Considerable research testifies to the harmful emotional effects that separation from parents and family can have on hospitalized children. Although institutions and HCPs have traditionally assumed that the child's eventual "settling in" is a healthful adjustment to separation, research in the past quarter-century indicates that this is not the case. Yet judging from their policies on sibling visits, rooming-in, and parental participation in the care of the child, one must conclude that a number of hospitals continue to fly in the face of the overwhelming evidence that militates against separation of parents and child.

True, some institutions do not have the room and resources to accommodate parental rooming-in. It is also true that parents who room in have the opportunity to ask far more questions than parents who do not and that answering these questions consumes time. And it must be admitted that parents rooming in have occasionally been known to make nuisances of themselves, actually impeding the efficient delivery of proper health care to their children. But such legitimate concerns must be evaluated in light of the verifiable positive effects on children of rooming-in and of unrestricted visiting policies.

What are the positive effects? For one thing, the child does not suffer separation anxiety, which makes the job of the HCP and the child's adaptation to the hospital that much easier. For another, the child gains a greater sense of security. This is very important in the child's response to illness and treatment. Still another effect is that parents are better able to absorb HCP instruction about their child's illness, and HCPs are better able to observe and evaluate family interactions. Given these advantages to rooming-in, liberal visiting hours, and sibling visits, it is hard to see how highly restrictive visiting policies can morally be justified. If the preceding analysis is valid, then HCPs in institutions with such restrictive policies would appear to be under a moral and professional obligation to work to change those policies.

There are additional ways that HCPs can help ensure family continuity during a child's hospitalization. One way is to provide family members with the opportunity to express their feelings, including their fears and anxieties about the child's illness. Another way is to help family members become aware of one another's needs, attitudes, and feelings in the context of the illness of the child. A third way is to ensure that the family has the necessary financial and emotional resources to deal with the stresses of the situation. Important here is the role of the HCP as a source of information about community agencies and social services that may be available for assistance. Notice that these suggestions not only allow HCPs to implement the assumption about the continuity of the parenting role but also the assumption that family members often harbor troubled feelings about the illness of the child.

As another example of how institutions and HCPs can implement these assumptions, consider the assumption that parents are a valuable source of information about the child. If more than lip service were to be paid this assumption, HCPs might collaborate with parents in setting goals for the care of the child. Specifically, HCPs and parents could jointly devise a plan of health care that would include opportunities for the parents to assist in the physical care of the child—bathing, feeding, preparing for bed, accompanying to diagnostic procedures, and participating in recreation. Again, this is not just a matter of frills, courtesies, or good public relations. Given the data available, such practices seem necessary for the total support of the child who is ill. *Not* offering these opportunities raises serious moral questions about the adequacy of the health care provided the child.

The Institutional Environment

Whether or not the health care that a child receives in a hospital can be considered adequate depends in part on the hospital's emotional environment. What is a proper pediatric emotional environment? This question, like several others raised in this section, cannot be answered with great specificity. But again, certain assumptions seem warranted.

First, because hospitalization can be a wrenching emotional experience for children, institutions and HCPs must meet the child's need for a feeling of security. Second, because of the unique qualities of children, the pediatric staff should be selected for their knowledge, sensitivity, patience, and warmth. Third, because communication directly affects every facet of health care, an atmosphere of open communication should exist; children should have ample opportunity to express their fears and anxieties and to become involved in their own treatment and care. Fourth, because parental participation is an essential part of child care, institutional policies should accommodate parental visits. Fifth, because hospitalization can seriously affect the child's progress in school and interactions with others, HCPs must help maintain the child's intellectual and social growth.[5]

This list of working assumptions could probably be expanded, but these are enough for a start. The key thing is that taking these propositions seriously

---

5. K. *Creason Sorensen and J. Luckman,* Basic Nursing: A Psychophysiologic Approach *(Philadelphia: W. B. Saunders, 1979), chapter 46.*

incurs an obligation to devise ways to implement them. This, in turn, entails scrutiny of operational procedures, with an eye on whether they advance or impede the institution's emotional atmosphere.

For example, take the claim that the pediatric atmosphere should meet the child's need for security. Certainly an integral part of the emotional security of children is their protection from overhearing discussions that may arouse fear and anxiety. Yet this protection is often violated during rounds, particularly in teaching hospitals. Too often HCPs intent on teaching and learning during rounds are unmindful of the negative effects of their comments.

Having a group of strangers surround one's bed can be frightening enough in itself to the child who is hospitalized for the first time. Furthermore, what occurs at one bed can affect children in other beds nearby. Thus, tarrying at the bedside of one child can cause other children to wonder why that child is getting so much attention, and children who are "passed by" can feel forgotten and neglected. Rounds also often include expressions of dissension among staff, which can alarm the child. Finally, and above all, much of the distorted information that children and parents obtain results from careless remarks during rounds. Here are a few examples:

> Eric, age 4, overheard his doctors discussing arrangements for a bone marrow aspiration. After overhearing this he asked his mother if he was to have a bow and arrow.
>
> At 15, Carol, a sophisticated and knowledgeable adolescent who was admitted for the treatment of subacute bacterial endocarditis, quickly lost her facade when she heard the pediatricians discussing the mortality rates of her disease.
>
> Jane's examining physician casually mentioned that she could not feel the child's ovaries. Not until much later did the staff learn from her father that this 13-year-old had interpreted the statement to mean that she had no ovaries.
>
> Nancy, age 10, was quite depressed as a result of several setbacks following cardiac surgery. Her feelings of hopelessness were reinforced when, during rounds, her pediatrician said in an effort to be reassuring, "Well, there is nothing more we can do for you." Her anxious expression was picked up by the head nurse who explained that the doctor meant to say all the diagnostic tests he had ordered were already carried out.
>
> Fourteen-year-old Carrie's anxious parents were present in the Intensive Care Unit when a surgeon checking the child's condition said to the nurse as he was making adjustments, "There is too much negative pressure in the chest tube." The parents became convinced that an error was made and that their child was incompetently managed.[6]

The point is that the conduct of medical and nursing rounds raises important moral concerns about the proper emotional environment. Conducted carelessly, rounds can undercut the security needs of the child, thereby causing harm. Thus, the obligation to meet the child's need for security entails an evaluation of such practices as rounds. It also entails a sensitive understanding of how these practices affect children and an effort to find less harmful ways of carrying out important institutional functions.

---

6. *Petrillo and Sanger, pp. 86–87. Reprinted by permission.*

As another example of the connection between a broad obligation and specific behavior, consider the obligation to help maintain the child's intellectual and social growth. One concrete and most important way to do this is through play. Children benefit from play in a variety of ways—from learning about themselves to releasing stress to developing a measure of control over a strange environment. In addition, play gives HCPs a beautiful opportunity to observe the child's level of development. Put in these terms, then, to play or not to play is no longer the question. Because play figures prominently in maintaining the child's intellectual and social growth and because we assume an obligation to help maintain these facets of the child's personality, it would seem that HCPs have a moral obligation to make some provisions for play. The question therefore becomes: How shall we provide opportunities for play?

The same sort of analysis can be made of the other assumptions about a proper emotional pediatric atmosphere. It would seem that HCPs need to ask a number of questions relevant to these assumptions. The fundamental question, of course, is whether or not the assumptions are to be endorsed. If they are, then we must ask whether these assumptions entail moral obligations. If they do, what is the nature of these obligations? Specifically, what must the HCP do to implement the assumptions? Answering this last question may require an inspection of institutional policy, a reassessment of procedures, and the development of imaginative and less harmful alternatives.

With these preliminary considerations in mind, we can now turn to specific issues that recur in pediatric health care. They can conveniently be clustered around questions of information, consent and autonomy, and privacy.

Truth and Information

The patient's right to truth and information is no less morally important for the ill child than for the ill adult. What should the child be told? When? How? By whom? Obviously such questions provoke introspection about sound pediatric practice. But they also raise moral concerns, because the rights of patients to truth and information about their conditions are at stake, even though the patients happen to be children.

It is tempting to minimize, even dismiss, a child's rights in this area. Sometimes it is claimed that children are rarely able to understand the nature of their illness and the treatment that is required anyway. As a result, providing information will only confuse, frighten, and intimidate the child. Another approach is to bypass the child entirely: HCPs communicate substantively only with parents or guardians. Presumably the parents are to inform the child in a way that satisfies curiosity and allays anxiety. Either way, the child is treated as not having a right to truth and information in the same way as an adult.

The child's inability to comprehend fully the nature of an illness and the need for hospitalization probably says more about the adult's communication skills than about the child's immature mind. More often than not, the child cannot understand simply because the adult does not know how to communicate on the child's developmental level. Consider the actual experience of five-year-old Paul, whose case was reported at the opening of this chapter. Recall that Paul was hospitalized for the correction of an obstruction in the neck of his bladder and reimplantation of a ureter. Paul's postoperative reactions could only be described as dreadful. He pulled out intravenous infusion tubes, tugged at his urinary drainage tubes, regurgitated medications, and resorted to frequent tantrums. Although he was told repeatedly about the purpose of the drainage tubes, it did not sink in; he just did not seem to understand. Finally, an explanation was delivered visually. What follows is a health team member's account of it:

> A body outline was sketched including the genitourinary system. This was followed by a simple interpretation of the anatomy and physiology, the congenital defects, surgical correction and the placement and purpose of the tubes; and most significantly in his case, we dealt with the mutilation and castration fantasies common to Paul's period of development. . . . This was done by playful repetition to clarify his understanding in this way: "This is where your operation is, right here in your belly. No other part of your body was operated upon." (Then, pointing to body parts) ". . . not your head, not your ears, not your eyes, nor your nose and mouth, not your arms and chest, not your 'peepee,' nor your legs, feet or toes."[7]

In conjunction with this explanation, Paul was given a stuffed doll on which the staff helped him place bandages and tubes in the appropriate areas for his type of surgery. He was also helped to pretend that he was administering oral and intramuscular medication to the doll and performing other procedures that were part of his treatment. In the words of the attending staff, "Paul's reaction was striking; he became cooperative with treatments, accepted the staff, and was interested in play."[8]

The point is not that all such difficulties are so happily or easily resolved. It is, rather, that children's supposed inability to comprehend is often more adults' failure to communicate effectively. Obviously, communication with ill children calls for considerable study about child development. But will anything less do if children's right to truth and information is to be respected? Without such background and the ability to apply it, HCPs risk injuring children in a most fundamental way—by rationalizing away their right to know and thus intensifying their anxiety and minimizing the control they have over their environment.

As for bypassing the child and dealing directly with parents, there is no question that the relationship between HCP and parents is vital to the well-being of the child. Given the importance of the relationship to meeting the child's needs, it follows that HCPs have a moral and professional obligation to do everything they can to ensure that the relationship is constructive. In part, this involves acknowledging and acting on the four assumptions about a proper

---

7. *Petrillo and Sanger, p. 11.*
8. *Petrillo and Sanger, p. 11. Reprinted by permission.*

HCP-parent relationship. In addition, a constructive HCP-parent relationship surely must be based on acknowledgment of the parental right to information about the child's hospitalization, illness, and progress.

At the same time, providing information to parents is not the same as informing the child. Ideally, the parents can, should, and will communicate with the child. But many factors impede such a fluid progression of events. First, generally speaking, parents have as many communication problems as HCPs. It is true that they may know the child better, but their abilities to communicate are nonetheless impaired by their own fears, embarrassments, attitudes, and guilt. In addition, communicating effectively in health care matters requires technical understanding and, as we saw in the case of Paul, an ability to pitch that understanding to the appropriate developmental level. In this sense, parents are not always the best communicators. Nor are parents always present when the child's need to know is most urgent.

In short, HCPs must take pains to establish a relationship with parents in order to maximize the welfare of the child. At the same time, HCPs cannot assume that, having informed the parents, they have thereby met their responsibilities to the ill child in the area of truth and information.

Consent and Autonomy

Because parents have almost complete legal control over children, their consent for health care treatment is usually solicited and accepted in place of the minor's. In effect, then, the autonomy of the child is overlooked. Even discussions of children's rights are really examinations of parental rights. And yet, the rights of the child and of the parent are not the same and are not always compatible. There are at least three kinds of cases that point up the tension between them: (1) cases of parents refusing to consent to treatment for their child, (2) cases of children consenting to medical treatment for themselves, and (3) cases of children refusing to undergo treatment to which parents have consented.

Parents Refusing Treatment for Children

Probably everyone has read or heard about at least one case of parents refusing medical treatment for their child, usually on grounds of religious conviction. For example, this chapter began with the case of a parent refusing to give consent for the blood transfusion needed for her child's operation. Such cases raise this central moral question: Is it ever morally permissible for a hospital and HCPs to treat a child despite the parent's wishes? Is HCP paternalism in such cases justifiable?

At stake are conflicts of rights and responsibilities held by various parties. On the one hand, parents have basic religious rights and legal and moral rights to control the lives of their children. Children, on the other hand, have rights to their lives and to state protection from threats to their health and safety. Precisely because they are minors, children's claim to state paternalism is regarded as more extensive than adults'. For its part, the state would seem to have the obligation to meet claims to conscience, religious freedom, and parental control over children and at the same time to meet the claims to health and safety of the minors. Obviously, in cases like the one mentioned, the state cannot fully meet the claims of both children and parents.

Health care institutions and HCPs find themselves right in the middle of these conflicts. Clearly, their primary concern is and should be the health and welfare of their patients, in this case minors. Still, for both moral and legal reasons, HCPs must remain sensitive to parental rights.

The distinction between cases that threaten life and those that do not threaten life is important in the moral calculation, as well as in the legal. When a child's life is endangered and the parent will not permit treatment, the parent's right can be honored only at great cost to the child. Thus hospitals and HCPs have traditionally come down squarely on the side of the child, usually by seeking legal authorization for the needed treatment. Generally speaking, they have been supported by the courts, which inevitably appoint someone (often a hospital staff member) as guardian of the child for the purpose of giving informed consent.

Cases that do not threaten life are more difficult to manage. When the child's life is not threatened, one cannot attempt to justify abridging parental rights by appeal to the child's greater claim to life. At the same time, however, it could well be that a child's physical, psychological, and social development, indeed her or his future happiness, hangs in the balance. Such was the case with the 15-year-old boy suffering from von Recklinghausen's disease, which caused a severe facial deformity. When this case actually occurred, the hospital thought it was morally obliged to act as an advocate for the child. So hospital officials asked the court to overrule the mother's refusal to consent to the transfusion. The court permitted the operation to proceed without the mother's consent.[9]

In most cases, however, the court will not interfere with parental choice in a nonmortal situation. Take, for example, the actual case of a child with a harelip and cleft palate whose father refused to consent to surgery to correct the condition. Again, a hospital requested the court to overrule the parent's refusal to consent. The court denied the request, and the child did not get the operation.[10]

Such cases always raise questions about personal and institutional responsibilities to advocate the rights of children. In fact, despite the weight of legal precedent upholding parental choice when life is not at stake, numerous institutions and HCPs are recognizing moral responsibilities to advance what they perceive as their child-patient's right to adequate health care. In the case of children, the question of whether to advocate must be considered in light of three important facts. First, ill children are especially vulnerable to physical and mental abuse because of their age and health status. Second, ill children are in no position to argue their own case; they are virtually powerless in our society to

---

9. *In re Sampson,* 317 N.Y. S (2d 641, 1970) aff'd. 29 N.Y. 2d 900, 328 N.Y. S 2d 278 N.E. 2d 918 (1972).

10. *Matter of Seiffreth,* 309 N.W. 80, 127 N.E. (2d 820, 1955).

advocate effectively for themselves. Third, and most important, HCPs have unique knowledge about the travail of ill children, as well as the long-range impact of illness; and of course, HCPs know what can be done to reduce the child's suffering. Therefore, it is safe to say that, whenever a parent refuses medical treatment for a child, there are always serious moral questions about the institution's and HCP's role as an advocate for the health care interests of the child.

Children Consenting to Treatment for Themselves

Is it ever morally permissible to allow children to consent to medical treatment for themselves? One way to approach this question is to raise some categories of cases that involve minors. There are four useful ways to categorize them: specific conditions, emergency situations, emancipated status, and unemancipated status.[11]

SPECIFIC CONDITIONS.   Minors can be afflicted with conditions of an extremely delicate sort. Venereal disease is one such condition; drug dependency is another. Requiring parental consent as a precondition of treatment might not only increase the psychological suffering of the minors but in some cases even discourage them from seeking treatment. There seem to be no substantive reasons to deny children with these conditions the right to consent to treatment for themselves. Indeed, some states explictly recognize the child's legal rights in these circumstances, even extending the right to include conditions of pregnancy, blood transfusions, alcoholism, contagious disease, and contraception. Whether the minor's right to consent to treatment should extend to all these conditions is a question for reasonable debate. Moreover, it may very well be that an HCP's moral obligation in a particular case would be to inform the parents of the child's condition. But to insist on parental consent as a precondition for treating minors can easily raise serious moral questions of harm and injury, invasion of privacy, and violations of autonomy.

EMERGENCY SITUATIONS.   It is generally acknowledged that HCPs legally can and morally should treat the adult patient who is brought to an emergency room unconscious or delirious and who needs immediate care. In such cases, the consent is considered implied, although in a strictly legal sense it is more accurate to describe the HCP as "privileged" due to the situation. Is there any morally significant reason to make an exception of minors? There does not seem to be. On the contrary, making an exception could obviously imperil their health and lives. This explains why minors are generally recognized to have the legal right to emergency treatment without consent. Some states have even codified the doctrine. Massachusetts, for example, permits a physician to treat a minor without parental consent if a delay in treatment will "endanger the life, limb, or mental well-being" of the patient. Surely no moralist could object to this principle.

EMANCIPATED STATUS.   The status of the child also bears on the morality of allowing minors to consent to treatment for themselves. Some minors are eman-

---

11. G. J. Annas, *The Rights of Hospital Patients* (*New York: Avon Books, 1975*)*, chapter 12.*

cipated from the care, custody, control, and services of their parents. In some cases, these minors are emancipated through marriage or by joining the armed forces. In other cases, for a variety of reasons, they simply find themselves on their own. Whatever the reason, to deny minors who are emancipated in practical fact the right to consent to treatment for themselves raises serious moral questions concerning violation of autonomy, invasion of privacy, and needless exposure to physical and mental danger.

UNEMANCIPATED MINORS. It is rather easy and uncontroversial to argue for the moral rights of minors to consent to treatment for themselves for certain conditions, in an emergency situation, or when emancipated. But what about unemancipated minors who seek ordinary treatment? It is tempting to assume that parental consent should be a precondition for medical treatment. Justifying that conclusion, however, presents real problems.

Simply because they are "minors"—because they have not attained the legal age to be considered adults—does not seem to warrant the conclusion. After all, how valid is such an arbitrary criterion as age in determining rights to autonomy, privacy, and consent? Presumably, the thinking is that, as a group, minors do not ordinarily have the maturity and sophistication to make prudent judgments about their medical treatment. In other words, it is not so much a matter of age as that, generally speaking, minors lack the requisite qualities to make informed judgments about their medical treatment. And so, taking a paternalistic tack, the state assigns the care and exercise of the child's rights to the parents.

Undoubtedly in many instances minors do not have the maturity to make prudent judgments about medical treatment for themselves. But the same can be said of many adults. What is more, some minors do have the maturity and intelligence needed to comprehend a medical treatment or procedure, including its consequences. The fact is that age in and of itself does not make a person mature, intelligent, or prudent.

What are we to say of such unemancipated minors? That they nonetheless should not be allowed to consent to treatment for themselves? This position would be very difficult to defend on moral grounds, because inevitably the appeal returns to the arbitrary, even irrelevant, criterion of age or to the stereotypes associated with youth.

It has been this sort of analysis that has led some states to adopt what is termed "the mature minor rule."[12] For example, Mississippi statutes state that "any unemancipated minor of sufficient intelligence to understand and appreciate the consequences of the proposed surgical or medical treatment or procedure" may consent to the treatment. The state of New York also recognizes this doctrine. But even where no laws expressly acknowledge rights for unemancipated minors, HCPs must still deal with such cases on a moral level. Thus, they must again evaluate their obligations to serve as advocates for unemancipated minors.

A final word: There have been no reported cases of HCPs or institutions being held liable for the performance of acceptable therapeutic procedures on

---

12. *Annas, p. 138.*

minors over fifteen. This last fact is particularly important to consequential analyses of cases involving unemancipated minors, for it addresses the pervasive fear of HCPs and health care institutions about incurring liability.

Children Refusing to Undergo Treatment

The central moral question when children refuse treatment is whether it is ever morally permissible to allow a child to refuse to undergo treatment to which the parent has consented. Recall the case raised at the outset involving a sixteen-year-old unmarried girl who gets pregnant. She wants to have the baby, but her mother insists on an abortion. Whose wishes ought to be respected, mother's or daughter's?

One way to approach this dilemma is to invoke the rather uncontroversial assumption that follows from the preceding discussion: Under carefully circumscribed conditions, children ought to be allowed to consent to treatment for themselves. For the sake of argument, let's assume that one of those carefully circumscribed conditions is the condition of pregnancy. Thus we are assuming that this sixteen-year-old should be allowed to consent for herself to medical treatment for her pregnancy. Specifically, we should recognize her right to choose to have the baby or have an abortion. But choosing one entails a rejection of the other: If she chooses to have the child, she refuses to have the abortion; if she chooses to have the abortion, she refuses to have the child. Evidently, then, the right to refuse is implied in the right to consent. Stated another way, if the right to refuse is denied, then so is the right to consent. Should the daughter be denied the right to refuse the abortion, then she is effectively denied the right to consent to treatment for herself in this matter.

The right to refuse, therefore, can be viewed as the "flip side" of the right to consent.[13] Recognizing a right to consent necessarily implies recognition of the right to refuse. If minors ought to be allowed the right to consent to treatment, then it follows that under certain conditions they ought to be allowed the right to refuse treatment as well. To recognize the right to consent while denying the right to refuse is logically inconsistent. To deny both, as we previously saw, raises serious moral questions about rights to autonomy, privacy, and adequate health care.

There is very little in the law that addresses cases of minors' refusing treatment, but there have been some cases suggesting that the preceding analysis, although intended to aid moral debate, has some legal merit. One case that occurred in Maryland involved precisely the circumstances outlined. According to Maryland law, minors have the same capacity to consent to medical treatment for pregnancies as adults. In finding that abortion was included under medical treatment for pregnancy, the court ruled that

> the minor, having the same capacity to consent as an adult, is emancipated
> from control of the parents with respect to medical treatment within the
> contemplation of the statute. We think it follows that if a minor may consent to
> medical treatment as an adult upon seeking treatment or advice concerning

---

13. *Annas*, p. 139.

pregnancy, the minor, and particularly a minor over 16 years of age, may not be forced, more than an adult, to accept treatment or advice concerning pregnancy. . . . The one consenting has the right to forbid.[14]

The right of minors to refuse treatment may be extended into the area of human experimentation. Under guidelines for the protection of human subjects proposed by the U.S. Department of Health and Human Services, researchers would have to obtain the consent of all subjects seven years of age or older, as well as their parents' consent, before they could participate in experiments. In effect, minors would have veto power over a parent's consent to their being used in an experiment.

Privacy

Privacy, as we have seen, is considered as much a basic patient right as are the rights to adequate treatment, truth and information, and autonomy. And, like the others, the right to privacy bears on the health care of children.

We have seen one example of this already. In the case of specific disease conditions, such as venereal disease or drug dependency, the minor's special need for and right to confidentiality make the argument for the morality of allowing minors to consent to treatment for themselves a most powerful one.

Frequently in such cases HCPs are faced with an additional moral dilemma: whether to divulge the child's condition or treatment to the parents against the child's will. Situations involving contraception, pregnancy, abortion, venereal disease, and drug use are among the most common that raise this issue of privacy and confidentiality.

According to the AHA's Patient's Bill of Rights, HCPs have a general moral obligation not to reveal information that they receive from patients or about the type of treatment provided. The key moral issue, therefore, is whether it is ever morally permissible to tell a child's parents about the child's condition or treatment against the child's will.

In answering this question, it is important to distinguish cases in which the parent is the consenting party and those in which the minor is (as in instances of specific conditions, emancipated minors, or mature minors). If the parents are the consenting party, then, like any patient, they need and have a right to all the information necessary for them to give informed consent. In cases like these, it is possible to argue that, since the parents stand in place of the child, informing parents presents no real breach of the patient's right to confidentiality. If such cases are viewed more strictly as involving a breach of confidentiality, these breaches could be justified as being necessary in order to elicit from the parents the informed consent that is necessary for initiating urgent medical treatment for the child.

On the other hand, if the child is the consenting party, then providing parents with information about the child's condition cannot be defended by appeal to a parent's inherent right to information, in the role of the child's surrogate, or by appeal to the necessity of obtaining informed consent in order to

---

14. *Matter of Smith, 16 Md. App. 209, 295 A. 2d 238–246 (1972), cited in Annas, p. 140.*

initiate urgent medical treatment. It does not necessarily follow, however, that HCPs should never disclose the child's condition or treatment to parents against the child's will—only that the foregoing defenses are inadequate to justify such a breach of confidentiality.

Some would argue that HCPs are never morally justified in breaching the confidence of consenting minors. Others, while respecting the consenting minor's right to privacy, would claim that their right to adequate health care can override their right to privacy. In some cases, adequate health care might necessitate involvement of parents in the treatment of the child, as with psychological and drug problems. Effective treatment for the child may simply require parental participation. But even those who hold this view would acknowledge that an overwhelming case must be made for going against the minor's will, because such an action would violate not only confidentiality but also autonomy. It might also result in unfortunate consequences: The child's trust or participation in the treatment might be impaired or even lost.

One of the most troubling situations that HCPs face in this area involves consenting minors who attempt to extract from the HCP an explicit promise in advance not to tell their parents anything about the examination or treatment. The consenting minor certainly has a legal right to make such a demand. But are HCPs morally obliged to meet it? At least two moral considerations make it impossible to answer such a question simply.

The first consideration involves the likely condition of the minor. Such requests are usually made in the presence of what is likely to be a serious medical problem, from both the personal and social viewpoints. Conditions relating to drugs and venereal disease come to mind. Rejecting the demand might delay urgent medical attention and, as a result, might jeopardize the health of the minor and others. On the other hand, HCPs could understandably be concerned about the propriety of making an explicit promise that, at some point, they may feel obliged to break. Therefore, such cases require a sensitive, compassionate, and thoughtful reading of all the nuances. But one thing is certain: They always raise serious moral questions about the minor's right to privacy.

The Dying Child

So far, the discussion of children as patients has focused on situations that involve consideration of basic patient rights as applied to minors. But no discussion about the moral aspects of pediatric health care would be complete if it did not consider cases involving minors who are dying. Chapter 9 takes up many of the moral aspects surrounding dying patients. But because the deaths of children raise a cluster of unique moral considerations, at least a brief discussion is warranted here.

It is hard to imagine a situation that more painfully taxes the resources of an HCP than the death or impending death of a child. Fully meeting one's obligations to child and family while continuing to meet routine duties requires an active awareness of the dynamic interplay of human needs and emotions in the face of death, a profound sensitivity to and compassion for the bereft, and a heightened sense of professional responsibility. This is no mean task.

The HCP's moral responsibility in this area probably begins with an honest and thorough examination of one's own attitude toward death in general and toward the death of the individual child in particular. As for the death of a particular child, it is not uncommon for an HCP to feel as aggrieved as the child's family. This reaction is understandable, given the attachments that can be formed during a long period of care, but it can easily undermine the HCP's overriding obligation: to provide emotional support for the child and family. At the same time, HCPs must be cautious about defending themselves from an onslaught of emotion by assuming a professional air of detachment that effectively undercuts their support role. Denied the emotional support from staff that they need, dying children can suffer the pain and loneliness that result from having no opportunity to discuss their impending death. In short, how HCPs feel about the death of a particular child and how they subsequently deal with those feelings carry serious moral implications for the care they give child and family.

Beyond this, a foremost moral issue in these situations concerns the child's right to truth and information. Inevitably, HCPs must decide when and how to inform the child. Notice, *when* and *how*—not *whether* to tell the child. Research indicates that dying children inevitably know the truth before they are told.[15]

Since the how and when decision that HCPs make can do the child great good or great harm, there is no doubt that these decisions raise moral concerns. No two situations are ever identical and each calls for a distinct analysis, but there is one most important determinant in informing the child, which HCPs must not overlook: how the parents feel about the death. In the final analysis, how children respond to their impending death will be determined largely by how their parents allow them to respond. And of course, the child's response will affect the suffering and pain he or she experiences. One could argue that the child's right to adequate health care obligates the HCP to investigate thoroughly the parents' feelings and to help them deal with those feelings in a way that is beneficial to parent and child.

Turning to specifics, parents—like anyone else facing death—generally pass through a variety of phases: shock, denial, disbelief, anger, guilt, acceptance. Each of these stages creates unique challenges for the HCP, but the denial stage raises perhaps the most serious moral questions that HCPs have to deal with.

Withdrawal is common to denial. Parents who refuse to accept the impending death of their child might simply stay away from the child or, more likely, might attempt to deceive the child by affecting a Pollyanna attitude. The potential harm to the dying child is enormous. First, the parental support the child needs is lost. Second, the child suffers conflict: On the one hand, the child knows something is wrong; on the other, the child is being told everything is okay. As a result, the child can easily end up protecting the parents. This turn of events seems an obscene violation of the child's right to his or her own grief.

---

15.   E. Waechter, "Death Anxiety in Children with Fatal Illness," Nursing Research Conference, *March 5, 1969, p. 83.*

One could argue, therefore, that the HCP's obligation is, in part, to help the parents be honest and truthful with the child. How to do this? As a start, HCPs might share with the parents an understanding of how the child's developmental level affects his or her understanding of death. For example, children under three associate death with separation from adults and a loss of love; children three through six are concerned with death as punishment and often view death as reversible. Children seven through ten seem to fear death itself, associating it with unexpected pain, mutilation, and mystery. Children who are older by and large understand the permanence and universality of death. They resent its impeding the realization of goals, opportunities for success, and chances for independence. Sharing information about developmental concepts of death is only a beginning, but it is vitally important because it often initiates a dialogue with parents that effectively dilutes the child's sense of loneliness, isolation, and confusion.

Although informing the dying child is of paramount concern, it is not the only moral consideration in these cases. Parental needs, including rights to privacy, sharing of grief, and empathy from staff, are very important for HCPs to recognize and try to meet. Of course, the extent of staff involvement is always a difficult problem. Ideally, HCPs provide the support that the family needs but maintain their effectiveness as professionals. They must always guard against overinvolvement, which can violate the family's right to their own grief and can end up calling attention to the HCP's needs rather than the family's. These problems can usually be avoided if HCPs take seriously their obligation to devise a plan of support for child and family in advance. This should include a determination of the family's usual coping mechanisms, which in turn is important in determining how much support they will need; an assignment of staff who will tell the parents; and a strategy for informing the child and siblings. It should be noted that siblings can suffer needless harm when denied information or an opportunity to express their feelings and work through them.[16] Since HCPs are in a position to educate parents about this problem and help them devise ways of meeting it, they can therefore also minimize the degree of suffering that siblings experience.

Cases of dying children, then, raise dramatic and unique moral questions about the HCP's obligations to child, parents, and family members. Because of the crucial role that parents play, it is a safe bet that HCPs who are serious about meeting their obligations in this area will ask themselves: How am I enabling the parents to provide the warmth, love, and support that the child needs, and how am I equipping them to provide the information that the child seeks?

Summary

This chapter focused on cases involving one special group of patients: children. The child's right to adequate health care was viewed in the context of

---

16.   *Sorensen and Luckman, pp. 1220–1223.*

certain factors that affect childhood, including sociocultural, family, and developmental factors; relationships among HCPs, children, and parents; and the institutional environment. Of utmost moral concern in pediatric health care are issues involving information, autonomy and consent, and privacy. Although there are many subcategories of children that raise special moral interest, none is of greater concern than the dying child. HCPs have distinct responsibilities to children who are dying, as well as to their parents and families.

Ethical Theories

Clearly there are a number of issues involving children to which the ethical theories we have examined could profitably be applied. We will consider just one: cases of mature unemancipated minors who want to consent to treatment for themselves. Are HCPs ever morally justified in honoring their request? We will also consider the theories' views on social policies or law aimed at acknowledging the minor's right to consent to treatment for himself or herself.

Egoism

Consistent with its position on rights generally, egoism would not acknowledge that minors have an inherent right to consent to treatment for themselves. In the final analysis, egoistic HCPs would determine whether or not to honor the minor's autonomy by appealing to their own self-interests. Of course, all sorts of considerations could arise in the calculation of self-interest. Foremost among them might be legal considerations, since HCPs have much to lose if they violate laws in this area. As for social policies aimed at acknowledging the rights of minors, egoists could support them if they seemed likely to produce the greatest amount of long-term self-interest.

Act Utilitarianism

Like egoists, act utilitarian HCPs would allow an evaluation of consequences to determine their interactions with children. Unlike egoists, however, their standard of proper conduct would be the interests of all concerned. In the case under discussion, the parties immediately affected would be the minor, the minor's parents or guardians, and the HCP. Indirectly affected would be the institution, the profession, and society generally. Although act utilitarians could or could not recognize the minor's autonomy, depending on the situation, they would always be very concerned about abridging it. Classic utilitarianism is emphatic on this point: Personal freedom should be limited only when, left unchecked, it would harm society. It is very difficult to see how, in exercising self-determination in medical matters, mature minors threaten the well-being of society. But of course, in some cases that may be the case. In any event, act utilitarianism does not provide a clear-cut position on the issue.

Act utilitarianism is no clearer in its stand on appropriate legislation. It could support a social policy that acknowledged the rights of mature unemancipated minors, if such a law maximized the common good. One positive effect would be that individuals who needed immediate medical attention probably would be more inclined to seek it out. Another might be that society would establish more control over the so-called social diseases. On the negative side, a systematic recognition of the rights of mature minors might encourage adolescents to be cavalier in their attitudes toward drugs and sex. Knowing that they could always be treated without suffering the humiliation and pain connected with parental censure, they might be more inclined to indulge themselves. In the end, it would be perfectly consistent for act utilitarian HCPs to approve of individual acts permitting mature unemancipated minors to consent to treatment for themselves but to oppose any systematic social policy acknowledging such a right.

Rule Utilitarianism

Like egoists and act utilitarians, rule utilitarians would allow the consequences to determine the propriety of HCP interactions with mature unemancipated minors. But unlike act utilitarians, rulists would evaluate the consequences of the rule under which the particular act fell. The appropriate rule here might read: Mature unemancipated minors should be permitted to consent to treatment for themselves. Would following this rule be likely to produce more total good than not following it? It is impossible to say for sure, but it seems that the social advantages would outweigh the disadvantages. The two paramount advantages are that needy minors and the public health would be served.

In any event, once rulists determined that the rule was to society's greatest advantage, then it seems they would be obliged to support appropriate social policies. After all, HCPs should not be in legal jeopardy for doing what they are morally obliged to do. But the only way to remove legal jeopardy is to pass legislation that expressly recognizes the rights of the mature unemancipated minor. Of course, rulists could always argue that even a limited recognition of such rights might lead to additional liberalizing of laws that could have negative social consequences. But although it is true that society might be predisposed to additional nonrestrictive legislation, such laws are not inevitable.

The Categorical Imperative

Kant stated clearly that rational, self-governing persons have certain basic rights. The question is, Do mature, unemancipated minors fall into the category of "rational, self-governing persons"? It seems that *mature* implies rational and self-governing. If so, then Kant would argue that such minors have the same right to consent to treatment for themselves that adult persons have and that others have an obligation to respect this right. Thus, by a Kantian account, HCPs encountering unemancipated minors have a clear duty to respect their freedom to consent, other considerations notwithstanding. Kantian ethics could provide a theoretical basis for social policy to this effect as well. After all, if mature unemancipated minors have this right and if, in the absence of law, they are or can be denied it, then society has an obligation to recognize it expressly in social policy.

Prima Facie Duties

If we again assume that *mature* implies rationality and the ability to be self-governing, then Ross, like Kant, would consider a paternalistic HCP action that overrode the minor's right to consent a serious breach of the prima facie duties of justice and noninjury. Of course, this belief in itself would not prohibit such an action, since consideration of other prima facie duties is always in order. Again, as with Kant, Ross could and likely would support appropriate legislation, once it was determined that the mature unemancipated minor had the right to consent.

The Maximin Principle

An application of Rawls's ethics suggests an endorsement of the rights of the mature unemancipated minor. This follows from a consideration of Rawls's original position. Rational people with a sense of their own interests would not be bound by a rule that denied them the freedom to consent to treatment for themselves. If *mature* implies the qualities that people in the original position have, then there is no sensible reason to deny these rights to mature unemancipated minors. After all, the only apparent reason for so doing would be age; but people in the original position would not permit the introduction of so arbitrary a criterion, since they could be victimized by it. Rawls's equal liberty principle would also argue for the rights of mature unemancipated minors, because permitting this freedom to consent is compatible with a greater amount of liberty for all than not permitting it. In addition, such a rule would improve the lot of the most disadvantaged in the area of consenting to medical treatment. If this analysis is correct, then Rawls's ethics would also support legislation aimed at acknowledging the rights of mature unemancipated minors to consent to treatment for themselves.

Roman Catholicism's Version of Natural-Law Ethics

Roman Catholic moral theology yields conclusions similar to those drawn by other nonconsequential theories. Again, if *mature* is roughly the emotional and psychological equivalent of being an adult, then mature unemancipated minors would have the right to consent to treatment for themselves as much as adults would. Looked at another way, given these assumptions about maturity, paternalism directed at the mature unemancipated minor would be no more defensible than paternalism directed at the adult.

It might be argued that the principle of double effect, at least in theory, could allow exceptions to this general obligation to respect the wishes of mature unemancipated minors. Perhaps so. But it is difficult to imagine how a violation of the "mature" minor's right to consent would be a "morally indifferent act." Certainly it would not be "morally good."

If this analysis is correct, then Roman Catholic moral theology would seem to endorse appropriate social policy. It could be argued that the absence of such legislation surely invites violation of the mature minor's rights and thus places

HCPs in legal and moral jeopardy. Since these evils can be rectified by appropriate legislation, then that legislation would be not only justified but obligatory.

The preceding applications are speculative, of course, and certainly incomplete. But they provide a framework in which serious students can begin to consider the important moral questions in pediatric health care. One other point: These analyses have implications for HCP advocacy of children's rights. Having determined that the mature unemancipated minor's freedom to consent must be respected and that appropriate legislation is needed, HCPs might have obligations to help advance, secure, and protect these rights. Although it is inaccurate to say that this obligation necessarily follows, given their unique role and moral positions, surely HCPs must seriously investigate their advocacy responsibilities in this area.

CASE PRESENTATION
Death and Redemption

It's strange how little you know about your own daughter, Mrs. Jeffries thought as she went through Beth's things. She had chanced upon a little brown leather book with *My Diary* embossed on the face. Mrs. Jeffries held the book lovingly. Like mother, like daughter, she thought. For years she herself had been keeping a diary, which by now had run to several volumes.

For a moment she thought that opening the little book might be a violation of trust. But then she decided it probably wouldn't make any difference now; Beth had been dead over a month. And so, with care bordering on reverence, Mrs. Jeffries opened her daughter's diary to glimpse again the person she had loved most in the world.

Sept. 15: Something really weird happened Tuesday. Felt weak and cold in school. Terrific pain in my left side. Blacked out in gym. Rushed to hospital. Not appendix (pain was on the wrong side, but they thought about appendicitis anyway). Feel great now.

Oct. 3: Another attack—this time at home. Mom and Dad really got scared. Rushed me to hospital. Dr. Reynolds thinks I'm anemic. Gave me three pints of blood. Makes you feel like a vampire! Took a lot of x-rays.

Oct. 5: Still having trouble. They don't know what to do. Dr. R. has called in a specialist. Whole bunch of tests.

Oct. 7: They say I'm aplastic—whatever that means. Dr. R. says it isn't so terrible. He doesn't have any idea what caused it. Told him I didn't care what caused it, just cure it! Fast! He just nodded—Old Stoneface! I'm worried about missing school. Doesn't anybody know this is the year I graduate from dear old Clinton High?

Oct. 9: Something's bothering Mom and Dad. They say no, but I can tell.

Probably the hospital bills. It must be costing them a fortune to have me in this place. I told them I'm okay to go home, but the doctors don't think I'm ready. Neither do I, really. More tests, x-rays. What are they looking for?

Oct. 12: Impossible to get any info from anybody around here. "How are you?" "Your hair looks nice today." "Don't worry." "Have a nice day!" They all sound like wind-up dolls! I'm sure my parents know something. I'm going to find out. This conspiracy of silence is killing me!

Nov. 15. Over a month since the last entry. Didn't feel like writing. But do now. A writer has to write, right? Mom says I will die. Nobody knows when, but soon. Certainly less than a year. Die. Die. Die. Some little words have big meanings. Mom and Dad are taking it awful. Feel worse for them than for myself. Wish someone would help them cope. Glad they've got their faith, though.

Nov. 17: It dawned on me: I'll never see how it turns out—my life. I wonder if I'd have been a great novelist. Of if I'd have been a not-so-great novelist. Wonder if I'd have married and had kids. Wonder what they would have been like? I wonder. . . . I guess that's all that's left—to wonder. I wish . . . wish I could tell Mom and Dad how unfair I think it all is. But that would only make things harder on them. Instead I just told Mom it must all be God's will and that I am in His hands. That made her feel better. I only wish I believed it. . . . Haven't seen Dad in a week. Mom says he has the flu, but I wonder. . . .

Nov. 23: Staff is unbelievable—won't tell me anything. "How are you?" "Have a nice day!" "See you later." *Ad nauseum!* Impossible even to mention death around these people. They react like I've just called them a dirty name! One nurse even said, "Now, now, let's not get maudlin." I figure I damn well have a right to be maudlin if I want to. It's me that's dying, not her! Boy, I'd like to put her in a story some day! I could have slugged her. Well, better I didn't. It would've hurt me more than her: She doesn't seem to have any feelings.

Nov. 27: Thanksgiving—irony of ironies. Dad came today. Looked awful. God, I wish somebody would help them deal with this thing. If there was only something I could do.

Dec. 11: Some of these doctors are really getting to me. They just come in and look you over and ask how you are today or something trivial like that. Too bad they don't stay around long enough for an answer. And they never give me any info. I guess it's tough on them, though, since I'm so young. Or maybe they just can't deal with death—or failure.

Dec. 20: The pain's pretty awful. The doctors were talking to my mom and dad in the hall today. God, I hate that, going behind my back. Makes me feel like a real wimp. I made Mom tell me what it was all about. Just discussing my progress she said, and then changed the subject.

Jan. 8: I don't feel good. I want my parents—my Mom.

Mrs. Jeffries slowly closed the book. Someday, she promised herself, she would read the rest of it. But now she reached for her own diary, hidden under her lingerie in the bottom drawer of her bedroom dresser. It didn't take her long to find the entries she was seeking.

Oct. 7: This was the worst day of my life. Dr. Reynolds says that Beth has aplastic anemia and that she will not recover. His words still throb in my ears: "She's not good, not good at all." With that he led me to a little room. It might as well have been a gas chamber. There he told me about this dreadful disease.

"Nothing can be done," he said. "We don't know what causes it, and we don't know how to cure it." Then he said, "If it's any consolation, lots of people have it. It's incurable, and that's all there is to it. Beth will just have to accept it." I couldn't believe it. After he left, I sat alone, numb. I wish Dr. Reynolds could have waited until Henry and I were together. Henry has been with me every other time we came. Dr. Reynolds should have waited. It would have been easier on me. I don't know how I bore up for Beth today. She must have sensed something. But I never let on. Why upset her?

Oct. 13: Beth made me tell her the whole truth. I don't know how I found the courage. We both had a good cry.

Oct. 20: Sometimes I think it was something we did or didn't do that's caused all this.

Nov. 15: Henry is not bearing up at all. He can't even bring himself to visit Beth, and he feels so guilty about it. I've told her he has the flu. Beth spoke today of her faith. I think she was speaking more for me than for herself. I wish she'd express what is really on her mind. She must be terribly troubled.

Dec. 20: The doctor says it won't be much longer. He wanted to give Beth some morphine to shorten the time and ease the pain. We said we wanted to think it over. He couldn't understand that. "If it were my daughter," he said, "I'd want to stop the pain." Then he walked off. We decided maybe he was right. So we told the floor doctor to tell him we'd agreed.

Jan. 12: Sometimes they make me feel that I'm just in the way around here, especially the resident doctor and intern. Sometimes I even hide out down the hall and try to avoid them. I hate this skulking about, but their eyes seem to be questioning my right to be here. Don't they realize that I'm here because my child asked me and she's never asked me to before? I try to be helpful. They seem so shorthanded. I don't know what Beth would have done if I hadn't been here the last two nights—she's been so sick. Throwing up half the night, she needed to be washed and cared for. And the older woman in the room, with the heart attack, can't even get on a bedpan. So I help her and Beth, because there doesn't seem to be anyone around when they need help the most. Maybe because I'm here the staff figures I should do it all. I don't know.

Mrs. Jeffries dropped the book to her lap. She sat on the edge of the bed for some minutes while something took hold inside her. At first a thought, then a belief, and finally a conviction gripped her. She felt compelled to act on it. Somehow, she thought I must share these unguarded moments of mother and daughter. With whom, she didn't know. But in some strange way she felt her redemption hung in the balance.[17]

Questions for Analysis

1. *Several issues raised in this chapter are evident in this case. Identify them.*

2. *Focusing on each physician-parent or physician-patient interaction, evaluate the behavior of Dr. Reynolds and the other doctors.*

3. *Do you think Mrs. Jeffries in any sense failed her daughter?*

---

17. *This case is based on one reported in E. Kübler-Ross,* On Death and Dying *(New York: Macmillan, 1969), pp. 200–215.*

4. *Taking the bureaucratic view, would you describe Beth as an "uncooperative" patient? Explain.*

5. *Mrs. Jeffries feels that she should share the diaries. Do you think that she has a moral obligation to do so? If yes, what is the basis of the obligation, and precisely how do you think she should meet it?*

6. *Do you think any of Beth's rights were violated?*

7. *State specifically how the hospital might have made things less painful and more enriching for everyone, including the staff.*

Exercise 1

In *Hulit* v. *St. Vincent's Hospital,* the Montana State Supreme Court decided that a hospital could reasonably deny fathers access to labor and delivery rooms. Some of the grounds cited were that this practice might increase the possibility of infection, increase the number of malpractice suits, disturb doctors in their locker room, increase costs, make nurses in the delivery room uneasy, invade the privacy of other women in the delivery room, and create disharmony between staff who favored and opposed natural childbirth. Commenting on this ruling, George Annas stated:

> In my view the court in this case took a far too narrow view of its role, gave the hospital far too much discretion, and did not adequately deal with the rights of the physician, the woman, or her husband.[18]

Do you agree with Annas or the court? Explain. Do you think a woman has a moral right to have the father of the child present during delivery if she so desires?

Exercise 2

Reconsider your responses under Values and Priorities. Having read the chapter, would you answer the same in all cases? Where you would, do you feel you have a firmer basis for your replies?

---

18. *Annas, p. 153.*

Children, Stress, and Hospitalization

James K. Skipper, Jr. and Robert C. Leonard

Can the social environment of a hospital produce a great deal of stress for child patients and their parents? Sociologists James K. Skipper, Jr. and Robert C. Leonard thought it might. They

---

From Journal of Health and Social Behavior 9 (December 1968): 275–287. Copyright ©1968 by the American Sociological Association. Reprinted by permission of the publisher and the authors.

conducted an experiment in order to find out, and their findings are presented in the article that follows.

In the field experiment that Skipper and Leonard conducted, children hospitalized for tonsillectomies were randomly placed in one of two groups. One group received routine care; the other received similar care but with the addition of a nurse contact who interacted with the mother and kept her informed about the surgery and what to expect. Skipper and Leonard demonstrated that the nurse's interactions with the mother reduced the mother's stress, which in turn seemed to alter the child's social, psychological, and even physiological behavior.

This study sheds light on how HCPs might better meet their moral obligation to provide adequate care to the child patient. More subtly, it shows how the nature of the HCP-parent interaction can affect the parent's perception of personal control over the health crisis of the child, which the parent then transmits to the child. The child's right not only to adequate care but also to autonomy is at stake in HCP-parent interactions, because feeling "in control" of a situation, the child enjoys at least the perception, if not the fact, of self-determination. This perception can positively affect the child's response to hospitalization and treatment.

This paper reports an experimental study concerned with the reduction of some of the effects of hospitalization and surgery—physiological as well as social and psychological—in young children. Usually much of children's behavior while hospitalized for surgery is presumed to be a response to psychological and physiological stress. This research offers evidence demonstrating the effects of *social interaction* on children's response to hospitalization for a tonsillectomy operation.

When illness is serious enough to warrant an individual's confinement to a hospital, the process of hospitalization may produce stress (for all concerned) independent of that precipitated by the illness itself. Illness may be a stress-provoking situation not only for the stricken individual, but also for the members of his immediate family. Of special interest here is the stress in the patient role resulting from discontinuities, ambiguities and conflicts in the network of role relationships in which the patient becomes involved when he enters the hospital care and cure system. Stress seems to be especially high for both the staff and the patient and his family in cases involving surgery on young children.

One of the most common causes of hospitalization and surgery in preteenage children is tonsillectomy. It has been estimated that over two million of these operations are carried out each year. They constitute about one-third of all operations in the United States. Often it represents the child's first admission to a hospital, his first separation from the security of his parents and home,

and his first real experience involving loss of consciousness and bodily parts. The stress produced by this type of hospitalization and surgery results from loneliness, grief, abandonment, imprisonment, and the threat of physical injury, as well as intense needs for love, affection, and maternal protection. The experience may even lead to grave psychological problems years after the child has been discharged.[1]

The data in this report are based on a field experiment designed to test the effects on the behavior of hospitalized children of nurses' interaction with the children's mothers. It was hoped that the experiment would develop a method of reducing the children's stress indirectly by reducing the stress of their mothers. The study is a logical extension of a series of small sample experiments used to measure the effects of nurse-patient interaction on the behavior of patients. Evidence from these studies indicates that interaction with a patient-centered nurse trained in effective communication often results in large reductions in the stress experienced by the patient and a large decrease in somatic complications. (Leonard et al., 1967)

We postulate that hospitalization for a tonsillectomy operation is likely to produce a great deal of stress for child patients and their mothers. For the children this stress is likely to result in: elevated temperature, pulse rate and blood pressure; disturbed sleep; postoperative vomiting; a delayed recovery period; and other forms of behavior which deviate from the medical culture's norms of "health" and normal progress of hospitalization and treatment.

We conceptualize these patient behaviors to be simply instances of individual human behavior, which therefore can be affected by the patient's attitudes, feelings, and beliefs about his medical treatment, hospital care, and those who provide it. This is not to disregard physical and physiological variables as stimuli for the patient's response, or to deny that the response may be "physiological." Rather, we reason that in addition the meaning the patient attaches to the stimuli will also affect his response. For instance, some (stressful) definitions of the patient role and the hospital situation may result in deviant patient behavior in spite of all attempts at control by medication, anesthesia, or variations in medical technology.

Past attempts at reducing children's stress in the hospital have not fully considered the effect that parents and especially the mother may have on the child's level of emotional tension. (Prugh et al., 1953) The mother is a prime factor in determining whether changes in the child's emotions and behavior will be detrimental or beneficial to his treatment and recovery. If she is affected by severe stress herself, her ability to aid her child may be reduced. Moreover, in her interaction with the child her feeling state may be communicated and actually increase the child's stress.[2]

If the mother were able to manage her own stress and be calm, confident and relaxed, this might be communicated to the child and ease his distress. Moreover, the mother might be more capable of making rational decisions concerning her child's needs and thus facilitate his adaptation to the hospital situation. An important means of reducing stress from potentially threatening events is through the communication of information about the event. (Janis, 1958) This allows the individual to organize his thoughts, actions, and feelings about the event. It provides a framework to appraise the potentially frightening and disturbing perceptions which one might actually experience. An individual is able to engage in an imaginative mental rehearsal in which the "work of worrying" can take place. According to Janis (1958) the information is likely to be most effective if communicated in the context of interest, support, and reassurance on the part of authoritative individuals.

Regular and Experimental Conditions

As has been noted in the literature in a variety of different contexts, the modern hospital is a notoriously poor organization for eliciting information, providing support or generating a reassuring atmosphere. From the patient's point of view, the lack of information and lack of emotional warmth from physicians and nurses are among the most criticized aspects of patient care. (Skipper, 1965; Mumford and Skipper, 1967) For whatever reasons, to the extent medical and nursing personnel do not engage in expressive interaction with patients and provide them with information, they contribute to, or actually become sources of stress.

We can describe the usual routine staff approach to tonsillectomy patients and their mothers from the experience of members of our research team who worked for several months in advance of the study on the ward where our experiments were conducted. Typically, the staff approached the patient as a work object on which to perform a set of tasks, rather than as a participant in the work process or an individual who needs help in adjusting to a new environment. The attending surgeon's interaction with the child was limited primarily to the performance of the operation and release from the hospital. The nursing staff tended to initiate interaction only when they needed some data for their charts or had to perform an instrumental act such as taking blood pressure, checking fluid intake, or giving a medication. The typical role was the bureaucratic one of information gatherer, chart assembler, and order deliverer. They offered very little information and were usually evasive if questioned directly. If the mother displayed stress, the staff tried to ignore it or to get her to leave the ward.

For our research purposes, actual practice made a good comparison condition against which to test the hypothesis that:

> the childrens' stress can be reduced indirectly by reducing the stress of the mothers.

The experimental approach began with the admission of the mother and child to the hospital. Although the child was present, the focus of interaction was the mother. No more attention was paid to the child than under routine (control) admission conditions. The special nurse attempted to create an atmosphere which would facilitate the communication of information to the mother, maximize freedom to verbalize her fear, anxiety and special problems, and to ask any and all questions which were on her mind. The information given to the mother tried to paint an accurate picture of the reality of the situation. Mothers were told what routine events to expect and when they were likely to occur—including the actual time schedule for the operation.

The experimental interaction may be characterized as expressive, yet affectively neutral, person-oriented rather than task-oriented, non-authoritarian, specific (not diffuse) and intimate. The special nurse probed the mother's feelings and the background of those feelings as possible causes of stress regardless of what the topic might be, or where it might lead. In each individual case the special nurse tried to help the mother meet her own individual problems.

With the experimental group, the process of admission took an average of about 5 minutes longer than regular admission procedures. In addition to the interaction which took place at admission, the special nurse met with the mothers of the first 24 experimental group patients for about 5 minutes at several other times when potentially stressful events were taking place. These times were: 6:00 and 8:00 p.m. the evening of admission; shortly after the child was returned from the recovery room the next morning; 6:00 and 8:00 p.m. the evening of the operation; and at discharge the following day. The remaining 16 experimental group mothers were seen only at admission. For purposes of analysis the first 24 patients and their controls constitute Experiment I, and the remaining 16 patients and their controls Experiment II.

Our theory predicts that: providing the experimental communication for the mothers of children hospitalized for tonsillectomy would result in less stress, a change in the mothers' definition of the situation, and different behavioral responses. This in turn would result in less stress for the child and, hopefully, a "better" adaptation to the hospitalization and surgery. If this could be demonstrated, not only might it be a practical means of reducing the stress of young children hospitalized for tonsillectomy, but it would also provide direct evidence on the effect of social interaction on be-

haviors often assumed to be responses to psychological stress.

Research Methodology

To test these hypotheses an experiment was conducted at one large teaching hospital, in a four-month period during the late fall and winter. The sample included all patients between the ages of 3 and 9 years admitted to the hospital for tonsillectomy and having no previous hospital experience. Patients were excluded from the sample if there were known complicating medical conditions, their parents did not speak English, or their mothers did not accompany them through admission procedures. A total of 80 patients qualified for the sample. Forty-eight of the children were male, and 32 were female. Thirty-six were between the ages of 3 and 5, and 44 between the ages of 6 and 9. Thirty-three of the mothers had more than 12 years of formal education, 45 between 10 and 12 years and 2 less than 10 years. All the families were able to pay for the cost of the operation and the hospitalization.

Children were admitted to the hospital late in the afternoon the day before surgery was performed. At admission each child received a physical examination which included securing samples of blood and urine and a check on weight, blood pressure, temperature and pulse rate. When the admission procedures were completed the child was dressed in his night clothes and taken to one of two four-bed rooms limited to children who were to undergo a tonsillectomy. Control and experimental patients were not separated, but placed in rooms with each other to eliminate any systemic peer influence. From midnight until their return to the room after surgery, the children were not allowed to take fluids. The next morning, starting at 8 o'clock, the children were taken to the operating room, one every half-hour. Each child voided before the surgery. Following the operation they were taken to the recovery room where they remained until awake. Then they were returned to their room where they stayed until their discharge late the following morning. Only six of the mothers gained permission to "room in" with their child overnight. Three of these were in the control group and three were in the experimental group. All but one of the remaining mothers was able to spend most of the operation day at the hospital.

However, a record was not kept of the actual amount of time spent with her child. In fact it was beyond the resources of this investigation to obtain systematic data on the mother-child interaction; that is, the actual differences in frequency, timing and quality of interaction between control and experimental group mothers with their children.

The study was experimental in the sense that R. A. Fisher's (1947) classic design was used. The children were randomly assigned to control and experimental groups. No significant differences were found in the composition of the groups on the bases of: sex, age or health of the children, age of the mothers, class background, religious affiliation, and types of anesthesia used during the operation. Since the children were randomly assigned, antecedent variations and their consequences are taken into account by the probability test.

Correlated measurement bias may be a much more important source of mistaken conclusion than bias in the composition of the groups. One way of gaining some control of this type of bias is a "blind" procedure in which the individual measuring the dependent variables does not know which treatment the subjects have been assigned. With one exception, blind procedures were employed in this research.

The independent variable in the experiment was interaction. The experimental manipulations were all communicative—affective as well as cognitive. They emphasized the communication of information and emotional support to the mothers. The dependent variable was the behavior of the children. Thus the experimental variation was the interaction under usual hospital conditions compared with what was added experimentally.

All patients and their mothers whether in control or experimental group were subjected to regular hospital treatment and procedures. In addition, experimental group patients and mothers were admitted to the hospital by a specially trained nurse. Admission is a crucial time to introduce the experimental communication. Entry into any new social situation can be a tense experience. Lack of attention to the patient's definition of the situation in the admission process not only does not relieve stress but may actually increase it. Previous experimentation (Leonard et al., 1967) has indicated the potential effect of providing such attention on immediate stress and also the patient's adjustment to the hospital experience.

The regular nursing staff was informed that a study was in progress and asked to complete a short questionnaire regarding the behavior of the child and mother, as well as making charts and records available. They did not know which patients were in the control and experimental groups. The study was conducted at a teaching hospital, and the staff was used to having all sorts of projects taking place on the ward. They had become immune to them and ignored them unless they seriously interfered with their work. The staff was also familiar with the research personnel, who had been working on various projects on the ward on and off for over a year.

At admission, regardless of group, each mother was asked if she would be willing to complete a short questionnaire which would be mailed to her 8 days after her child was discharged, and would concern the hospital experience and its aftermath. None of the mothers refused. The mothers were not aware of whether they were in the control or experimental group. The questionnaire asked for the mother's perception of: her own level of stress before, during and after the operation, as well as her possible distress about a future similar operation; her desire for information during the hospitalization and her feeling of helpfulness; her trust and confidence in the medical and nursing staff; and her general satisfaction with the hospital experience. By means of a second questionnaire administered to the regular nursing staff, an independent measure of each mother's level of stress and general adaptation to the hospital experience was secured. To discover the effects of hospitalization on the child after discharge, a section of the mailback questionnaire to the mother also concerned aspects of the child's behavior during his first 7 days at home. Items were related to such matters as whether the child ran a fever, whether it was necessary to call a physician, and whether the child recovered during the first week at home. In addition, mothers were asked if their child manifested any unusual behavior which might be regarded as an emotional reaction to the operation and hospitalization such as disturbed sleep, vomiting, finicky eating, crying, afraid of doctors and nurses, etc.

Based on previous research several somatic measures of children's stress in the hospital were selected. Each child's temperature, systolic blood pressure and pulse rate were recorded at four periods during the hospitalization: admission, preoperatively, postoperatively, and at discharge. The normal variability of these vital signs is not great in children between the ages of 3 and 9. Children at this age have not developed effective inhibiting mechanisms, so that an increase in excitement, apprehension, anxiety and fear, etc. will be reflected in the level of these indicators. Inability to void postoperatively and postoperative emesis also may be responses to stress over which a child has little conscious control. The time of first voiding after the operation was recorded as well as the incidence of emesis from the time the child entered the recovery room until discharge. Finally, the amount of fluids a child is able to consume after the operation may be related to the mother's understanding of its importance and her ability to get the child to cooperate. Fluid intake was recorded for the first 7 hours upon the child's return from the recovery room. This period represented the shortest time that any mother in the study stayed with her child after the operation.

Systolic blood pressure was measured and recorded by the special nurse. Checks on the objectivity and reliability of the special nurse were made periodically. Data on pulse rate, temperature, postoperative vomiting, ability to void postoperatively, and oral intake of fluids were collected and recorded by staff nurses who had no knowledge of which children had been assigned to the control and experimental groups.

Data were complete on all patients and mothers with two exceptions. First, since reliability checks were not made on the special nurses' measurement of systolic blood pressure for several patients in Experiment II, these were not used. Second, the regular nursing staff's estimate of the mothers' stress and adaptation was not available

for two mothers in Experiment II. The response rate to the mailback questionnaire was over 92 per cent, 74 of the 80 mothers returning the questionnaire. Four of the nonreturns were control group mothers (2 in Experiment I and 2 in Experiment II) and 2 experimental group mothers (Experiment I).[3] All hypotheses predicted the direction of differences between control group mothers and experimental group mothers and children.

Findings

In a previous paper (Skipper et al., 1968) the effect of the special nurse's interaction with the mothers was presented in detail. In summary, according to the mothers' reports on the mailback questionnaire, experimental group mothers suffered less stress than control group mothers during and after the operation. This finding was substantiated by the independent evaluation of the regular nursing staff. The regular nursing staff also estimated each mother's difficulty in adapting to the hospitalization. Experimental group mothers were rated as having less over-all difficulty in adaptation. This agreed with the mothers' own self-evaluation. Experimental group mothers, as compared to control group mothers, reported: less lack of information during the hospitalization; less difficulty in feeling helpful to their child; and a greater degree of satisfaction with the total hospital experience. Taken together these measures provide evidence in support of the hypothesis that social interaction with the special nurse was an effective means of changing the mothers' definition of the situation to lower stress levels, thus allowing them to make a more successful adaptation to the hospitalization and operation.

In this paper we are concerned with the effect of the nurse-mother interaction on the children. Tables 1–3 compare the mean systolic blood pres-

Table 1. Mean Systolic Blood Pressure of Control and Experimental Group Children at Four Periods during Hospitalization

	Admission	Preoperative (8:00 P.M.)		Postoperative (8:00 P.M.)		Discharge		Total N
	$\bar{x}$	$\bar{x}$	t^*	$\bar{x}$	t^*	$\bar{x}$	t^*	
Experiment I Experimental	111.5	109.1	4.81 $P<.0005$	109.7	7.73 $P<.0005$	104.7	6.81 $P<.0005$	24
Control	110.4	120.3	....	127.8	....	120.9	....	24

*One-tailed test.

Table 2. Mean Pulse Rate of Control and Experimental Group Children at Four Periods during Hospitalization

	Admission	Preoperative (8:00 P.M.)		Postoperative (8:00 P.M.)		Discharge		Total N
	$\bar{x}$	$\bar{x}$	t^*	$\bar{x}$	t^*	$\bar{x}$	t^*	
Experiment I								
Experimental	103.6	95.8	5.10	101.6	6.31	95.2	5.08	24
			$P<.0005$		$P<.0005$		$P<.0005$	
Control	104.6	110.8	....	122.2	....	110.8	....	24
Experiment II								
Experimental	105.6	100.2	1.38	117.1	.83	105.4	2.13	16
			$P<.10$		$P>.10$		$P<.025$	
Control	104.9	107.5	....	123.1	....	116.8	....	16

*One-tailed test.

sure, pulse rate, and temperature of control and experimental sets of children at four periods during hospitalization—admission, preoperatively, postoperatively, and discharge.

At admission, the differences in systolic blood pressure were, of course, random, with the experimental mean actually slightly higher than the control (Table 1). This difference was reversed after the experimental treatment, and the control children continued to have higher average blood pressure throughout their hospital stay. In Experiment I the mean for experimental group children at admission, 111.5, dropped preoperatively to 109.1, remained relatively the same postoperatively, 109.7, and then dropped sharply at discharge to 104.7. The discharge mean was lower than the admission mean. The mean for control group children at admission, 110.4, rose to 120.3 preoperatively, and continued to rise to 127.8 postoperatively, before falling to 120.9 at discharge. The discharge mean was much higher than the admission mean. The mean differences between the control and experimental group children reached a

level of statistical significance beyond .005, preoperatively, postoperatively, and at discharge.[4] As mentioned previously, the data for Experiment II are not presented since reliability checks on the special nurses' measurement of systolic blood pressure were not available for several patients. However, the data that were available followed the same patterns as described in Experiment I.

We see in Table 2 that in both Experiments there was little difference at admission between the mean pulse rate of control and experimental group children. In Experiment I the mean for experimental group children at admission, 103.6, dropped to 95.8 preoperatively, rose to 101.6 postoperatively, and then fell to 95.2 at discharge. The discharge mean was lower than the admission mean. The control group mean at admission, 104.6, rose preoperatively to 110.8 and continued to rise to 122.2 postoperatively, before falling only to 110.8 at discharge. The discharge mean in the control set was much greater than the admission mean. The mean difference between the two groups reached a statistical level of significance

Table 3. Mean Temperature of Control and Experimental Group Children at Four Periods during Hospitalization

	Admission	Preoperative (8:00 P.M.)		Postoperative (8:00 P.M.)		Discharge		Total N
	$\bar{x}$	$\bar{x}$	t^*	$\bar{x}$	t^*	$\bar{x}$	t^*	
Experiment I								
Experimental	99.4	99.1	1.13	100.1	2.48	99.2	1.68	24
			$P>.10$		$P>.01$		$P<.05$	
Control	99.5	99.8	....	100.7	....	99.8	....	24
Experiment II								
Experimental	99.3	98.9	2.65	99.3	1.93	99.3	.85	16
			$P>.01$		$P<.05$		$P>.10$	
Control	99.3	99.4	....	99.9	....	99.7	....	16

*One-tailed test.

Table 4. Incidence of Postoperative Emesis for Control and Experimental Group Children

	None		Once		More than Once		Total N	X^2
	N	%	N	%	N	%		
Experiment I								
Control	14	58	3	12	7	29	24	8.40
Experimental	21	88	3	12	0	0	24	$P<.01$
Total	35	73	6	12	7	15	48	
Experiment II								
Control	12	75	1	6	3	19	16	1.15
Experimental	14	88	1	6	1	6	16	$P<.10$
Total	26	81	2	6	4	12	32	

beyond .005 at each of the periods. Exactly the same pattern appeared in Experiment II, but the differences between the group means were considerably less and did not reach as high a level of statistical significance.

Table 3 shows that in both Experiments I and II at admission there was little difference between the mean temperature of control and experimental children. In Experiment I the experimental group mean at admission, 99.4, fell to 99.1 preoperatively, rose to 100.1 postoperatively and dropped to 99.2 at discharge. Again, as in the case of systolic blood pressure and pulse rate, the discharge mean was lower than the admission mean. The control group mean at admission, 99.5, rose to 99.8 preoperatively and continued to rise to 100.7 postoperatively before falling to 99.8 at discharge. Again the mean discharge figure for the control group children was higher than the admission mean. The same pattern appeared in Experiment II.

In addition to systolic blood pressure, pulse rate, and temperature the childrens' postoperative emesis, hour of first voiding, and oral intake of fluids were checked. Tables 4 and 5 present these data. Table 4 shows that in Experiment I, 10 of the children vomited after the operation, 7 of them more than once, while only 3 experimental group children vomited, none of them more than once. Although the incidence of postoperative emesis was not as great in Experiment II as Experiment I, the same pattern appeared. Control group children experienced more emesis than experimental group children. As can be seen from Table 5, control group children did not void as rapidly after the operation as experimental group children. In Experiment I the mean hour of first voiding for control group children was well over 7½ hours compared to 4½ for experimental group children. In Experiment II the corresponding figures were: control group children approximately 6¾ hours and experimental group children 5¾ hours. Moreover, in both Experiments control children consumed much less fluid during the first 7 hours after the operation than experimental group children (Table 5).

Taken together these physiological measures indicate that the level of stress among experimental children was much lower. This was true for both Experiments. Experimental children had

Table 5. Mean Number of Hours from the End of the Operation to the Hour of First Voiding and Mean Intake of Fluids Postoperatively for Control and Experimental Group Children

	Mean Hours before First Voiding		Mean Fluid Intake (no. of c.c. after 7 hours)		Total N
	$\bar{x}$	t^*	$\bar{x}$	t^*	
Experiment I					
Experimental	4.54	5.94	629.17	4.81	24
Control	7.63	$P<.0005$	413.13	$P<.0005$	24
Experiment II					
Experimental	5.75	.58	520.00	3.62	16
Control	6.81	$P>.10$	351.56	$P<.005$	16

*One-tailed test.

Table 6. Regular Nursing Staff's Evaluation of Control and Experimental Group Children's Adaptation to the Hospitalization

	High Adaptation		Averge Adaptation		Low Adaptation		Total N	X^2
	N	%	N	%	N	%		
Experiment I								
Control	4	17	15	62	5	21	24	6.00
Experimental	12	50	9	38	3	12	24	$P<.02$
Total	16	33	24	50	8	17	48	
Experiment II								
Control	5	31	5	31	6	38	16	3.14
Experimental	9	56	5	31	2	12	16	$P>.10$
Total	14	44	10	31	8	25	32	

lower mean levels of systolic blood pressure, pulse rate and temperature preoperatively, postoperatively, and at discharge than control group children. Experimental group children had less postoperative emesis, voided earlier, and drank more fluids than control group children. These data lend support to the hypothesis that the experimental nurse-mother interaction would reduce mothers' stress and increase their ability to adapt rationally to the hospitalization, which, in turn, would have profound effects on their children. The hypothesis is further supported by the regular nursing staffs' evaluation of the children's general over-all adaptation to the hospitalization. By means of a short questionnaire each staff nurse who had the most contact with a child was asked to judge whether she considered the child's adaptation to be high, average or low. The staff nurses had no knowledge of whether a child was in the control or experimental group. Table 6 presents these data. In Experiment I, 50 per cent of the experimental group children were judged as making a high adaptation to the hospitalization compared to only 17 per cent of control group children. The corresponding figures for Experiment II were: experimental group children 56 per cent high adaptation and control group children 31 per cent high adaptation.

The mailback questionnaire to the mothers provides data on the children's condition and behavior at home during the first week after discharge (Table 7). In both experiments over 50 per cent of control group mothers reported that their child ran a fever during the first week at home while less than one third of the experimental group mothers reported this. None of the experimental group mothers reported their child vomiting, but this was reported by four control group mothers, two in Experiment I and two in Experiment II. Almost 41 per cent of control group mothers in Experiment I and almost 29 per cent in Experiment II indicated that they were worried enough about their child's condition to call a physician. Less than 14 per cent of experimental group mothers in Experiment I and less than 19

Table 7. Fever, Emesis, Condition Requiring Mother to Call Physician, and Recovery Time during First Week at Home, for Control and Experimental Group Children

	Fever Reported			Emesis Reported			Called Physician			Recovery Time within One Week		
	N	%	$*X^2$	N	%	$*X^2$	N	%	$*X^2$	N	%	$*X^2$
Experiment I												
Control	11	50	2.39	2	9	.51	9	41	2.86	11	50	12.12
Experimental	6	27	$P<.07$	0	0	$P>.10$	3	14	$P<.05$	22	100	$P<.0005$
Total	17	39		2	4		12	27		33	75	
Experiment II												
Control	8	57	1.07	2	14	.22	6	43	1.08	5	36	8.83
Experimental	5	31	$P>.10$	0	0	$P>.10$	3	19	$P>.10$	15	94	$P<.002$
Total	13	43		2	7		9	30		20	67	

*Corrected for continuity, one-tailed test.

per cent in Experiment II indicated it was necessary to call a physician. Finally, and perhaps most significantly, 100 per cent of experimental group mothers in Experiment I and 94 per cent in Experiment II reported that their child had recovered from the operation before the end of the first week after discharge. Only 50 per cent of control group mothers in Experiment I and 36 per cent in Experiment II claimed their child recovered during the first week. In other words, based on mothers' reports all but one of the experimental group children recovered from the operation during the first week after discharge, in contrast to less than half of the control group children.

These data indicate the experimental group children seemed to experience, physiologically, less ill effects from the operation and hospitalization and made a more rapid recovery than control group children. In addition, there were differences in the social and psychological behavior of the two groups. Major differences were found in three areas: excessive crying, disturbed sleep, and an unusual fear of doctors, nurses, and hospitals (Table 8). In both experiments, twice as many control as experimental mothers reported their child cried more than usual during the week after his discharge. Over 68 per cent of control group mothers in Experiment I, and over 78 per cent in Experiment II indicated their child suffered unusual sleep disturbances at night. This was compared with just 14 per cent of experimental mothers in Experiment I and 25.0 per cent in Experiment II. Of all the effects of the operation and hospitalization at home, disturbed sleep appeared to be the most common and the most severe.

Although only one experimental mother (Experiment I) reported her child seemed to have an unusual fear of the hospital and its personnel, 36 per cent of control mothers in Experiment I and 50 per cent in Experiment II reported that their child did. Often fear of the hospital, disturbed sleep, and excessive crying occurred in combination with one another. A written comment by one of the control group mothers aptly illustrates this:

> My child has had nightmares ever since he left the hospital. This is very unusual for him. He wakes up in the middle of the night yelling and screaming and crying his heart out. He is afraid someone will put him back in the hospital and leave him forever.

In addition to excessive crying, disturbed sleep, and fear of the hospital, slight differences were discovered in a number of other behavioral areas. According to mothers' reports, control group children had greater difficulty than usual in eating, drinking, and relating to others, as well as in manifesting more regressive behavior (thumb sucking, bed wetting, etc.) than experimental group children.[5]

Conclusion and Implications

The control group data confirms our hypothesis that under prevailing conditions the social environment of the hospital is likely to produce a great deal of stress for child patients and their mothers. For the children this stress is likely to result in elevated temperature, pulse rate and blood pressure, disturbed sleep, fear of doctors and nurses, a delayed recovery period, and other

Table 8. Unusual Fear of Hospitals, Physicians and Nurses, Crying More Than Usual, and More Disturbed Sleep Than Usual during the First Week after Discharge, for Control and Experimental Group Children

	Unusual Fear			Crying More than Usual			Disturbed Sleep More than Usual			Total N
	N	%	$*X^2$	N	%	$*X^2$	N	%	$*X^2$	
Experiment I										
Control	8	36	5.02	10	46	2.63	15	68	11.38	22
Experimental	1	4	$P<.05$	4	18	$P<.06$	3	14	$P<.0005$	22
Total	9	20		14	32		18	41		44
Experiment II										
Control	7	50	7.81	8	57	2.00	11	79	6.56	14
Experimental	0	0	$P<.0003$	4	25	$P<.08$	4	25	$P<.005$	16
Total	7	23		12	40		15	50		30

*Corrected for continuity, one-tailed test.

forms of behavior which deviate from the medical culture's norms of "health" and normal progress of hospitalization. The experimental group data indicate that a change in the quality of interaction between an authoritative person such as the experimental nurse and the hospitalized child's mother can lower the mother's level of stress and produce changes in the mother's definition of the situation. Due to the mother's intimate relationship and interaction with the child, a reduction in her level of stress and changed definition of the situation alters a salient component of the child's social environment. The data support the hypothesis that this may result in less stress for the child and consequently a change in his social, psychological, and even physiological behavior.

In Experiment I the special nurse interacted with the experimental group mothers at admission and at several other times during the hospitalization. In Experiment II the interaction was limited to the admission process. The observed effects of the experimental interaction on the childrens' behavior were in the predicted direction for both experiments, although the magnitude of the relationship was generally slightly higher in Experiment I. Although this finding highlights the effectiveness of the initial interaction and/or suggests that admission may be the crucial time and place to begin stress reducing interaction, it also suggests that further interaction throughout the hospitalization has important effects.

According to general sociological theory much of the important variation in individual human behavior is explained by variation in the culture and structure of the group to which the individual belongs. Additional variation is explained by the individual's status and position within the group. On occasion sociologists implicitly or explicitly specify intervening psychological states and processes that mediate group effects on individual behavior. When psychological variables are included their source is usually hypothesized in the socialization process or simply in social interaction. Indeed, sociology is often defined as the study of human interaction. However, many times sociologists do not find it convenient in their research to observe interaction, or the actual behavior that is supposedly affected by interaction. Self-reported values, statuses, role definitions, individual psychological states and behavior have been more accessible for study. Thus sociologists

have accumulated data suggesting that status inconsistency or low status crystallization is likely to result in strong liberal political attitudes, voting for the Democratic party, or, depending on the type of inconsistency under discussion, higher frequency of self-reported psychosomatic symptoms. Although this line of research does not appear to have been explicitly linked to the "social structure and anomie" theories of deviant behavior stemming from Durkheim, it has been linked to the "status integration" suicide research which also derives from Durkheim and with the psychological theories of cognitive dissonance.[6] Most previous research has relied on static macro-level correlations using census-type statistics for infrequent events such as suicide, or on survey analyses of self-reported physiological stress symptoms correlated with various indexes of individual status consistency. The intervening social process activating the psychological inconsistencies has not explicitly figured in the research, nor has the research been experimental. Obviously, it is extremely difficult in most cases to manipulate these structural variables. The research reported above, by focusing on the effects of the immediate social environment rather than on more permanent social structural determinants or long-term personalities changes points the way to nonlaboratory experimental tests of social environmental stress theory.

In addition to its potential contribution to social psychological theory, this type of research can form an interesting chapter in applied sociology. It has immediate implications for the control of stress, since control lies in the dyadic interaction which can be manipulated by individual practitioners. The results of this research suggest that even just one such practitioner out of the dozens with whom a patient may come in contact may be able to have a major effect. In contrast, manipulation of either relatively permanent statuses, major structural features of the organization, or deep-seated personality traits must be more difficult.

Specifically, if suported by further, more extensive research, the data suggest that some of the after-effects of hospitalization and surgery in young children, physiological as well as social and psychological, may be alleviated through a relatively simple and inexpensive social process. An authoritative figure, by establishing an expressive relationship with the mother of a child, and providing her with information, may reduce the mother's stress and allow her to make a more rational adaptation to the child's problems and take a more active role in aiding him. The change in the mother's behavior may then have a profound effect on the child's behavior. We suggest that this process might be an effective and efficient procedure which could easily be added to the arsenal of ways and means which health professionals may have at their disposal for combating the stress of hospitalization and surgery on both mother and child.

Acknowledgments

This research took place at the Child Study Center, Yale University, in cooperation with the Yale School of Nursing and the Yale New Haven Community Hospital. It was supported in part by a grant to the senior author from the Yale Medical School. The authors would like to express their appreciation to Julina Rhymes, Perry Mahaffy, Jr., Margaret Ellison and Powhatan Woolridge for their helpful assistance during the data collection stage of the research.

Notes

1. Lipton (1962) summarizes much of the literature concerning the nature, extent, and psychological effects of tonsillectomy operations. However, recent evidence is leading many physicians to question the need for tonsillectomy at all, especially in routine cases. (McKee, 1963).

2. Escalona (1953) points out that the communication of feeling states between a mother and her child may take place on a non-verbal as well as verbal level, may occur at even a very early age in the life of the child, and may not be fully subject to the voluntary control of the mother.

3. The actual design of the questionnaire and the return rate is described and discussed in (Skipper and Ellison, 1966).

4. The reader should keep in mind that statistical significance does not necessarily indicate practical significance. For the most part variations in the somatic indicators are within what might be considered the normal range. Their importance lies in the fact that they are symptomatic of the degree of stress suffered.

5. When all the results from the mailback questionnaire were controlled for the age and sex of the child, and the education of the mother, one

important association was discovered. Regardless of treatment (control or experimental) children age 6 and under suffered more from disturbed sleep during the first week after the operation than those age 7 and over.

6. Much of the literature on this topic is summarized in Martin's (1965) cogent review of theories of stress.

References

Escalona, S. 1953. Emotional Development in First Year of Life. pp. 11–92 in Milton J. E. Senn (ed.), Problems of Infancy and Childhood. Packanack Lake, N.J.: Foundation Press.

Fisher, Ronald. 1947. The Design of Experiments. Edinburgh: Oliver & Boyd.

Janis, Irving. 1958. Psychological Stress. New York: Wiley.

Leonard, R., J. Skipper, and P. Woolridge. 1967. "Small sample field experiments for evaluating patient care." Health Services Research 2 (Spring): 46–60.

Lipton, S. 1962. On the Psychology of Childhood Tonsillectomy. pp. 363–417 in Ruth Eissler et al., (eds.), The Psychoanalytic Study of the Child. New York: International Universities Press.

Martin, W. 1965. "Socially induced stress: some converging theories." Pacific Sociological Review 8 (Fall): 63–69.

McKee, W. 1963. "A controlled study of the effects of tonsillectomy and adenoidectomy in children." British Journal of Preventive and Social Medicine 17, 2:46–69.

Mumford, Emily, and James K. Skipper, Jr. 1967. Sociology in Hospital Care, New York: Harper & Row 117–139.

Prugh, D., E. Staub, H. Sands, R. Kirschbaum, and E. Lenihan. 1953. "A study of the emotional reactions of children and families to hospitalization and illness." American Journal of Orthopsychiatry, 23:70–106.

Skipper, J. 1965. Communication and the Hospitalized Patient. pp. 61–82 in James K. Skipper, Jr., and Robert C. Leonard (eds.), Social Interaction and Patient Care. Philadelphia: Lippincott.

Skipper, J., and M. Ellison. 1966. "Personal contact as a technique for increasing questionnaire returns from hospitalized patients after discharge." Journal of Health and Human Behavior 7 (Fall): 211–214.

Skipper, J., R. Leonard, and J. Rhymes. 1968. "Child hospitalization and social interaction: An experimental study of mothers' stress, adaptation and satisfaction." Medical Care (in press).

For Further Reading

Anthony, J. A., and Koupernik, C. *The Child in His Family: The Impact of Disease and Death.* New York: Wiley, 1973.

Boulding, E. *Children's Rights and the Wheel of Life.* New Brunswick, N.J.: Transaction Books, 1979.

Brody, S., and Axelrad, S. *Mothers, Fathers, and Children: Explorations in the Formation of Character in the First Seven Years.* New York: International Universities Press, 1978.

Cook, S. Sheets. *Children and Dying: An Exploration and Selective Bibliographies.* New York: Health Sciences Publishing, 1974.

DeLone, R. H. *Small Futures: Children, Inequality, and the Limits of Liberal Reform.* New York: Harcourt Brace Jovanovich, 1979.

Esson, W. M. *The Dying Child.* Springfield, Ill.: Charles C Thomas, 1970.

Gross, B., and Gross, R., eds. *The Children's Rights Movement.* Garden City, N.Y.: Anchor Books, 1977.

Haller, A. J., ed. *The Hospitalized Child and His Family.* Baltimore: Johns Hopkins University Press, 1967.

Hamovitch, M. B. *The Parent and the Fatally Ill Child: A Demonstration of Parent Participation in a Hospital Pediatrics Department.* Duarte, Calif.: City of Hope Medical Center, 1964.

Klein, A. F. *The Professional Child Care Worker: A Guide to Skills, Knowledge, Techniques, and Attitudes.* New York: Association Press, 1975.

Koocher, G. P. *Children's Rights and the Mental Health Professions.* New York: Wiley, 1976.

Noland, R. L. *Counseling Parents of the Ill and Handicapped.* Springfield, Ill.: Charles C Thomas, 1971.

Paul, J. L., Neufeld, R. G., and Pelosi, J. W., eds. *Child Advocacy within the System.* Syracuse, N.Y.: Syracuse University Press, 1977.

Sarkissian, W. *The Design of Medical Environments for Children and Adolescents: An Annotated Bibliography.* Monticello, Ill.: Vance Bibliographies, 1980.

Sheleff, L. Shaskolsky. *Generations Apart: Adult Hostility to Youth.* New York: McGraw-Hill, 1981.

Stacey, M. *Hospitalized Children and Their Families.* London: Routledge & Kegan Paul, 1970.

Westman, J. C. *Child Advocacy: New Professional Roles for Helping Families.* New York: Free Press, 1979.

8
THE ELDERLY PATIENT

Mr. Winter had for some years single-handedly run an operation for his company that no one else fully understood. As Winter approached his sixty-fourth birthday, the company assigned him a young man to learn Winter's job, so that he could replace the elderly worker on retirement. Winter did not want to retire; company policy simply decreed retirement at sixty-four. Winter complied reluctantly.

Not long after his retirement, a significant change occurred in Winter: a withdrawal from people and a general decline in his zest for life. Within a year, he was hospitalized with a diagnosis of senile psychosis. Winter had become a vegetable.

As chance would have it, two years after assuming Winter's job the young worker died of a sudden heart attack. The company found itself with a vacancy and at a loss to find a competent employee. So it was decided to approach Mr. Winter, to see if he could pull himself together to resume his job long enough to train someone else.

Four of his closest co-workers were sent to the hospital. After hours of trying, one of the men finally got through to him. The idea of going back to work brought a sparkle to Mr. Winter's eyes—the first in two years. Within days, this "vegetable" was back at work, operating at full steam, and interacting with people as he had all his working life.

The preceding case actually occurred.[1] It is not atypical; many of the elderly find themselves in similar circumstances. Such situations trigger numerous questions about the plight of the elderly in our society. What pressures do they face? What stresses do they feel? What are their needs? What is society's response? What is the health care system doing? The questions are many, the concerns profound.

This chapter considers another category of special patients: the elderly, defined here, chronologically, as any person sixty-five years of age or older. In general, it will try to throw some light on HCP responsibilities to the elderly patient. Since questions of adequate health care and autonomy inevitably arise in discussions of moral responsibilities to aged patients, this chapter will focus

---

1. B. Margolis and W. Kroes, "Work and Health of Man," *in* Work in America, Special Task Force (Cambridge, Mass.: MIT Press, 1972).

on these issues. But it will also take a close look at the place where more and more of us are spending our final days: the old-age institution. The social organization of these facilities raises moral concern and therefore warrants our attention.

But since many of the moral problems involving HCP interactions with the elderly have their roots in how individuals and society generally view the aged, the chapter starts with a consideration of our perceptions of the aged, the myths that surround aging and the elderly, and the discrimination that is often directed at the aged.

Values and Priorities

The following questions are designed to help you identify your own feelings about aging and the elderly before you start the chapter.

1. Do you think the company was right to make Winter retire?

2. Many people think mandatory retirement is good because it makes room for young people. Do you agree?

3. How important do you think work is to a person?

4. What value does our society place on work?

5. From what important values, besides work, are most of the elderly separated?

6. "I wish my grandparents were still alive," young adults sometimes say. Why do you think they feel this way?

7. Do you think an old person has anything unique to offer others?

8. Assuming old people do have something unique to offer, do you see evidence that society values it and tries to exploit it or evidence that society does not?

9. How would you like to be treated when you're old?

10. How do you think you will be treated when you get old?

11. How do you feel about old-age institutions?

12. Have you ever been inside an old-age institution? If so, what were your feelings?

13. Ideally, where would you like to end up in your old age?

14. What's your greatest fear about growing old? Why does this frighten you?

15. How many people over sixty-five do you interact with on a fairly regular basis?

16. Do you think society gives the aged about as much assistance as they deserve?

17. As a taxpayer, do you think you'd be more inclined to support social welfare programs for children rather than the aged?

18. Would you ever put an aged relative in an old-age institution?

19. How do you feel about the possibility that your children will one day put you in a "home"?

20. Do you think that most old people are unable to manage their affairs?

Do you agree or disagree with the following?

21. Society tends to ignore the elderly.

22. The elderly remind us of our own mortality.

23. Old people generally are senile, hard of hearing, and cranky.

24. Communities planned specifically for the needs of the elderly are the best places for old people to live.

25. Society should pay for the food, shelter, clothing, and medical services of any elderly person who can't afford them.

26. If people weren't prudent enough to plan for their old age, then they have no right to expect society to maintain them.

27. Old people don't have very much to live for.

28. Paternalism directed at the aged is pretty much like paternalism directed at children.

29. About all HCPs can do for the institutionalized elderly is to provide custodial care.

30. As people get older, autonomy means less and less to them and security more and more.

Perceptions of Old Age

Ask yourself this question: What do I associate with old age? Your answer will give you a pretty good idea of your perceptions of the old and, more importantly, insight into the basis of your interactions with the elderly ill.

If your answer included such descriptions as *nonproductive, out-of-touch, sick, lonely, poor* and similar negative impressions, your perceptions correspond with how most of the young *and* old see old age. Most of us believe old age is bleak. This largely explains our general dislike for aging. And a dislike for aging can lead to a kind of selective perception, in which we notice what confirms and ignore what denies our expectations. Curiously, there is some evidence for believing that the elderly themselves engage in this selective perception.

A study conducted in 1975 by the American Council on Aging illustrates the point.[2] According to this study, more than half of the people interviewed— including many of the elderly—thought of the elderly as being in poor health, but only 21 percent of the people age sixty-five and over actually found themselves with health problems; 60 percent of the respondents considered old people lonely, but only 12 percent in that age category actually felt lonely; 62

2. The Myth and Reality of Aging *(Washington, DC: National Council on Aging, 1975).*

percent of the public thought old people's income too low to live on, but only 15 percent of the aged actually felt this financial pinch; half of those asked thought crime was a very serious problem for the elderly, but only 23 percent of the elderly actually considered themselves threatened. On each subject, elderly people shared the public's characterization of old age but exempted themselves from any stereotype.

Clearly, the study points up a stereotype of old age and proves how pervasive it is. More interesting is how intractable this stereotype appears to be. After all, even those whose personal experience of old age belied the stereotype did not seem able or willing to change their view of old age in general. Former Secretary of Commerce Juanita M. Kreps offers this explanation:

> There is surely some tendency for the older person to play down his age-related problems; to want to spare children and friends any additional burdens; to want to retain a feeling of independence and competence as long as possible. In short, *we may refuse to acknowledge the onset of problems that would confirm us as elderly. One would prefer not to identify with his age group when his concept of what it is like to be old leads him to view that phase of life as totally unrewarding.*[3]

It is not difficult to understand why people have negative perceptions of old age. For one thing, television, newspapers, magazines, comic books, jokes, and songs often dwell on the painful, empty, dangerous, and humiliating aspects of old age. Add to this a value system that glorifies youth, prizes the future, devalues the past, and equates human worth with productivity and financial earning, and it is small wonder that we labor under numerous myths about old age and the elderly. What makes such myths of moral concern is that they can undercut the elderly's right to health care.

Myths about Old Age

It is impossible to catalog all the myths that circulate about old age and the elderly, for they are legion. But we can and should isolate those that affect the elderly's right to adequate health care, since they have particular moral implications.[4]

1. *Old age is an illness.* The myth is that old people get sick and die because they are old and worn out. Such thinking is based on the assumption that, since old age is inevitable, everything we dislike about it is also inevitable. But people do not die of old age; they die of disease. Missing the point can lead to action or inaction that would be of moral concern.

    For one thing, the elderly and their families might resign themselves to conditions that in younger people would be diagnosed, treated, and perhaps even cured. Thus, by accepting this myth, the elderly may set

---

3. *J. M. Kreps, "Human Values and the Elderly," in* Aging, Death and the Completion of Being, *ed. D. D. Van Tassel (Philadelphia: University of Pennsylvania Press, 1979), p. 18. Italics added.*

4. *C.* Epstein, Learning to Care for the Aged *(Reston, Va.: Reston Publishing, 1977); and H. Downs,* Thirty Dirty Lies about Old *(Niles, Ill.: Argus Communications, 1979).*

themselves up for unnecessary suffering. What is more, this myth can lead to a less than ambitious social commitment to research on aging, an easy justification for questionable allocations of tax monies and societal tolerance of an array of psychosocial problems faced by many of the elderly, some of which could be avoided or alleviated. Most important, endorsing this myth can predispose HCPs to resignation, even defeatism, in the care of the elderly. As one HCP put it:

> The attitude is apparent every time a physician says, "What can you expect? You're seventy years old!" Every time an orthopedist says, "How much walking do you have to do? You're 75 years old!" Every time a treatable illness is overlooked because the expectation is that the observable symptoms are merely signs of old age.[5]

2. *Old people cling to old values and resist change.* Put in homelier terms, this myth reads: You can't teach an old dog new tricks. Although it is true that some things, such as language learning, might more easily be mastered at a young, impressionable age, old people are not ineducable. Learning is intimately connected with motivation, preferably with the prospect of positive benefit. People learn best when they are convinced that the change the learning will bring is desirable. But what incentives does society hold out to the aged as the reward for change? Not many. In fact, a case could be made for the opposite: Often we isolate the aged from other age groups, deny them involvement in activities that require change, and demand that they behave in a way that "befits their age." It is not surprising, then, that some of the elderly appear to live more in the past than the present.

    In health care, the "old dog" myth can seriously undermine care by fostering a "basic maintenance" attitude toward treatment. Thus, when the reasoning is that one cannot expect people to change their ways after seven decades, there is no point in even trying. As a result the HCP's only obligation is to do what is minimally needed to maintain these individuals.

3. *Certain personality quirks are attributable to old age.* In keeping with our general perceptions of old age and the elderly, we frequently write off displays of irritability, impatience, and general deterioration to growing old. As a result, needed health care can go begging. A good example is "recruitment deafness," a condition in which normal speech cannot be heard, although shouting can be. Sufferers seem forever to be complaining that people are either whispering or yelling at them. The apparent inconsistencies can lead one to the erroneous conclusion that these people are faking it. Thus, the stereotype is reinforced that the old are ornery and difficult to live with simply because they are old. When evident in the old, such traits are more likely the results of untreated illnesses or the effects of improper drug use.[6]

4. *Old age means many negative behavioral patterns.* This myth assumes that preoccupation with body functions, irritability, and a general disinterest in

---

5. *Epstein, p. 60.*
6. *Epstein, pp. 61–62.*

things are integral parts of aging. The danger of such a belief is that it can impair the development of strategies, techniques, and activities to involve the aged in the world around them. Actually, the negative behavioral patterns the aged sometimes show are more properly attributable to "bereavement overload."[7] Faced with the sudden deaths of friends and loved ones, the loss of valued possessions, forced relocation, and diminished physical faculties, the aged who can no longer lose themselves in work can easily be overwhelmed. The results may indeed be negative, even bizarre or self-destructive, patterns of behavior. But assuming that such behavior is simply part of being old without considering the impact of multiple bereavements can lead to tragically inappropriate responses. Thus, whereas elderly persons suffering from bereavement overload need an environment that permits their personalities to remain integrated, they often receive treatment that completely overlooks the values and characteristics that have been central to their personalities.

5. *Old people are not interested in the kinds of marital relationships that are of concern to younger people.* If strictly adhered to, this myth can deprive older people of the emotional security that they need as much as any other age group does. Although little research has been done in this area, one study involving 227 older husbands and wives identified twenty-four factors by which they measured marital need satisfaction:

    1. Providing a feeling of security in me.
    2. Expressing affection toward me.
    3. Giving me an optimistic feeling toward life.
    4. Expressing a feeling of being emotionally close to me.
    5. Bringing out the best qualities in me.
    6. Helping me to become a more interesting person.
    7. Helping me to continue to develop my personality.
    8. Helping me to achieve my individual potential (becoming what I am capable of becoming).
    9. Being a good listener.
    10. Giving me encouragement when I am discouraged.
    11. Accepting my differentness.
    12. Avoiding habits which annoy me.
    13. Letting me know how he or she really feels about something.
    14. Trying to find satisfactory solutions to our disagreements.
    15. Expressing disagreement with me honestly and openly.
    16. Letting me know when he or she is displeased with me.
    17. Helping me to feel that life has meaning.
    18. Helping me to feel needed.

---

7. R. Kastenbaum, *"Death and Bereavement in Later Life," in* Death and Bereavement, *ed. A. Kutscher (Springfield, Ill.: Charles C Thomas, 1969), pp. 28–54.*

19. Helping me to feel that my life is serving a purpose.
20. Helping me to obtain satisfaction and pleasure in daily activities.
21. Giving me recognition for my past accomplishments.
22. Helping me to feel that my life has been important.
23. Helping me to accept my past life experiences as good and rewarding.
24. Helping me to accept myself despite my shortcomings.[8]

Clearly, the elderly's needs correspond with those of young married people. Thus, despite one's age, the need for personal growth, for meaning, purpose, and self-esteem, and for honesty in human relationships all remain constant.

6. *Old people have no interest in sex.* It is still common to think that older people have no interest in sex or that, if they do, such an interest is unnatural. Not only is sexual interest perfectly normal and natural at any age, but research definitely indicates a continuing interest in sex among older people. Furthermore, the considerable publicity usually given to instances of older men marrying younger women and subsequently fathering children gives lie to this belief. Yet the myth persists, sometimes with unfortunate consequences for the older person.

In his book *Thirty Dirty Lies about Old,* television communicator, lecturer, and consultant Hugh Downs gives an illustration of how physicians can be as misguided as anyone else when dealing with the sexuality of older patients:

> Dr. Robert Butler tells of a physician who in consultation described the condition of a vigorous sixty-nine-year-old, listing several disorders but neglecting a genital problem—Peyronie's disease—which is rare, but treatable and sometimes self-correcting. The malady causes pain during intercourse. "He's too old for that to matter," the doctor offered as an explanation for ignoring the condition. But it did matter a great deal to the patient.[9]

Here a physician's prejudice leaves a treatable medical problem untreated. The moral implications need no comment.

Dr. Butler, whom Downs mentions in his example, is perhaps best known for his investigations of prejudice directed at the aged. "Ageism," as this is termed, is an outgrowth of our perceptions and myths about the aged.

Ageism

The term *age-ism* was coined by Robert N. Butler in 1968. Butler, now director of the National Institute on Aging, was then practicing psychiatry in Washington, DC. So striking were the similarities he saw between discrimination against the aged and racial and sexual discrimination that he developed the concept of ageism as the senescent counterpart of racism and sexism.

---

8. N. Stinnett, J. Collins, and J. E. Montgomery, "*Marital Need Satisfaction of Older Husbands and Wives,*" Journal of Marriage and the Family, *November 1968, pp. 428–434.*

9. *Downs, p. 75.*

According to Butler, ageism is another form of bigotry. Specifically, ageism is a process of systematically stereotyping and discriminating against people because they are old.[10] Like racism and sexism, which accomplish the same thing by skin color and gender, ageism insists that traits of personality and character are determined by inherent biological factors. And like racism and sexism, ageism raises serious moral concerns about human dignity and worth, about potential damage to the self-image of the elderly, and about our very definition of what it means to be human. It also raises fundamental questions about justice. For example, is it fair to evaluate a person for a job on any criteria not directly related to job performance? Or on a broader social level, is it fair to society to disqualify someone on the basis of age even though that person may still be a valuable resource?

Just as important is another consideration Butler observes: stereotyped role expectations. One of the great handicaps of aging is the reduction in the range of choice. In part, this reduction obviously results from physiological and economic limitations. But just as significant are the restrictive norms of a biased culture. Much as society's role expectations for blacks and women persist, age-graded categories abound, with old age set in an especially derogatory mold. Indeed, some commentators, such as physician Gerald J. Gruman, have detected a striking similarity in the roles and characteristics attributed to blacks, women, and the elderly. Stereotyped "good" attributes ascribed to blacks have included: skill in handling children and animals, musical and dancing ability, loyalty as servants and as menial employees, and a feeling of religion.[11] Gruman sees these same traits, with the exception of the song and dance abilities, linked to women and the elderly. In the same vein, prominent gerontologist Robert Kastenbaum has spoken of the "aging mystique." "The gracefully aging person," Kastenbaum writes satirically, "has the knack of making us [professionals] feel better. This type of elder is . . . content with his or her lot . . . is non-competitive, non-complaining . . . knows his place . . . also eats watermelon and fried chicken and has a natural sense of rhythm."[12]

The bottom line is that ageism touches the lives of the elderly. It is a prejudice, as Butler observes, that is both insidious and destructive. For these reasons, ageism is of utmost moral consequence. To HCPs it should be a matter of intense concern, because it can violate the elderly's right to adequate health care by precluding an accurate and thorough understanding of their needs.

Just as important as recognizing ageism and its moral implications is the recognition that ageism has its roots in cultural values. Two of these bear particular mention: time and work.

---

10. R. Butler, Why Survive? Being Old in America *(New York: Harper & Row, 1975), p. 12; for Butler's earliest development of the concept of ageism, see "Age-ism: Another Form of Bigotry,"* The Gerontologist, *1969, pp. 243–246.*

11. G. J. Gruman, "Cultural Origins of Present-Day 'Age-ism': The Modernization of the Life Cycle," *in* Aging and the Elderly, *eds. S. F. Spicker, K. M. Woodward, and D. D. Van Tassel (Atlantic Heights, N.J.: Humanities Press, 1978).*

12. R. Kastenbaum, "Should We Have Mixed Feelings about Our Ambivalence toward the Aged?" *Journal of* Geriatric Psychiatry 7 (1974): 97.

Cultural Values: Time and Work

Addressing the subject of growing old, a recent United Nations study declared: "Strongly competitive societies in which too much emphasis is given on an individual's worth in terms of productive work and achievement, in which inactivity is somewhat suspect and leisure is highly commercialized and therefore expensive, are not congenial environments in which to grow old."[13] Few would argue with this or with the claim that the United States is a strongly competitive society in which an individual's value more often than not is viewed in terms of productivity. In brief, there is a decided tendency in our society to equate human worth with earnings.

Underlying this tendency are two interlocked core cultural values. One has to do with time and its utility, the other with work and the importance of productivity.

The common view of time in our society relates its value to productivity. The value of time, then, depends on how productively it is used. Productivity, in turn, is generally associated with earning. Time is money. Individuals who spend their time providing some marketable good or service are thought to be making "valuable" use of their time. Time is of value to them and to those observing them. Following this line of thought, we go on to make rather subtle value assessments about the allocation of our time. For example, it is not uncommon for professional people to decide that it is a better use of their time to pay someone to cut the grass, look after the kids, and cook the meals than to spend their time doing these things. Stated in other terms, we often associate the value of time with efficiency, another word for productivity. The value we attach to work cannot be divorced from this view of time.

In stressing the worth of time spent productively, we confer on the worker a useful and valuable role in society. At the same time, by implication, we devalue the importance of nonworkers, even consign them to a useless existence. Small wonder, then, that society's attitude toward the nonworking elderly is something less than wholesome. After all, what are they contributing? What are they producing? In a society as strongly competitive as ours, the answer must be nothing, and their value is thus reduced to little more than nothing.

What is particularly curious about this set of value assumptions is that the nonworking elderly internalize them so easily. Quite predictably, they often view themselves as no longer of any value or worth. The sinister impact of these negative feelings on health needs no documentation. Yet the broader, social implications for the elderly often go overlooked. We fail to realize that the enthusiasm with which we set up research, programs, and services for the elderly is often sapped by the equation between time and productivity, human worth and earnings. Kreps has stated the matter vividly:

> By current-productivity standard a retiree merely consumes the output of others just as a child does; both must depend on others for their support. The important difference between retiree and child, this line of reasoning continues,

---

13. *United Nations, Department of Economic and Social Affairs,* The Aging: Trends and Policies *(New York: United Nations, 1975), p. 11.*

lies in their potential. Whereas investments in the young person will "pay off" because he will become a producer, spending for the older dependent yields no future return. Hence, it is easy to develop an economic rationale for heavy investments in the education of youth. But the cost of supporting the aged is not recouped, and there is some tendency to view these expenditures as poor investments.[14]

In short, the value that we attach to time and work bears directly on how we view the elderly. It also goes a long way toward explaining ageism as it functions in our society.

With this common understanding in mind, we may now turn to a crucial moral issue involving the elderly who are ill: paternalism practiced to protect the aged from outside threats or from themselves. This issue raises profound concerns about autonomy.

Paternalism and Autonomy

Certainly no group of adults is more the object of paternalism, both personal and state, than the elderly. In understanding the moral questions that paternalism for the elderly raises, it is useful to distinguish two separate and distinct categories of "dangers" from which paternalistic acts are intended to protect the elderly: external and internal dangers.[15]

External Dangers

When people think about paternalism and the elderly, it is generally in the context of protecting the elderly from dangers outside themselves, such as the threat of being destitute or of being enfeebled. Because society views the elderly as generally incapable of caring for themselves, it feels obligated to intervene on their behalf, *for their own good.* Thus, society over the years has enacted all sorts of social welfare programs, from Social Security to Medicare, to protect the elderly from various physical, social, and psychosocial harm. Similarly, numerous private charities have sprung up that intervene in the lives of the elderly, again *for their own good.*

Behind paternalistic acts that attempt to ensure the safety and well-being of the elderly there is an irreproachable moral impulse: the desire to help others. What is more, through acts of paternalism the state and family meet their obligations to the elderly to reduce their suffering and protect them from outside threat. The only problem, of course, is that the help is imposed, not just offered; a helping hand is not so much extended as laid on. Thus paternalistic acts, although altruistic, always carry such self-conceits as "We know better than you do" and "We know what is best for you" and such zealous imperatives as "We must save you, even from yourself if necessary." Nevertheless, paternalism for

---

14. *Kreps, p. 12.*

15. *This discussion is indebted to points made by T. Halper, "Paternalism and the Elderly," in* Aging and the Elderly, *ed. S. F. Spicker, K. M. Woodward, and D. D. Van Tassel (Atlantic Heights, N.J.: Humanities Press, 1978), pp. 321–339.*

the elderly involves a conflict between the state's duty to protect its citizens from harm and the citizen's right to autonomy.

ARGUMENTS AGAINST PATERNALISM. Several arguments have been advanced against paternalism for the aged. First, it is claimed that, although some elderly cannot or will not perceive their own interests, the vast majority can and will if given the opportunity. Indeed, some claim that the elderly themselves are in a better position to perceive their interests than anyone else. No matter how altruistic individuals or the state may think they are, they do not have the personal knowledge that the aged have of themselves. This supposedly explains why paternalistic policies have had such mixed results. A good example is Social Security. Although in some cases Social Security may have removed the threat of destitution in old age, it has also contributed to the transfer of responsibility to provide for the aged from family to government. As a result, today it is much easier for a family to shunt aside its moral obligations to aged relatives, thus hastening their isolation from the family. Furthermore, to some extent Social Security has quickened elderly people's withdrawal from the economy, for the law limits the amount of money one can earn without severe restrictions on Social Security benefits. So, at least in the case of Social Security, what is intended as altruism might in fact be separatism.

Related to this argument against paternalism is a second one: Paternalism often masks self-interest, plain and simple. In other words, authorities and individuals may help the elderly not for the sake of the elderly but for their own. A simple example can be seen in the family that simply does not want to be bothered with tending to an elderly relative. Captalizing on the existence of a public or private institution that ostensibly exists for paternalistic reasons, the family institutionalizes the older person. In this way, both the authorities and the family can delude themselves that they acted for the person's own good. In fact, they acted to unload a problem.

A third argument contends that, by attempting to sanitize the lives of the aged, paternalistic acts may deprive them of the very challenges they need to enrich their later years. It is curious how glibly we concede the importance of challenging people at every earlier stage of life but then make old age the exception. In fact, talk of challenging the elderly more often than not is received with skepticism, even incredulity. The implication, of course, is that the challenges of adjusting to change, developing talents and skills, meeting new people, and effecting different lifestyles are suitable for the young and middle-aged but not for the elderly. But why not? Are not challenges a fundamental part of continued functioning? Don't all people need challenges in order to keep growing and to find meaning in life? If yes, then no matter how well-intentioned our paternalism, it may do considerably more harm than good to the elderly. Just as bad, it may be reinforcing us in an invalid and pernicious concept of what is proper to old age.

These three arguments are essentially consequential; they object to effects of paternalism. There is another common argument against paternalism for the elderly that is strictly nonconsequential. This view objects that paternalism violates the principle of political equality by subordinating the individual to the state in matters that are strictly personal. Moreover, by blurring, sometimes even obliterating, the distinction between public and private spheres, paternalistic policies undercut privacy. Thus paternalism, where it effectively denies elderly people's right to manage their personal affairs, is thought to infringe on their status as rational, moral, and autonomous beings.

ARGUMENTS FOR PATERNALISM.   First, paternalists concede that some old people may know better than anyone else what is in their best interests. But many do not. What is more, many who do are in no position to act on that knowledge. For example, old people who realize precisely what they need to survive may not have been prudent enough to ensure the necessary resources in their younger years. Worse, given our labor structure, they may no longer be able to find work to satisfy their needs. Thus, programs like Social Security's Old Age and Survivors' Insurance Program are geared to ensure that the older person's maintenance needs will be met. Although such programs do infringe on autonomy, paternalists ask what good maximum self-determination is if it does not lead to happiness. Paternalists who argue this way clearly value freedom insofar as it promotes happiness. They conclude, therefore, that paternalism is justified, even obligatory, when it will maximize the happiness of the elderly. Determining precisely when it will is, of course, problematic.

Second, paternalists again concede that in some instances families and society do use paternalism to mask selfish interests. But they claim that these isolated occurrences should not obscure the fact that countless pitfalls lie in the path of any citizen in a society as highly complex as ours. The dangers are compounded for the aged, who may not have provided for old age or whose knowledge and skills have become obsolete. Such realities, claim paternalists, justify paternalistic policies.

Third, paternalists claim that critics draw too sharp a line between the private and public spheres. Again, given the complexity of our society, it is not easy to distinguish the private sphere, from which government coercion presumably would be excluded, from the public sphere, within which some coercion presumably would be allowed. As a result, it is not always easy to separate the selfish act from the altruistic one. For example, a Medicare regulation requires three days of hospitalization as a precondition for reimbursement of subsequent home health services. Such a policy is designed in part to protect the aged from the consequences of their ailment being underestimated. At the same time, the regulation serves the interests of other parties. Physicians, for one, are protected from the legal consequences of too hasty a dismissal of the complaint; other HCPs are protected from spotty and uncertain employment resulting from an underuse of their services; family and friends are protected from the disruption of personal lifestyles that may result from having to provide close home care and supervision. The point is that, given society's pluralistic interests, it is very difficult to separate a policy's self-serving aspects from its altruistic purposes. Indeed, the challenge to justice is to draft policies that protect the elderly while harmonizing with the needs and interests of other parties.

Fourth, paternalists argue that, generally speaking, old age is so fundamentally different from other stages in life that the argument to allow the elderly continued challenges loses its force. The combination of physical and mental impairments, the frequent social displacement, and the diverse psychosocial crises that inevitably beset the old impair the elderly person's capacity to deal with challenges the way that the more youthful ordinarily can. And it is claimed that, within the context of enlightened paternalistic policies, the aged still retain a large measure of autonomy. In fact, when freed from debilitating financial and medical worries, the old can pursue enrichment activities that might not have been within reach before.

Finally, although the rights to equality, autonomy, and privacy are fundamental, paternalists claim that these rights do not mean much to those plagued by a variety of physical, mental, and psychosocial problems they are ill equipped or unable to manage. Often such conditions so enshroud the elderly that they are powerless to pursue loftier ideals.

Before turning to the internal paternalistic policies from which paternalists often seek to protect the elderly, it is important to crystallize the broad philosophical question in the debate between paternalists and antipaternalists: Just what should be the limits of society's influence and authority over the individual? As we saw in Chapter 6, answering this question calls for drawing a line between individual and societal authority. Where that line should be drawn is a basic question of social philosophy. No attempt will be made here to settle the question, but it seems safe to say that we must draw the line anew according to the tugs and pulls of historical circumstance and changing social conditions. If for no other reason than their uniqueness as source people, gerontologists and HCPs who work with the aged would seem to have a special obligation to help us draw the line between authority and the elderly. The moral dimensions of this challenge loom even larger when the issue involves protecting the elderly from internal dangers.

Internal Dangers

So far, we have considered paternalism as a way to protect the elderly from external threats, such as destitution. But the elderly are also vulnerable to internal threats that result from mental deterioration or disability—what is commonly termed *mental incompetency*. It is one thing for the state to withhold a specified amount of people's earnings in order to protect them from destitution in old age; it is quite another to consign them to a mental institution or nursing home because they are considered mentally incompetent. What makes this later kind of intervention so serious morally is that it strikes at psychic privacy, the privacy of the inner sphere, of one's thoughts, ideals, ambitions, and feelings.

Paternalistic acts that aim to protect the elderly from the dangers of mental incompetency take numerous forms. In the private sector, as adults grow older, younger people are inclined to treat them as not fully capable of making decisions and managing affairs as they once did. Thus it is not uncommon for the adult children of the aged to impose their own will on their parents in such things as where, how, and with whom their parents will live, what they will continue to own or divest themselves of, and what kind of medical treatment they will have. The rationale behind this intervention is that, left to their own devices, the elderly would make imprudent decisions and thus harm themselves.

This tendency to restrict the autonomy of the aged can also be seen in the behavior of HCPs who are inclined to deal almost exclusively with the adult children in determining what is in the best health care interest of the aged parents. Just as they deal with minors through the parents, they deal with elderly parents through the adult children. Often such HCPs join forces with the children to pressure elderly relatives into compliance. In particular, HCPs and adult children sometimes make decisions about life-sustaining medical technology without even consulting the elderly patient.

In addition to these sorts of interventions, there are legal proceedings intended to challenge the elderly person's competence to write a will or care for person or property. If someone believes that an old person can no longer properly care for self or property, that person may file a petition with the court to have the old person declared incompetent and to have a guardian appointed to act on the person's behalf. Such hearings always bring into stark conflict the individual's right to self-determination and the state's duty to protect the vulnerable—that is, autonomy versus paternalism.

The use of expert testimony is crucial in presenting evidence of incompetency at a trial. Generally the testimony of HCPs is sought, especially that of physicians. In order to show mental incompetency, a medical expert has to establish not only that some kind of mental incapacity exists but also that it is the cause of previous "unreasonable" conduct or may be the cause of future "unreasonable" conduct. But because legal competency tests are not concerned with the physical condition of the brain, physicians are allowed, even encouraged, to state their nonmedical opinions about the behavioral competency of the person.

At this point serious questions arise about the relevance of medical testimony. Do physicians have any greater claim to such expertise than a lay person? Legal expert Ronald Leifer puts the problem this way:

> The psychiatrist is trained primarily as a medical physician. His expert status is based on two factors: first, his training in the sciences of anatomy, physiology and chemistry as they are specifically applicable to the detection and alteration of physico-chemical events in the bodily machine; second, his authorization by a civil authority to practice these skills. The legal sanctioning of the physician-psychiatrist is based on his knowledge of the methods of physics and chemistry. However, the methods of communication and not the physico-chemical methods of medicine are utilized in the investigation and description of human affairs. It is interesting to note that no legal test of competency mentions the physical condition (of the brain) as a criterion. . . . It is clear that the psychiatrist's primary claim to expert status, his medical training, is irrelevant in the determination of competency.[16]

As a concrete example of Leifer's point, suppose a physician were asked to judge whether an aged person had the mental capacity and ability to handle an estate of considerable size. To answer that question, the physician must deter-

---

16. R. Leifer, "The Competence of the Psychiatrist to Assist in the Determination of Incompetency: A Skeptical Inquiry into Courtroom Functions of Psychiatrists," Syracuse Law Review 14 (1965): 572.

mine what capacity or ability it takes to manage the estate and whether the person has that capability. But what is there in the physician's medical experience that qualifies him or her as an expert in answering such a question?

A number of factors prompt questions about the appropriateness of medical testimony at incompetency hearings. For one thing, if the medical testimony is irrelevant, then the old person is being evaluated on spurious grounds. At the very least, this practice raises serious questions of justice. More important, much is at stake for the person—not only from the view of losing autonomy but also of losing needed protection. An elderly person certified as competent may all the same need protection in certain areas. Since the potential exists for great harm, the issue is of serious moral concern. Finally, and most important from the view of HCPs, there is the question of personal obligations in such cases. These obligations must be viewed in light of the preceding discussion. HCPs may be morally justified, even obligated, to offer expert testimony in competency cases. But where they are asked to offer testimony based on nonmedical standards, their real obligation might be to refrain from offering such inexpert opinion lest it bias the evaluation of the person's competency.

Determining one's moral obligations in competency hearings requires more than just considering the relevance of one's testimony. It also calls for a deep understanding of and sensitivity to the impact of an incompetency declaration. To illustrate, suppose a son has reason to suspect that his aged father is about to squander away a fortune on some misbegotten venture. He files for and is granted a declaration of incompetency. Precisely what has occurred?

From the paternalistic viewpoints of the son and the court, the old person is now protected from foolishly frittering away his fortune. But there is more. Someone else's judgment about how the father will dispose of his holdings has been substituted for the father's own. If the case involved a minor, then one could argue consequentially that such paternalism maximizes the child's potential and preserves the child's future opportunities. In other words, the intervention is justified because eventually there will be a payoff for the child. But this argument does not apply to the elderly. If anyone's potential and future opportunities have been served, it has been somebody else's—the son's, in this case. Certainly there is no payoff for the father. Thus a declaration of incompetency, although perhaps protecting the elderly, also substitutes someone else's judgment for their own at a time in their lives when self-determination in such matters as how they want to spend the money they have worked hard for may be one of the few pleasurable prerogatives left to them.

So, although paternalism applied to the external realm raises moral questions, paternalism applied to the internal realm raises far more serious ones. In applying paternalism to protect the elderly from outside threats, we make judgments about what the elderly will and what they should will and end up exercising control over a relatively small and possibly inconsequential portion of their lives. But by applying paternalism to the internal, psychological realm, we deny that the elderly person has even the capacity to reason, which implies that the person is unable to will anything meaningful. Thus we exercise complete control over the person, effectively denying the opportunity and right to change lifestyles, alter lifelong behavior patterns, or live by different sets of values. This raises as basic a moral question as imaginable: the denial of the existence of a reasoning self and the substitution of another's judgment.[17] In determining obligations in competency cases, then, HCPs have much to consider that is of moral significance. Such concerns provoke additional questions. An important one relates to the institutions in which many of the elderly willingly or unwillingly live out their final years.

Old-Age Institutions

No discussion of morality and the elderly would be complete without some mention of old-age institutions (OAIs), which include nursing homes, rest homes, and homes for the aged. Although only 4 percent of the elderly are in a convalescent care home at any given time, the number of people over sixty-five in OAIs has increased dramatically in the past twenty years. Several factors help to account for this: the impact of federal welfare programs, the growth in number of the elderly population, and the decline of the extended family and other changes in family living arrangements.[18] Moreover, public welfare programs for the aged will probably help to increase the numbers of institutionalized elderly further. For one thing, having more money, the elderly will be able to obtain relatively acceptable institutional care. For another, with a supplementary allowance, the economically marginal elderly will be tempted to live alone. Current provisions of the Supplementary Security Income program, in guaranteeing a minimum income for the elderly, reduce the allowance by one-third if the elderly live in the household of a relative. Probably, then, the pool of economically marginal elderly living apart from families will increase. At the same time, the loosening of family ties and their economic marginality make these elderly high risks for future institutionalization.[19] Given this scenario, the numbers of the institutionalized elderly are likely to increase, which is all the more reason to be concerned with their plight.

Many of the moral problems that surround OAIs grow out of their social organization. Individual institutions do differ, but they seem to share a basic social organization in terms of the socioeconomic status of patients, the prognosis for patients, and staff makeup and relationships. A consideration of these factors will reveal how easily the elderly can be abused and denied basic rights.[20]

The Socioeconomic Status of Patients

Studies done on OAIs overwhelmingly indicate a correlation between patient socioeconomic status and (1) the likelihood of being institutionalized and (2) the quality of care received if institutionalized. Specifically, elderly white

17. *Halper, p. 330.*

18. *U.S. Department of Health, Education, and Welfare,* Medical Care, Expenditures, Prices, and Costs: Background Book, *pub. no. (SSA) 74–11909 (Washington, DC: Government Printing Office, 1973), p. 68.*

19. *B. B. Manard, C. S. Kart, and D. W. L. van Gils,* Old Age Institutions *(Lexington, Mass.: D. C. Heath, Lexington Books, 1975), pp. 107–131.*

20. *C. I. Stannard, "Old Folks and Dirty Work: The Social Conditions for Patient Abuse in a Nursing Home,"* Social Problems *20 (Winter 1973): 329–342.*

people in relatively good health who have money, friends, and family are far less likely to be institutionalized than those who do not fall into these categories. If elderly people with these social and economical advantages are institutionalized, chances are they will end up in a private nursing home rather than in a public mental hospital or other institution. Of those in private nursing homes, the majority are in those homes with the best resources and thus receive the most attention.

Given this connection between the quality of OAIs and the socioeconomic composition of the residents, serious questions arise about the adequacy of health care for the socially disadvantaged elderly. For example, nursing homes certified by Medicare as extended-care facilities are generally thought to be better in terms of professional staff and provisions for intensive nursing care than homes that do not meet Medicare requirements. As a result, the elderly poor needing institutional care often find themselves in small, old, and substandard OAIs. Furthermore, they are frequently placed in facilities located farther from their families and previous dwellings than the affluent elderly are.[21] Compounding matters is the staff's attitude, which often reflects a general sociocultural bias against the poor and powerless. Thus, far too often, staff unconsciously equate poverty with lack of worth, which can lead to high-handed and abusive interactions with patients.

The Prognosis for Patients

A second important social factor in the structure of most OAIs is the prognosis for patients, which is inevitably pessimistic. There is no question that the elderly in OAIs are dying, at least in the sense that they will never get out of the institution. Thus, for the elderly who must be institutionalized and for their families, OAIs represent the end of the line, the ultimate defeat in the battle to maintain social independence. Staff generally shares this perception. More often than not, they receive an admission to a nursing home as they would a terminal disease. Their job is to preside over the ebb of life. Looked at biologically, this view is altogether realistic. But by taking a strict biological approach, staff can easily misinterpret one of the psychological stages that many people facing death pass through: bargaining.

Chapter 9 investigates the psychological phases that the dying often pass through. Here our concern is that, in facing their own imminent death, individuals sometimes engage in "bargaining," which is an attempt to prolong life by striking an agreement, a bargain, with God or fate. The terminal cancer patient might bargain, "If I can just live until Christmas" or "If I can just live to see my daughter's graduation from high school." As some gerontologists and students of the social organization of OAIs have observed, bargaining is often the main source of social life for the elderly in OAIs.[22] But in this context, bargaining is not

---

21.   J. I. Kosberg, "Differences in Proprietary Institutions Caring for Affluent and Nonaffluent Elderly," *Gerontologist* 13 (1973): 304.

22.   E. Gustafson, "Dying: The Career of the Nursing-Home Patient," *Journal of Health and Social Behavior* 13 (1972): 226–235.

so much for the preservation of biological life as for the social and psychological self. Thus bargaining in OAIs generally centers around such internal matters as one's self-image and self-identity. Specifically, the elderly in OAIs bargain for moral and social support for their self-esteem. They want authorities and peers to reassure them, by word and deed, that they continue to be vital and of importance to other people. The case of Mr. Winter is a graphic example. The institutionalized elderly want to feel that social death will not precede their physical death.

It is common for staff who fail to distinguish the social phase of dying from the biological to remain unresponsive to the real meaning of the institutionalized elderly person's bargaining. Worse, if they misinterpret the bargaining, staff members can misinform, lie, and offer outlandish hope about the patient's biological condition. The irony here is that by misreading the meaning of the bargaining, staff members impale themselves on the horns of a needless dilemma. They feel they must either be ruthlessly frank and thus, perhaps, not compassionate or compassionate and thus, perhaps, not truthful. But when the bargaining is seen for what it is—a plea to be reassured that the social self will not die before the biological self—then HCPs can take measures within the context of institutional care to reassure the elderly of their social worth. They can design a variety of activities, exercises, and challenges that will provide the reassurances the elderly need, to be offered when they need it and for as long as they need it.

During the social stage of dying, which can consume most of the elderly person's institutionalized time, the aged are fighting an understandable and justifiable battle against the tendency of relatives, staff, and society to force them into a premature social death. Rather than supporting the elderly in this fight, OAIs frequently thwart them by segregating them from highly trained professional staff, labeling their conditions irreversible, and denying them opportunities for mingling with other patients. Such rejection really amounts to socially structured dying. And as some authors have noted, "socially structured dying may be as deadly to the mind and emotion of a patient as irreversible biological deterioration is to the body."[23] In the last analysis, then, whether or not staff in OAIs meet their obligations to provide adequate health care will largely depend on the efforts they make to preclude the elderly person's premature social death.

The Staff

There are a number of observations about the staff of OAIs that bear on social structure. Like all generalizations, the following allow exceptions for the many dedicated, compassionate, and effective individuals who work in institutions for the aged.

Over the years, a number of studies have reported that the people who work in OAIs, including the professionals, tend to occupy marginal positions in the

---

23.   W. H. Watson and R. J. Maxwell, Human Aging and Dying (New York: St. Martin's Press, 1977), p. 132.

labor market. Indeed, the ones with the most client contact, aides and orderlies, occupy some of the lowest positions. Also, given their lower social-class origins, they often share a latent culture that regards force and aggression as effective and acceptable means of resolving conflicts.[24] Similarly, because of their limited formal and professional education, they lack the understanding of the elderly, of human motivation and mental illness, that is needed to make OAIs responsible and effective health care centers for the aged. Probably, then, staff members with the most patient contact will interpret and respond to patient behavior in ways that reflect an amalgam of misconceptions and stereotypes more than any sound health care philosophy.[25] Consequently, these aides or orderlies are likely to rely on what is for them a well-established means of meeting difficulties: force. Given this picture, one can understand how patient abuse occurs.

The problem of staff incompetence is made worse by the insulation of professional from nonprofessional staff. For example, it is common for the registered nurses who run the daily affairs of OAIs to provide little or no direct patient care themselves. Given their burdensome administrative chores, they have to rely on others, usually aides and orderlies, to provide treatment. Too often, as a result, the institution takes on a custodial ideology: Patients are assumed mentally enfeebled and in need of control by drugs and restraints; care is equated with tending to body needs, with keeping patients and institutions clean. One author has succinctly described how this condition arises and how HCPs in OAIs become estranged from the noblest ideals of their profession:

> The professionals and semiprofessionals who work in these institutions are the less successful members of their professions. Work in custodial mental hospitals and nursing homes does not bring professional recognition and is regarded as a step down by their professions in general. Once in these institutions, they find themselves with patients they cannot help, confronted by staff problems which make it difficult or impossible to achieve the goals expounded by their professions. The lofy goals of help and service learned during their professional training give way to more realistic goals of custody and order maintenance. Rather than taking active leadership in caring for patients, they withdraw from this aspect of their role, become cynical, and concentrate their attention and energy on activities which reduce their contact with patient and lower-level employees. Patient care becomes the almost exclusive province of the lower-level employees to whom the professionals delegate a great deal of discretionary power. This insulates the lay perspectives of the lower-level employees from the more sophisticated and potentially ameliorative ideas of the professionals.[26]

The real danger in such a context, as this gerontologist points out, is that supervisory and professional staff will seldom witness abusive behavior by other

---

24. H. S. Becker and B. Geer, "Latent Culture: A Note on the Theory of Latent Social Roles," Administrative Science Quarterly 5 (September, 1970): 304–313; and M. D. Blumenthal, R. L. Kahn, F. M. Andrews, and K. B. Head, Justifying Violence (Ann Arbor: University of Michigan, Institute for Social Research, 1971).

25. A. L. Strauss, L. Schatzman, R. Bucher, D. Ehrlich, and M. Sabshin, Psychiatric Ideologies and Institutions (New York: Free Press, 1964).

26. Stannard, p. 341.

employees. Worse, they frequently develop a "culture of accounts" to deal with reports of abuse. *Culture of accounts* refers to the social structure that prevails in OAIs, characterized by a mind-set about the basic attributes of the work force and the patients. Thus, from the viewpoint of the professional, both employees and patients show many discrediting attributes that make them untrustworthy and unreliable—indeed, attributes that explain why they are in the institution to begin with.[27] Given this definition of institutional reality, professionals can conveniently meet reports of abuse with counterclaims about the person making the report. It has even been suggested that OAIs require this culture of accounts. Professionals who remain in the institution are socialized into the culture. Those who cannot or will not accept the prevailing concept of institutional reality must deal with a torturous dissonance between their self-image as professionals and the realities of institutional life. Unable to preside over the corruption of their professional ethics, they usually quit.

There are undoubtedly many selfless, committed, and unheralded individuals working in OAIs, but the social structure described here is apparently pervasive enough to cause moral concern. For one thing, it raises serious questions about how well such institutions are delivering or are capable of delivering health care to the elderly. Not only may the atmosphere be inhospitable to proper care, but it may be downright conducive to patient abuse. Then there is the issue of professional ethics. How can the well-intentioned, conscientious, and principled employee maintain ethical standards in a climate of suspicion, distrust, and hostility?

Postscript: Reevaluating the Structure

What are the health problems of the elderly? HCPs who are serious about meeting their obligations to the elderly must at some point ask this question because the answer to it will largely determine HCP-elderly interactions. As we saw in Chapter 4, medicine has traditionally allowed its preoccupation with disease to determine its conception of a proper health care delivery system. Thus, "health problems of the elderly," as with health problems for anybody, have been equated with disease. But there are a number of other problems the elderly face that, at least according to some gerontologists, fall under the heading of "health problems." Such problems might be termed functional as opposed to medical. Failure to consider functional problems as health problems may result in a system that does not meet the real health care needs of the elderly at all. If so, then HCPs and society in general have profound obligations to reevaluate the whole structure of our health system as it applies to the elderly.

Gerontologist Stanley Brody has been one critic of the health care system as it now exists.[28] A decade ago, he delivered a paper at the twenty-fourth annual

---

27. *Stannard, p. 335.*

28. *S. Brody, "Comprehensive Health Care for the Elderly: An Analysis," in* Social Problems of the Aging, *ed. M. M. Seltzer, S. L. Corbett, and R. C. Atchley (Belmont, Calif.: Wadsworth, 1978), pp. 195–203.*

meeting of the Gerontological Society in which he argued that the elderly have a number of functional problems that are an integral part of their health problems. By "functional problems," Brody means physical limitations, mental impairments, and environmental hazards that limit the elderly person's ability to function.

For example, studies indicate that physical disability is a correlate of aging. In other words, the older a person is, the more likely that person is to suffer some kind of limitation on mobility. Thus many of the noninstitutionalized elderly are housebound or report difficulty walking on stairs. Mental impairment also immobilizes a large number of the elderly. As with physical disabilities, rates of psychosis, somatic illness, and organic mental disorders increase with advancing age. Similarly, certain environmental hazards can impair the function of the elderly. Lacking adequate income, many of them must occupy substandard housing located in areas with limited transportation. As a result, they do not have ready access to the medical and social services they need. In some cases, fear of robbery and attack further isolates them from the sources of care and service.

Clearly, then, the elderly in our society are subject to high risk stemming from various functional disabilities. Whether physical, mental, or environmental, these disabilities seriously impair the quality of the aged person's life. In Brody's view, a proper health care system must address these quality-of-life problems. But this is unlikely to happen as long as we remain disease-oriented, for disease-orientation stresses the quantity of life to the exclusion of the quality of life.

Brody's plea for restructuring of the entire health care delivery system takes root in the World Health Organization's definition of health, which is characterized by the ideals of "complete physical, mental and social well-being," not merely the absence of disease or infirmity. Following this line, Brody has argued for a "continuum of health care" that would include social services equal to medical services. Brody envisions five categories of health-social services. One would consist of services keyed to personal hygiene. A second would include supportive or extended medical services, which recognizes the need for various HCPs to implement physician orders, observe patients, and report back to physicians. A third category would be maintenance services: housekeeping, food preparation, environmental hygiene, and the like. A fourth would consist of counseling, which would include listening skillfully, extending help, mobilizing existing resources, and enabling the use of these services. Fifth, Brody lists a category termed "linkages," which would include any service that helps connect the elderly to the needed assistance. He feels that these services are vital to adequate health care for the aged.

The basic issue here is one we encountered earlier: the proper concept of health. Whatever the concept, it largely determines the nature of the health care delivery system and dramatically affects the care provided the elderly. What health care system is proper for the elderly? This is a question that HCPs who take their duty to provide adequate health care seriously must surely ask themselves. And of course, that question implies another: What is health?

The issue of HCP advocacy for the elderly ties into the answers to these key questions. If a thorough investigation leads one to conclude that our working definition of health, and the resulting assumptions imbedded in the health care system, do not in fact allow us to meet the health care needs of the elderly, then we have an obligation to help change that system. Presumably this would entail advocacy not only at the professional level but also at the legislative level, for as Brody points out, "the nature and number of [the elderly person's] problems are beyond individual and family resources, thus requiring public coordination and support through services and programs."[29]

Summary

This chapter was concerned with the HCP's responsibilities to the elderly. We saw how many of the moral problems involving HCP interactions with the aged grow out of misconceptions about aging and the aged. These false perceptions and myths give rise to and are further reinforced by ageism, which is discrimination against the elderly. Undoubtedly, paternalism directed toward protecting the elderly raises fundamental questions about the proper limits of personal and state interference in the lives of the elderly. Paternalism aimed at walling off the elderly from internal threats raises the most serious moral concerns, for at stake is the individual's very capacity to make any decision. In addition, we encountered three factors of social organization that raise many moral issues in the operation of old-age institutions. These factors are patient socioeconomic status, patient prognosis, and staff quality and relationships. Finally, this chapter suggested that the question of adequate health care for the elderly is inseparable from certain assumptions about health embedded in the health care system. If the elderly are not receiving adequate health care, then the problem is likely to be systemic. If so, changes need to be made, and presumably HCPs have an obligation to advocate them.

Ethical Theories

Is ageism ever morally justifiable? No question that deals with morality and the elderly is more basic than this one. If ageism is never morally justifiable, then individually and collectively we should never deliberately discriminate against a person solely on the basis of age.

To put the issue concretely, let's suppose that an automobile accident has occurred. Two victims are rushed in an ambulance to the emergency room of a nearby hospital. One is a thirty-five-year-old man, the other a man seventy-three years old. Both are unconscious and show signs of head injuries. A staff member barks, "Take the younger one. He's got more to live for." Assuming that the proposed action is a form of ageism, is it morally justifiable?

---

29. Brody, p. 197.

Egoism

In theory, egoism could defend ageist practices on the basis of best long-term self-interest. On the other hand, the same consideration might lead to a condemnation of a specific ageist act. Whatever their position in an individual case, egoists would find nothing inherently objectionable about ageism.

As for the case under discussion, egoists could not approve or disapprove staff action without introducing the interests of the particular moral agent in question. But who would this be? The old man? The younger one? Staff? An individual staff member? In the last analysis, egoism lacks the theoretical structure to determine whether or not staff acted rightly. Evidently, all egoists can say is that the morality of a case like this depends on whose viewpoint one takes. Egoism appears powerless to arbitrate the issue.

Act Utilitarianism

Like egoism, act utilitarianism would not support an unqualified position on ageism. If an ageist practice would be likely to produce more social good than any other alternative, then it would be good and right. If not, then it would be bad and wrong. Where act utilitarians objected to a discriminatory act, it would not be because the act violated a basic human right or because it was inherently unfair. Remember, the act utilitarian calculus is based on social utility and not on any inherent characteristic of the act. Human rights and unfair treatment would enter the calculation only with regard to the likely consequence of supporting or abridging them. Thus any appeal to principles of justice would be based on efficiency, not fair play.

Was the staff's action right or wrong then? The answer would depend on a calculation of the consequences. If the action produced more total good than any other alternative, then it would be right; if not, then it would be wrong. Whatever the ultimate verdict, act utilitarianism clearly provides a theoretical basis for acts of ageism.

Rule Utilitarianism

Rule utilitarianism would also base its moral decision about ageism on determination of the social good. But it would do so by appealing to the rule under which the particular practice or act fell. Thus the action of denying the old man immediate attention only because he was old might fall under such a rule as "Never deny a person an equal chance for emergency medical treatment on the basis of age." If following this rule would be likely to produce more favorable consequences for all concerned than breaking it, then the staff acted immorally. But would it? It is impossible to say with certainty, although a powerful consequential argument undoubtedly could be made for such a rule. As for ageism itself, again rulists would inquire about the social utility of the rule under which ageist acts fell. Would a rule like "Never engage in ageist acts" be likely to produce more social good than its contradiction, which would permit at least some ageist acts? On first look, an affirmative answer may seem apparent, but on closer inspection we might see that endorsing such a rule would presumably prohibit all preferential programs for the aged, since these victimize members of other age groups—for example, by making fewer resources available to them. So it appears that rule utilitarianism does not provide a clear-cut answer to the issue of ageism or to the morality of staff's action. But again, in theory, rule utilitarians undoubtedly could provide a basis for ageist practices.

The Categorical Imperative

Kant would consider ageist practices incompatible with the inherent worth, dignity, and equality of all rational creatures. Any maxim that would legitimize such practices would not be universalizable, because it would be inconsistent with the nature of rational, autonomous creatures. It seems, then, that staff acted immorally in discriminating against the old man. It is important to clarify what Kant would object to here. He would not necessarily question staff's choice but would question the way it was made. Had staff chosen the young man on the basis of some kind of lottery, for example, that would have given each candidate a roughly equal chance to be chosen, then Kant could approve the determination. But the facts indicate that the selection was strictly ageist. It is this ageist criterion that Kant would find immoral. (We will encounter more about determining the allocation of resources in Chapter 14.)

Prima Facie Duties

Undoubtedly Ross would condemn ageism as injuring individuals and groups, as well as insulting the individual's basic right to fair and equal treatment. This means that, at least in theory, ageist practices would not be justifiable. However, other prima facie duties might override one's duty not to engage in ageist practices. Whether those duties prevailed in the case under discussion is uncertain. If staff did not even consider other prima facie duties, then it acted immorally. But at least in theory, staff could have acted morally if a consideration of all the prima facie duties indicated that the action taken was an actual duty. But then its action could hardly have been termed ageist.

The Maximin Principle

There is no doubt that Rawls's maximin principle would condemn ageism. Just consider the original position. Knowing that they could be victimized by such practices, rational, self-interested creatures would hardly condone them. Thus ageism appears to be a clear violation of Rawls's equal liberty principle. If this interpretation is correct, then staff acted immorally.

Roman Catholicism's Version of Natural-Law Ethics

Roman Catholic moral theology would disapprove of ageism as incompatible with the equal treatment persons deserve. All humans are equal in the sense of having been made in God's image, of having inherent worth and dignity, of deserving respect and basic rights. One of these is the right to be given equal consideration under similar circumstances, of not being discriminated against unfairly. Ageism, then, would be viewed as unfair discrimination and therefore a violation of God's will as expressed in natural law. If this analysis is correct, then it follows that staff acted immorally.

Clearly the nonconsequential theories are more decisive in their views of discrimination against the aged than consequential views are. However, no attempt should be made to extend a theory's condemnation of ageism as traditionally practiced to a condemnation of "reverse ageism"—that is, preferential programs for the elderly at the expense of other age groups. Indeed, it is entirely possible that, although a nonconsequential theory unequivocally opposes discrimination against the aged, it may support preferential treatment of the sort described. Rawls's ethics is an example.

One could argue that, in Rawls's original position, reverse ageism would be condemned as a species of ageism. However, Rawls's difference principle allows for unequal treatment as long as everyone benefits by it or at least as long as the most disadvantaged people benefit. The elderly surely qualify as being among the most disadvantaged in the scramble for health care. Since they would be likely to benefit from preferential health programs, it seems that at least in theory Rawls's maximin principle could condone the practice. This conclusion is strengthened when one introduces Rawls's principle of paternalism: Those in a position of authority are obliged to introduce the interests of those unable to support their own.

This general caution about glib extensions of nonconsequential positions on ageism to positions on issues involving preferential treatment for the aged applies with equal force to Kant, Ross, and Roman Catholic moral theology. We need not consider more about this point here, since we will consider it fully when we examine the plight of the poor who need health care.

As always, the preceding analyses are incomplete and tentative. They are intended to chart general directions of ethical theories on the issue of ageism. Certainly much more needs to be done before the full force of these views can be brought to bear on this issue or others involving the health care of the elderly.

CASE PRESENTATION
A Tragic Flaw

It's hard to know where to begin telling about a tragedy. Perhaps with the tragic hero. In this instance that would be Ralph Waldo Anderson, brilliant jurist and legal scholar. Maybe you've seen his name in the papers or heard it mentioned on the news. He's been associated with an array of celebrated cases over the years. Take my word for it that, in the legal profession, Ralph Waldo Anderson is a giant.

I had the profound good fortune to learn the law under Professor Anderson.

A sharper, more incisive man I've never known in my own thirty years of practice. And to be sure, I've never known a man more true to his namesake, for Ralph Waldo Anderson was nothing if not self-reliant. Indeed, his sense of independence bordered on a conceit.

After leaving law school, I lost personal touch with "R. W.," as we used to call him behind his back. Losing contact like that hurt, because I always felt a mixture of adulation and affection for the man and wanted very much to see him again. Today that yearning was quelled—but the pain remains and, I expect, will grow.

It seems that in the years since I last saw him, through his eighty-second birthday, R. W. remained extraordinarily active. Then he began to suffer the complications of diabetes, something he'd evidently had for many years. By his eighty-fourth year he had lost his sight and hearing. As a result, he retreated to his upstairs study, withdrawn from the outside world, and from Kate, his wife of over fifty years, who herself was hobbled by severe arthritis and the effects of a mild stroke.

About two years ago, R. W. developed total heart block. His pulse dropped to about thirty beats a minute, and he couldn't even get up in a chair for meals. He was totally bedridden. His doctor recommended a pacemaker, but Kate refused permission. Apparently the doctor pressed her. Probably lacking the strength to resist, she reluctantly agreed. Had there been children, it might have been different, but the Andersons were childless.

So R. W.—the man who had once prided himself on being the last of a vanishing species of "autoless birds"—was taken by ambulance to a hospital and admitted for cardiac monitoring of his heart block. After several days of observation, the pacemaker was implanted. Sure enough, R. W.'s pulse rate picked up to a normal seventy. Otherwise, his condition remained unchanged.

R. W. was then moved to a nursing home. Initially Kate objected, insisting that R. W. would be dead-set against the idea. But again, professional wisdom triumphed, and Kate acquiesced. Perhaps it was for the best, for within the year Kate died of another stroke.

You're probably wondering how I know all this. Well, this afternoon I chanced upon a mutual friend of R. W. and mine who filled me in. As it happened, he was visiting R. W. at the very nursing home where I was visiting my mother.

My friend told me that I would never recognize R. W.: "They have him diapered and turn him from side to side every so often on this waterbed mattress to prevent bed sores." He went on in such graphically ugly detail that I felt faint by the time he concluded, "But I'll be damned if his pulse isn't just humming along at a nice, normal seventy."

I didn't have the courage to look in on R. W. Maybe some day I will. But now I'm beset with thoughts of the tragic flaw that brought Ralph Waldo Anderson to this pathetic end and of those who conspired in his undoing.[30]

---

30. *This account was inspired by a true case reported by L. S. Baer,* Let the Patient Decide: A Doctor's Advice to Older Persons *(Philadelphia: Westminster Press, 1978), pp. 69–70.*

Questions for Analysis

1. *How do you feel when you hear of something like this case? Angry? Fearful? Depressed? Despairing?*

2. *List the various values embedded in the happenings reported here. Does any single value seem to have won out? If so, how do you feel about that, and why?*

3. *An earlier chapter mentioned some of the negative side effects of the medical commitment to technology. Are any evident here? Do you think that technology may pose a special threat to the elderly? Explain.*

4. *Do you think that R. W.'s autonomy was in any way directly or indirectly violated?*

5. *The narrator strongly hints that R. W. shares responsibility for his plight. Do you agree?*

6. *To what extent do all of us exercise control over the conditions that will surround us should we end up institutionalized in our old age?*

7. *Suppose you were part of the pacemaker team. Would you try to convince Kate that the procedure should be done? Suppose most of your colleagues were for it. Would you acquiesce, or would you feel obliged to advocate Kate's right to decide?*

8. *Suppose you were the physician trying to decide whether to implant the pacemaker. Apply each of the ethical theories to your decision. Show what each theory would emphasize and likely prescribe as a moral course of action.*

Exercise 1

The following is a statement of the conditions for nursing home admission taken from a typical form:

> Medical and Surgical Consent: The Patient is under the control of his attending physicians and the facility is not liable for any act or omission in following the instructions of said physicians, and the undersigned consents to any medical treatment or services rendered the patient under the general and special instructions of the physician.

At least one physician, Louis Shattuck Baer, clinical professor of medicine emeritus at the Stanford University School of Medicine, considers such a statement far too broad and permissive. In its place he recommends a document that, in part, includes the following directives:

> If at some time in the future I am admitted to a nursing home, *and in the absence of my ability to give directions concerning my care,* it is my intention that this directive shall be honored by my family and physician(s) as the final expression of my legal right to refuse medical and surgical treatment and accept the consequences of such refusal.

I consent only to nursing care, pain medication, and needed sedation.
NOTHING ELSE!

All other medicines of any type are STRICTLY FORBIDDEN!

I prohibit any type of medical or surgical treatment UNDER ANY
CIRCUMSTANCES!

If I sustain a fracture or any other injury, I will allow sedation, *ample* and
frequent pain medicine, immobilization, and ABSOLUTELY NOTHING ELSE!

Diagnostic tests, except the required chest x-ray, ARE NOT PERMITTED!

I may be assisted to drink, food may be offered, BUT SPOONFEEDING IS
FORBIDDEN![31]

Would you favor such a revision? Would you sign it? How would the various
nonconsequential theories evaluate the morality of signing such a document?

Exercise 2

Reconsider your responses under Values and Priorities. Do you remain
satisfied with them?

---

31. *Baer, p. 150.*

Nursing Homes for the Aged: The Human Consequences of Legislation-shaped Environments

Jeffrey Wack and Judith Rodin

---

*The following article, authored by two psychologists, attempts to show that nursing homes are not
set up to meet the social and psychological needs of the elderly. On the contrary, they seem to foster
psychological and economic dependency.*

*Wack and Rodin point out that current federal legislation has established nursing homes as
part of the medical health care system rather than as community-based institutions. As a result,
federal regulations allocate resources and provide services in a way that emphasizes meeting the
medical needs of the elderly, often at the expense of their social and psychological requirements. In
addition, the authors indicate that the quality of life in nursing homes receiving public funds has
been determined largely by the kind of bureaucracy mandated by federal legislation and the
supporting state standards of operations required for certification. The resulting organization, in
the authors' view, constricts the range of options open to the elderly in such settings.*

*Granting the accuracy of this account, serious moral questions arise about whether nursing
homes are meeting or can meet the elderly person's rights to adequate care and autonomy. Just as
important are questions about whether HCPs working in nursing homes can meet these patient
rights, given the concept of the nursing home as part of the medical health care system. If such a
concept precludes meeting the needs and rights of the institutionalized elderly, then HCPs must
consider their obligations to advocate changes.*

From Journal of Social Issues *34, 4 (1978): 6–21.*

The social and economic opportunities available to any group of people in this society depend not only on their own resources, capabilities, and aspirations but, as importantly, on the resources, capabilities, and aspirations that the public attributes to them and thus permits. Unfortunately, the public at large views most older people as passive, sedentary types who have lost the openmindedness, mental alertness, and efficiency of the young and who are beset with economic problems, poor health, and loneliness (Harris, 1975). However, while the young may prefer to picture older people off by themselves, spending a good deal of their time sleeping, sitting and doing nothing, or nostalgically dwelling upon the past, many elderly people do not want to be excluded from happenings around them. Like the young, three out of four people 65 and over say they would prefer to spend most of their time with people of all different ages, rather than with people their own age only (Harris, 1975).

Fortunately, the lives of most elderly people can continue in the community where they can interact with people of all ages. However, more than 1.5 million of the elderly population are institutionalized in 20,000 nursing home settings across the United States; for most of them it is the last place that they will live. This paper explores the legal and socioeconomic factors that influence the development and operation of these institutions and examines their impact on the psychological and physical condition of those who live there.

To consider this issue we must acknowledge that whether elderly persons wind up in a nursing home or not, the transition from adulthood to old age is often perceived as a process of loss, physiologically and psychologically (Birren, 1958; Gould, 1972). However, it is as yet unclear just how much of this change is biologically determined and how much is a function of the environment. The ability to sustain a sense of competence in old age may be greatly influenced by environmental factors which in turn may affect one's physical well-being. In this connection one of the most debilitating environments available may be the current type of nursing home.

Nursing homes, broadly defined, have undergone profound changes over the past few decades. They have changed not only in size and number but also in the perception of their purpose

and function. Many of these changes merely parallel the altering social and demographic profile of the United States. Prior to 1930, the few ill or needy aged not cared for by their families were sustained by almshouses or poor farms, supported by local civic or religious organizations. As a result of the Depression, the federal government became involved in public relief for the aged for the first time and the Social Security Act, passed in 1935, made provisions to pay for care in proprietary boarding homes for the elderly. With increasing longevity, there was increasing need for more facilities to care for the aged. Further pressure for these facilities arose because urbanization made it less likely that families had sufficient room for elderly family members to live at home. In response to these changes, the first major legislation to provide substantial federal money for nursing home care was passed in 1951. With the legislation and the money came numerous regulations regarding these institutions and their structure, regulations largely intended to prevent abuses in payment and thus concerned with mandating which institutions were qualified to give care rather than how persons should be treated. This intent was explicit in the Medicare and Medicaid legislation passed in 1965, which created incentives for a vast and sudden increase in the number of nursing homes because the allocation of large sums of federal money made them lucrative businesses.

Before Medicare and Medicaid legislation, the nursing home was viewed as a residence for the elderly person who, primarily for economic and social reasons, could not live independently in the community. While government data indicate that this type of demand is still substantial,[1] current federal legislation has established nursing homes (or in federal jargon "long term care facilities") as a part of the medical health-care system rather than as community-based residences. The associated regulations allocate resources and provide services in a manner that emphasizes meeting the medical requirements of the individual, often at the expense of his social and psychological needs. The result is a setting that is optimal for those whose medical problems make them physically or emotionally dependent upon medical assistance but unnecessarily restrictive for the vast numbers of nursing home residents capable of greater degrees of independence.

The poor conditions of some long term care institutions qualifying for public funds received widespread publicity in the early 1970s. As a result, the nursing home is now one of the most closely watched and regulated institutions. There are currently at least five federal agencies, ten regional offices of Long Term Care Standards Enforcement, and six congressional committees and subcommittees having the concerns of the elderly as their charge. The executive branch has organized White House Conferences and Presidential Commissions concerning aging and institutionalization. Recent years have seen the passage of three major pieces of legislation and numerous amendments to legislation as well as frequent additions, revisions, and reinterpretations of standards governing nursing home regulation.

Despite all of the human energies directed at them, few are satisfied with the quality of life experienced by the aged in these settings. Federal and state administrators complain of the patchwork development of recent laws and the lack of any omnibus bill covering the aged in general and the role of the nursing home and support systems for the elderly in particular. Nursing home administrators complain that their staffs are swamped by the paperwork required to keep the institution in compliance with regulations and thus cannot adequately care for the residents. We share the concern that the quality of life lived by those in many nursing homes is not what it could and should be. In part this is due to the very nature of institutionalization, the effects of which have been discussed at great length elsewhere (Goffman, 1961, 1963; Sarason, 1972). However, the quality of life that can be provided in nursing homes receiving public funds has also been determined in large part by the kind of bureaucracy mandated by federal legislation and supporting state standards for certification.

To understand the nature of the legislation and its impact on the institution and the individual, we will consider how the system of care and nursing home environment shaped by federal legislation restricts the range of options open to the individual. Next we will discuss the consequences of these constraints for his or her psychological and biological well-being. Finally, we will consider an alternative public policy that addresses the needs of the individual rather than institutional efficiency.

The Legislation-shaped System

The Medicare and Medicaid programs were, in part, motivated by concern for those persons who would require a long period of convalescence to regain their health. Health policy planners recognized that people at various stages of recovery would not only overload and inefficiently utilize the full-scale medical services that hospitals provided, but that hospital care would be expensive as well. The decision was to reimburse the costs of care in nursing homes, which would provide less sophisticated medical care. It is thus not surprising that the resulting eligibility standards for nursing homes were a modified version of the standards that had been in use in the accreditation of hospitals.[2] The personal consequences of this move for the individuals' nonmedical needs were never considered. Instead, licenses were granted to nursing homes that made available the required services and met guidelines established by the Code of Federal Regulations. Further legislation in 1972 (Public Law 92-603) developed the concept of "level of care" by allowing Medicaid to cover care in a new category of institutions of still less skilled medical and nursing care, known as the intermediate care facility. The concept is that people are matched to the type of facility that best meets their assessed medical needs.

The difference between the skilled nursing facility (SNF) and the intermediate care facility (ICF) is in the sophistication of medical care they are qualified to deliver. The SNF provides services most like those found in an acute-care hospital, including seven-day-a-week nursing, and emphasizes rehabilitative services. The ICF provides "on a regular basis, health-related care and services to individuals who do not require the degree of care and treatment which a hospital or skilled nursing facility is designed to provide but who because of their mental or physical condition require care and services (above the level of room and board) which can be made available to them only through institutional facilities."[3]

Overall, the balance of services that must be provided by the SNF and ICF weighs heavily towards a view that none who enter a facility will ever improve significantly. Standards for the ICF require that services must be provided such that "residents are transferred promptly to hospitals, skilled nursing facilities or other appropriate facilities" (Code of Federal Regulations, 1976, p. 82) as changes occur in their physical or mental condition necessitating service or care that cannot be adequately provided by the facility. There is no explicit specification for movement in the other direction, no mandated mechanism for returning residents to the community if they improve. Furthermore, the legislation determining services in SNFs and ICFs conspicuously lacks the active treatment and explicit release conditions, which are an important element in the regulations governing intermediate care facilities for the mentally retarded, for example. Indeed, the tone and language of the legislation and Code of Federal Regulations is overly fatalistic, reflecting an attitude that entering a nursing home signals the beginning of the inevitable end for the elderly person.

Most disturbing, however, is that the focus of the regulation is on the institution and not on the individual, and the dictated standards have resulted in a rigid environment appropriate only for persons with medical requirements and insufficiently flexible to accommodate individual differences in need and independence. In fact, the regulations may have unwittingly minimized the possibility for planning and independent choice that could be as critical to recovery rate and health as medical care per se (Janis & Rodin, in press; Johnson, 1975).

Thus the Medicare and Medicaid legislation has had a major effect in reconceptualizing the functions to be served by the nursing home. No longer an extension of the family home for those who lacked such assistance, the nursing home has become an extension of the medical delivery system and participates in its structure of incentives. The implicit demands made by the licensing legislation have created profound changes in the quality of life of nursing home residents now caught in the health-care delivery system. How does the elderly person get "caught" in this system to begin with?

Routes of Entry into a Nursing Home

There are two means by which people 65 years or older in the community may find their way into a nursing home. The first is to be struck by an acute illness necessitating hospital care that then qualifies them for a Medicare-covered stay in the nursing home. Between 1964 and 1974, the percentage of nursing home residents entering the facility from a hospital increased from 12% to 35%. A second route of entry opens for individuals in the community whose physical condition declines over a period of weeks, months, or years, so that at some time they meet the Medicaid criteria.

The Medicare program requires three days in an acute-care hospital before the costs of a nursing home stay will be covered. After that period, the physician may decide that a patient no longer requires the intensive care of a hospital and may authorize moving the patient to an SNF. Medicare then provides assistance for up to 100 days of inpatient care in the SNF. After the first 100 days, patients may be discharged if they are well or need some minimal additional care that could be given at home. Medicare will cover a designated number of home health care visits by nurses, assuming that there are agencies in the person's community able to provide these services, that the physician and discharge planners are aware of them, and that such services are judged to be reasonable alternatives to institutional medical care. However, mostly as a result of restrictive state eligibility requirements and inadequate financial incentives, home health care is not yet a viable option;[4] it constituted only 1% of total Medicaid expenditures in 1975.[5] Thus the simplest alternative is to make patients eligible for Medicaid coverage and to keep them in an SNF or ICF.

To obtain Medicaid coverage for a nursing home stay, federal law mandates that both need for medical care and financial eligibility conditions be met. The financial eligibility determination is the source of the infamous "spend down" requirement. In assessing eligibility for Medicaid benefits, any income (e.g., Social Security benefits, Supplemental Security income) and all assets including the savings and home of the individual are taken into account. The individual must then cover his/her own costs of institutional care until he/she has "spent down" all resources to the state-determined level—in general, an amount adequate to cover burial costs.[6] With an estimated average annual cost of $7300 for a nursing home stay in 1975,[7] the spend down may not take long. Statistics indicate that over 47% of nursing home patients whose costs were paid by Medicaid were not initially poor by state definitions but depleted their resources and became qualified as "medically needy."[5] Thus the legislation, which at least acknowledges the possibility of discharge from the facility, also requires that recipients spend themselves into poverty. While they may become sufficiently independent medically to be capable of maintaining themselves in the community, they are no longer in a financial position to do so. They have become trapped by the system presumably set up to protect them.

There are other disincentives to returning individuals to living in the community. One is the simple economic fact that the proprietary nursing home, where over 70% of the institutionalized aged reside,[1] requires a certain occupancy rate to operate profitably. While it is likely that a resident discharged could be replaced by one admitted, there are disincentives in terms of staff time and paperwork in making arrangements for the discharge and the new admission. It is easier, and more profitable, to keep a resident in the home as long as possible.

Some Effects of the System on the Individual

To provide a background for understanding the bureaucratic structure of the present-day nursing home system we have discussed one of two main converging forces that result in keeping the elderly in nursing homes once they have entered: the eligibility requirement itself. The second factor is that the system created by the legislation fosters both dependence and a loss of individuality, so that an elderly nursing home resident comes to view himself as helpless and unable to return to the community.

It is significant that rather than having the beneficiary paid directly by the fiscal intermediaries and in turn billed by the nursing home for services delivered, the basic costs of care are paid directly to the facility on the basis of the number of beds occupied by Medicare and Medicaid beneficiaries during the payment period, regardless of who these individuals are. Similarly, ancillary costs are generally broken out and reimbursed as costs per service and, hence, do not provide information as to which residents are being provided particular services. While these

procedures expedite the cost-payment cycle, they bypass the actual consumer, eliminating any control residents might exert on the character of the nursing home through monetary incentives. The reimbursement methods have further served to slant the view of nursing homes toward monetary issues rather than the life of each individual. Thus the resident has been relegated to a passive and dehumanized role in the system.

The nursing home environment unwittingly accelerates the processes of deindividuation, leading to feelings of loss of personal identifiability (Zimbardo, 1970). As greater numbers of people require treatment, names are replaced by ID numbers or, even worse, by ailments. Residents of nursing homes are usually called patients and are often referred to as "the stroke at the end of the hall" or "the coronary who came in yesterday." Persons come to be thought of as their medical problems rather than as unique individuals. In addition, most problems related to health arouse some degree of fear and uncertainty. Thus most contacts with health care professionals in the nursing home occur in an emotionally charged context. High degrees of arousal have been shown to increase feelings of deindividuation (Zimbardo, 1970). Moreover, the aged person who is ill may have a distorted sense of the present, as the future seems bleak and the past, when he or she was well, no longer seems relevant. A narrowing of temporal perspective is another factor that increases feelings of deindividuation (Zimbardo, 1970). As forces in the system contribute to a feeling of lost identifiability and individuality on the part of the aged resident, stress may be added to the original medical problem.

A major source of the helplessness and hopelessness one senses in the nursing home stems from the medical model of care from which the services provided in the SNF and ICF are derived. The model is based on assumptions that are sometimes inappropriate for the care of the aging person and result in a mode of nursing home life that can be detrimental. One assumption is that the person is suffering from an aberrant condition that is curable. The debilitation occurring with normal aging, however, is often not curable in the typical sense, and the duration of stay in the nursing home is generally years. Because of this, medical practitioners, trained to expect cures, come to feel that their treatments are largely ineffective.

Hence, most professionals view work in nursing homes as unrewarding and demoralizing. They then come to treat the aged residents as though they were helpless and in a hopeless situation.

Adoption of the medical model of care for the nursing home has had many other ramifications. One is that the nursing home is continually having to justify its belonging to part of the medical care and reimbursement system. Consequently, there already exists a tendency to overdiagnose and overmedicate. Furthermore, the criteria for discharge are biased toward the convenience of medical care delivery and protection from injury, rather than taking a broader view of psychosocial needs and feasibility of care via other services in the community.

The medical model even determines perceptions of what constitute medical problems as well as the methods for their resolution. Social psychological problems are often misinterpreted as medical symptoms requiring a medical solution. For example, the origin of a bout of crying may be an acute environmental cause, such as an argument with a good friend. However, medical personnel in the nursing home are trained to see crying as a symptom of depression and, therefore, to use drugs as the appropriate treatment. Similarly, in a medical setting, the death of a patient is viewed as a failure and the concern is to maintain the image of infallibility by keeping information of the death from other residents. However, for the aged, death is a natural and inevitable outcome and should not be treated with shame. In contrast to the medical model, the social psychological perspective holds that the most appropriate method for dealing with the death of a person is to acknowledge the event with mourning and respect.

Perhaps the most disturbing ramification of the medical model is that the setting fosters and maintains the dependency of the nursing home resident. Within the treatment mode of the medical model, the person is expected to relinquish much personal control to those medical personnel who provide treatment. The role of patient requires the person not to engage in any activity without the permission of the physician. Information and decisionmaking, even over day-to-day activities, is the domain of doctors and nurses rather than the individual and his or her social network. The result is increased passivity allowing

little opportunity for, and indeed discouraging, the residents' further growth. Life in the nursing home thus represents a radical departure for most people. Even the most humane institutions constrict the individuals' choices by treating people as ill, effectively isolating them from the community, and providing a relatively uniform, inflexible environment within which to function. In learning to deal adaptively with this setting, elderly people may begin to lose self-management skills, or at least to believe that they have (Felton & Kahana, 1974). Once they no longer feel able to care for themselves, they may be reluctant to leave the nursing home.

The bureaucratic structure and staffing patterns of nursing homes further increase deindividuation and dependence. Most of the residents' contact with staff is with low-paid, poorly educated paraprofessionals who perform custodial functions such as bathing, dressing, and assistance in using the toilet. They see the residents at those times when they are indeed most dependent and develop attitudes and behavior toward them that are often authoritarian and condescending. These responses then generalize to the rest of the day's activities as well. The professional staff, such as nurses and social workers, are reduced largely to administrative functions by the federal code, which requires them to keep lengthy medical and social service records on each resident for inspection by federal and state teams. They therefore not only have little time to interact with the residents, but are more familiar with them as cases and records than as people. Finally, we must note that, as in many other institutions, nursing homes are simply easier to run when the residents are docile and dependent. Hearing complaints and finding solutions takes time, giving choice creates the need for more options and is potentially disorganizing. Thus there are clear bureaucratic incentives for induced dependency.

Reversing the Trend to Dependency

We believe that structural changes could be made that would increase independence among elderly nursing home residents without creating bureaucratic havoc. Working with a Connecticut nursing home, Langer and Rodin (1976) designed an intervention to encourage elderly nursing home residents to make a greater number of choices and to feel more control and responsibility for day-to-day events. The study was intended to determine whether the decline in health, alertness, and activity that generally occurs among the aged in nursing home settings could be slowed or reversed by choice and control manipulations that have been shown to be beneficial in other contexts (Ferrare, 1962; Schulz, 1976; Seligman, 1975). The data indicated that residents in the group given more control and responsibility became more active and reported feeling happier than a comparison group of residents who were encouraged to feel that the staff would care for them and try to make them happy. Patients in the "responsibility-induced" group also showed significant improvement in alertness and increased behavioral involvement in many different kinds of activities, such as movie attendance, active socializing with staff and friends, and contest participation.

In addition to multiple questionnaire and behavioral measures at the time, long-term follow-up data were collected on several variables, including mortality (Rodin & Langer, 1977). A doctor on the nursing home staff evaluated the medical records of each resident for two periods: the six months prior to the first study in 1974 and the six months that immediately preceded the follow-up. While there was no significant difference between the groups in the preintervention health evaluations, the responsibility-induced group showed a significant increase in health at the end of two years. There were also death rate differences between the treatment groups: By the time of the follow-up, twice as many residents in the comparison group, as compared to the responsibility-induced group, had died. Because these results were so startling, other factors were assessed that might have accounted for the differences. Unfortunately one simply cannot know everything about the equivalence of these subjects prior to intervention, but it is clear that, when the study began, those who died did not differ significantly from those who lived in age, length of time they had been institutionalized, or overall health status.

We are not suggesting that it is universally beneficial for people in nursing homes to feel increased responsibility and independence. Feelings of control may be stress-inducing, especially when the individual believes that there are responses that he or she can and should be making but is not (Averill, 1973). This may be especially true if the legislation-mandated structure of the nursing home prevents or restricts the opportunity for choice and control: Enhancing the residents' sense of independence without creating real changes and options for exerting control can be expected to have profoundly debilitating effects. Thus it should be clearly noted that the Rodin and Langer studies were conducted in a setting that was extremely open and responsive to change, to the extent that it was possible. We believe, however, that the legislation-mandated system restricts the flexibility of even the most progressive nursing homes and thus the real focus of initial change must be in the structure and not in the individuals—residents or personnel—who are simply caught in the system.

Considerations for a Person-Focused Public Policy

Before proposing specific changes, we must acknowledge that, at the most basic level, what need to be changed are many of society's values and attitudes. This is critical since ultimately social attitudes pervade the policy-making process and shape legislation, and legislation often determines the nature of institutional settings that service a large segment of American society. Like all legislation-shaped environments, nursing homes are structured to provide services according to mandated regulations based on economic considerations rather than psychosocial factors. The bureaucracy functions in the most cost-effective way in terms of dollars and perhaps the least effective way in terms of human needs and values. Since the legislation establishes means for supervising the institution and evluating outcomes at the level of institutional operation, rather than individual benefit, it provides bureaucratic incentives for induced helplessness and dependency.

Legislation should be changed to evaluate the performance of the setting in terms of the gains of those serviced by it.[8] This would implicitly reflect the modal aged person as one capable of and desiring to make decisions for himself or herself, an active consumer of services rather than a passive recipient of welfare. Such changes would result in alterations beneficial to the aged at all levels of the system. It would focus the legislation, services, and finances on the individual rather than on the institution. The nursing home would become more responsive to the individual needs of its residents. Finally, aged residents' control over their environment would be enhanced and their sense of competence and well-being maintained.

Current legislation also reflects society's sense of hopelessness for those in the nursing home. It even makes the perception of hopelessness reasonable by giving little consideration to the needs of those who have the potential to return to the community. Indeed, in mandating a system of payment that results in poverty and a setting that fosters psychological dependency, it guarantees that a return will not happen.

Bureaucratic structure would have to change if legislation supported a view that persons entering the nursing home may once again return to their homes in the community. It would have to establish mechanisms to keep the person financially secure so that return to the community would remain an option. This would mean eliminating spend-down requirements or making grants available to those who could return to living in the community. In addition, the nursing home would no longer view itself as merely a caretaker facility but instead would be active in maintaining the residents' community contacts, preparing them psychologically for a return, and becoming more involved in managing moves both into and out of the nursing home. It would also have to reorient its intra-institutional incentive structure to encourage residents to remain as independent as possible. For example, nursing home personnel could be trained to reinforce those who exhibit initiative and/or creativity.

Due to the complexity of federal regulations, it is often easier for institutions serving the elderly to obtain federal money than it is for community-based service agencies. However, recently the Administration on Aging, the Department of Housing and Urban Development, the Social Security Administration, and the Public Health Service have independently financed programs and services such as home health care, adult day care, congregate housing, foster care, and meals on wheels. Similar programs have been developed by state agencies and community governments. But these services are poorly organized and not well publicized. As a result, the elderly currently wishing to stay or return to the community face an overwhelming array of fragmented services and financing arrangements. Too few communities

have agencies that might attempt to coordinate these programs into a coherent system that could adequately meet the needs of the elderly. As a consequence, physicians, discharge planners, and offspring have found that placing and maintaining the individual in a nursing home is no longer a necessity but just less complicated. Thus again, a major barrier to independence and personal control is bureaucratic.

Another problem with the current bureaucratic structure is that it is based on a medical rather than social model of health care delivery. This too is a situation fostered by current laws since the incentives for supporting nursing homes derive from an attempt to find alternatives to acute hospital care. Thus to some extent, the helplessness, dependency, and hopelessness produced by the intra-institutional norms and rules of a nursing home are manifestations of the medical model of care adopted for these long-term care facilities. A conception of the nursing home as an extension of the medical setting is appropriate for that minority requiring a relatively brief period of convalescent care following treatment in a hospital. However, the nursing home has also become a long-term residence for many elderly people, for whom the medical model of acute care becomes inappropriate. In planning change, policy makers must recognize that the nursing home today is attempting to serve two very different populations and perhaps find ways for these functions to be divided into different types of facilities.

Moreover, the medical model, which stresses immediate and crisis-based care, decreases the likelihood of following adequate decisionmaking and orientation for the elderly person at the point of entry. While some people come from a hospital and need to be moved quickly, most have the time for more careful planning and staging of the entry process. Since perceived choice and decisionmaking are crucial to the outcome of moves for the elderly (see Rowland, 1977, for review), more explicit attention should be paid to the preparation and participation of the aged person.

If they are able, most older people desire to remain active and contributing members of their families and communities (Harris, 1975). It is thus disturbing that current legislation has resulted in systems that so easily subvert the satisfaction of this desire by unnecessarily isolating the aged from the mainstream of society. Economically, the legislation mandates that they spend themselves into poverty so that a return to the community is no longer an option. Psychologically, the legislation shapes a nursing home environment that fosters helplessness and dependency and makes people feel that they are unable to leave. Not only is a reexamination of the legislation affecting the aged needed but also an entire reconceptualization of the role of the nursing home within a comprehensive and coherent system of services. Enlightened legislation will recognize aging as a process and reflect this process in a system of community-based support that will not force the individual to choose between independence and dependence.

Notes

1. Department of Health, Education, and Welfare. *National nursing home survey, 1973–1974* (DHEW Publication No. 77-1778). Rockville, Md.: Health Resources Administration, 1974.
2. Barker, A. Personal communication, July 8, 1977.
3. Department of Health, Education, and Welfare. *Five years of accomplishments of the office of long term care,* 1971–1976, October 1976, p. 5.
4. General Accounting Office. *Home health: The need for a national policy to better provide for the elderly* (Report No. HRD 78-19). Washington, D.C.: December 1977.
5. Congressional Budget Office. *Long-term care: Actuarial cost estimates,* February 1977.
6. In Connecticut, for example, this level is $850, $600 of which is a "burial reserve." In Maryland, state law requires the person "spend down" to $2500 before he is eligible for Medicaid.
7. Congressional Budget Office. *Long-term care for the elderly and disabled,* February 1977.
8. While this paper was in press, Kane and Kane (1978) proposed a similar idea.

References

Averill, J. R. Personal control over aversive stimuli and its relation to stress. *Psychological Bulletin,* 1973, *80,* 286–303.

Birren, J. Aging and psychological adjustment. *Review of Educational Research,* 1958, *28,* 475–490.

Code of Federal Regulations (Title 45 Public Welfare Chapter II Social and Rehabilitation Service (Assistance) Part 249.12 (2) (1) (ii) (A) (1) (ii). Washington, D.C.: U.S. Government Printing Office, 1976.

Felton, B. A., and Kahana, E. Adjustment and situationally-bound locus of control among institutionalized aged. *Journal of Gerontology,* 1974, *29,* 295–301.

Ferrare, N. A. *Institutionalization and attitude change in an aged population.* Unpublished doctoral dissertation, Case Western Reserve University, 1962.

Goffman, E. *Asylums: Essays on the social situation of mental patients and other inmates.* Chicago: Aldine, 1961.

Goffman, E. *Stigma.* Englewood Cliffs, NJ: Prentice-Hall, 1963.

Gould, R. The phases of adult life: A study in developmental psychology. *American Journal of Psychiatry,* 1972, *129,* 521–523.

Harris, L. *The myth and reality of aging in America.* Washington, D.C.: National Council on Aging, 1975.

Janis, I. L., and Rodin, J. Attribution, control and decision-making in health situations: A social psychological perspective. In G. C. Stone, F. Cohen, and N. E. Adler (Eds.), *Health psychology.* San Francisco: Jossey-Bass, in press.

Johnson, J. Stress reduction through sensation information. In I. Sarason and C. Spielberger (Eds.), *Stress and anxiety* (Vol. 2). New York: Wiley, 1975.

Kane, R. L., and Kane, R. A. Care of the aged: Old problems in need of new solutions. *Science,* 1978, *200,* 913–919.

Langer, E., and Rodin, J. The effects of choice and enhanced personal responsibility for the aged: A field experiment in an institutional setting. *Journal of Personality and Social Psychology,* 1976, *34,* 191–198.

Rodin, J., and Langer, E. Long-term effects of a control-relevant intervention with the institutional aged. *Journal of Personality and Social Psychology,* 1977, *35,* 897–902.

Rowland, K. F. Environmental events predicting death for the elderly. *Psychological Bulletin,* 1977, *84,* 349–372.

Sarason, S. *The creation of settings.* San Francisco: Jossey-Bass, 1972.

Schulz, R. Effects of control and predictability on the physical and psychological well-being of the institutionalized aged. *Journal of Personality and Social Psychology,* 1976, *33,* 573.

Seligman, M. E. P. *Helplessness: On depression, development and death.* San Francisco: W. H. Freeman, 1975.

Zimbardo, P. G. The human choice: Individuation, reason, and order versus deindividuation, impulse, and chaos. In W. J. Arnold and D. Levine (Eds.), *Nebraska Symposium on Motivation* (Vol 18). Lincoln: University of Nebraska Press, 1970.

For Further Reading

Achenbaum, W. A., and Kusnerz, P. A. *Images of Old Age*. Ann Arbor, Mich: Institute of Gerontology, 1978.

Aldridge, J. W. *In the Country of the Young*. New York: Harper Magazine Press, 1969.

Birchnell, J. and Streight, M. E. *Care of the Older Adult*. Philadelphia: J. B. Lippincott, 1973.

Birren, J. E., and Schaie, K. W. (eds.). *Handbook of the Psychology of Aging*. New York: Van Nostrand Reinhold, 1976.

Curtin, S. R. *Nobody Ever Died of Old Age*. Boston: Little, Brown, 1972.

De Beauvoir, S. *The Coming of Age*. New York: G. P. Putnam's, 1972.

Fischer, D. H. *Growing Old in America*. New York: Oxford University Press, 1977.

Huttman, E. D. *Housing and Social Services for the Elderly*. New York: Praeger, 1977.

Kett, J. *Rites of Passage*. New York: Basic Books, 1977.

Kimmel, D. C. *Adulthood and Aging*. New York: Wiley, 1974.

Neugarten, B. L., and Havighurst, R. J. (eds.) *Social Policy, Social Ethics and the Aging Society*. Washington, DC: U.S. Government Printing Office, 1977.

Rosenfeld, A. *Prolongevity*. New York: Knopf, 1976.

Smith, B. Kruger. *The Pursuit of Dignity*. Boston: Beacon Press, 1977.

Woodruff, D. S., and Birren, J. E. (eds.). *Aging: Scientific Perspectives and Social Issues*. New York: Van Nostrand, 1975.

9
THE DYING PATIENT

What follows are extracts from the recorded impressions of a doctor dying of heart disease in the north of Scotland:

> I have no fear of Death itself, but how long can I bear the dying? To be useless, and lonely to the point of agony, that's the fear. I know I'm going down and there is no return. But before this time, I was never bitter. Due to my illness, I have, per force, to spend my time alone, as I can't go out alone; and the loneliness is devastating. I loved my work, and being no longer of any use to anyone is bitter, above all. . . . I get no comfort, no peace in the thought of the God my mother trusted so implicitly. In all her suffering, He never failed her, and she found comfort and sustenance in that belief. I envy that. But I regret deeply, actively, the cruelty of the God who's supposed to be kind. . . . The loneliness is still the worst thing to bear, so far. Total isolation, cold, still, far away from the world now, and encompassed in the silence. I am looking back to that world I used to live in, and observing everything going on normally out there. There is no emotion in that silent steady drop of ice. But where am I? Where am I going? I don't know, I hardly even care. . . . My loneliness may be hard to bear, but who would want to visit such a misery? Those who are well can only enjoy *fun,* it seems. . . . They're busy; maybe they want to come, but they can't find the time. . . . I have just lived too long; I have been discarded. I'm not good company so why should they come? . . . My new General Practitioner doesn't help. . . . I am in continuing pain, but I am non-curable, and so of no interest to him. . . . They have told me that the only possible treatment for my heart would be a cardiac transplant, but my general condition would not allow me to survive the operation. I would gladly take the risk, but I accept their decision. . . . I believe that the fear could be eliminated, if the doctor would only talk it over with us. Little by gently little we could lose the fear of the unknown, and could work *with* him, to face the end. Just some common-sense talk about my needs. Have some regard for my occupation, my life, my activities. . . . Ask my opinion, my views; and let me talk about my death when I want to.[1]

---

1. *From the book* The Facts of Death *by Michael A. Simpson, pp. 50–53. Copyright ©1979 by Prentice-Hall, Inc. Published by Prentice-Hall, Inc., Englewood Cliffs, N.J. 07632.*

Even in these few words, the dying doctor communicates what appear to be main currents in the experience of the dying: loneliness, fear, bitterness, self-doubt. Feelings of uselessness and desertion. Loss of self-governance, control, and participation. Indignity. Her correspondence stands as a touching testimony to human need and fragility; a poignant reminder that one need look no farther than the corner bed to see man's inhumanity to man.

This person's experience is far from an isolated one. The growing literature on death and dying testifies to the near-universality of her experience. Nevertheless, it was not until the mid-1970s, at a workshop called "The Terminally Ill Patient and the Helping Person," that the following Bill of Rights for the dying person was created:

I have the right to be treated as a living human being until I die.

I have the right to maintain a sense of hopefulness however changing its focus may be.

I have the right to be cared for by those who can maintain a sense of hopefulness, however changing this might be.

I have the right to express my feelings and emotion about my approaching death in my own way.

I have the right to participate in decisions concerning my care.

I have the right to expect continuing medical and nursing attention even though "cure" goals must be changed to "comfort" goals.

I have the right not to die alone.

I have the right to be free from pain.

I have the right to have my questions answered honestly.

I have the right not to be deceived.

I have the right to have help from and for my family in accepting my death.

I have the right to die in peace and dignity.

I have the right to retain my individuality, and not be judged for my decisions which may be contrary to beliefs of others.

I have the right to discuss and enlarge my religious and/or spiritual experiences whatever these may mean to others.

I have the right to expect that the sanctity of the human body will be respected after death.

I have the right to be cared for by caring, sensitive, knowledgeable people who will attempt to understand my needs and will be able to gain some satisfaction in helping me face my death.[2]

Almost all these rights could be viewed as subcategories of the four basic categories of patient rights detailed in Chapter 6. They fall largely under the

---

2. *This Bill of Rights is reprinted in M. I. Donovan and S. G. Pierce,* Cancer Care Nursing *(New York: Appleton-Century-Crofts, 1976), p. 33.*

rights to adequate health care, truth and information, and autonomy. As is true of the general categories, unless these rights of the dying are examined within the context of specific health assumptions and care transactions, procedures, and policies, they will remain little more than platitudes.

This chapter will therefore consider still another category of special patients: the dying. It examines HCP responsibilities to the dying within the context of the rights of the terminally ill. Specifically, the chapter engages three issues that raise moral questions: disclosing terminal diagnoses, choosing the place of death, and administering narcotics for pain relief. But before we look at any of these, we should examine two basic conceptual issues. One concerns the definition of death, the other a psychological model of dying. It is also important to note that many other important subjects fall under the rubric of death and dying, among these the "right to die" and issues involving euthanasia. These topics are reserved for Chapter 12, which deals with these broader bioethical matters.

Values and Priorities

Before beginning this chapter, respond to the following questions, which are intended to help you identify your attitudes toward death and dying.

1. If somebody told you that you had just a year to live, how would your life be different?

2. What feelings do you get when contemplating your own death or that of a loved one?

3. Would you rather die before or after the person you love most in the world?

4. Make a list of the various ways people look on death.

5. Our culture has numerous euphemisms for *death*. Does that say anything of our feelings about death?

6. Do you think patients should be told they are dying?

7. Where would you rather die, at home or in a hospital?

8. As an HCP, how do you feel about caring for the terminally ill?

9. Do you think the care of the dying should be left to specialists in the study of death and dying?

10. What meanings do you attach to the phrase "dying with dignity"?

11. Do you think the way you learned or were told about death is probably the best way?

12. Would you want to be told if you were dying?

13. As an HCP who knows that a patient is dying, how would you respond to the person's question, "Am I going to die?"

14. Do you think terminal patients should be allowed powerful painkillers, even though they may become addicted?

15. Would you prefer to die at home rather than in a hospital? Rank the following options in order of your preference. *I would like to die:*
    a. *In my sleep.*
    b. *At home.*
    c. *With my family around me.*
    d. *At a "ripe old age."*
    e. *When I want to.*
    f. *Painlessly.*
    g. *Alone.*
    h. *Suddenly.*
    i. *Only after attaining my life's goals.*
    j. *With all my affairs in order.*
    k. *Only after every measure was taken to sustain my life.*
    l. *Without ever knowing that I was dying.*
    m. *Without any guilt.*
    n. *Only after accepting my imminent death.*
    o. *With religious faith and conviction.*

Defining Death

A terrible auto accident has occurred. One of the cars was occupied by a husband and wife. Authorities on the scene pronounce the man dead and rush the unconscious woman to a hospital, where she spends the next seventeen days in a coma due to severe brain damage. On the morning of the eighteenth day, she dies. Some time afterward a relative, contesting the couple's estate, claims that the two people died simultaneously. Did they?

Not too long ago the answer from most quarters would have been a resounding yes. In fact, an identical case went to the Supreme Court of Arkansas in 1958. The court ruled, however, that since the woman was breathing, although unconscious, she was alive.[3] In making its decision, the court relied on a time-honored definition of death as "the cessation of life; the ceasing to exist; defined by physicians as a total stoppage of the circulation of blood and a cessation of the animal and vital functions consequent thereon, such as respiration, pulsation, etc."[4] By this definition, death occurs if and only if there is a total cessation of respiration and blood flow.

Using heart-lung functioning as a criterion for death served well enough until recent developments in biomedical technology made it questionable. One of these developments is the increasing and widespread use of devices that can sustain respiration and heartbeat indefinitely, even when there is no brain activity. If the traditional heart-lung criterion is applied in cases like the one above, then these individuals are technically still alive. Yet, to many—including relatives of the comatose and those who must treat them—such people are, for all intents and purposes, dead.

---

3. Smith v. Smith, (229 *Arkansas 579, 317 S.W.* (2d 275 1958).
4. Black's Law Dictionary, *rev. 4th ed., 1968, p. 488.*

Another development has complicated further reliance on the traditional definition of death: the need for still-viable organs in transplant surgery. Generally speaking, a transplant is most successful if organs are removed immediately after death. Thus there is substantial pressure on transplant teams to harvest organs as soon as possible. The moral implications of this pressure are very serious, as we will see.

But these developments are only part of what makes the whole issue of defining death such a complex one. Three distinct categories of concerns that can be identified in any discussion of death also figure prominently: the philosophical, the physiological and the methodical.[5]

The philosophical level refers to one's basic concept of death, which inevitably springs from some view of what it means to be human. For example, if we believe that it is the capacity to think and reason that makes a human, then we will be inclined to associate the loss of personhood with the loss of rationality. Similarly, if we consider consciousness as the defining characteristic, then we will likely consider a person to have lost that status when a number of characteristics—such as the capacity to remember, enjoy, worry, and voluntarily will—are gone. Although the absence of rational or experiential capacities would not necessarily define death, they would incline us toward such a definition, since we are already disposed to accept the absence of personhood in the absence of those criteria. In this way, there is some interplay between our concept of what it means to be a person and our definition of death.

A second category of concerns in discussions of death is the physiological, which refers to standards related to the functioning of specific body systems or organs. The traditional physiological standard for recognizing death has been irreversible loss of circulatory and respiratory functions. More recent physiological standards focus on the central nervous system—the brain and spinal cord. Specifically, these standards are the irreversible loss of reflex activity mediated through the brain or spinal cord, electrical activity in the cerebral neocortex, and/or cerebral blood flow. Whether traditional or recent, these physiological standards can be used individually or in combination.

The third category, the methodical, refers to specific methods or means for determining the fulfillment of physiological standards. The method used in determining traditional heart-lung standards has been taking the pulse or reading an electrocardiogram or both. Where the standards focus on the central nervous system, electroencephalographs can be used to measure electrical activity in the neocortex, and radioactive tracers can be injected into the circulatory system for detecting cerebral blood flow.

Brain-Death Criteria

In an attempt to devise a definition of death that speaks to our times and needs, an ad hoc committee of the Harvard Medical School was formed in the 1960s. In 1968 the committee issued four criteria for death that focus on the

---

5. T. L. Beauchamp and L. R. Walters, Contemporary Issues in Bioethics (Belmont, Calif.: Dickenson, 1978), pp. 254–256.

physiological level. The first criterion is unreceptivity and unresponsivity. This means that the individual shows a total unawareness and unresponsiveness to applied stimuli and inner need. The second criterion is lack of movement or breathing. The third criterion is lack of reflex action, as determined, for example, by a fixed and dilated pupil that will not respond to a direct source of bright light. The fourth criterion is a confirmation procedure using an encephalograph to assure the absence of cerebral function. This final criterion can be distinguished from the first three in that the first three define death, whereas the fourth ascertains that death has occurred. Thus, the first three fall into the physiological category, and the fourth into the methodical.

The committee's skillful synthesis of previous proposals has met with widespread acceptance. Most transplantation units and many health care facilities have adopted it. At the same time, some basic assumptions underlying the criteria have proven controversial. One assumption is that, when irreparable brain damage is more or less total to the whole brain, individuals cannot possibly return to spontaneous, respirator-free body activity. This is why the committee's criteria are often termed the "brain-death" definition. Another assumption that gives this one its significance is that spontaneous body activity is a necessary condition for a person to be considered alive.

These assumptions have led to criticisms of the committee's criteria. Some critics reject the committee's reliance on spontaneous respiration, claiming that artifically sustained life is nonetheless life. Others consider the loss of the central nervous system, even of brain function, as irrelevant to the task of defining death. They argue that respiration and circulation are not subsystems that, like the growth of hair or nails, function locally and display biochemical activity for themselves. Rather, they are activities whose function extends throughout the total system and ensures the preservation of other parts. This would make circulation and respiration at least as important as brain activity—maybe more important, since brain activity depends on them. Still others argue that a sharp line between life and death cannot be drawn. Such criticisms usually are part of an attempt to revitalize the heart-lung criteria.

On the other hand, there are those who think that the committee's criteria do not go far enough in making brain state the overriding criterion for defining death. By this account, the primary physiological standard for recognizing death is the irreversible loss of functioning in the cerebral neocortex, the highest region of the brain, wherein resides the capacity for conscious life, which is taken to be the hallmark of personhood. Other functions of the central nervous system, such as the mediation of spinal or brainstem reflexes or the activation of spontaneous respiration, are considered wholly irrelevant in recognizing life and death.

Moral Implications

What makes defining death so important in the study of bioethics is the interplay between definitional and moral considerations. For one thing, definitional problems of death can rarely be discussed apart from moral considerations about what it means to be human and about the implications of a particular definition. And moral considerations about death cannot be discussed without laying to rest the definitional problems.

To illustrate the interplay between definitional and moral problems, consider this example. An attacker has clubbed a woman into a comatose condition. She is rushed to a hospital, determined to have profound and irreversible brain damage, and put on an artificial respirator. Efforts to identify her fail. As the team tending her debates whether to remove her from the respirator, one member, using one of the brain-death criteria, claims that she is already dead. Therefore, withdrawing the respirator poses no special problems. Another member, using the heart-lung criterion, insists that she is still alive and that they have an obligation to sustain her life. What should they do?

An easy answer is to let the law decide. Unfortunately, many states lack an adequate definition of death. Worse yet, some states allow either of the two alternative definitions—one based on heart-lung functioning, the other on brain functioning. Even where the law decisively comes down on the side of one set of criteria or another, moral problems remain about the propriety of the standard itself. Similarly, when the law sanctions a brain-death criterion, it still does not force an HCP or anyone else to implement it. Thus, although brain-death law may legally protect HCPs, it does not obligate them. HCPs, presumably in consultation with others, must still wrestle with moral obligations in cases of irreversible coma.

In addition, there is the phenomenon of organ transplants, which promises to become of even greater concern as technology increases and techniques improve. A number of interests are identifiable in such cases. First there are the interests of the recipients whose welfare depends on the availability of quality organs. Then there are the interests of health teams who are obliged to provide adequate health care, which may include appropriate quality organs. There are also the interests of the donors, who may fear that their organs will be pirated before death or that their own health care providers will perform less than adequately in trying to maintain their lives. Similarly, there are the duties and obligations of health teams to guard donors against physical violations as well as the psychological threat of violations. Compounding these problems are the interests of the health care community at large, which must be concerned about developing a cannibalistic image. Finally, society must be watchful that the rights of its citizens to protection are not flouted; but at the same time it must ensure that its ill citizens are not denied needed medical care and treatment, which may involve transplants.

So the definition of death, which remains to be refined and agreed on, is clearly fundamental to many of the moral questions raised by discussions of death. Given their intimacy with death and the dying, HCPs should be aware of this conceptual issue, crystallize their own positions, and help society articulate a view that comes to grips with the interests of all the parties involved.

Kübler-Ross's Psychological Model of Dying

Defining death is not the only conceptual issue that bears on HCP interactions with the dying. Another of more immediate impact is the psychological understanding that HCPs bring to their dealings with the terminally ill. One model that has received widespread attention within and outside health care has been advanced by Dr. Elisabeth Kübler-Ross.

Many scholars have written about the process of dying, especially about stages that the dying seem to pass through. An understanding of these stages, and the capacity to respond meaningfully, will in part determine whether HCPs meet their obligation to provide adequate health care to the dying. This is particularly true when "adequate health care" is seen to relate to the dying patient's right to be cared for by caring, sensitive, and knowledgeable people. Thus HCPs must try to understand the needs of the dying and be able to gain some satisfaction in helping others face death. This discussion is based on the assumption that understanding the needs of the dying means understanding the stages they pass through, since these stages grow out of needs for assistance in preparing for death. Furthermore, it is assumed that, in gaining awareness, acceptance, and understanding of each stage, HCPs improve their chances of becoming comfortable with death and thus with the dying. As a result, they stand to gain some personal satisfaction in helping the dying face their death.

A good source of information about the stages of death are the works of Kübler-Ross, who has broken ground in isolating people's reactions to impending death. On the basis of extensive interviews with dying patients, Kübler-Ross has identified five psychological stages that dying people frequently pass through: denial, anger, bargaining, depression, and acceptance.[6]

Denial

In the denial stage, people will not accept the death verdict. "No, not me; it can't be me" typifies the reaction. Almost all people use denial in the first stage of a terminal illness. The extent to which patients will deny death depends on many factors, such as how they are told, how much time they have to acknowledge the inevitable, and how they have been prepared throughout life to cope with stressful situations.

For most patients, the period of denial is generally short-lived, but for others it can be lengthy. For example, upon receiving word from her physician that x-rays indicate a terminal pathology, one woman might declare that someone has made a mistake. "You must have mixed me up with someone else," she insists. Assured that there is no mistake, she sets off on a costly odyssey to find someone who will provide her with reassuring answers. In a case like this, the denial stage may be protracted. What is more, even when she accepts the inevitable, such a person—like other terminally ill people—may continue to engage in denial to protect others: relatives, friends, even HCPs.

As to the terminally ill person's right to be cared for by knowledgeable people, it is important to recognize that people in the first stage of a fatal illness are particularly vulnerable. Indeed, this is why they are defending themselves through denial. As a result, HCPs are in a formidable position to help or hurt. They can injure patients further by insisting on, almost browbeating them into,

---

6.   E. *Kübler-Ross*, On Death and Dying *(New York: Macmillan, 1969).*

an acknowledgment of their fate. But if HCPs can allow these patients their defenses, they might be able to help them adopt less radical defense mechanisms. Being present when needed, allowing patients to talk freely, listening carefully, avoiding judgment, and being warm, supportive, and understanding are all essential in providing care to the dying patient who is in the denial stage. It is just as important to coordinate one's own approach with those of other health team members, in order to avoid transmitting inconsistent information.

Anger

When the stage of denial cannot be maintained any longer, it is usually replaced by feelings of anger, rage, envy, and resentment. Thus, after patients have exclaimed "It can't be me!" and then realize that it is, they often ask, "Why me?" The woman in the previous example—a decent, hard-working widow— might look at an elderly street-corner beggar and think, "As far as I can tell that man is of no value to anybody." And then she wonders, "So why couldn't it have been him instead of me?"

In the anger stage, people frequently project their feelings onto others, such as the family or staff. They blame, complain, and find fault easily. It is tempting for HCPs to personalize such outbursts and to retaliate in kind, thus increasing the patient's distress and frustration. Just as likely, HCPs might react judgmentally, with expressions like "You shouldn't say things like that!" or "You shouldn't behave that way." Such reactions come close to denying patients the right to their feelings—that is, to psychic privacy. Certainly these reprimands effectively deny the patient's need to express feelings of hostility. Caring for the terminally ill in the anger stage, then, requires great patience and tolerance of rational and irrational anger. HCPs can provide such care by overcoming their own fears and dropping their own defenses; by, again, careful listening; and by recognizing the legitimate source of the patient's anger.

Bargaining

Having faced up to the inevitable and unloaded their anger and frustration, the terminally ill then often try to effect an agreement whereby the inevitable may be postponed. Conditioned as we all are to be rewarded for good behavior, to be granted a wish for special services, the dying often engage in bargaining that almost always involves a wish for an extension of life or temporary surcease from pain. Thus the woman in the example might wish to live "just long enough" to see her son graduate from high school the following year or promise to devote her life to God for the comfort of a few painless days. Such bargaining is an attempt to postpone the unavoidable. It includes a prize offered for "good behavior," a self-imposed "deadline" (the son's graduation), and the implicit promise that the person will not ask for more if this one postponement is granted. Furthermore, such bargaining is usually made with God and kept secret, although occasionally it is implied in conversations or revealed in confidence to a chaplain.

Regarding care, it is important for HCPs to recognize bargaining as just as

helpful and appropriate a coping mechanism as denial and anger. More important, since psychological promises may be associated with quiet guilt, HCPs who
brush them away risk harming patients. Speaking to this point, Kübler-Ross
notes an opportunity, even need, for an interdisciplinary approach in the care of
such patients:

> If a sensitive chaplain or physician elicits such statements, he may well wish to
> find out if the patient feels indeed guilty. . . . It is for this reason we found it
> so helpful to have an interdisciplinary approach in our patient care, as the
> chaplain often was the first one to hear about such concerns. We then pursued
> them until the patient was relieved of irrational fears or the wish for punishment
> because of excessive guilt, which was only enforced by further bargaining and
> more unkept promises when the "deadline" was past.[7]

Depression

When the bargaining period ends, depression often sets in. Patients experience the deep psychological pain of grieving, which takes various somatic,
psychological, and behavioral forms. Among the common somatic forms of
grieving are excessive sighing, nausea, loss of strength. Psychologically, grieving can take the form of a loss of reality, feelings of guilt, or emotional distance
from people. Some behavioral forms are restlessness and aimlessness in movement, loss of warmth toward others, and disoriented behavior.

Kübler-Ross differentiates two kinds of depression: reactive and preparatory. *Reactive depression* is related to what has already been lost—a body part
that has been removed surgically, a job lost because of excessive absence, or
symbols of self-esteem that are gone. *Preparatory loss*, the kind the terminally ill
must undergo, relates to the anticipation of losing all love objects and life itself.

This distinction is important, because the appropriateness of care depends
on the kind of depression the patient is experiencing. Suppose the woman in the
example is worried about her family and household and whether her children
are managing without her. There may be good reason for such concern. If so,
health team members such as chaplains and social workers can help lift this
woman's sense of loss by helping to reorganize the household. This is especially
true when children or people who cannot fend for themselves are involved.

In response to preparatory depression, HCPs, like anyone else, are inclined
to try to cheer up patients, to tell them there is nothing to worry about, to urge
them to look on the bright side. Although well-intentioned, such reassurances
may be woefully inadequate for patients who are seriously trying to prepare for a
severance from everything they love. Indeed, the reassuring word often masks
the implicit value judgment that people should not think about their impending
death. Such a judgment, tendered as it is to the most vulnerable under the most
trying of human conditions, raises serious questions about the invasion of
psychic privacy.

Furthermore, because preparatory depression generally is suffered silently,
HCPs who insist on verbally interfering with patients may end up hindering

---

7. *Kübler-Ross, p. 84.*

their emotional preparation for dying rather than enhancing it. "In the prepar-atory grief," writes Kübler-Ross, "there is no or little need for words. It is much more a feeling that can be mutually expressed and is often done better with a touch of the hand, a stroking of the hair, or just a silent sitting together."[8]

Acceptance

When patients have had enough time and help in working through the first four stages, they often reach a stage during which they are neither depressed nor angry about their fate. They will have vented their anger, envy, and bitter-ness and will have sufficiently mourned the impending loss of all in the world that they love. Now they are able to contemplate their death with a degree of quiet expectation. In short, they are ready to accept death.

Patients in this final stage are usually weak and quite tired. As a result, they have a need to doze off for short intervals. The sleep of acceptance is not the sleep of avoidance or of escape from pain that characterizes depression. "It is a gradually increasing need to extend the hours of sleep very similar to that of the newborn but in reverse order."[9]

HCPs sometimes invade this profoundly personal stage of dying by misin-terpreting it. They may view the dying person's behavior as resignation or hopeless "giving up." Consequently, they exhort the person not to despair, to continue to fight the good fight.

Acceptance can just as easily be interpreted as a time when the patient needs to be surrounded by friends and loved ones. In fact, however, acceptance is almost devoid of feelings. The pain has ceased, the struggle is over, the time has come for the final rest before the long journey.[10] Having found some inner peace and acceptance, patients generally wish to be left alone. They do not wish to be agitated by the concerns of others. As a result, in this final stage the dying generally do not desire visitors; nor do they feel like talking when they do have visitors. Sensitive, knowledgeable HCPs recognize such behavior for what it is and respect the dying person's right to the privacy that this stage of acceptance entails. HCPs can also contribute to the comfort and care of the dying by helping loved ones to understand the nature of what is happening, to reassure them that their silent presence and tender touch are enough to communicate the only message that the dying may need: "I'll be with you until the end."

Before this discussion concludes, several obervations need to be made to avoid oversimplifying the experience of dying. First, it is by no means conclusive that there are only five stages that the dying pass through. As Dr. Michael A. Simpson puts it, "There are not five stages, there are three, fifteen, ninety-two, and five hundred."[11] Second, it is not known whether these stages are univer-sally experienced, and certainly they do not have to occur in the sequence

---

8. *Kübler-Ross, p. 87.*

9. *Kübler-Ross, p. 112.*

10. *Kübler-Ross, p. 113.*

11. *Simpson, p. 44*

described. Many people do experience all stages in the order presented here, but the stages can also occur in any order, disappear, and recur. Indeed, it is not uncommon for the dying to move in and out of stages during a very brief time, in the course of a conversation, for instance. Third, there is nothing necessary about these stages, except possibly for some varieties of denial.[12] In other words, these stages are not an inevitable part of dying or of dying "successfully."

In short, Kübler-Ross's model describes five normal and rather common responses to any important loss. They represent very complex patterns of states of knowing and of emotional response. Whether they last moments or weeks, or even occur, depends in part on the individual's personality, approach to life, and behavior during previous periods of crisis and in part on the HCPs understanding, capacity to respond, and skillful management. Furthermore, when these stages do occur, there is no guarantee that the dying will progress smoothly through them toward acceptance. Rather, there is often marked fluctuation, characterized by retreats to denial. The woman in the example might at one moment convince the staff of her open and clear acknowledgment of the nearness of her own death but in the next astound them with chatter that borders on pure fantasy. As Simpson points out, "The variance is merely the surface sign of deeper debate within, of the shifting balance between knowing and needing not to know."[13]

HCPs who miss these points will oversimplify the experience of dying and consequently stereotype the terminally ill. Consequently, they may misinterpret behavior or force it to fit a predetermined model. It is most important to realize that any model, including Kübler-Ross's, is subsidiary to the individual's right to die in a style that typifies his or her way of life. People may die angrily, noisily, melodramatically, even spitefully. Consider this example:

> Maureen, an actress—expressive and flamboyant—did it her own way. She treated Death like an impatient and unreasonable agent whose contract had no escape clause. Accordingly, she'd go if he insisted, but was damned if she'd abandon her usual style. She greeted me each ward round with "Darling! *Do* have whisky! Why ever not?" She prepared for her new role with panache. ("You may not think it's much of a part, darling," I assured her, "but it'll run for *ages!*") She was the center of attention when she died, dead drunk, after a cheerful and amusing afternoon, scandalizing some other people's visitors. It was unexpected in its timing—she could never quite resist jumping her cue if it would steal the scene. A sudden exit. Stage Left.[14]

The point is that it is very easy to compel the dying to conform to our own script for dying, to satisfy our own norms of what constitutes a "proper" death. All HCPs must be aware of this tendency, for it raises serious moral concerns about several rights of the dying, especially the right to autonomy.

So far, the discussion of dying patients has been largely conceptual. We have examined the meaning of death and one psychological model of the stages that

---

12. *Simpson, p. 44.*

13. *Simpson, p. 44.*

14. *Simpson, pp. 44–45.*

the dying often pass through. We have seen in a general way how definitions of death and understanding of the psychological process involved in dying bear on the rights of the dying, especially on the rights to adequate health care and autonomy. Apart from these conceptual matters, there are numerous issues unique to professional interactions with the dying that raise moral concerns. Three of these are disclosing diagnoses to the dying, determining the location of death, and treating the dying with drugs for pain relief. The remainder of this chapter deals with these issues and their moral implications.

Disclosing the Diagnosis: The Right to Information

One central moral question is whether the dying should be told they are dying. What makes this question troublesome from a moral viewpoint is that it raises twin obligations for HCPs in meeting the claims of the dying. On the one hand, HCPs presumably have a general obligation to provide patients with truth and information. On the other hand, they supposedly have an obligation to provide adequate health care.

To grasp the dilemma, one must understand that adequate health care for the dying would be likely to include psychological management. Psychological factors partially determine the measure of comfort that the dying experience in their final days and whether they are able to face death with dignity. But some people cannot successfully cope with the unvarnished truth about their conditions; telling them about the diagnosis may harm them. Since the obligation to provide adequate health care embraces the obligation to refrain from harming patients, it seems that under certain circumstances patients should not be told they are dying. Thus, HCPs often find themselves on the horns of a dilemma: Either they tell the person he is dying and therefore risk injury to him (and thus deny him adequate health care), or they don't tell the person, and therefore deny him the right to truth and information. At the outset, it is important to determine whether this dilemma is a genuine one. If it is not, then perhaps HCPs should be asking some other question, one that better focuses the moral analysis.

To determine whether the dilemma is a real or false one, it is necessary to examine the reasons that are commonly given for not telling patients their diagnoses. There are three chief ones: (1) Patients do not want to know, (2) they will not be able to handle the truth, and (3) they will lose all hope if they know. In some cases, any or all of these claims may be correct, but evidence does not support these as generalizations.

Do patients who are dying want to know? The evidence suggests that they do. For example, when the question is posed in a theoretical way, 80 to 90 percent of those asked say they would want to know if they had a terminal disease.[15] Of course, it could be argued that the affirmative reply can be ac-

---

15. V. A. Gilbertson and O. H. Wangensteen, "Should the Doctor Tell the Patient the Disease is Cancer?" Cancer 12 (1962): 80–85.

counted for by the good health the respondents currently enjoy. But ask them when they are ill, when the jury is still out about the precise nature of their disease, and the result will be the same. Indeed, one study asked 500 people attending a tumor clinic whether they would want to know of a potential fatal diagnosis. About 80 percent said yes.[16] Another study compared 100 patients who had cancer with 100 patients who did not. Over 80 percent of both groups wanted to know if they had inoperable cancer.[17] And on it goes—the evidence indicates overwhelmingly that people do want to know about a terminal diagnosis.

At the same time, evidence indicates that the majority of doctors do not want to tell them. In most surveys, well over 80 percent of the doctors interviewed said they never or seldom inform patients of terminal diseases. And yet, 80 percent of the physicians indicated that they themselves would want to be told if they were similarly ill.[18] Why the discrepancy? The most common and likely explanation is that HCP inability or unwillingness to speak frankly with patients reflects their own unresolved feelings about death more than it does concern for the patient. All the more reason for HCPs to do some hard thinking about their own attitudes toward death as a precondition for interacting adequately with the dying.

Can people handle a terminal diagnosis? Again, evidence suggests that generally speaking they can. For example, some researchers have compared patients who had been told about their incurable illness with those who had not. They found that the ones who had been told were better able to maintain their emotional balance and showed a greater capacity to deal with the stress and tension of progressive deterioration.[19]

Do patients abandon hope when told of a terminal diagnosis? They easily can inasmuch as "terminal" is tantamount to "hopeless." The issue, then, is how to inform patients of the gravity of their condition without undercutting their hope. Perhaps the best and only way to do this is to tell patients not that they are terminal or that the situation is hopeless but that they are very ill and may not recover. The latter locutions leave room for the hope dashed by a statement like "You are terminally ill." In addition, such expressions are more honest. HCPs, after all, are only in a position to confirm the gravity of a disease; when they transmit such valuations as "hopeless," they exceed the limits of their knowledge, authority, and position. Ironically, they may end up violating a fundamental principle of all health care professions: Never do harm to patients. Furthermore, locutions such as "very ill" and "you may die" honestly recognize

---

16.   C. J. Hinton, "Facing Death," Journal of Psychosomatic Research 10 (1960): 22–28.

17.   J. Aitken-Swan and E. C. Easson, "Reactions of Cancer Patients on Being Told Their Diagnosis," British Medical Journal 1 (1959): 779–783.

18.   D. Oken, "What to Tell Cancer Patients," Journal of the American Medical Association 175 (1961): 1120–1128; and B. M. Mount et al., "Death and Dying—Attitudes in a Teaching Hospital," Urology 4 (1974): 741–748.

19.   Simpson, p. 98.

the possibilities of spontaneous remission, an eleventh-hour cure, and a mistaken diagnosis.

Some might argue that merely telling patients they are very ill and may die will also undercut hope. There is no substantial evidence to support this contention. Indeed, kept ignorant, patients seem more likely to be seized with hopelessness under the onslaught of progressive deterioration, endless treatment, and growing incapacity. Despite the conspiracy of silence around them, these events stand as incontrovertible proof that they are gravely ill and getting worse. Indeed, if kept ignorant, patients can find themselves in a double bind. Everything they are going through tells them they are seriously ill, but the HCPs treating them insist that nothing of consequence is happening. It is hard to imagine conditions more fertile for despair.

So the common defenses that many HCPs erect to avoid telling patients of terminal diagnoses simply do not square with the evidence, which suggests that patients generally want to know and can handle the knowledge when it is properly communicated. In fact, there are several good reasons for patients wanting to know, which bear on the HCP's determination of moral obligation in these matters. For one thing, honesty can provide patients and their loved ones with greater mental ease during their final days together. Without disclosure, tension can build between the dying and those around them. Communication can become strained, with an uneasy sense of playing games and of all sorts of subterfuges being erected. The result for the dying can be a terrifying sense of isolation and despair, from which they surely have a right to be preserved. Also, honesty can give the terminally ill an emotional handle in their struggle to accept imminent death. If they know what to expect, they can emotionally brace themselves for it; they can seek emotional help where they need it. Finally, there is the sense of maintaining control over their lives that honesty can bring the terminally ill. Surely the dying have a fundamental right to determine how they wish to spend the time that remains for them. They may want to take care of unfinished business, clear up relationships with others, tend to domestic or professional affairs, seek consolation in philosophy and religion. Whatever the particular need may be, if deprived of the knowledge of their condition, patients are denied the right to manage the time left to them. Their autonomy is undercut.

In concluding this discussion, several facts stand in sharp focus. First, people want to know about their terminal condition. Second, people do not break down or lose control when told. Third, people only fall into despair if they are told that their condition is hopeless rather than serious or grave. Fourth, a number of factors argue for the wisdom of honesty. Furthermore, all information amassed on the dying indicates that, whether informed or not, people know intuitively at some point that they are dying. The question that emerges is when—not if—the terminally ill should be told. The answer clearly involves a thorough and sensitive reading of situational nuance and patient idiosyncrasy. By focusing on these aspects, HCPs probably stand their best chance of satisfying their twin obligations to provide truth and information while providing adequate health care.

Deciding Where to Die: The Right to Autonomy

The central moral question about deciding where to die is whether the terminally ill should be permitted to die at home. In a theoretical sense, the answer must be a simple yes. The reason is that, as autonomous agents who presumably have initiated their own admission into a health care facility, competent patients have both the moral and legal right to sign out of the facility. It can also be argued that, since the terminally ill have the right to decide where to die, others therefore have a duty to respect that right and to permit them to die at home.

But as a practical matter, such a decision is not so simple. For one thing, our society makes it very difficult for people to die at home. For another, other people's rights are involved as well, especially the rights of the dying person's family. Still another reason is that there may be duties that are incompatible with honoring the dying person's right to die at home. For example, a health care facility presumably has a duty to provide the dying with adequate health care. But what if it cannot adequately meet this duty to the patient who chooses to die at home? Clearly, the question of permitting the dying to choose where to die deserves more than a glib endorsement of their right to decide.

Over the years, dramatic changes have occurred in the location of death. At the turn of the century, about 80 percent of all deaths in the United States occurred at home. By 1949 half of them occurred in a hospital, and by 1958 the figure had reached 60 percent. Currently about 60 to 75 percent of all deaths occur in a health care facility, and the proportion is slowly rising.[20] Despite this pattern, most people today say that they would prefer to die at home. At the same time, they apparently do not think others should die at home. Again, as with telling the diagnosis, there is a contradiction between what is thought to be right for oneself and right for others. Why this ambivalence?

Dying at Home

Two reasons not only help explain the ambivalence but also constitute the most common attempts to justify denying the terminally ill an at-home death. First, it is argued that home cannot provide the care they need, that only health care facilities can provide that. HCPs show this institutional bias at least as much as lay people do. In fact, HCPs frequently try to persuade the family that the institution is the proper place for their dying loved ones. Considerations about the best health care interests of the dying thus lead many to view the institution as the appropriate place to die.

Although such considerations are honorable, HCPs and family alike should ponder a number of factors before blindly deciding that the health care institution can provide the best care for the dying. To begin with, in many cases the family can easily acquire the skills necessary for maintaining the comfort of the

---

20. *Simpson, p. 17, and E. S. Sneidman,* Death: Current Perspectives *(Palo Alto, Calif.: Mayfield, 1976), chapter 3.*

dying person in the final stages of a terminal disease. For example, they can learn how to give an injection and how to move and position a person for maximum comfort. They can also rent equipment for home care, such as wheelchairs and beds. Furthermore, they can capitalize on the expertise of HCPs who can provide such vital support services as counseling, teaching, administering parttime care, and tending to more difficult aspects of care. The effort involved in caring for the dying at home should not be minimized, however. Obviously, a commitment to home care does not come cheap, either emotionally or financially. It requires great personal sacrifice and love on the part of the family. Failing to realize this, outsiders can easily judge the reluctant family as cruel and selfish. On the other hand, it is just as easy to overstate the problems that are involved, thus prejudicing a potentially willing family.

Just what are the HCP's and facility's obligations when faced with a patient who wishes to die at home and a family that wants to accommodate their loved one but feels inadequate to the task? It seems that the obligations to provide adequate health care and to honor the autonomy of the patient make it necessary for HCP and facility to assist the family on the patient's behalf. The help might consist essentially of three things: (1) realistically determining whether care can adequately be administered at home; (2) if it can, assuring the family that they can provide that care; and (3) helping the family provide the care, which includes teaching and providing support services. Anything less would seem to constitute inadequate health care for the dying, as well as violate several subcategories of this general right.

There is a second reason why people resist allowing others to die at home. It is claimed that children should not be exposed to the dying and to death, that somehow they will be psychologically injured by the experience. Like the first reason, this argument bears scrutiny.

In reality, any emotional injury sustained by children is probably more attributable to the attitudes, values, and reactions of the people around them than to the fact of the dying itself. Indeed, it can be argued that having the loved one die at home minimizes the emotional damage a child may incur, because there is far less chance that children will be isolated in their own grief. Rather, they will probably be privy to the talks and fears of the rest of the family, which gives them the comfort of shared responsibility and shared mourning. Just as important, a home death can help them view death as a natural and necessary part of life and thus may enhance their own emotional maturity.

In addition to the reservations about care and children, there is another consideration, the central one, in determining the proper place to die. This relates to the advantages, if any, of dying at home. It is small wonder that most people indicate they would prefer to die at home. Home, after all, is a beloved and familiar environment, a bastion of security. Generally speaking, it is the respository of everything that is near and dear to us. Thus dying at home requires far less adjustment for the terminally ill than dying in a health care facility. For some, home may be the only setting that provides enough emotional security to help them find a separate peace before they die.

Surely, these potential benefits should remain central in any determination of the proper place to die. It is easy, however, for family and professionals to forget these in their preoccupation with efficiency and in their tendency to relegate life's "dirty details" to the "experts." Given their background and training, HCPs would seem to have special obligations to avoid falling into such a mind-set and to assist in the education of others, particularly families, so that they can make informed, intelligent, and thoughtful decisions. On a broader level, the question of where to die raises considerations about professional and moral obligations to help investigate alternative ways of providing terminal care.

Alternative Care: The Hospice Approach

Of all the alternative care systems for the dying, the hospice approach has received more attention in recent years than any other. Hospices are special settings devoted exclusively to care of the terminally ill. Some are separate agencies unaffiliated with hospitals; others are special units within hospitals. The most publicized hospice is St. Christopher's in London, which was founded in 1967 by Dr. Cicely Saunders. Home care teams in the United States, such as one that operates in New Haven, Connecticut, often operate in association with St. Christopher's and St. Joseph's Hospice in Britain.

The hospice approach to care of the dying differs from conventional care in several important ways.[21] First, the hospice approach stresses comfort and care. It recognizes that a time comes when curing and healing the patient is no longer appropriate or possible. Comfort and care include pain and symptom control and assistance at all levels to patients and their families during and after death. Note that this is an enlarged concept of what constitutes proper care for the dying.

A second characteristic is the hospice's team approach. In recognition of the dimensions beyond the physical that make up any human being, the hospice team includes physicians, nurses, clergy, social workers, psychologists and psychiatrists, and various therapists and volunteers, as well as patient and family. This feature is based on a concept of death that involves the full range of an individual's total personality: physical, spiritual, social, and emotional.

A third characteristic involves the hospice approach to pain and symptom control. Hospice practitioners acknowledge that pain, especially chronic pain, is a complex phenomenon that involves the emotional, social, and spiritual dimensions of the patient, as well as the physical. Drawing on this assumption, hospice treatment takes a preventive rather than a reactive approach to pain and symptom control. Thus every attempt is made to preclude the patient's having pain. This means that in some cases drugs are administered not only when patients are already in pain but also before. Contrast this approach with the prevailing reactive attitude toward pain control that characterizes conventional treatment. In many traditional hospitals, the dying patient's requests for pain relief are met with such reactions as "It isn't time for your pill yet" or "You just

---

21.   *J. Thiroux,* Ethics: Theory and Practice *(Encino, Calif.: Glencoe, 1980), pp. 187–189.*

had a shot" or the monstrous "You can't be in pain; you just had your medication!" Obviously, the hospice approach to pain and symptom control takes a rather liberal attitude toward the use of powerful drugs to control the pain of the dying. We will come back to this point shortly.

A fourth characteristic is that hospices may have both inpatient and outpatient services. Outpatient service is, in effect, home care service. One of the goals of such service is to prevent the isolation of the dying from significant others. Some hospices have round-the-clock on-call teams to help people die at home. Inpatient service, or hospital care, tries to prevent patient isolation by allowing unlimited visiting hours, not only for adults but for children and pets as well. Also, loved ones are invited to participate in the care of the dying. Whether inpatient or outpatient, the care is ideally provided on a nonprofit basis. Those who can afford to pay should contribute, but those who cannot should be free from the burden of financial worry by having costs met by existing medical insurance programs.

Finally, the hospice approach attempts to help patients and their families adjust to death before, during, and—most important—after its occurrence. This is where the team approach is particularly beneficial. Throughout the dying process, the team is intimately involved in caring for the needs of patient and family alike. And when death occurs, the family is not left in isolated bereavement. On the contrary, the team continues to support it however it can.

Whether or not the hospice approach is appropriate for every dying person, or even for the majority, is less important here than what it represents: an alternative way of providing care for the dying that is based on a profoundly humane enlargement of the general obligation to provide adequate health care. In this it reminds us all that care for the dying should be evaluated in reference to what we know about the needs of the dying and their loved ones and what we believe to be their rights.

Administering Drugs: The Right to Adequate Care

The central moral question about drugs and the dying is whether the dying should be permitted narcotic drugs for pain relief. Those with reservations about providing the dying whatever drugs they need inevitably express fears of addicting people to powerful painkillers. To react intelligently to this reservation, and thus to the moral question at issue, requires placing the drug issue within the larger frame of symptom control.

Everyone would agree that the obligation to provide adequate health care certainly includes an attempt to control the unpleasant symptoms that accompany terminal illness. The only question is how far the obligation extends. Some would draw the line at the point of addiction.

It is a fact that although most of the symptoms connected with terminal disease can be treated, often they are not. Indeed, studies indicate that 50 percent of the terminally ill have unrelieved pain; 26 percent have severe or serious pain; and 16 percent have continuous severe pain.[22] Yet in almost every case the pain can be controlled by the use of drugs and techniques that have been available for years.

Why this state of affairs? Apparently part of the answer is that professional staff have not received proper training in symptom conrol.[23] As a result, management of symptoms is often not even attempted. When an attempt is made, problems may arise because the drug that is used is not long-acting enough, is used too seldom, or is administered in too small a dose. The point is that fears of addicting the patient can mask ignorance about the management of symptoms. In other words, in many cases skillful medical care can provide pain relief without any risk of addiction.

But what of those cases where painkillers have been skillfully managed and HCPs then face the risk of addicting a dying person to a powerful drug? Is it ever permissible to run this risk? First of all, it is crucial to understand addiction. *Addiction* refers to a physiological dependence on a drug, as evidenced by withdrawal symptoms when the drug is discontinued. Drug addiction is an extremely complex phenomenon and apparently requires a particular kind of personality using the drug under certain circumstances for certain reasons. In other words, as far as we can determine, an appropriate mix of personality, circumstantial, and motivational factors are necessary to produce addiction. Rarely, if ever, does this mix of conditions characterize the terminally ill. Indeed, what makes their situation unique is that they are taking the drug only to kill physical pain. There is no evidence to believe that, once the pain is deadened, terminal patients desire more of a powerful drug.

Thus the fear of addicting the dying to powerful drugs is based on a faulty premise: that cases of terminal illness are analogous to cases of drug addiction among other members of society. This just is not so. On the contrary, evidence indicates that proper drug management, including the management of powerful painkillers, can relieve the symptoms of the dying with no risk of drug addiction.

But even if there is the slightest chance of addicting the dying patient, so what? If a person is dying, what difference does it make if he or she becomes addicted? Which is the greater evil: subjecting patients to excruciating pain when they are begging for relief, or addicting them to a drug? It is hard to categorize a judgment that addiction is worse as anything but an objectionable form of paternalism.

Of course, one could always argue that the drugs will only make the patient fuzzy-headed, disoriented, even delusionary. Thus the patient will stand little if any chance of expressing informed consent or exercising autonomy. But is a patient in unspeakable pain any more capable of expressing informed consent and exercising autonomy? In fact, in opting for a death free from powerful

22. J. M. Hinton, "The Physical and Mental Distress of Dying," Quarterly Journal of Medicine 5 (1963): 1–21; W. D. Rees, "The Distress of Dying," Nursing Times 68 (1972): 1479–1490; and M. A. Simpson, "Planning for Terminal Care," Lancet 1 (1976): 192–193.

23. Simpson, p. 46.

drugs, are we not usurping the very autonomy we so zealously pretend to be guarding?

In the last analysis, the obligation to provide adequate health care entails an obligation to relieve pain. How far should professionals go in meeting this obligation? In answering this question, HCPs might be well advised to dwell on the dying patient's Bill of Rights, which includes the right to be free from pain. Underlying this right is the recognition that pain not only tortures patients but also pollutes their interactions with family, visitors, and staff. Pain also impedes the patient's struggle to come to grips with imminent death. Thus allowing pain to go unrelieved raises serious questions about the rights of the dying and the duties of those who minister to them.

Summary

This chapter was concerned with the rights of the dying. After indicating the various definitions of death and their moral implications, it discussed Kübler-Ross's psychological model of dying. HCPs who are unaware of the stages that the dying often pass through can violate their obligations to deliver adequate health care to the terminally ill. Similarly, concerns about proper health care and patient rights to information and autonomy arise in connection with disclosure of diagnoses, selection of the place to die, and the administration of powerful drugs for pain relief. Each of these issues raises moral concerns about which reasonable people will differ, but strong arguments can be made for disclosing diagnoses, allowing people to die at home, and permitting the use of drugs as needed.

Ethical Theories

In applying the ethical theories to cases involving the terminally ill, we will concentrate on one issue: drugs. We will consider how each of the theories would be likely to answer this question: Should the dying be permitted narcotic drugs as needed for pain relief? The remarks that follow are based on two assumptions. First, although there is always a chance of addicting the dying patient, the chance is remote. Second, the dying patient is suffering excruciating pain and has requested a painkiller.

Egoism

From the viewpoint of egoism, one would be justified in permitting the dying to have powerful drugs if and only if doing so served one's own best interests. In part, one's interests would be determined by law and institutional policy, since violating either could seriously harm one's professional status. These and other factors would play a role only insofar as they affected self-interest. In short, egoism would not offer any decisive position on the issue. Permitting the drugs may or may not be justified and obligatory, depending on the interplay of circumstances and best long-term self-interests. This ambiguity notwithstanding, one thing is clear: Egoism can provide a theoretical basis for permitting the dying to have narcotics.

Act Utilitarianism

Act utilitarians would also base their decision on a comparative evaluation of the consequences of permitting as opposed to refusing the drugs. Unlike egoists, act utilitarians would consider the impact of each of these alternatives on everyone concerned. Certainly, by this calculation, the person with the most to gain or lose would be the dying patient. The pain that the person would suffer without the drugs would figure prominently in weighing the effects of not permitting the drugs. Similarly, freedom from pain would figure prominently in calculating the effects of permitting the drugs. In addition, act utilitarians would consider the impact of each alternative on other interested parties, especially the family, attending team, other patients, and the institution. In the last analysis, act utilitarians would make their decision on a case-by-case basis. Certainly in theory act utilitarianism could condone permitting the dying to have powerful painkillers in at least some cases.

Rule Utilitarianism

The rule under which the proposed act seems to fall: The dying should be permitted narcotics for pain relief. The likely benefits of following such a rule appear enormous. First, the dying are guaranteed a pain-free death. Second, their families are spared the torment of watching their loved ones driven mad by pain. Third, HCPs are spared the agony of having to deny the dying what, in effect, are last requests. Fourth, all of us can rest easier that we ourselves will not have to endure excruciating pain in similar circumstances. The list could undoubtedly be expanded, but even these likely consequences argue formidably for such a rule. All of which leads to the conclusion that rule utilitarians probably would endorse permitting narcotics to the dying for pain relief. Moreover, if this analysis is correct, then any instance of a dying patient's being denied such drugs would be unjustifiable and immoral. By this account, much of the conventional approach to pain and symptom control is morally indefensible. Indeed, by a rulist calculation, HCPs would apparently have an obligation to help change this approach as well—that is, to advocate for the right of the dying to be free from pain.

The Categorical Imperative

Kantian ethics holds that each of us has a duty to refrain from intentionally and needlessly injuring someone else. To injure others would be inconsistent with the way rational, self-governing moral agents deserve to be treated. What is more, such behavior is incompatible with the behavior that is proper to such beings. Presumably, one can injure someone as much through an act of omission as through an act of commission. Thus physicians can harm patients just as much by neglecting them as by poisoning them. By this analysis, then, HCPs who deny dying patients pain-relieving narcotics do harm through omission and thereby act immorally.

The fly in the ointment of this application is the issue of patient rationality and autonomy. Both will probably suffer, perhaps even be destroyed, as a result of administering powerful drugs. Given this possibility, and the emphasis Kant places on rationality and autonomy, one wonders whether Kantian ethics could unequivocally countenance the use of drugs.

One way to tackle this problem is to consider the status of pain-ridden dying patients. Are one's rational facilities not impaired as much by pain as by drugs? Most of us have experienced the pain of a toothache or a headache and realize how devastating it can be. Often we seem unable to think about anything except how much it hurts; we exercise very little control over our decisions, moods, frame of mind, or interactions with others. In brief, we are obsessed with one thought: killing the pain. If the pain of a migraine headache, for example, produces such constraints on our rational faculties and self-determination, how much more constraining must be the torture of a terminal disease like stomach cancer? Not only are patients thus afflicted in unspeakable agony, but unlike the migraine sufferer, these patients have little hope of remission from either pain or condition. Who can say for sure that under these conditions rationality and autonomy are less impaired than under the influence of powerful drugs? Indeed, there is some evidence to suggest that when relieved from pain, dying patients are more clearheaded, more capable of rational choice and making decisions for themselves.

Therefore, concern about rationality and autonomy, although always important in a Kantian analysis, does not undercut the original conclusion that Kant would probably permit narcotics for the dying.

Prima Facie Duties

Ross's primary concern in such cases would be the duty of noninjury. By refusing to permit dying patients pain-relieving drugs, HCPs stand to injure these patients seriously. Looked at another way, HCPs are in a position to render the dying profound beneficent services. Not only can they relieve pain, but they can also situate the dying so that they are able to continue to enjoy relationships with the family and tend to unfinished business. Furthermore, HCPs certainly have a prima facie duty of fidelity to do all they can to relieve the pain of their patients. It is always possible, of course, that given the mix of prima facie duties in any case, one's actual duty would be not to permit the drugs. But in the light of the aforementioned duties, in most cases one's actual duty would apparently be to permit the drugs. In any event, Ross's ethics provides a theoretical basis for drug use.

The Maximin Principle

Given the original position, rational creatures with a sense of their own interests surely would not wish to be bound by a rule that would prohibit pain relief in their most desperate hours. Looked at another way, nonrestrictive drug policies for the dying would be compatible with both of Rawls's principles of justice. On the one hand, allowing rather than denying painkillers to the dying in effect allows a greater amount of freedom that is compatible with a like freedom for all. On the other hand, permitting the dying such drugs but denying them to the nondying is an action that benefits the most disadvantaged in the class termed "patients." Furthermore, since any of us could ultimately find ourselves in pain on our deathbed, everyone could be said to benefit from a permissive use of drugs for the dying. This analysis leads to the conclusion that Rawls would not only permit drugs for the dying but would probably discern a moral obligation to advocate for it.

Roman Catholicism's Version of Natural-Law Ethics

An application of the principle of double effect would probably lead Roman Catholic moral theologians to endorse narcotics for the dying. First, administering such drugs is in itself neither good nor bad. Its morality depends largely on why drugs are used, under what circumstances, and with what likely effect. Second, the bad effect—possible loss of rationality as well as possible addiction—is not the means by which the person is relieved of pain. Thus, the dying could just as easily be relieved of pain if their faculties were preserved and they were not addicted. Third, the good effect—certain pain relief—surely is at least the equivalent of the possible bad effect. Finally, it is assumed that the motive in such cases is exclusively the relief of the patient.

As always, these applications are speculative and tentative. If they are correct, then clearly a theoretical basis can be found in each of these theories for permitting the dying narcotics for pain relief and perhaps a concomitant obligation for HCPs to advocate for it.

CASE PRESENTATION
Independence Day

The psychiatric resident put down her pen. Across the river, in the distance, she could hear the last explosions of another Independence Day. *Independence Day.* She remembered what it used to be like when she was a kid. Then she reread what she'd just written about the final Independence Day in the life of Deborah, age eighteen:

> I interviewed this patient in April, when she was admitted for rape. She left the hospital shortly thereafter but returned on the Fourth of July, this time with a stab wound. She was not expected to live through the night.
> When I stopped to see the parents outside the intensive care unit before going inside to visit the girl, the father appeared inconsolable. He was unable to relate either to his wife or to me. The mother told me that they had had trouble

reaching the hospital because they did not own a car. I knew from the April admission that the family lived in a ghetto area. When the emergency call came from the hospital this time, the parents apparently called a neighbor for a ride. The neighbor said his car was broken, but later, as they were hurrying to the bus stop, they saw him drive past.

Then, as they were waiting for the bus, they were accosted by two young men bent on robbery. Apparently the father had been sobbing, and the mother was trying to console him. The thugs, assuming they had frightened the man to tears, began to ridicule him. Fortunately, a uniformed police officer appeared, and the hoodlums ran.

The woman's anecdotes made me angry, but I didn't understand why she was relating them while her child lay dying. The point, as I soon realized, lay buried in my own anger. Both incidents represented people as insensitive and cruel.

I asked the mother if she thought something cruel and insensitive was happening to her daughter right now. She began to cry and pointed in the direction of the girl's room.

When I entered the unit, I understood the mother's feelings. Deborah lay on her bed half-naked, hooked up to infusion tubes and a respirator. Her eyes darted around the brightly lit room with panic and desperation.

I started to cover her with a bedsheet, but a nurse said, "She'll only push it off again."

I held Deborah's hand and asked if the light was bothering her. She indicated that it was. I asked that the light be turned down but was told they had to keep a close watch on the monitors, so it was not practical.

Then I asked for a chair so the mother could sit with her child. I was informed that the mother had overstayed her previous visit by five minutes. The chair had been taken away to encourage compliance with the ten-minutes-per-hour visiting rule next time she came in.

Deborah died, six hours after her parents had been informed that her death was imminent. She died with the light in her eyes, with tubes in her mouth and veins, and with her parents sitting outside in the waiting room, weeping.

The psychiatric resident closed Deborah's file. Independence Day, she thought. Somehow it's not the same any more.

Questions for Analysis

1. *What are your feelings about what occurred at the hospital?*

2. *Do you think Deborah died with dignity?*

3. *Do you think any of Deborah's rights as a dying person were violated? Explain.*

4. *Do you think the nurse's behavior was justified? Explain.*

5. *If you had been the nurse, how would you have acted? Explain.*

6. *Would you say the psychiatric resident's behavior was beyond reproach? Could she have behaved other than she did?*

7. *Do you think HCPs have an obligation to work actively to change institutional rules like the ones evident here? Under what conditions, if any, do you think an HCP is morally justified in violating such institutional rules?*

8. *Do you think the psychiatrist has any moral obligations to the grieving parents? If so, what are they?*

9. *Do you think a health care facility has a duty to help grieving loved ones work through their grief? Apply each of the ethical theories to this issue and determine what moral directions they suggest.*

Exercise 1

Pretend you are an HCP who is confronted by the following situations. How would you respond? Do your responses correspond with how you think you *should* respond?

1. The family of a patient who is going to die do not want the person to be informed. One day the patient says to you, "I just know I'm not going to live long, I know it! That's true, isn't it?"

2. A harried colleague comes to you seeking help. She is having trouble dealing with a patient who persists in believing that she is not going to die, although the prognosis is terminal.

3. You are a nurse who deals with terminal cancer patients. One of the patients, who has not been told he is terminal, insists on being told the truth. His doctor does not think it prudent that the patient be informed.

4. A physician who does not feel up to the job asks you to discuss a patient's impending death with the person.

Exercise 2

Reconsider your responses under Values and Priorities in light of this chapter's considerations.

Death, Dying, and the Biological Revolution

Robert M. Veatch

Robert M. Veatch is one of the foremost scholars on bioethical issues. In the following selection, Veatch contends that a philosophical view of both humans and human values is necessary for an adequate medical definition of death. Veatch provides a framework for his claim by isolating four levels in the definition of death: the formal definition of death, the concept of death, the locus of death, and the criteria for the determination of death. In Veatch's view, the Harvard committee confused these four levels and, in fact, did not go far enough in updating the definition of death. Indeed, Veatch argues that the irreversible loss of functioning in the cerebral neocortex, rather than in the whole brain, should be the basic physiological criterion for determining death. He thereby distinguishes himself from the Harvard committee as well as from those who espouse more conventional criteria, including spinal and brain stem reflexes. Veatch's position has significance for all HCPs who care for the dying, especially for organ transplant teams, because his definition permits a more liberal determination of death than is currently permissible.

While we philosophize about the meaning of death in the age of the biological revolution, people are being pronounced dead (or alive) by physicians who choose one definition or another. The philosophical discussion becomes literally a matter of life and death. You may be pronounced dead by a randomly available physician even if you and your family believe (or have believed) you are still alive and even if you would be considered alive at another hospital down the block. Or you may be considered living by a physician who has chosen to reject the newer notions of death centered on the brain or some part of it, even if you have thought about the issue and decided in favor of a brain-oriented concept.

Doctors in the forty states that have not adopted specific legislation are taking it upon themselves to use a brain-oriented concept of death, although the laws in these states do not authorize them to. Other doctors are reluctant to use newer concepts of death, fearing they may offend the patient's family or some district attorney. The fact is, "There is currently no way to be certain that a doctor would not be liable, criminally or civilly, if he ceased treatment of a person found to be dead according to the Harvard Committee's criteria, but not according to the 'complete cessation of all vital functions' test presently employed by the courts."[1]

Some order must be brought out of this confusion. A public policy must be developed that will enable us to know who should be treated as alive and who should be treated as dead. . . .

Laying the responsibility on the individual physician or the profession as a whole for deciding what the definition of death should be is the result of inadequate analysis, of failing to distinguish adequately between the levels of the debate. The medical professional undoubtedly has special skills for determining and applying the specific criteria that measure whether particular body functions have irreversibly ceased. Whether the Harvard criteria taken together accurately divide those who are in irreversible coma from those who are not is clearly an empirical question (although the important consideration of just how sure we want to be takes us once again into matters that cannot be answered scientifically). But the crucial policy question is at the conceptual level: should the individual in irreversible coma be treated as dead? No medical answers to this question are possible. If I am to be pronounced dead by the use of a philosophical or theological concept that I do not share, I at least have a right to careful due process. Physicians in the states that do not authorize brain-oriented criteria for pronouncing death who take it upon themselves to use those criteria not only run the risk of criminal or civil prosecution but, in my opinion, should be so prosecuted.

A Statutory Definition of (Whole) Brain Damage

When the *Tucker* case (in which the Virginia physicians defended their use of criteria for death not sanctioned by state law) reached court, it was the first case to test a *public* policy for defining death. Judge A. Christian Compton was not willing to have such a major question resolved in his court, saying, "If such a radical change is to be made in the law of Virginia, the application should be made therefore not to the courts but to the legislature wherein the basic concept of our society relating to the preservation and extension of life could be examined, and, if necessary, reevaluated."[2]

The Kansas Proposal

In 1968 Kansas was the first state to pass a law permitting the procuring of organs for transplantation. . . . Maryland next passed an almost identical bill.[3] Subsequently Alaska, California, Georgia, Illinois, Michigan, New Mexico, Oregon, and Virginia passed such legislation. States now considering changes include Florida, Minnesota, and New York. Others have legislators interested in new death definitions.

These statutory proposals have not gone without opposition. Probably the best focused and most widely known criticism has come from British law professor Ian McColl Kennedy.[4] "Let us have guidelines by all means. They are essential," he argues. "But let them be set down by the medical profession, not by the legislature." That the medical profession, as a profession, may have no special competence to set such guidelines is a possibility he completely misses. Like many others, he confuses medical and policy expertise. He goes on to outline six specific criticisms of the Kansas bill. Some of them seem to me more valid than others.

The first is probably the most critical and the most valid. The act, he observes, "seems to be drafted only with transplantation surgery in mind." Indeed, the bill incorporates explicit directions on this matter: "death is to be pronounced before artificial means of supporting respiratory and circulatory function are terminated and before any vital organ is removed for the purpose of transplantation." As Dr. Taylor has revealed in his 1971 article, the University of Kansas Medical Center was concerned about transplants when staff members began promoting the change in the law.

The relation between a new definition of death and transplantation is complex,[5] and Ian Kennedy's first critical point identifies a major cause of worry: "To draft a statute on death inspired apparently by the desire to facilitate what must still be considered experimental surgical procedures must serve to disturb the man in the street. . . . The Act in its present form does not serve to reassure the person who may fear that during his last hours on earth his doctors will be less concerned with his condition than with the person earmarked to receive one of his vital organs."[6]

Don Harper Mills, physician and lawyer, does not agree that the statute is so closely associated with transplant policy. He claims that it intentionally extends to questions of when the physician can terminate resuscitative efforts or discontinue artificial maintenance.[7] Whatever the intentions of the bill's authors, both the authors and Mills may be wrong in their assumptions of what purposes such a statutory definition should serve. It is dangerous to propose a statutory definition solely for the purpose of obtaining organs, but it is equally dangerous to confuse the issue of when resuscitation should be stopped with the one of when a patient is dead. Neither considers that a statutory definition may be needed to prevent the basic indignity of treating a corpse as if it were alive—of confusing a living human with one who has lost essential humanness. Kennedy is right in recognizing that the link between transplantation and the definition of death should not be as close as in the Kansas bill.

Second, Kennedy objects that the Kansas bill seems to propose two alternative definitions of death, implying a person may be simultaneously dead according to one criterion and alive according to the other. In a law review article agreeing with Kennedy on this point, Alexander Capron, law professor at the university of Pennsylvania, and Leon Kass, professor of bioethics at Georgetown University, pose a bizarre problem.[8] A patient who meets the brain-oriented criteria for death and is a good tissue match with a potential organ recipient is pronounced dead under a special "transplant definition." What would the patient's status be if the potential organ recipient died before the donor organs were removed? The donor would be alive according to the heart- and lung-oriented definitions but pronounced dead according to a definition no longer applicable. If it is the person who dies and not some organ or cells or function, then we need a single definition that can apply to all of us, independent of what someone may want to do with our parts. These two problems raised by Kennedy—the dangerous link with transplantation and the implication of alternative definitions of death—should be taken into account in any future bills dealing with a new definition of death.

Third, Kennedy senses something wrong with the requirement that death be pronounced before artificial means of supporting respiration and circulation are stopped. Here his instincts may be sounder than the reasons he uses to support them. The proposal that death be pronounced first is taken from the Harvard Committee report. Kennedy seems to agree with the policy but feels it should not be written into the legislation. He writes that the dilemma faced by physicians is "more imagined than real" and declares that "doctors" do this every day without legislative fiat and will continue to do so with impunity. . . . I don't follow his reasoning. Does he mean that physicians declare death every day before turning off resuscitation equipment? The cry for legislative protection seems to contradict that. Or does he mean that physicians decide to stop supportive maintenance on dying patients every day? That is probably true, but an entirely different issue. Kennedy goes on to argue that the requirement that death be pronounced before stopping life support is "entirely redundant." He says, "Once the doctors decide that the conditions specified in the Act exist, and 'further attempts at resuscitation or supportive maintenance will not succeed,' death has already occurred." Indeed it has, according to the new definition, but to say that "death must be pronounced" is something else. If nothing more, this makes clear that the concept of death being used is radically new.

There is a more serious problem, which Kennedy does not mention. To say that death should be pronounced before supportive maintenance ceases (on a corpse) might imply to the less careful reader that it is never appropriate or legal to decide to stop life support on a dying individual. If anyone were to read that from the Kansas legislation, it would be a serious problem. The question of stopping treatment of the dying is a separate issue to be taken up later.

Kennedy's fourth objection to the Kansas bill is that it does not require a confirmatory judgment of a second physician before pronouncing death according to brain-oriented criteria. He criticizes others who find this "commendable."[9] Whether the requirement of a second judgment is reasonable will depend upon the purpose and context of the legislation. In the context of organ-transplant practices, a second judgment may indeed control aggressive transplanters. But if Kennedy is also right that the redefinition shuld not be limited to the transplantation context, then a confirmatory judgment seems less crucial. Is his position that the brain criteria are so much more complicated than the older heart and lung criteria that two technically competent individual judgments are

necessary? I doubt that this is true now, and surely it will become completely unlikely, as experience is gained during the life of the legislation. There seems no plausible reason to have two experts involved in the general task of pronouncing death unless the techniques used are so complex that one cannot handle them adequately.

Kennedy's fifth criticism is that the act should require the physician pronouncing death to be a different one from the transplant physician. He calls for "safeguards" to protect the patient from potential conflict of interest. This is important and valid, particularly in the context of legislation explicitly for transplantation. Even better would be a more general ban on conflict of interest as part of a more general redefinition of death. No physician who has any interest beyond the patient's own welfare should be permitted to pronounce death.

Kennedy's final criticism is the most confusing. He claims that the act implicitly incorporates "the detailed clinical procedures that serve to determine 'brain death,' " and he is rightly concerned that the law is no place to spell out in great detail the technical procedures for measuring whether a death has occurred. But it is impossible to read any such specification into the Act, which simply says that the diagnosis of absence of spontaneous brain function is to be "based upon ordinary standards of medical practice." These standards will vary from place to place and from time to time. New technical innovations or empirical data will change the tests to be used or the way they are used. The length of time an electroencephalogram has to be flat may change. Virtually all others who have criticized the Kansas bill[10] have thought that it does avoid the trap of overspecificity. The problem seems to be one of confusing the levels of the definition debate. Whatever Kennedy is taking exception to, "the absence of spontaneous brain function" certainly seems a rather general term. It specifies a function or a "locus" in the body, not empirical criteria or tests.[11]

A Better (Whole) Brain Statute

Capron and Kass are not happy with the Kansas statute for some of the same reasons as Kennedy: they do not like the close link with the transplantation issue, and they are particularly distressed at the implication that there are alternative forms of death appropriate for different situations. But they are still in favor of legislation. The questions at stake, in their opinion, are crucial matters that call for public involvement. "Physicians *qua* physicians are not expert on these philosophical questions nor are they expert on the question of which physiological functions decisively identify the 'living, human organism'."[12] The legislative route, they argue, would permit the public to play a more active role in decision making. It would also dispel both lay and professional doubt and provide needed assurance for physicians and patients' families that the new definition could be used without fear of a legal suit. They propose five "principles governing the formulation of a statute."

1. The statute should concern the death of a human being, not the death of cells, tissues, or organs, and not the "death or cessation of his role as a fully functioning member of his family or community."
2. It should move incrementally, supplementing rather than replacing the older cardiopulmonary standards.
3. It should avoid serving as a special definition for a special function such as transplantation.
4. It should apply uniformly to all persons.
5. It should be flexible, leaving specific criteria to the judgment of physicians.[13]

On the basis of these guidelines they propose a new draft statute as an alternative to the laws in Kansas and Maryland:

> A person will be considered dead if in the announced opinion of a physician, based on ordinary standards of medical practice, he has experienced an irreversible cessation of spontaneous respiratory and circulatory functions. In the event that artificial means of support preclude a determination that these functions have ceased, a person will be considered dead if in the announced opinion of a physician, based on ordinary standards of medical practice, he has experienced an irreversible cessation of spontaneous brain functions. Death will have occurred at the time when the relevant functions ceased.[14]

Capron and Kass have captured all of the virtues and none of the problems of the Kansas statute. Their bill fails to meet two of Kennedy's objections—it does not require two physicians to participate in determining death and it does not provide that the death-pronouncing physician be separate from the physician interested in the potential cadaver's organs—but these requirements seem superfluous for a general public policy for determining when we are dead. Nevertheless, in holding to the principle of making the definition independent of transplantation concerns, Capron and Kass may have missed an important protection for the patient potentially dead because his brain has completely and irreversibly ceased functioning. They argue, "if particular dangers lurk in the transplantation setting, they should be dealt with in legislation on that subject, such as the Uniform Anatomical Gift Act."[15] That is reasonable, but it is also reasonable that there be observed a general requirement that the physician pronouncing death should be free of significant conflict of interest (whether interest in a respiring "patient," research, continued treatment fees, or transplantation). That there must be no such conflict is obviously essential, whether or not it should be banned by the statute itself.

Critics of the proposals for statutes setting out new standards for determining death have either dealt with technical wording difficulties or made misguided appeals for vesting decision-making authority in physicians or medical professional groups. These, however, are not the only problems. In order to accept the Kansas statute or the preferable Capron-Kass revision, it is first necessary to accept the underlying policy judgment that irreversible destruction of the brain is indeed death—that individuals should be treated as dead when, and only when, their brains will never again be able to function. Some of us continue to have doubts about that basic judgment.

A Statutory Definition of Cerebral Death

There has been great concern that statutes designed to legalize and regularize the use of brain-oriented criteria may not be sufficiently flexible to keep up with changes in this rapidly developing area. Kennedy and others who place their faith in medical discretion fear that a statute would not permit adoption of new techniques and procedures. For the most part they are wrong, since none of the proposed statutes specifies any particular criteria, techniques, or procedures. Techniques and procedures are changing rapidly; with that the proposed laws can cope. But our concepts, our philosophical sophistication, are evolving rapidly, too. Even today most people writing in the field, including competent scientists and physicians, are careless in distinguishing between the whole brain and the cerebrum and the functions of each. Here may arise a significant problem, for under even the highly generalized statutory proposals it may not be possible to make wanted distinctions between lower brain functions, such as those that control spontaneous respiration, and those giving rise to consciousness and individual personality.

If it is decided that a person without the capacities which are thought to reside in the higher brain (cerebral) centers should really be considered dead, then an amendment to the brain-death statutes might be in order. The change could be a simple one: simply strike the word *brain* and replace it with *cerebral*. This change in specifying the locus or the general standards for determining death may or may not have practical significance to the clinician who pronounces death. The question of criteria is an empirical one and the answer will change periodically. It may be that the only way of knowing for sure that the cerebrum has irreversibly lost its ability to function is to use exactly the same tests as for determining that the whole brain has lost its power to function, that is, the Harvard Committee criteria or something similar. But it may also be that other tests—such as EEG alone—could predict with certainty when individuals have irreversibly lost cerebral function even if they retain some lower brain functions, even if, say, they are still breathing spontaneously. The question of criteria can and must be left to the neurological experts.

There may be reasons for sticking with the old-fashioned statutes based on whole-brain conceptions of death. Only a few people will be dead according to a cerebral concept but alive according to a whole-brain concept. There may be some risk of making an empirical error in applying cerebral criteria and pronouncing someone dead who could still regain some form of consciousness. Some moral doubt may remain about the legitimacy of pronouncing someone dead who retains lower brain function. But these same problems

arise with the whole-brain-oriented statutes as well. Once the judgment has been made that false positive diagnoses of life are a serious problem, serious enough to overcome any empirical or moral doubts, there is a strong case for moving on from the whole brain to a cerebral focus.

A Statute for a Confused Society

There is still another option. Part of the current confusion reflects sincere and reasonable disagreements within society over which philosophical concept of death is the proper one. As with many philosophical questions, the conflict will not easily be resolved. In a democratic society, however, we have a well-established method for dealing with a diversity of religious, moral, or philosophical perspectives. It is to allow free and individual choice as long as it does not directly infringe on the freedom of others and does not radically offend the common morality.

When dealing with a philosophical conflict so basic that it is literally a matter of life and death, the best solution may be individual freedom to choose between different philosophical concepts within the range of what is tolerable to all the interests involved. There have been rare and tentative hints at this solution in the literature. In 1968 proposed by the general definition of human death Halley and Harvey had an apparent option clause:

> Death is irreversible cessation of *all* of the following: (1) total cerebral function, (2) spontaneous function of the respiratory system, and (3) spontaneous function of the circulatory system.
>
> Special circumstances may, however, justify the pronouncement of death when consultation consistent with established professional standards have [sic] been obtained and when valid consent to withhold or stop resuscitation measures has been given by the appropriate relative or legal guardian.[16]

They abandoned this "consent" formula, however, in later versions of their proposal.[17]

Halley and Harvey have been criticized for their "mistake in making the state of being dead (rather than the acceptance of imminent death) depend on the 'consent' of a relative or guardian."[18] It seems likely that they did indeed confuse the state of being dead with the state of being so close to death that a decision could justifiably be made by a relative or guardian to stop resuscitation. But I do not see that their perhaps naive formulation makes "the state of being dead" dependent upon consent of a relative. It makes the state of being *pronounced* dead dependent upon consent. Being dead or alive may be quite independent of the wishes of relatives, but the treatment of persons as if they were dead or alive can logically still be a matter of choice of a relative or even a prior choice of the individual. For those who believe that metaphysical states are to some extent independent of personal choice (as I do), this will mean that in some cases we shall continue to treat corpses as if they were alive or living people as if they were corpses, but we run that risk under any public policy alternative, whether or not it permits freedom of philosophical choice.

More recently Michael Sullivan, county probate judge in Milwaukee, had to make two critical legal decisions concerning whether patients have the right to refuse treatment. He has explained the basis of his decisions in the *New England Law Review*.[19] He writes in his article that he does not believe legislation defining death to be advisable "in this context." Since he is discussing whether dying patients have the right to refuse treatment, this attitude is perfectly plausible. But, although it is also irrelevant to his context, he goes on to state his opinion on who should decide what definition of death should be used: "The individual should decide whether he will employ the Harvard criteria, or some other definition for his death." According to Sullivan, it is the individual, not the physician, the medical society, or the state, who should have the "right to prescribe his death style" including the person's own definition of death. This obviously raises some problems, as in the cases of individuals in irreversible coma who have not recorded an opinion while conscious and competent. Some provision will have to be made for these cases.

There are two possibilities: (1) shifting decision making to the individual (or the next of kin or other legal guardians) and (2) setting up a definition to be followed unless otherwise instructed. As a practical matter both can probably be used. The law could specify a given general standard—oriented to heart or the whole brain or the

cerebrum—with the proviso that the individual has the right to leave explicit instructions to the contrary. Further, as with the Uniform Anatomical Gift Act, the law could provide that, in those cases where the individual has left no instructions while conscious and competent, the right would be exercised by the next of kin or guardian appointed for the purpose. Many of these issues also arise in setting up mechanisms for refusing further medical treatment for the still living patient.

There is another problem, however. Has individualism run amok? Do we really want to be so antinomian, so anarchical, that any individual no matter how malicious or foolish can specify any meaning of death which the rest of society would be obliged to honor? What if Aunt Bertha says she knows Uncle Charlie's brain is completely destroyed and his heart is not beating and his lungs are not functioning, but she still thinks there is hope—she still thinks of him as her loving husband and does not want death pronounced for a few more days? Worse yet, what if a grown son who has long since abandoned his senile, mentally ill, and institutionalized father decides that his father's life has lost whatever makes it essentially human and chooses to have him called dead even though his heart, lungs, and brain continue to function? Clearly society cannot permit every individual to choose literally any concept of death. For the same reason, the shortsighted acceptance of death as meaning whatever physicians choose for it to mean is wrong. A physician agreeing with either Aunt Bertha or the coldhearted son should certainly be challenged by society and its judicial system.

There must, then, be limits on individual freedom. At this moment in history the reasonable choices for a concept of death are those focusing on respiration and circulation, on the body's integrating capacities, and on consciousness and related social interactions. Allowing individual choice among these viable alternatives, but not beyond them, may be the only way out of this social policy impasse.

To develop model legislation, we can begin with the Capron-Kass statutory proposal and make several changes to avoid the problems we have discussed. First, a cerebral locus for determining if a person is dead can be incorporated by simply changing the word *brain* to the narrower *cerebral.* Second, it seems to me a reasonable safeguard to insist, in general terms appropriate for a statutory definition, that there be no significant conflict of interest. Finally, wording should be added to permit freedom of choice within reasonable limits. These changes would create the following statute specifying the standards for determining that a person has died:

> A person will be considered dead if in the announced opinion of a physician, based on ordinary standards of medical practice, he has experienced an irreversible cessation of spontaneous respiratory and circulatory functions. In the event that artificial means of support preclude a determination that these functions have ceased, a person will be considered dead if in the announced opinion of a physician, based on ordinary standards of medical practice, he has experienced an irreversible cessation of spontaneous cerebral functions. Death will have occurred at the time when the relevant functions ceased.
>
> It is provided, however, that no person shall be considered dead even with the announced opinion of a physician solely on the basis of an irreversible cessation of spontaneous cerebral functions if he, while competent to make such a decision, has explicitly rejected the use of this standard or, if he has not expressed himself on the matter while competent, his legal guardian or next of kin explicitly expresses such rejection.
>
> It is further provided that no physician shall pronounce the death of any individual in any case where there is significant conflict of interest with his obligation to serve the patient (including commitment to any other patients, research, or teaching programs which might directly benefit from pronouncing the patient dead).

Notes

1. Alexander Morgan Capron and Leon R. Kass, "A Statutory Definition of the Standards for Determining Human Death: An Appraisal and a Proposal," 121 *University of Pennsylvania Law Review* 97 (1972).

2. *Tucker* v. *Lower*, No. 2831 (Richmond, Va., Law and Equity Court, May 23, 1972), p. 10.

3. Maryland Sessions Laws, ch. 693 (1972). The phrase "in the opinion of a physician" was deleted from the first paragraph and the phrase "and because of a known disease or condition" was added in the second paragraph following "ordinary standards of medical practice." It is not clear why the irreversible loss of brain function must be caused by a known disease or condition unless this is thought to be a protection against falsely diagnosing irreversibility in cases where a central nervous system depressant is present, unknown to the medical personnel.

4. Kennedy, "The Kansas Statute," pp. 946–50.

5. See the discussion in chapter 1 [of *Death, Dying, and the Biological Revolution* (New Haven, Conn.: Yale University Press, 1976)].

6. Kennedy, "The Kansas Statute," p. 947.

7. Don Harper Mills, "The Kansas Death Statute: Bold and Innovative," *New England Journal of Medicine*, 285 (1971), 968–69.

8. Capron and Kass, "A Statutory Definition," p. 197 n. 70.

9. William J. Curran, "Legal and Medical Death—Kansas Takes the First Step," *New England Journal of Medicine*, 284 (1971), 260–61. See also Capron and Kass, "A Statutory Definition," pp. 116–17.

10. See Mills, "The Kansas Death Statute," p. 969; Capron and Kass, "A Statutory Definition."

11. In order to clarify the problem of what can and should be legislated, Capron and Kass ("A Statutory Definition," pp. 102–3) have outlined four possible levels for legislative action. These parallel to some extent those in chapter 1 [of *Death, Dying, and the Biological Revolution*]. While they also specify a purely formal definition ("the transition, however abrupt or gradual, between the state of being alive and the state of being dead"), the *basic concept* is the most general level of the four on the list. Not unlike my use of the term *concept*, they mean a philosophical specification of what it is that is the essential change in a person who is no longer considered alive. This, they argue, should not be legislated. I would agree, provided it is recognized that certain assumptions at the basic conceptual level will have to be made in order to move to the next level, which they call "the general physiological standard." They mean here something like what I called the locus: an area of the body whose functioning is critical. Here, we all agree, is the prime area for legislation. The third and fourth levels outlined by Capron and Kass are the operational criteria (e.g., absence of cardiac contraction and movement of the blood) and "specific tests and procedures" (e.g., pulse, heartbeat, blood pressure, etc.). All agree that there is no lace in legislation for something as ephemeral

as specific empirical tests. I also concur with Capron and Kass that "operational criteria" should not be incorporated into the law.

12. Capron and Kass, "A Statutory Definition," p. 94.
13. *Ibid.*, pp. 104–8.
14. *Ibid.*, p. 111.
15. *Ibid.*, p. 116.
16. M. M. Halley and W. F. Harvey, "Medical v. Legal Definitions of Death," *Journal of the American Medical Association,* 204 (1968), 423–25.

17. Halley and Harvey, "Law-Medicine Comment: The Definitional Dilemma of Death," *Journal of the Bar Association of the State of Kansas,* 39 (1968), 179.
18. Capron and Kass, "A Statutory Definition," p. 105 n. 66.
19. Michael T. Sullivan, "The Dying Person—His Plight and His Right," 8 *New England Law Review* 197–216 (1973).

For Further Reading

Aries, P. *Western Attitudes toward Death.* Baltimore: Johns Hopkins University Press, 1974.

Bakan, D. *The Duality of Human Existence.* Chicago: Rand-McNally, 1966.

Bakan, D. *Disease, Pain and Sacrifice.* Chicago: University of Chicago Press, 1968.

Bell, T. *In the Midst of Life.* New York: Atheneum, 1961.

Bettelheim, B. *The Empty Fortress.* New York: Free Press, 1967.

Bowers, M. K. *Counseling the Dying.* New York: Thomas Nelson & Sons, 1964.

Cutter, F. *Coming to Terms with Death.* Chicago: Nelson-Hall, 1974.

De Beauvoir, S. *A Very Easy Death.* New York: Warner Communications, 1973.

Donovan, M. I., and Pierce, S. G. *Cancer Care Nursing.* New York: Appleton-Century-Crofts, 1976.

Grollman, E. A. (ed.). *Concerning Death: A Practical Guide for the Living.* Boston: Beacon Press, 1974.

Krant, M. J. *Dying and Dignity.* Springfield, Ill.: Charles C Thomas, 1974.

Kübler-Ross, E. *Death: The Final Stage of Growth.* Englewood Cliffs, N.J.: Prentice-Hall, 1975.

Lifton, R. J., and Olson, E. *Living and Dying.* New York: Praeger, 1974.

Saunders, C. *Care of the Dying.* London: Macmillan, 1959.

Scherzer, C. J. *Ministering to the Dying.* Englewood Cliffs, N.J.: Prentice-Hall, 1963.

Scott, N. A., Jr. (ed.). *The Modern Vision of Death.* Richmond, Va.: John Knox Press, 1967.

Shneidman, E. S. (ed.). *Death: Current Perspectives.* Palo Alto, Calif.: Jason Aronson, 1976.

Sudnow, D. *Passing On.* Englewood Cliffs, N.J.: Prentice-Hall, 1967.

10
THE MENTALLY ILL PATIENT

Dr. Myerson had heard patients make threats before. What therapist hadn't? Like other therapists, he realized that most threats are idle, that just expressing them seems enough for most patients. In any event, in his fifteen years of practice at the university, Dr. Myerson had never felt obliged to warn the target of a threat made by a patient during a therapeutic session. In fact, he regarded such action as a breach of confidentiality.

So when Peter Prudhomme, a student who was in psychotherapy on an outpatient basis, began making threats to kill fellow student Tanya Trippett, Dr. Myerson didn't feel obliged to inform the young woman. Even if he had wanted to, he couldn't have, because Trippett was away on summer vacation.

Nevertheless, Dr. Myerson believed that in Prudhomme's case the threats constituted a possible danger to Tanya. So he asked the campus police to take Prudhomme into custody incident to pursuing the legal machinery for involuntary hospitalization. After a seventy-two hour observation, however, the police found no sign of abnormality or dangerousness in the young man and consequently released him.

For his part, Prudhomme immediately terminated treatment with Dr. Myerson. Furthermore, he became the roommate of Tanya's brother. In the month they roomed together, he showed no signs of disturbance or potential harmfulness. Certainly he expressed no threats on the life of Tanya Trippett. Nonetheless, shortly after her return in the fall, Peter Prudhomme sought out Tanya Trippett and brutally murdered her.

This fictionalized case is based on a real one. In the actual case (*Tarasoff* v. *The Regents of the University of California*), the parents of the young woman charged the therapist with failure to warn them of the danger to their daughter. The lower courts dismissed the action without trial because of the legal immunity of the defendants and the overriding need to maintain confidentiality in psychotherapy. But the parents appealed to the California Supreme Court, which reversed the lower court's ruling on December 23, 1974 and ordered the case to trial.

Professional societies bristled at this turn of events. They demanded a rehearing, which they were granted. At the rehearing, numerous mental health professionals (hereafter MHPs) argued that violating patient confidentiality under such circumstances would undercut effective patient treatment. They pointed out that in almost every case a professional judgment of "dangerous" would be invalid, and they backed up their claim with statistics. Nonetheless, on July 1, 1976 the court filed a revised decision that in no way altered the California Supreme Court's recognition of an obligation to warn.

Many features of this case, and its fictional counterpart, make it an appropriate departure point for this chapter. First, it deals with yet another group of special patients: the mentally ill. Second, it invites a consideration of patient rights as they apply to the mentally ill. Third, it elucidates the web of relationships that ensnare MHPs and the opposing obligations that they sometimes entail. Fourth, it reflects the friction that exists on the interface between law and mental health. In addition, the Tarasoff case also provokes speculation about the moral aspects of involuntary commitment, the inadequacy of MHPs to predict dangerous behavior, the nature of paternalism in mental health care, and the nature of the MHP-patient relationship in general. In a word, the case abounds with issues that raise moral concerns.

This chapter engages most of these issues and the moral concerns they raise. Of course, there are other issues worthy of consideration: the nature and extent of society's obligations to the mentally ill, the institutional environment, staff competency, and the nature of psychological interventions, to name just a few. Indeed, many more complex moral issues surround the mentally ill than any other group of patients. That is why you should approach this chapter, more than any other, as a mere introduction to the staggering array of knotty moral problems that beset HCPs.

Values and Priorities

As usual, before beginning, you are encouraged to respond to the following questions, which are intended to initiate introspection about your beliefs, attitudes, and values toward the mentally ill and the mental health fields.

1. Do you think Dr. Myerson acted responsibly?

2. Do you think Dr. Myerson should have made some attempt to warn Tanya Trippett or at least her family?

3. Do you think Dr. Myerson was right in asking the police to take Peter Prudhomme into custody?

4. If you were Tanya Trippett's mother or father, would you hold Dr. Myerson responsible?

5. Do you think the university and the police bear any responsibility for what happened?

6. Would you be inclined to think that people who are mentally ill are potentially dangerous?

7. Recall your earliest contact with mental illness. What feelings did you have?

8. Do you think society's attitude toward mental illness today is enlightened.

9. Would you say that mental illness is different from other illnesses—that someone who is mentally ill is different from someone who is physically ill? If so, how?

10. Would you have any reservations about working with the mentally disturbed?

11. Do you think it should be easier or more difficult to institutionalize people against their will?

12. Which of the four general patient rights would you give priority in dealing with the mentally ill—adequate care, information and consent, privacy, or autonomy?

13. Do you think patients should have legal counsel at an involuntary commitment hearing?

14. Who do you think is potentially more dangerous—a criminal released from prison after serving a sentence or a patient released from a hospital after being certified sane?

15. With whom would you feel less comfortable working: an ex-convict or a formerly institutionalized mental patient?

16. Do you think insurance companies should have access to the health records of their insured mental health patients?

17. Do you think individuals who have been found mentally ill and in need of treatment should be involuntarily committed?

18. How do you feel about the fact that more and more mental patients are being treated in outpatient settings?

Values in Mental Health Care

As we have seen throughout this book, values play a key role in the structure and delivery of health care and in interactions between HCPs and patients. But nowhere in health care is the fundamental role of values more apparent than in the care of the mentally ill.

One reason that helps to account for the centrality of values in mental health care relates to the nature of the field itself. Unlike other health care occupations, mental health work deals with human sickness that does not readily lend itself to scientific analysis, description, prevention, and cure. As numerous commentators have pointed out, the MHP deals with a "sickness of the soul." Rarely can this sickness be seen through a microscope, cultured in a laboratory, or cured by injection, medication, or operation. People in the field even dispute the meaning of the terms *mental illness* and *mental disorder.*

Indeed, in the third edition of its *Diagnostic and Statistical Manual of Mental Disorders* (DSM-III), the American Psychiatric Association faces up to these conceptual problems. Although it concedes that there is no satisfactory definition for *mental disorder,* it does explain why certain conditions were included in DSM-III as mental disorders and others were excluded. Each of the mental disorders in DSM-III is "a clinically significant behavioral or psychological syndrome or pattern that occurs in an individual and that is typically associated with either a painful sympton (distress) or impairment in one or more important areas of functioning (disability)." In addition, there is an inference that there is a behavioral, psychological, or biological dysfunction and that the disturbance is not only in the relationship between individual and society.[1] Still, one might ask what constitutes "a clinically significant behavioral or psychological syndrome or pattern" and who is to decide. This is not to minimize the contribution of DSM-III to both clinical practice and research. Indeed, its authors are to be commended for laying to rest common misconceptions about mental illness such as (1) that a classification of mental disorders classifies individuals, when actually what are being classified are disorders that individuals have, and (2) that all individuals described as having the same mental disorder are alike in all important ways, when in fact they may differ in ways that can affect clinical management and outcome. Nevertheless, conceptual problems remain.

Beyond these conceptual problems, there is another sort of problem: The methods that MHPs use are not the concrete ones observable in the practice of medicine. As one author puts it, "[The MHP] bandages no cuts, administers no pills, and carries no needle. He cures by talking and listening."[2] In short, MHPs are not scientists but clinicians. As such, they must make use of a vast array of techniques and knowledge, make studied selection according to the case before them, and make careful application of what are often highly speculative theories and procedures. Also, they must monitor results that are by no means predictable and they must always be prepared to readjust, even to scrap, their approach.

Granting the accuracy of this occupational description, a profile begins to emerge that is dominated by value concerns. For one thing, given the nature of the clinician role that they play, MHPs must make interpretations and select techniques in a highly idiosyncratic way. Inevitably embedded in these professional judgments will be personal presuppositions, attitudes, biases, and beliefs. Values, both personal and professional, will figure prominently. Personal and professional values color the interactions of any HCP, but they probably affect the MHP more, because of the nature of mental health care.

A related factor that accounts for the impotance of values in mental health care relates to the kind of patients being treated. It seems fair to characterize patients in therapy as suffering from disturbed thoughts and feelings. The source of the disturbance is often a conflict between the conceived and actual self-image. This means that there are discrepancies among the way the mentally disturbed see themselves, the way they want to be seen, and the way they actually appear to the outside world. But neither how we see ourselves nor how we would like to be can be divorced from our values, especially our moral

---

1.   Diagnostic and Statistical Manual of Mental Disorders, 3rd ed. (Washington, DC: American Psychiatric Association, 1980), p. 6.

2.   P. London, The Modes and Morals of Psychotherapy (New York: Holt, Rinehart & Winston, 1964), p. 3.

values. Indeed, the very nature of the conflict bespeaks a value conflict, as does any perception of discrepancy between "is" and "ought."

If this characterization of patients in therapy is even roughly accurate, then it follows that part of the therapist's job, probably a large part, involves helping patients to discover, think about, and rank values in their lives. But again, in doing this, therapists and other MHPs inevitably draw from their own value systems. Thus, therapeutic interaction with patients involves a moral confrontation that makes communication of the MHP's moral values an integral part of mental health work.[3] Moreover, value considerations largely dictate how MHPs define patient needs, how they operate in a therapeutic situation, and how they define treatment and cure—even reality.

It is true that some MHPs ardently believe that therapists should not make moral judgments, that they should avoid entanglement in the web of patient beliefs. By this account, the work of the MHP is to provide a service, to restore the human being to some measure of normal functioning. In short, the closest MHPs should get to values and morality in their practice is in their attempt to help human beings function as they were meant to.[4] This is not the place to debate the merits of this view of therapy and mental health care, but we can observe that this view is itself laden with value judgments, which seem to belie the very claim being made. After all, the act of adopting this position inevitably influences therapeutic interactions with patients. The fact is that therapists cannot avoid making value judgments in assessment, diagnosis, and treatment; nor can they avoid transmitting some of these value judgments to patients, since assessment, diagnosis, and treatment are integral parts of patient care.

At the same time, there are those who would distinguish selecting values and goals from the technical procedures for achieving them. Presumably, the patient always does the selecting; the therapist always decides the procedures. Thus, it is concluded, the claim that therapists are inescapably involved in values transmission has little merit. Two points can be made to counter this analysis. One is that the selection of procedures, the therapy and treatment, involves a value judgment that inevitably affects the care of patients. At the very least, selection raises a question about the weight that should be given to patient autonomy—to participation in one's own care. But beyond this, rarely does a patient come to a therapist with clearly thought-out values and goals. The task of helping patients clarify their goals and values, then, and of helping them set priorities, generally falls to the MHP. In support of this view, one therapist has written:

> At first glance, a model psychiatric practice based on the contention that people should just be helped to learn to do the things they want to do seems uncomplicated and desirable. But it is an unobtainable model. Unlike a technician, a psychiatrist cannot avoid communicating and at times imposing his own values upon the patients. The patient usually has considerable difficulty in

---

3. *London,* p. 11. *See also J.C. Hoffman,* Ethical Confrontation in Counseling *(Chicago: The University of Chicago Press, 1979).*

4. *E. Goffman,* Strategic Interaction *(Philadelphia: University of Pennsylvania Press, 1969).*

finding the way in which he would wish to change his behavior, but as he talks to the psychiatrist his wants and needs become clearer. In the very process of defining his needs in the presence of a figure who is viewed as wise and authoritarian, the patient is profoundly influenced. He ends up wanting some of the things the psychiatrist thinks he should want.[5]

The point is simple but crucial: MHPs' personal and professional values are an integral and highly influential part of their interactions with patients. Evidence of underlying values can be seen in the particular school of therapy that MHPs embrace, in their concepts of what is best for the patient, in their diagnosis and selection of treatment, and in their day-to-day patient interactions. Little wonder, then, that psychotherapy has been described as "an intimate dialogue about moral issues."[6]

If the preceding analysis is fair and accurate, at least two important moral concerns emerge. The first and most obvious deals with MHP responsibility to be clear about their own values and to recognize how these values affect patient care. If they are indifferent to these issues, MHPs will be unaware of what is perhaps the most profound level on which they interact with patients, and thus they will remain oblivious to a vital aspect of the delivery of adequate health care.

A second concern relates to the patient's general right to information and autonomy. Specifically, the following question arises: Should the patient be informed by the MHP of the MHP's working value assumptions? Some say no. They argue that patients rarely are in any position to grasp such concepts, that MHPs should exercise a large measure of paternalism in this area. Others disagree. Indeed, they press for a "therapeutic contract" that, in part, would spell out precisely the values of the therapist. Although in some cases this could not or should not be done, contractualists insist that in most cases it is feasible. Whatever the proper course, one thing is certain: Given the importance of the patient's general right to autonomy and the potential for undermining this right in the therapeutic setting, disclosure of the therapist's values emerges as an important moral issue in mental health care.

The debate that surrounds values and value transmission in mental health care is only one of numerous issues with moral overtones that confront MHPs. One thing that rules out an easy resolution to these issues is the moral plight in which MHPs find themselves.

The Moral Plight of Mental Health Professionals

A plight is an unfortunate or distressing state. In terms of the constraints on their ability to act, the divided loyalties they feel, the roles they are asked to

5. S. L. Halleck, The Politics of Therapy (New York: Science House, 1971), p. 19.

6. N. Hobbs, "Ethics in Clinical Psychology," in Handbook of Clinical Psychology, ed. B. B. Woolman (New York: McGraw-Hill, 1965).

assume, and the delicate balance of patient rights they are expected to strike, the moral state of MHPs can truly be termed distressing. There are several reasons that account for the moral plight of MHPs, two of which bear special note. One is that more and more MHPs are being asked to serve as "agents of the state." The second is that MHPs easily get caught up in a web of relationships that pose conflicting loyalties. Examining each of these issues will not only highlight the moral constraints under which MHPs work but will also carve out key areas of moral interest in the delivery of mental health care.

Social Function

Probably more than any other group of health professionals, MHPs are called on to function as "agents of the state." An occupational reality explains this: For young MHPs, most employment opportunities are not in private therapy but in the area of social control. Dr. Myerson is a good example.

Dr. Myerson spent the early years of his practice employed by the state of Massachusetts. There his duties had little to do with what he was trained in: clinical therapy. On some occasions he was asked by the state to determine whether a person was competent to stand for trial. (Indeed, his colleagues were undoubtedly asked to do precisely the same when the Prudhomme case came to trial.) On other occasions, he was asked to administer a battery of tests, on the basis of which he was supposed to decide whether a person was mentally ill. On still other occasions, Dr. Myerson was asked to distinguish between the drug user and the addict, between the social drinker and the chronic alcoholic. And quite often he was charged with determining who should be institutionalized "in the interests of society" and who was well enough to be turned loose. In each instance, the state was trying to exercise control over "unstable" elements in the society—sometimes "for their own good," more often for the "good of society." And the state engaged the services of Dr. Myerson and other MHPs to help in these control efforts, even though the MHPs were trained to dispense therapy.

For his part, Dr. Myerson felt conflict. On the one hand, by training and inclination he was directed to helping people who were mentally disturbed. On the other, he was functioning as an agent of the state—a role for which he had no training and felt no affinity. Dr. Myserson, like other MHPs, was being drawn increasingly into the problems of society. Despite his inadequacies to do so, he was being asked to render decisions with far-reaching consequences for many individuals.

To make Dr. Myerson's conflict more concrete, consider the events that led up to the passage of his state's "sexually dangerous" law. In the late 1950s, two young boys were sexually molested and killed in Boston. As it happened, the man charged with the crime had been released from prison only months before, after serving a sentence for another sexual offense. Understandably, the citizenry was outraged and demanded action. In response, the Massachusetts legislature immediately enacted a law whereby persons convicted of sexual offenses may be committed for "indeterminate treatment." In other words, they could be hospitalized from one day to a year, in lieu of a criminal proceeding and incarceration.[7]

Clearly, under such a law sexual offenses and offenders would be treated differently from other crimes and criminals. This, of course, presumes that such offenses and offenders are in fact different. In order to establish difference, the legislature summoned a number of MHPs as expert witnesses, among them Dr. Myerson. Although Myerson felt totally inadequate to make such a judgment, he and his colleagues nevertheless were expected to do so. In addition, after the law was in place Dr. Myerson and other MHPs were called on occasion to give expert testimony as to the "sanity" of the offense and offender. And in the event that both were found insane, Dr. Myerson and other MHPs were expected not only to treat the offenders but to decide when the offenders no longer posed a threat to society.

Numerous people in the field are beginning to focus on the problems associated with the MHP's role as an agent of the state. One of them, associate professor of psychology Norman J. Finkel, characterizes the situation this way:

> The new community psychologist is asked to be an active participant in the problems of society, rather than maintaining the passive receptive stance of an age just passing. He is asked to be a "social change agent," a "mental health quarterback," a "social engineer," a "participant-conceptualizer," and a "political activist." As the professional attempts to intervene earlier in problems rather than belatedly, as he attempts to *detect* disorders at their earliest stages and *prevent* them from worsening, or prevent them from occurring at all, he runs grave ethical risks of doing too much, and of intruding too far on rights that are properly left to the individual.[8]

Increasingly, then, MHPs are expected to perform control functions, despite inadequate training. In meeting this obligation, they risk unjustifiable invasion of individual rights. Where does their real duty lie?

Perhaps some political action is necessary to relieve MHPs of this facet of their moral plight. For example, MHPs and their representative bodies might lobby for additional training requirements that would better equip them to function as agents of the state. Were such requirements mandated, MHPs with inadequate training would be protected from illegitimate expectations. Moreover, the state would have an identifiable subgroup of MHPs to turn to for control functions. Of course, this solution assumes that the problem can be remedied by training. But even if it cannot, political action still might be appropriate for relieving MHPs of control functions and placing these functions squarely with another entity—for example, the criminal justice system. In any event, MHPs and their professional organizations seem well-advised to remem-

---

7.   N. Kittrie, The Right To Be Different: Deviance and Enforced Therapy (Baltimore: Johns Hopkins University Press, 1971), p. 176.

8.   N. J. Finkel, Therapy and Ethics: The Courtship of Law and Psychology (New York: Grune & Stratton, 1980), p. 8.

ber that politics is a branch of ethics and that in some instances political action may be the moral course to pursue to remedy a problem.

Multiple Relationships

Because MHPs are involved in multiple relationships, they often get caught between conflicting obligations. As a result, their duty is not clear. Among the many relationships that MHPs get involved in, five stand out: the relationship to patient, to community, to law, to the patient's family, and to "third parties."

PATIENT. In theory, the MHP's primary relationship is to the individual patient. This relationship entails MHP recognition of certain patient rights. One right might be the right of consent and choice about the treatment and direction of any proposed behavioral change. Another might be the patient's right to terminate treatment. Specific patient rights do not concern us at this point so much as the fact that the relationship between MHP and patient is based on the assumption that the individual patient's interests come first. This is a fine assumption, but it does not always hold true in practice, because responsibilities stemming from other relationships intrude.

COMMUNITY. Another relationship involves the MHP with the community. This is especially true of MHPs outside private practice. Thus MHPs like Dr. Myerson must often balance the legitimate interests of the patient against the legitimate interests of the community. In their relationship to community, MHPs like Myerson presumably must work as agents of social regulation. In part this means that they must be concerned primarily with the interests of the group, which may not parallel those of the individual patient. In some instances, MHPs may even have to impose coercive measures on individuals in order to force compliance. The point is that the MHP's relationship to community often involves a set of responsibilities that conflict with the set imposed by the relationship with an individual patient. When these dual responsibilities conflict, ethical dilemmas inevitably arise.

LAW. Although the relationship between law and medicine has always been uneasy, today it bristles with controversy, especially in the mental health field. The reason is that in recent years we have seen an upsurge of interest in the civil liberties of the mentally ill. Various laws have been enacted, bills submitted to legislatures, and public debate begun concerning the mentally ill person's rights to treatment, to a humane environment, to the least restrictive alternative to hospitalization, and to services tailored to their individual needs. The upshot is that the patient-MHP relationship has been greatly altered. More to the point, MHPs now find themselves with legal obligations that are not always clear, often conflict, and sometimes seem to undercut adequate health care.

A number of courts and legislatures have attempted to assure procedural safeguards for patients prior to commitment. For example, in one case a court ruled that patients have a right to be present at the commitment proceedings, to be represented by counsel, to cross-examine witnesses, and to call witnesses of their own.[9] This ruling has serious implications for health care and MHPs. For one thing, it means that MHPs will probably have less time available for direct patient care. For another, it means that they will probably think twice, even beyond, before forcing treatment—*even in cases where they think treatment is needed.* In addition, it seems safe to speculate that, with increasing frequency, patient-MHP relationships will be strained by the adversarial role MHPs are expected to play. Thus, although such a ruling attempts to buttress the rights of patients to due process, it also threatens their right to adequate health care and the MHP-patient relationship. The situation can be characterized as a conflict between legal and medical priorities. The law's primary interest is the patient's civil liberties; medicine's primary interest is the delivery of adequate health care. Obviously both civil liberties and adequate health care are patient rights. The question is: How are MHPs to resolve conflicts between the two? Where does their primary responsibility lie?

Among the most troublesome areas is the patient's right to refuse treatment. On face value this right appears to be basic, but implementation is fraught with problems for the MHP. For one thing, short of allowing every patient this right, MHPs presumably have to make judgments about which patients are capable of exercising it. More likely than not, MHPs will be rather liberal in their judgments lest they later be charged with violating civil liberties. Undoubtedly the recognition of such a right would eliminate many abuses in the mental health field, but it would also effectively deny some patients needed care.

Another problem looms as large but for a different reason. Suppose a patient judged competent refuses treatment. MHPs treating the patient then face an agonizing decision. If they discharge the patient, they may later be found liable for malpractice should the patient harm self or others. If they retain the patient but give no treatment, they may be found to have violated the patient's constitutional right to liberty.[10] Either way, the MHPs lose. This double bind undercuts moral accountability by constraining the MHP's capacity to choose. Even worse, it encourages preoccupation with law and malpractice rather than with health care and patients.

PATIENT'S FAMILY.   It is a psychological truism that functioning families are the building blocks of mental health. Even the greenest of MHPs is aware of the overwhelming evidence linking troubled families to psychological disturbances in its members. Moreover, for many MHPs the first patient contact is the indirect one that comes as a result of a harried plea for help made by a family. "My teenage son hasn't left the house in a month!" a frantic mother reports. "Our father is in a terrible black funk again and is threatening to kill himself!" a couple of worried children say. "My mother is drunk again," a daughter sadly phones in, her voice trailing off in despair. Obviously these families are not merely reporting their domestic trouble to a perfect stranger. Rather, they do not know

---

9. *Lessar v. Schmidt, 349 F. Supp. 1078, (E.D. Wis. 1972).*

10. *L. E. Kopolow, "Patients' Rights and Psychiatric Practice," in* Law and the Mental Health Professions, *ed. W. E. Barton and C. J. Sanborn (New York: International Universities Press, 1978), pp. 264–265.*

what to do about their problem, and they hope that the person on the other end of the telephone or the other side of the desk has the answer.

This does not mean, however, that families do not harbor and express expectations. They do, and those expectations present problems for MHPs. "Give him some medication," says the mother. "Stop him before he hurts himself," plead the children. "Put her in a hospital," demands the daughter. Such family requests for psychological intervention are common. If MHPs acquiesce to a family's demands, they might easily jeopardize their relationship with the patient. But ignoring the family's expectations or treating them lightly can undermine the sound family-MHP relationship that is so invaluable in treating the disturbed member.[11] The MHP's relationship with the family can become even more complex when the two do not share a common view of diagnosis, assessment, and treatment. In that instance, MHPs must juggle duties to respect the family's prerogative with duties to deliver adequate health care to the patient.

Similar problems arise when the principals are a married couple, one of whom, by the account of the other, is mentally disturbed. "My husband is dangerous and should be committed," a troubled wife tells an MHP. Should the MHP act on the claim? If so, how? In the same vein, MHPs are often asked to give testimony about the mental competency or potential dangerousness of a spouse. Should they?

THIRD PARTIES.   Third parties are individuals other than the patient who are paying for mental health services. MHP relationships with third parties can raise a real question about who is the client—patient or bill payer.

For example, a colleague of Dr. Myerson's, Jeannine Mittner, an M.S.W. (master of social work) and mental health specialist, is employed in industry. Often she confesses to Dr. Myerson that she does not really know where her primary loyalties lie—with the company that is paying her salary or with its employees whom she is treating. As she puts it, "I'm not sure who my client really is." Indeed, where a third party is involved, such a question inevitably arises.

It is vital to determine who the client is in the analysis of the ethics of psychological interventions because patients presumably should participate in establishing the goals and treatment, and MHPs should be accountable to them. When the patient and bill payer are not the same, serious problems involving informed consent, privacy, and autonomy can easily arise.

Take Mittner's case, for example. Since the company is paying the lion's share of the services she provides its workers, it feels entitled to the information she is collecting on them. Mittner understands the employer's claim to the data, but she nevertheless feels conflict. "After all," she tells Dr. Myerson, "the patient also has a right to privacy and confidentiality." Myerson agrees. Then she asks him pointedly, "How am I supposed to balance these conflicting interests?" How indeed, especially when insurance coverage depends on

---

11. *Finkle, p. 4.*

the submission of detailed records concerning diagnosis, assessment, and treatment?

We need not examine further the web of relationships in which MHPs like Myerson and Mittner often find themselves. Suffice it to say that each relationship involves a client who has certain rights that entail MHP responsibilities. Because the interests of the various parties do not always run parallel, MHPs must juggle competing claims and decide where their primary responsibility lies. Taken together with the social control function that MHPs are increasingly asked to perform, this network of relationships makes moral decision making very difficult for MHPs.

So far, this chapter has focused on two general observations about mental health care. One is that the practice of mental health care is embedded with values; the other is that MHPs operate within a network of relationships. Both of these points bear directly on key issues in the delivery of care to the mentally ill. This chapter cannot touch on all of these issues, but it will now turn to three of profound importance: involuntary commitments, competency and consent, and privacy.

Involuntary Commitments

Involuntary commitment will be used to refer to cases where people are institutionalized against their will. Since by definition involuntary commitments violate the patient's right to consent and autonomy, they always raise moral concerns. To understand the current status of involuntary commitments and the moral problems surrounding this issue, it helps to know something about the modern development of mental health care.

In the late nineteenth century, the prevailing view in mental health stressed the biological factors in the development of mental *disease*. Thus the focus was on observation, description, and classification. Sigmund Freud (1856–1939), however, viewed psychological factors as paramount in mental disorders. As he gained prominence, the individual's psychodynamic process received primary attention.

During the first half of the twentieth century, psychiatrists and mental health professionals had time to reevaluate their assumptions, do considerable research, apply new knowledge to patient treatment, and develop new organizational arrangements. As a result, by the late 1950s and early 1960s, a series of changes revolutionized mental health care.[12] "Open" hospitals, work-release programs, informal admissions, and the aftercare concept all became familiar and integral parts of health care. New treatments, such as electroconvulsive treatment and chemotherapy, shortened hospital stays and allowed for treatment in a general hospital setting, even in MHP offices. The emergence of community mental health centers in the mid-1960s provided early entrance into

---

12. W. E. Barton and C. J. Sanborn, Law and the Mental Health Professions (*New York: International Universities Press, 1978*), part VII introduction.

therapy, crisis intervention, and ambulatory care. Underlying these and other changes was a new awareness of the role that social factors play in mental illness. Increasingly, disturbed individuals came to be seen as integral parts of society, indeed as products of their environment. Social factors came to be seen as contributing conditions of mental illness, together with biological and psychological considerations.

Given these changes in care and the growing emphasis on social factors, one could anticipate that mental hospitals, which had been exanding in the first half of the century, would begin to decrease in resident population. They did, in some cases by as much as 50 percent. People who would have been institutionalized just a few years earlier no longer were. They were treated primarily in some outpatient setting. It was at this point, when more* and more mental patients were being reintegrated into society and treated in the least restrictive environment, that legal activism and judicial involvement sought to regulate the public practice of psychiatry.[13] There followed a spate of court decisions dealing with the rights of the mentally ill.

Common to many of the court decisions in the 1970s was an emphasis on personal freedom over a determined need for treatment. For example, in *Suzuki v. Quisenberry,* a federal court judge in Hawaii ruled that people could not be hospitalized involuntarily solely because they were mentally ill.[14] Similarly, the year before this case, the U.S. Supreme Court held in *O'Connor v. Donaldson* that nondangerous patients who are capable of surviving in freedom by themselves or with the help of a willing and responsible support system may not be confined.[15]

On first look, such rulings appear to be beyond reproach. After all, given that individuals stand to lose their liberty and have their will violated, it seems that only the most pressing reasons warrant involuntary confinement. At the same time, however, these rulings turn up a number of moral problems for MHPs. In approaching these problems, it is helpful to understand the two involuntary commitment procedures: emergency and nonemergency procedures. States vary widely in their emergency and nonemergency procedures, but cautious generalizations can be made that crystallize the moral issues involved.

Emergency Procedures

Recall that, when Dr. Myerson thought Peter Prudhomme was a potential danger to Tanya Trippett, he requested that the police detain Prudhomme for observation. This action would be termed an emergency commitment. In most cases the individual is not detained in a jail but in a hospital. In any event, the purpose of an emergency procedure is the same: to protect citizens from mentally ill people who might be dangerous. Furthermore, emergency procedures are always enforced ostensibly for observation and diagnosis. In Prudhomme's

---

13.   *Barton and Sanborn, p. 311.*

14.   *Suzuki v. Quisenberry, 411 F. Supp. 1113, D. Haw, 1976.*

15.   *O'Connor v. Donaldson, 422 U.S. 563, 45 L. Ed. (2d 396 at 407, 1975).*

case, and in the actual Tarasoff case, the individual was released after seventy-two hours. States differ in their regulations, but emergency commitment is always brief, rarely exceeding fifteen days.

The first thing to note about an emergency commitment is that it is a misnomer.[16] In almost all states, the so-called emergency procedure is in fact the *standard* procedure. In New York City, for example, almost all involuntary patients are hospitalized initially under the emergency procedure. The chief moral significance of this practice does not lie in its noncompliance with the spirit of the law but in its potential for violating patient liberties. After all, under emergency procedures patients generally enjoy fewer rights and safeguards than they would in nonemergency procedures. Indeed, this is precisely why the emergency procedure is supposed to be used *only* in exceptional cases.

The second thing to note about emergency procedures is that they are based on the immediate protection of society from potential harm. This concern for public safety, termed the *harm principle,* was stated most clearly and forcefully by John Stuart Mill in his essay *On Liberty.* In that most influential work, Mill stated:

> The only purpose for which power can rightfully be exercised over any
> member of a civilized community against his will is to prevent harm to others.
> . . . His own good, either physical or moral, is not a sufficient warrant. He
> cannot rightfully be compelled to do or forbear because it will be better for him to
> do so, because it will make him happier, because, in the opinion of others, to do
> so would be wise, or even right. These are good reasons for remonstrating with
> him or reasoning with him, or persuading him, or entreating him, but not for
> compelling him.

Clearly Mill condoned paternalistic intervention *only* to prevent harm to others. In effect, this is the underlying rationale for most emergency commitment procedures. It is why Dr. Myerson intervened and why the psychiatrist in the Tarasoff case did. In each instance the physician judged the patient potentially dangerous.

But are these psychiatrists, or any other MHP for that matter, in a position to say who is potentially dangerous? Granted they are qualified to judge who is mentally ill, but mentally ill is not the same as potentially dangerous— misconceptions to the contrary notwithstanding.[17] The fact is that there is no support for the assumption that the mentally ill are more dangerous as a group than the general population.[18] Furthermore, if the record is any indication, MHPs are not very good forecasters of dangerous or aggressive patient behavior. Indeed, they seem less accurate in their predictions than police officers and social workers are.[19] An example often marshaled to support this observation is the so-called Operation Baxstrom.

16.   B. J. Ennis and L. Siegel, The Rights of Mental Patients *(New York: Avon, 1973), p. 17.*

17.   *J. M. Livermore, C. P. Malmquist, and P. E. Meehl, "On the Justification for Civil Commitment,"* University of Pennsylvania Law Revue, *117 (1968): 75–96.*

18.   *B. J. Ennis and T. R. Litwach, "Psychiatry and the Presumption of Expertise: Flipping Coins in the Courtroom,"* California Law Revue, *72 (1974): 716.*

19.   *Ennis and Siegel, p. 21.*

In New York in 1966 there were about a thousand mentally ill patients whom psychiatrists had certified as too dangerous to be housed in regular civil mental hospitals. Psychiatrists claimed that only the Department of Correction's high-security hospitals could handle such "potentially dangerous" patients. But because of a Supreme Court decision in the same year, all were transferred to civil mental hospitals.[20] As it turned out, the psychiatrists' predictions were almost totally wrong. To be exact, their forecasts proved right in only seven cases.[21]

The main problem facing MHPs in predicting dangerous behavior, then, is one of definition. What constitutes "dangerous"? There is no agreement. In some states, an individual who is a threat to property is considered dangerous. But in the Suzuki case, the court ruled that involuntary commitment on grounds of being dangerous to property violates the due process clause of the Fourteenth Amendment. Other states identify dangerousness with threat to others. But precisely how is this to be determined? Is someone babbling on a street corner to be considered a threat to others or just a public nuisance? Despite these conceptual problems, MHPs are expected to and indeed do make judgments about potentially dangerous behavior.

But emergency procedures raise not only legal concerns. Court rulings are indeed important for understanding the nature and level of the debate that enshrouds this issue. But there are other vital moral issues that emerge as a result of invoking Mill's harm principle for involuntary emergency commitment. First, individuals who, by their own admission, are not qualified to make judgments about who is potentially dangerous are being asked to do just that. Surely this raises a question of fairness from the viewpoints of both MHP and patient. Second, given that MHPs, as much as the lay public, tend to equate "mental illness" with "potentially dangerous," it is safe to surmise that MHPs will base their judgment of dangerousness on their diagnosis of mental illness. Indeed a judgment of "potentially dangerous" would seem an integral part of a judgment of "mentally ill." Such a facile but erroneous correlation raises questions about the legitimacy of denying patients their freedom and liberties. Third, the danger standard, in effect, is a form of preventive detention. It deprives one of liberty on the basis of a *potential, future* act. Thus, the danger standard invites grave misgivings about the moral legitimacy of a social policy that denies a segment of the population, the mentally ill, equal protection under law.[22]

Nonemergency Procedures

In contrast to emergency procedures, nonemergency procedures for involuntary commitment generally are invoked to *help* the patient. As such they focus on treatment and the patient's best interests and not on the protection of society from potentially dangerous behavior. Usually, nonemergency procedures follow a diagnosis of "mentally ill and in need of care, treatment, or custody." The

---

20.   *Baxstrom v. Herold, 383 U.S. 107, U.S. Supreme Ct. (1966).*

21.   *Ennis and Siegel, pp. 21–23.*

22.   *Finkel, pp. 50–51.*

length of institutionalization under these procedures varies but is always longer than under emergency procedures. In some states, in fact, confinement can be indefinite. Because the patient's freedom can be denied so long, nonemergency procedures warrant special moral scrutiny.

One moral issue immediately apparent in nonemergency procedures involves paternalism. In the case of emergency procedures, intervention is based on the harm principle, on the belief that the individual is potentially dangerous to others. But in nonemergency cases, intervention is based on the claim that the individual is mentally ill and should be helped or treated. Thus the intervention is justified by appeal to the person's own good. Whether or not that defense is compelling depends largely on the validity of the general arguments for and against paternalism outlined earlier. At the same time, the questionable competency of the mental patient introduces a unique feature into the paternalism-autonomy debate, which we will explore later.

Beyond the paternalism issue, nonemergency intervention raises other moral concerns. The most important ones cluster around the inevitable nonemergency hospitalization commitment hearing. In his treatment of this subject, Norman Finkel points out that, because commitment hearings are considered civil, not criminal, the patient's full due process guarantee may be diminished.[23]

One guarantee is "notice." In many states, notice of the purpose, time, and place must be given patients prior to judicial action. But in other states notice is ignored. Apart from the legal proprieties of denying patients notice, moral questions arise about the patient's general right to information. It is difficult to imagine information more relevant to patients than why, when, and where the hearing will be held that will determine whether or not they remain free.

A second guarantee involves counsel. Although in criminal cases the accused who cannot provide counsel for themselves automatically receive it, there is no such automatic consideration granted patients at commitment hearings. More often than not the patient must demand legal representation. And yet there is considerable reason to think that the patient's right to adequate health care and autonomy depends on counsel. There may be, for example, scheming relatives to be warded off, incompetent and lax medical judgments to be combated. In addition, there always looms the possibility that ability to pay will determine the care and treatment patients get or that patients will get lost in the bureaucratic shuffle.

A third diminished guarantee is the right to be heard and to cross-examine. Finkel feels that, because the loss of liberty can be just as great as in criminal cases, there is no sound reason why patients should be denied the right to cross-examine MHPs, family members, or anybody else who volunteers evidence in support of hospitalization.

A fourth guarantee refers to protection against self-incrimination as guaranteed in the Fifth Amendment. In a criminal proceeding, the defendant may remain silent and suffer no penalty. But in involuntary commitment hearings,

---

23. *Finkel, pp. 51–62.*

silence is often construed as a symptom, not a right; as an indication that the patient is hiding some mental defect. As a result, the silence is taken as confirmatory evidence for mental illness and confinement.

A fifth guarantee involves proof beyond a reasonable doubt. Again, because the person's liberty is involved, a serious question emerges about the amount and quality of evidence submitted for institutionalization. Some states require the same standard as a criminal proceeding—that is, proof beyond a reasonable doubt. But many states require something less: a preponderance of evidence, sufficient evidence, or a majority of evidence. Cases such as *Suzuki*, however, suggest that the courts recognize that the transcendent value of an individual's liberty requires as much evidence as demanded in a criminal case.

Clearly the concerns in nonemergency commitments are with the patient's autonomy and due process, which presumably stand a better chance of being respected the more the hearing becomes like a criminal proceeding. But the other side of the coin is the patient's right to adequate health care. Although "criminalizing" the hearing may afford patients greater guarantees of civil liberty, in some instances it may undercut health care.

Perhaps the most striking examples of this effect can be seen in California, whose Lanterman-Petris-Short (LPS) Act of 1969 made involuntary hospitalization for the mentally ill beyond seventeen days extremely difficult to achieve. By this law, need for hospitalization was no longer the standard for involuntary detention. Thus, even if MHPs feel that someone is desperately in need of mental health treatment, institutionalizing them beyond the prescribed limit will prove an undertaking of major legal proportions. What has been the effect of LPS? Some California MHPs believe that in many cases it has undermined health care for the mentally ill. One psychiatrist, who works with a county courts and corrections unit, puts it this way:

> There is no easy way to know how many mentally disordered persons are being routinely processed by the criminal justice system into jail and prison, persons who before LPS could have been detained and treated in mental hospitals. Those diverted to hospitals by being considered incompetent to stand trial may be only a tiny fraction. A California prison psychistrist said in a recent newspaper interview, "We are literally drowning in patients, running around trying to put our fingers in the busting dikes, while hundreds of men continue to deteriorate psychiatrically before our eyes into serious psychoses. . . . The crisis stems from recent changes in mental health laws allowing more mentally sick patients to be shifted away from the mental health department into the department of corrections. . . . Many more men are being sent to prison who have serious mental problems.[24]

In short, a conflict has developed between the patient's right to autonomy and due process on the one hand and adequate health care on the other.

Apparently, judicial concern with the patient's civil rights has interfered with quick access to treatment on an involuntary basis. Increasingly, emphasis is

---

24. *M. F. Abramson, "The Criminalization of Mentally Disordered Behavior: Possible Side Effect of a New Mental Health Law,"* Hospital Community Psychiatrist 23 (1972): 105.

put on outpatient treatment and the least restrictive alternative. Many MHPs are beginning to feel that, in its enthusiasm to protect individual freedom, society may in some cases be undercutting health care. MHPs find themselves squarely in the middle of this clash of rights and values.

Compounding the problem is the whole issue of competency. Some MHPs argue that a preoccupation with patient civil rights is misguided because many patients simply do not have the capacity to understand and use the rights that courts have gone to great lengths to delineate. But patient rights activists reply that this is all the more reason to secure rights through legal counsel. In order to respond intelligently to these claims and counterclaims, one must have some understanding of the issues that envelop competency, the subject to which we turn next.

Before leaving the topic of involuntary commitments, however, we should realize that more than legalities are involved. Embedded in the legal concerns are fundamental value judgments about patient autonomy, rights of due process, and health care. Furthermore, as we have seen elsewhere, even when the law is clear and precise on an issue, moral questions remain about (1) whether the law is fair and just, (2) whether one ought to follow the law, and (3) whether one ought to work to change the law—and if so, in which direction. Furthermore, laws are inadequate to cover the wide range of cases that fall under them and that require individuals to make momentous decisions. Of course, the law is far from being clear, precise, and consistent in respect to involuntary commitment and the procedures that surround it. Furthermore, many rulings speak directly to patients' civil rights but not to their treatment—which MHPs have a basic obligation to provide. There are glaring exceptions, of course, as in Alabama's *Wyatt* v. *Stickney* (1971), a momentous ruling that went far to spell out the rights of mental patients, including their right to adequate treatment.

One of the chief issues in the Stickney case was whether the state could involuntarily commit a mentally ill person and provide mere custodial care. This practice had evolved in Alabama and other states as a means of giving relief to families burdened with caring for the mentally ill. The court rejected this justification for involuntary commitment on the grounds that personal liberty was far too precious to be sacrificed for the comfort or convenience of others. The court did, however, accept the view that patients could involuntarily be committed provided they had the right to treatment aimed at restoring them to freedom. Such a ruling was a long-overdue righting of an unjust set of conditions that had led to individuals being unnecessarily imprisoned for their entire lives. At the same time, in ordering that mental hospitals be refurbished to meet court standards, the ruling appeared to subordinate the development of ambulatory care as the primary focus of the health delivery system. Was its focus a proper one? Did the court overstep its judicial charge? Such questions are not intended to detract from the practical merit of the Stickney ruling or to question its philosophical or practical need. Rather it is meant to point up that even judicial rulings that are most commendable for raising the level of debate about the rights of the mentally ill and for providing a framework within which to consider these rights are nonetheless inadequate to resolve the underlying moral issues evident in the delivery of mental health care. Indeed, in many cases these rulings have raised additional issues or made the analysis of perennial ones more complex.

Competency and Consent

Competency is a legal concept that deals with a person's mental capacity to perform an act.[25] *Mental capacity* generally is determined by appeal to the thought-feeling-behavioral patterns of a person, depending on the specific legal act at issue. *Performance of an act* has numerous meanings, depending on the context. Thus a person may be declared incompetent to make a will, enter into a contract, or manage business affairs; that is, the person is not considered to have the mental capacity to perform these specific acts. Most frequently, however, competency deals with the ability to assent or consent.

Where competency deals with the ability to consent, the act is "one of adequate intellectual comprehension unaffected by significant factors which would interfere with decision-making powers."[26] Regarding informed consent and the mentally ill, then, a fundamental issue concerns patients' mental capacity to comprehend a proposed psychological intervention being directed toward them. If they have this capacity, then they can give informed consent; if they do not, then they cannot consent.

No MHP would object to consent as a concept. After all the MHP-patient relationship is based on open communication, on the belief that the patient should assume an increasing measure of responsibility, and on the assumption that patients should participate in the decisions that affect their health and lives. But accepting this as a concept and trying to implement it in the case of the mentally ill are two different things.

Clearly, the feature that gives the issue of informed consent its unique twist is the diminished mental capacity of the mentally ill. A patient's attention span and concentration may be limited, and memory recall may be impaired. Furthermore, a patient may be in the throes of depression, aggression, or anxiety. If in an acute phase of mental illness, patients may make decisions that they would not make were they free of disabling mental symptoms. Granted that symptoms of mental illness do not necessarily render a patient incompetent to give consent, they may do so all the same. One of the mental health expert's most important duties, then, is to help determine whether the patient has a mental disease or defect that so affects judgment, decision making, and behavior that consent is not possible.

A determination of incompetence—although strictly a legal matter—has all sorts of moral ramifications. One of the most serious is that patients who are found incompetent in effect forfeit their autonomy, at least for a period of time. This forfeiture means that someone else will exercise control over the decision

---

25.   I. N. Perr, *"The Many Faces of Competence," in* Law and the Mental Health Professions, *ed. W. E. Barton and C. J. Sanborn (New York: International Universities Press, 1978), p. 211.*

26.   *Perr, p. 212.*

about hospitalization, the environmental conditions under which they will live, and the treatment they will receive. Even though those making these decisions may make them in a most commendable paternalistic mode, the preclusion of individual autonomy always raises moral concern.

Of course, it could be and is argued that, by definition, the incapacity to consent means that the person is incapable of exercising autonomy. But this is not necessarily so. For example, patients who may not be capable of consenting to institutionalization may be quite capable of determining whether they want to submit to electroconvulsive treatment. The problem is that, once the court, in consultation with MHPs, renders a decision of incompetency to consent, the ruling has a ripple effect, at least temporarily, over the facets of psychological intervention. Furthermore, there is no guarantee that, shortly after being hospitalized, the patient will not regain the mental capacity to consent—or that it will be acknowledged and respected. Undoubtedly, this is in part why professor of law Henry H. Foster includes the following right in his proposed Declaration of Interest and Rights of Patients: "A patient retained in a medical facility without his informed consent shall have the right to periodic medical review of his case which shall be made at least at the end of every six-month period or before the end of any period stipulated by court order."[27]

In addition to this basic moral concern, there are others that raise questions about adequate health care, informed consent, and autonomy. Many of these arise when MHPs are summoned by courts as expert witnesses in competency hearings.

Traditionally the law has been quite liberal in allowing a wide variety of people to offer opinions about a person's mental functioning. But the court, which ultimately decides competency, is generally influenced by the background of the person offering the opinion. Indeed, in some cases the law may place limitations by statute in this regard. For example, for some purposes and in some states, the testimony of a physician is required; in others, a psychiatrist may be the only one legally allowed to offer an opinion; in still others, opinions may be restricted to psychiatrists and psychologists. Many of the issues involved in competency hearings are based exclusively on medical-psychiatric parameters, thus making only medical information appropriate. But a particular case may involve intellectual functioning, which psychologists can measure; or it may involve social-environmental data that MHPs other than psychiatrists are best qualified to offer.[28] By relying exclusively on a medical frame of reference, then, courts introduce a bias into their competency evaluations that may lead to erroneous judgments with serious consequences for the patient's right to adequate care, informed consent, and autonomy. It seems that at the very least MHPs should be aware of this problem and apprise others of it, especially the judiciary. In this way, they not only meet obligations as patient advocates but help to ensure that patients receive proper care.

---

27. H. H. Foster, Jr., "Informed Consent of Mental Patients," in Law and the Mental Health Professions, ed. W. E. Barton and C. J. Sanborn (New York: International Universities Press, 1978), p. 91.

28. Perr, p. 214.

In addition to these problems, there are some that pertain almost exclusively to the MHP-patient relationship—although they, too, involve striking a reasonable and moral balance between patient rights to adequate care and informed consent. MHPs are often called on to secure patient cooperation for needed therapy without at the same time so alarming them as to vitiate informed consent. This is a recurring problem in the HCP-patient relationship, but it is especially troublesome in dealing with patients who are mentally disturbed.

Similarly, many of the treatments and medications used in mental health therapy do not have the highly predictable effects of their nonpsychological counterparts. Drugs used in psychiatry in identical doses may induce a remission of symptoms in one case, have no effect in another, make another worse, and possibly induce a fatal blood disorder in still another.[29]

Comparable problems arise in psychotherapy generally. In some instances, patients have been known to develop a dependence on or to get emotionally involved with their therapist. Perhaps more serious, psychotherapy always involves risks to significant aspects of the patient's life: marriage, job, religion, and so on. Indeed, in some isolated cases psychotherapy seems to have catalyzed suicidal urges. As a result, many would argue that patients are entitled to know the risks they run in psychotherapy, that anything short of a complete disclosure is insufficient for informed consent.

The patient's general rights to information and informed consent, then, seem particularly pronounced in psychological interventions. Against this must be balanced their right to adequate health care. The moral challenge for MHPs is therefore to strike a reasonable balance between the rights to adequate care and informed consent. To repeat, this challenge does not appear to have been eased by recent court decisions. Although in many instances these rulings have spelled out the civil rights of patients, they have left many in the mental health field choosing to practice defensive care—sacrificing treatment rather than risk judicial censure. As well-intentioned as such social policies may be, they are morally suspect if they have the effect of (1) torpedoing care for the mentally ill, (2) making MHPs more preoccupied with "guesstimating" judicial responses to psychological intervention than with patient treatment, or (3) so constraining MHPs' capacities to make decisions that they cannot rightly be held accountable.

Privacy

So far, we have seen that MHPs face unique problems in trying to meet the rights of the mentally ill to adequate care, information, and autonomy. The problems are no less unique when they try to honor the patients' right to privacy. Earlier we saw that MHPs are often in conflict when they try to respect the patient's civil rights and provide treatment at the same time. A similar problem arises with respect to privacy, as the Prudhomme case well illustrates. Recall that Dr. Myerson felt obliged to invoke emergency procedures for com-

---

29. *Barton and Sanborn, p. 313.*

mitting Prudhomme because he considered the patient potentially dangerous. Clearly, Myerson attempted to comply with the law. But he could not possibly do this without violating Prudhomme's right to privacy and confidentiality. Indeed, had he taken additional measures to warn the victim, he would further have sullied Prudhomme's privacy. And yet, judging from the actual Tarasoff case, the court apparently expects MHPs to do precisely this. Again, then, a fundamental tension between conflicting rights underlies this issue. On the one hand, like any other patients, the mentally ill have a general right to privacy; on the other hand, individual citizens have a right to be protected from harm. Thus, MHPs like Dr. Myerson and Dr. Moore (in the Tarasoff cased) find themselves having to arbitrate the interests of two claimants. Furthermore, they must decide to what lengths they should go to prevent harm. Myerson thought he had gone far enough; any further action, he thought, and Prudhomme's privacy would be unjustifiably violated. Dr. Moore probably reasoned similarly. Yet in its Tarasoff decision, the court disagreed; it felt Moore had an obligation to warn the victim. Legalities aside, the moral issues remain: MHPs must still rank in order competing duties that derive from the rights of two claimants.

By the same token, competing duties involving privacy can sometimes derive from the opposing rights of the same claimant. The Prudhomme case also illustrates this point. Dr. Myerson and all MHPs have a duty to preserve in confidence the particulars of professional transactions with patients. But they also have an obligation to respect the autonomy and civil rights of their patients. There was no way that Dr. Myerson, or Dr. Moore, could have respected the due process rights of his patient without at the same time violating the patient's confidences. Whereas the first point concerns a conflict between the patient's right to privacy and the citizenry's right to protection from harm, here the conflict is between two patient rights: privacy and due process. In the actual Tarasoff case, then, Dr. Moore had to juggle three eggs: the patient's right to privacy, the patient's right to due process, and the citizenry's right to protection from harm. The court felt he had dropped the third. Maybe so, but one cannot help wondering what the court might have ruled if (1) Moore had informed the victim and (2) Tarasoff had not killed her but instead had sued Dr. Moore for breach of confidentiality. In any event, the legalities concern us only insofar as they help crystallize the issues and show the moral constraints under which MHPs operate today. In the last analysis, moral issues remain—including the conditions under which patient privacy may be violated, if any.

For simplicity, the remaining problems involving privacy can be grouped around three relationships in which MHPs find themselves: relationships with outside parties, with the patient, and with the profession.

Outside Parties

Earlier we looked at the implications of MHP relationships with third parties. The topic deserves to be recalled here because third parties often expect access to the confidential files of patients. Recall Jeannine Mittner's dilemma. She understood the employer's desire for, and perhaps right to, some confidential information; at the same time she felt an overriding duty to preserve her client's privacy. Such situations again underscore the importance of defining *client.*

Beyond this concern, third-party interests raise another moral issue, a more subtle one. Inevitably MHPs react to third-party requests for confidential data with the same sort of righteous indignation that Mittner felt. Often they guard the records of the mentally ill with far greater diligence than other HCPs exhibit in regard to nonmental patients. Indeed, MHPs together with mental patients and the general public, seem preoccupied with hiding psychological interventions in as much secrecy as possible. We still consider mental illness and the mentally ill as different, a view that makes any enlightened attitude about mental illness and treatment very difficult. The question is: Do MHPs unwittingly retard enlightened attitudes and undermine health care by wrapping the records of the mentally ill in such profound secrecy and by insisting that those records be treated differently from those of nonmental patients? Indeed, they may.

For example, MHPs call out for equal insurance coverage and treatment for their patients and for the abolition of discrimination—each of which is necessary to assure proper care. Yet at the same time MHPs are often among the strongest exponents of special consideration for the privacy of the mentally ill and resist insurance company requests for patient information. Given the potential for abuse of highly sensitive material, MHP concerns are understandable. Still, it seems that MHPs must recognize that, as well-intentioned as their defense of patient privacy may be, ultimately it may have the effect of slowing down the development of more enlightened attitudes and of precluding the quality care that patients need. In this instance, MHPs must face up to the fact that the choice may reduce to insurance coverage without strict confidentiality or strict confidentiality without coverage.[30] In either event, since patients have the most to lose, it seems that they should be intimately involved in decisions relating to the disposition of their records.

The Patient

A second group of privacy issues clusters around the MHP's relationship with patients themselves. Often patients request to see their records. The result poses as much a dilemma for MHPs as for any other HCP in a similar situation. On the one hand, MHPs have a general obligation to respect the patient's right to the material; on the other, they must guard themselves against inflicting needless harm that patients may suffer from seeing the records. This problem was considered in Chapter 6, which discussed the patient's right to information. What was said there applies here with equal force.

There is, however, another aspect of granting mental patients access to their records that speaks directly to a privacy right, not of the patient but of outside parties. For example, it is rather common for the reports of relatives, friends, and associates of the patient to find their way into a psychiatric record. When that

---

30.   J. Donnelly, ''Confidentiality: Myth and Reality,'' *in* Law and the Mental Health Professions, *ed.* W. E. Barton and C. J. Sanborn (New York: International Universities Press, 1978), pp. 193–194.

happens, protecting the privacy of these individuals becomes a moral issue. This does not mean that the outsider's right to privacy necessarily overrides the patient's right to access; but the outsider's is a legitimate privacy claim that MHPs must be aware of. This is especially true since most of such anecdotal testimony is at best hearsay, at worst falsehood. Exposure to such information, then, not only raises questions of potential harm to the patient but of serious violations of privacy to those who submitted the information. A further complication is the question of whether the patient's right to confront accusers overrides the informant's right to privacy.

The Profession

We have seen how moral problems involving privacy can arise for MHPs in their relationships with outside parties or patients themselves. Additional privacy problems can evolve from the MHP relationship with the profession. The most frequently addressed problem involves privilege, which is a variety of confidentiality.

Psychiatrist John Donnelly provides a useful definition of *privilege:* "A specific exception to the customary rules of evidence in the administration of the judicial system whereby one individual may prohibit the testimony of another with whom he has had a strict professional relationship recognized by law."[31] As Donnelly further points out, the basic reason for privilege is that it advances the ultimate public welfare and is rooted in the prohibition against self-incrimination. The central moral question concerning privilege is who should claim it. Should privilege belong to the patient or to the MHP? This question is most important to answer, for if privilege is ultimately a professional right, then even in cases where patients waive privilege, MHPs can invoke it.

As an illustration, consider a case that was heard in the California Supreme Court in 1970.[32] The case involved a teacher named Housek who allegedly was assaulted by a man named Arabian. Housek sued for compensation for physical injuries, pain, suffering, and severe mental and emotional distress. In a deposition he stated that he had received psychotherapy for about six months, ten years earlier, from Dr. Joseph E. Lifschutz. When Dr. Lifschutz was subpoenaed, he refused to testify, even as to whether he had treated Housek—although Housek had waived privilege. Lifschutz claimed that the disclosure would interfere with the practice of psychotherapy generally and with his livelihood in particular. The court insisted that privilege was the patient's right, not the therapist's; since the patient had waived privilege, the therapist was obliged to testify. In the end Lifschutz did testify.

Although this court case provides a convenient frame for considering the issue of privilege, it does not resolve the moral issues embedded in it. (Indeed, it does not even resolve the legal ones, for history reveals that the Lifschutz

---

31. *Donnelly, p. 188.*

32. *In re Lifschutz on Habeas Corpus, Cal. S. Ct. Crim 14131; 2 Cal. 3d 415, 467 P.2 557, 85 Cal. Reptr. 829, 1970.*

decision was not held up in subsequent court practice.[33]) From the viewpoint of the mental health profession, what is at stake is the relationship between MHP and client. To a large extent, the quality of the patient-therapist relationship depends on the trust that the patient has in the therapist. Some in the profession, like Dr. Lifschutz, believe that trust will be destroyed if MHPs break confidence, even if patients authorize it.

In the debate over who ought to have privilege, this position has merit. However, it is relevant to note that privilege is based on the privacy of the individual. Therefore, the legal and moral right would appear to be the patient's or client's, not the MHP's. Furthermore, it is by no means certain that breaking confidence under such conditions will, in fact, undermine the MHP-patient relationship. It might as easily strengthen it. For one thing, MHPs who are willing to break confidence when a client waives privilege are investing clients with a large measure of autonomy. In effect, they seem to be saying, "I trust you to make judgments about what you think are your best interests, and I will abide by those decisions. Thus, if you want me to maintain silence, I will; but if you want me to break silence, I will do that also. The decision rests with you." It is difficult to see how this can have anything but a positive effect on the therapist-patient relationship. In contrast, where MHPs always maintain veto power over the patient's waiving of privilege, they seem to imply either distrust of the patient's decision-making capacity or the priority of something other than the welfare of the individual patient. Furthermore, if privilege belongs to MHPs, then it follows that they, not the patient, can waive it whenever they choose. It is difficult to see how this could foster a constructive MHP-patient relationship.

But even if MHPs share in the right to privilege, they must answer a couple of questions before exercising that right over the protests of patients. One question concerns competency. MHPs must determine whether the client is competent to waive privilege. If not, then MHPs may not only be violating the patient's right to privacy but may well be doing the person great personal harm. Another issue they must consider are the interests of the patient and what priority to give them. Evidently Lifschutz subordinated client Housek's interests to the interests of other patients, profession, and self. In so doing, he highlighted the web of relationships that MHPs find themselves in, as well as the need to clarify values, set priorities, and balance competing interests.

Additional Issues

The preceding discussion has raised important issues for MHPs in their interactions with the mentally ill, but it has hardly exhausted all the concerns. There are many additional issues that unfortunately cannot be covered here. Before concluding this chapter, however, we should at least become aware of some of them.

---

33. M. Grossman, "Right to Privacy vs. Right to Know," in Law and the Mental Health Professions, ed. W. E. Barton and C. J. Sanborn (New York: International Universities Press, 1978), p. 177.

One issue involves psychosurgery (also known as psychiatric neurosurgery, mental surgery, functional neurosurgery, or sedative neurosurgery). *Psychosurgery* may be defined as brain surgery that has as its primary purpose the alteration of thoughts, behavior patterns, personality characteristics, emotional reactions, or some similar aspect of subjective experience in human beings. Convinced of a relationship between brain and mind, proponents of psychosurgery claim that significant therapeutic results may be obtained for some behavioral disorders by the surgical destruction of particular brain regions.[34] But given the irreversibility of many psychosurgical techniques and the fact that their effects inevitably exceed those intended, is it ever moral to impose such treatment on a patient? And if so, under what conditions?

A second issue concerns casting mental patients in a sick and dependent role. Is such casting conducive or detrimental to the patient's recovery?[35]

A third issue deals with psychiatric intervention in schools as a means of dealing with developmental problems (for example, poor performance by a bright child or hyperactivity). Just as MHPs are expected to act as agents of the state, so are they being drafted as "educational facilitators"—agents for serving the values of educators and parents but not necessarily of children. Just what rights do schoolchildren have in these situations? Specifically, what is to be said about the child's right to consent?

Finally, a fourth issue relates to the treatment of various complaints that do not even fall into the category of mental illness. For example, is it legitimate for people to seek the services of MHPs to achieve changes in personality or their quality of life, or both? Do MHPs have an obligation to serve the pursuit of the higher degree of pleasure and happiness implied in such topics as job satisfaction and marital discord? Or does the public that creates the schools that train MHPs have a right to demand that those so trained not squander that training on persons who do not qualify as mentally ill?

Summary

This chapter dealt with the moral aspects of delivering health care to the mentally ill. After considering the central place of values in the mental health professions, it examined the moral plight of MHPs in terms of the conflicts they face in trying to meet professional and social functions and in managing relationships with the patient, the community, the patient's family, the law, and various third parties. Particularly troublesome for MHPs is the apparent ambivalence in the law regarding the rights of the mentally ill and civil rights even at the expense of adequate care. Finally, we saw that three issues predominate in discussions of morality and mental illness: involuntary commitment, competency and consent, and privacy. Embedded in these issues are concerns about

34.   *S. L. Chorover, "Psychosurgery: A Neuropsychological Perspective,"* Boston University Law Review 54 (March 1974): 231–248.

35.   *L. H. Cutter, "Operant Conditioning in a Vietnamese Mental Hospital,"* American Journal of Psychiatry 124 (July 1967): 123–128.

basic patient rights to adequate care, truth and information, confidentiality, and autonomy.

Ethical Theories

The ethical theories presented in Chapter 3 can be applied to nonemergency intervention—intervention generally invoked to help the mental patient rather than to protect society. The question under discussion will be: Is nonemergency intervention ever morally permissible? The following applications are intended to stimulate further debate. They stand as modest attempts to show how important health care issues can be framed within ethical theory. As such, they pose tentative judgments that should be scrutinized and used to initiate, not close, moral evaluation.

Egoism

Egoists, of course, would evaluate each nonemergency situation in terms of self-interest. Taking the viewpoint of the patient, there may be times when the individual stands to benefit in the long run by the intervention of others. From the egoistic viewpoints of MHPs, they too would benefit from nonemergency intervention, if in no other way than having opportunities to practice their professions and to derive personal satisfaction from helping to restore the mentally ill to some functional level. Of course, it is conceivable that in some situations egoistic interest may not be served. Much of the egoist position, as usual, turns on whose egoistic interests one considers. But one thing is certain: Nonemergency intervention would be compatible with the egoistic outlook.

Act Utilitarianism

Under the right circumstances, where the probable results would benefit society, nonemergency intervention would be consistent with act utilitarianism. At the same time, Mill's harm principle and the weight that utilitarians traditionally have given it should make one cautious in calculating possible results. In the utilitarian view, society does not benefit by violation of individual freedom in the absence of a threat to society. By the same logic, however, if an individual act of intervention promised to yield the most total good, then presumably that calculation would override considerations associated with the harm principle.

Suppose, for example, that a brilliant scientist of enormous potential benefit to society developed a mental condition that, although of no threat to society, rendered the scientist professionally inert. MHPs believe that they can restore the scientist to her former level of professional competence if they take nonemergency measures and initiate appropriate therapy. In such a case, it is at least conceivable that a nonemergency intervention could be justified on ultilitarian grounds. Of course, the case admittedly would be extraordinary; most instances of nonemergency intervention could not so glibly be demonstrated to have unquestionable social benefit. If intervention meant merely the restoration of the individual, the act utilitarian would presumably not sanction it. Such cases notwithstanding, nonemergency intervention is compatible with act utilitarianism when it is likely to yield the greatest amount of good.

Rule Utilitarianism

Rule utilitarians would scrutinize the consequences of the rule under which an act of nonemergency intervention falls. The rule might read: Society should not intervene solely on behalf of the mentally ill. Is this rule likely to produce more total good than its contradiction, which would permit at least some acts of nonemergency intervention?

In determining which of the rules is preferable, rulists would certainly consider the potential abuses of allowing nonemergency intervention, as well as the possibility that such paternalistic acts could be extended into other areas. They would also be mindful of the negative effects that would surely follow on the difficulty of defining one's own good and then implementing that standard. Against this, rulists would weigh the social consequences of such interventions. For one thing, society would stand to benefit from the mental restoration of its citizens. For another, it may in some cases be taking action that precludes individuals from deteriorating to the point where their conditions may pose a threat to society.

In the final analysis, rulists must decide whether the unqualified prohibition against all nonemergency intervention would likely produce the most total good. Although the issue is open to debate, given the utilitarian's historical endorsement of the harm principle, rulists might endorse a prohibition against nonemergency intervention. But since such rules are always subject to change as historical circumstances and social conditions warrant, such a prohibition would be open to periodic reevaluation and thus subject to change.

The Categorical Imperative

Of foremost importance in applying Kant's ethics to nonemergency intervention is the preeminence Kant gives to the inherent worth and dignity of rational creatures. Accordingly, Kant would oppose the subversion of individual rights and of the liberty of rational persons who pose no threat to society. Problems arise, of course, when the rationality of individuals is questionable, as it is in cases involving the mentally ill. In the absence of rationality, the perfect duty to refrain from violating a person's rights and liberties cannot so easily be invoked.

Having noted this, it could be argued that individuals have an imperfect duty to help the needy. How far any imperfect duty extends, including this one, is always problematic for Kant. But given their professional expertise, MHPs are in a unique position to provide health care for those in need. Thus they would feel the press of this imperfect duty more than others might. Furthermore, assuming that the mentally ill do not qualify as rational creatures, if MHPs intervene they could not be charged with insulting the worth and dignity of rational creatures. On the contrary, they might be commended for attempting to meet an imperfect duty to restore those individuals to a state of rationality. Indeed, it could be argued that to ignore those who are in need but not dangerous would be inconsistent with the way that rational creatures (MHPs and society generally) ought to behave toward their fellow beings—even those who lack rationality.

In any event, if we presuppose that the person being committed is not rational, then nonemergency intervention could be compatible with Kant's ethics. If the person is rational and not dangerous, then the intervention could not be justified within a Kantian frame. Obviously, judgment must depend on the meaning of *rational* and whether the mentally ill qualify as rational.

Prima Facie Duties

Several prima facie duties surface in applying Ross's ethics to nonemergency intervention. One is a professional duty of fidelity that MHPs have to the mentally ill in general to provide them with mental health care. Another is the duty of justice, whereby MHPs and society are obliged to provide services for those who are in need. Duties of beneficence are also evident here, since MHPs and society are in a position to help the mentally ill. By the same token, MHPs and others can perhaps meet a duty to improve themselves morally by intervening on behalf of those who are needy and helpless.

On the other hand, individuals are entitled to freedom and autonomy as long as they do not threaten anyone else in exercising these rights. Also, any intervention might be viewed as injuring individuals by detaining them against their will. But this objection is questionable, because of the apparent limitations on the individual's competency and will. Furthermore, the issue of rationality is as relevant in Ross's ethics as in Kant's. Thus, if the mentally ill cannot be considered rational to begin with, then objections to interventionism based on principles of justice and noninjury lose their force.

Undoubtedly, then, Ross would consider the various prima facie duties operating in intervention decisions and decide which were the overriding ones. If the preceding analysis is accurate, nonemergency intervention could be justified under Ross's provisions. When it is, MHPs would be justified, even obliged, to take appropriate measures. Moreover, society would be justified and obliged to provide the social machinery to permit such interventions.

The Maximin Principle

Rawls's natural duties could be applied to nonemergency procedures much like Ross's prima facie duties, with the same result. In addition, Rawls could justify acts of intervention when they were calculated to enlarge the freedom of the individual. Taking the view of those in the original position, rational creatures with a sense of their own interests would probably be willing to bind themselves to a rule that permitted nonemergency interventions under carefully circumscribed conditions, such as when individuals desperately needed help and could not provide it for themselves. If they were to prohibit such intervention, inhabitants of the original position would be sanctioning a rule that, in effect, denied them the possibility of getting the help they needed. It is difficult to see how this would ultimately be in their best interests. In short, non-emergency intervention finds theoretical justification in Rawls's ethics.

Roman Catholicism's Version of Natural-Law Ethics

Roman Catholicism would endorse nonemergency intervention. By the principle of double effect, what is intended in such cases is the provision of help for individuals who need it. Although intervention would deny these people a certain freedom, that is not its intention. Indeed, by restoring those in need and providing them with mental health care, MHPs and society act to increase their freedom and autonomy and thus to return them to a state that presumably reflects the Creator's design for human beings. In any event, the denial of the person's freedom is not the means by which the individual is restored to health; the therapeutic intervention is. Thus, given (1) that there is no intention to undercut individual rights and (2) that the patient stands to benefit enormously from the intervention, nonemergency procedures could be justified under Roman Catholicism's version of natural-law ethics.

CASE PRESENTATION
The Odyssey

"The trouble with you lawyers is that you're too concerned with legal triumph and too little with psychiatric catastrophe." Strong words, but Preston Rogers meant them. The lawyer in this instance was his fiancee Carolyn Bassey, and the young couple were making their way out of the courthouse after a grueling commitment hearing.

Preston's remark amused Carolyn. "Spoken like a true psychiatric social worker," she said, with a laugh that echoed through the marble corridor.

"Hey," said Preston, freezing in his tracks, "it's not just me." He pointed in the direction of the hearing room. "You heard the testimony in there—dozens of people in the field felt that Clayton Pierce should be committed."

"*Involuntarily* committed."

"All right, involuntarily."

Clayton Pierce was a forty-five-year-old man for whom the court, at Carolyn's urging, had just recommended treatment as an outpatient at a local community mental health center. In so doing, the court had decided the immediate fate of a man who had been severely schizophrenic for nearly thirty years. But it certainly hadn't put an end to the debate that surrounded the case nor written an ending to the story, whose already well established plot did not promise a happy finish.

Clayton Pierce's odyssey in the land of law and medicine began three decades earlier when, at the onset of his mental disorder, he had spent several years in and out of various psychiatric hospitals. At one point he was seen daily for nearly eighteen months in a specialized private hospital by a psychoanalyst, even though Clayton remained mute and catatonic. After that he was transferred to another hospital, where he improved slightly with medication and electroconvulsive therapy.

The cost of such intensive treatment ultimately became too much for the Pierce family. After spending nearly a quarter of a million dollars on medical care over a period of ten years, they transferred Clayton to another institution, where attempts were made to extend his privileges and give him the "least restrictive alternative." Unfortunately, these efforts always ended in disaster, once with a suicide attempt. So despite years of treatment, Clayton Pierce remained delusional and withdrawn and continued to hallucinate.

Several years after the suicide attempt, Clayton was placed on a trial home visit. But when his symptoms worsened, he refused to go back to the hospital. The police had to be summoned to return him to the institution. Not long after this episode, Clayton was again permitted to go home, with the same unfortunate results. This time the police refused to take action, and ultimately Clayton was dropped from the hospital rolls.

It was at this point that the services of Preston Rogers were engaged. It didn't take Preston long to realize that the Pierces were in no position to manage Clayton. Both were in their seventies, enfeebled by various ailments. They felt threatened by Clayton, and for good reason: He was a chain smoker and spent most of the time in his room puffing one cigarette after another. The Pierces were desperately afraid that he would start a fire. But whenever they cautioned him about his smoking, Clayton would react with a fury that left the Pierces fearing for their lives. Given this grim picture, Preston advised the Pierces to seek legal counsel.

The Pierces took his advice. They engaged a lawyer who, in consultation with psychiatrists and only after great effort, succeeded in getting Clayton to agree voluntarily to enter the psychiatric unit of a large hospital. Two days after admission, a petition for involuntary commitment was prepared, accompanied by the affidavits of psychiatrists. At this point, public defender Carolyn Bassey entered the case.

Carolyn contended that involuntary commitment proceedings could not be instituted against a person who had entered a hospital on a voluntary basis. Psychiatrists objected that such a ruling defies medical experience and common sense. They pointed out that the condition of hospitalized patients often deteriorates to such an extent as to preclude voluntary choice. They argued that in such cases the sensible thing is to file for involuntary hospitalization on a temporary basis while the patient is still in the hospital, which was precisely what they were attempting to do in Clayton's case. For her part, Carolyn pointed out that current law forbade such a procedure.

The upshot was that Clayton had to be discharged from the hospital before the commitment petition could be filed. Immediately on release this was done, but a hearing was not set until a month later.

Preston Rogers was thunderstruck by the ruling. He couldn't believe that the court could be so indifferent to the plight of the Pierces. In his view, the court simply had no idea of the problem involved in caring for an acutely psychotic patient who is waiting reluctantly to be committed.

At the hearing, Carolyn Bassey did her job of representing Clayton Pierce extremely well. So well in fact that, although the court concluded that the patient was likely to injure himself and others if allowed to remain at liberty, it recommended he be committed as an outpatient to a local community mental health center.

Outside the courthouse, Carolyn and Preston continued their debate over the ruling, Carolyn emphasizing the importance of Clayton's civil liberties, Preston his need for treatment and the rights of the Pierces. There was no telling what passersby thought of the heated exchange. Certainly they couldn't guess that at issues was something that might very well touch their own lives someday. someday.

Questions for Analysis

1. *What are your feelings about the court's ruling?*

2. *What are the dangers of allowing involuntary commitment of patients who have been hospitalized voluntarily?*

3. *Do you think that Preston's concern for the Pierces is well founded? What rights, if any, do you think the families of mental patients have?*

4. *Do you think the principle of the least restrictive alternative dominated Carolyn Bassey's and the court's thinking? What are the implications of such an emphasis?*

5. *In this case, all the principles of criminal law were applied. Do you think they were appropriate?*

6. *What obligations, if any, do you think Preston Rogers has now?*

7. *Suppose Carolyn Bassey had argued: "As a lawyer my primary responsibility is to ensure the legal rights of my client, just as your primary responsibility as a mental health worker is to ensure the best care for your patient. In the Pierce case I met my moral and professional responsibility the only way I could. If I had done anything less, I would have been irresponsible." How would you respond?*

Exercise 1

You have been seeing a client in therapy for six months. You know a considerable amount about the person—sex life, love life, needs, wants, desires, strengths, and weaknesses. The person admits being very attracted to you, in fact in love with you. This makes you feel good. It is always nice to feel that one is attractive, admired, and desired. What complicates things is that you feel an equally strong attraction toward your client. In fact, you see in the person characteristics and qualities that you prize and search for in a mate. You are torn between expressing your own feelings—which you know the client would respond to—and stifling your emotions. Make a list of your alternatives. What do you think you *would* do? What do you think you *should* do?

Exercise 2

Reconsider your responses under Values and Priorities. Would you alter any of them? Have any been reinforced?

On Being Sane in Insane Places

D. L. Rosenhan

Can psychiatrists distinguish the sane from the insane in psychiatric hospitals? Or is the traditional psychiatric classification of mental disorders unreliable, invalid, and harmful to the welfare of patients? Not too long ago, the official journal of the Association for the Advancement of Science reported a study that strongly challenged basic psychiatric concepts and practices.

The study, reported here, was conducted by professor of law and psychology D. L. Rosenhan. A group of pseudopatients presented themselves at various psychiatric hospitals complaining of hearing voices that said "empty," "hollow," and "thud." What happened to these pseudopatients, as well as the diagnosis and treatment they received, is reported in the study. In Rosehan's view, his findings indicate that we cannot distinguish sanity from insanity—a point made also by some other MHPs, such as Thomas Szasz and Ronald Laing.

Since the study appeared, it has received both support and criticism. Critics claim that Rosenhan's study of pseudopatients is really irrelevant to the issue of the validity of psychiatric diagnoses—that all it shows is that clever people can fool psychiatrists. Maybe these critics are right. But even if the report does not prove that psychiatric diagnoses are invalid, it does raise concern about other important aspects of care for the institutionalized mentally ill. As sociologist David Mechanic points out, the concerns cluster around such evident facts as the following: (1) psychiatric diagnostic practices are often sloppy; (2) patients are sometimes institutionalized too hastily; (3) observation of patients in hospitals is inefficient and casual; (4) MHP preconceptions about patients and their illnesses guide professional interactions with patients; and (5) patients are often overmedicated, sometimes dangerously so.[36]

If sanity and insanity exist, how shall we know them?

The question is neither capricious nor itself insane. However much we may be personally convinced that we can tell the normal from the abnormal, the evidence is simply not compelling. It is commonplace, for example, to read about murder trials wherein eminent psychiatrists for

---

From Science 179 *(January 19, 1973): 250–258. Copyright ©1973 from the American Association for the Advancement of Science. Reprinted by permission.*

36. *David Mechanic,* Readings in Medical Sociology *(New York: Free Press, 1980), pp. 83–84.*

the defense are contradicted by equally eminent psychiatrists for the prosecution on the matter of the defendant's sanity. More generally, there are a great deal of conflicting data on the reliability, utility, and meaning of such terms as "sanity," "insanity," "mental illness," and "schizophrenia" (1). Finally, as early as 1934, Benedict suggested that normality and abnormality are not universal (2). What is viewed as normal in one culture may be seen as quite aberrant in another. Thus, notions of normality and abnormality may not be quite as accurate as people believe they are.

To raise questions regarding normality and abnormality is in no way to question the fact that some behaviors are deviant or odd. Murder is deviant. So, too, are hallucinations. Nor does raising such questions deny the existence of the personal anguish that is often associated with "mental illness." Anxiety and depression exist. Psychological suffering exists. But normality and abnormality, sanity and insanity, and the diagnoses that flow from them may be less substantive than many believe them to be.

At its heart, the question of whether the sane can be distinguished from the insane (and whether degrees of insanity can be distinguished from each other) is a simple matter: do the salient characteristics that lead to diagnoses reside in the patients themselves or in the environments and contexts in which observers find them? From Bleuler, through Kretchmer, through the formulators of the recently revised *Diagnostic and Statistical Manual* of the American Psychiatric Association, the belief has been strong that patients present symptoms, that those symptoms can be categorized, and, implicitly, that the sane are distinguishable from the insane. More recently, however, this belief has been questioned. Based in part on theoretical and anthropological considerations, but also on philosophical, legal, and therapeutic ones, the view has grown that psychological categorization of mental illness is useless at best and downright harmful, misleading, and pejorative at worst. Psychiatric diagnoses, in this view, are in the minds of the observers and are not valid summaries of characteristics displayed by the observed (3–5).

Gains can be made in deciding which of these is more nearly accurate by getting normal people (that is, people who do not have, and have never suffered, symptoms of serious psychiatric disorders) admitted to psychiatric hospitals and then determining whether they were discovered to be sane and, if so, how. If the sanity of such pseudopatients were always detected, there would be prima facie evidence that a sane individual can be distinguished from the insane context in which he is found. Normality (and presumably abnormality) is distinct enough that is can be recognized wherever it occurs, for it is carried within the person. If, on the other hand, the sanity of the pseudopatients were never discovered, serious difficulties would arise for those who support traditional modes of psychiatric diagnosis. Given that the hospital staff was not incompetent, that the pseudopatient had been behaving as sanely as he had been outside of the hospital, and that it had never been previously suggested that he belonged in a psychiatric hospital, such an unlikely outcome would support the view that psychiatric diagnosis betrays little about the patient but much about the environment in which an observer finds him.

This article describes such an experiment. Eight sane people gained secret admission to 12 different hospitals (6). Their diagnostic experiences constitute the data of the first part of this article; the remainder is devoted to a description of their experiences in psychiatric institutions. Too few psychiatrists and psychologists, even those who have worked in such hospitals, know what the experience is like. They rarely talk about it with former patients, perhaps because they distrust information coming from the previously insane. Those who have worked in psychiatric hospitals are likely to have adapted so thoroughly to the settings that they are insensitive to the impact of that experience. And while there have been occasional reports of researchers who submitted themselves to psychiatric hospitalization (7), these researchers have commonly remained in the hospitals for short periods of time, often with the knowledge of the hospital staff. It is difficult to know the extent to which they were treated like patients or like research colleagues. Nevertheless, their reports about the inside of the psychiatric hospital have been valuable. This article extends those efforts.

Pseudopatients and Their Settings

The eight pseudopatients were a varied group. One was a psychology graduate student in his 20's. The remaining seven were older and "established." Among them were three psychologists, a pediatrician, a psychiatrist, a painter, and a housewife. Three pseudopatients were women, five were men. All of them employed pseudonyms, lest their alleged diagnoses embarrass them later. Those who were in mental health professions alleged another occupation in order to avoid the special attentions that might be accorded by staff, as a matter of courtesy or caution, to ailing colleagues (8). With the exception of myself (I was the first pseudopatient and my presence was known to the hospital administrator and chief psychologist and, so far as I can tell, to them alone), the presence of pseudopatients and the nature of the research program was not known to the hospital staffs (9).

The settings were similarly varied. In order to generalize the findings, admission into a variety of hospitals was sought. The 12 hospitals in the sample were located in five different states on the East and West coasts. Some were old and shabby, some were quite new. Some were research-oriented, others not. Some had good staff-patient ratios, others were quite understaffed. Only one was a strictly private hospital. All of the others were supported by state or federal funds or, in one instance, by university funds.

After calling the hospital for an appointment, the pseudopatient arrived at the admissions office complaining that he had been hearing voices. Asked what the voices said, he replied that they were often unclear, but as far as he could tell they said "empty," "hollow," and "thud." The voices were unfamiliar and were of the same sex as the pseudopatient. The choice of these symptoms was occasioned by their apparent similarity to existential symptoms. Such symptoms are alleged to arise from painful concerns about the perceived meaninglessness of one's life. It is as if the hallucinating person were saying, "My life is empty and hollow." The choice of these symptoms was also determined by the *absence* of a single report of existential psychoses in the literature.

Beyond alleging the symptoms and falsifying name, vocation, and employment, no further alterations of person, history, or circumstances were made. The significant events of the pseudopatient's life history were presented as they had actually occurred. Relationships with parents and siblings, with spouse and children, with people at work and in school, consistent with the aforementioned exceptions, were described as they were or had been. Frustrations and upsets were described along with joys and satisfactions. These facts are important to remember. If anything, they strongly biased the subsequent results in favor of detecting sanity, since none of their histories or current behaviors were seriously pathological in any way.

Immediately upon admission to the psychiatric ward, the pseudopatient ceased simulating *any* symptoms of abnormality. In some cases, there was a brief period of mild nervousness and anxiety, since none of the pseudopatients really believed that they would be admitted so easily. Indeed, their shared fear was that they would be immediately exposed as frauds and greatly embarrassed. Moreover, many of them had never visited a psychiatric ward; even those who had, nevertheless had some genuine fears about what might happen to them. Their nervousness, then, was quite appropriate to the novelty of the hospital setting, and it abated rapidly.

Apart from that short-lived nervousness, the pseudopatient behaved on the ward as he "normally" behaved. The pseudopatient spoke to patients and staff as he might ordinarily. Because there is uncommonly little to do on a psychiatric ward, he attempted to engage others in conversation. When asked by staff how he was feeling, he indicated that he was fine, that he no longer experienced symptoms. He responded to instructions from attendants, to calls for medication (which was not swallowed), and to dining-hall instructions. Beyond such activities as were available to him on the admissions ward, he spent his time writing down his observations about the ward, its patients, and the staff. Initially these notes were written "secretly," but as it soon became clear that no one much cared, they were subsequently written on standard tablets of paper in such public places as the dayroom. No secret was made of these activities.

The pseudopatient, very much as a true psychiatric patient, entered a hospital with no foreknowledge of when he would be discharged. Each was told that he would have to get out by his own devices, essentially by convincing the staff that he was sane. The psychological stresses associated with hospitalization were considerable, and all but one of the pseudopatients desired to be discharged almost immediately after being admitted. They were, therefore, motivated not only to behave sanely, but to be paragons of cooperation. That their behavior was in no way disruptive is confirmed by nursing reports, which have been obtained on most of the patients. These reports uniformly indicate that the patients were "friendly," "cooperative," and "exhibited no abnormal indications."

The Normal Are Not Detectably Sane

Despite their public "show" of sanity, the pseudopatients were never detected. Admitted, except in one case, with a diagnosis of schizophrenia (10), each was discharged with a diagnosis of schizophrenia "in remission." The label "in remission" should in no way be dismissed as a formality, for at no time during any hospitalization had any question been raised about any pseudopatient's simulation. Nor are there any indications in the hospital records that the pseudopatient's status was suspect. Rather, the evidence is strong that, once labeled schizophrenic, the pseudopatient was stuck with that label. If the pseudopatient was to be discharged, he must naturally be "in remission"; but he was not sane, nor, in the institution's view, had he ever been sane.

The uniform failure to recognize sanity cannot be attributed to the quality of the hospitals, for, although there were considerable variations among them, several are considered excellent. Nor can it be alleged that there was simply not enough time to observe the pseudopatients. Length of hospitalization ranged from 7 to 52 days, with an average of 19 days. The pseudopatients were not, in fact, carefully observed, but this failure clearly speaks more to traditions within psychiatric hospitals than to lack of opportunity.

Finally, it cannot be said that the failure to recognize the pseudopatients' sanity was due to the fact that they were not behaving sanely. While there was clearly some tension present in all of them, their daily visitors could detect no serious behavioral consequences—nor, indeed, could other patients. It was quite common for the patients to "detect" the pseudopatients' sanity. During the first three hospitalizations, when accurate counts were kept, 35 of a total of 118 patients on the admissions ward voiced their suspicions, some vigorously. "You're not crazy. You're a journalist,

or a professor [referring to the continual note-taking]. "You're checking up on the hospital." While most of the patients were reassured by the pseudopatient's insistence that he had been sick before he came in but was fine now, some continued to believe that the pseudopatient was sane throughout his hospitalization (11). The fact that the patients often recognized normality when staff did not raises important questions.

Failure to detect sanity during the course of hospitalization may be due to the fact that physicians operate with a strong bias toward what statisticians call the type 2 error (5). This is to say that physicians are more inclined to call a healthy person sick (a false positive, type 2) than a sick person healthy (a false negative, type 1). The reasons for this are not hard to find: it is clearly more dangerous to misdiagnose illness than health. Better to err on the side of caution, to suspect illness even among the healthy.

But what holds for medicine does not hold equally well for psychiatry. Medical illnesses, while unfortunate, are not commonly pejorative. Psychiatric diagnoses, on the contrary, carry with them personal, legal, and social stigmas (12). It was therefore important to see whether the tendency toward diagnosing the sane insane could be reversed. The following experiment was arranged at a research and teaching hospital whose staff had heard these findings but doubted that such an error could occur in their hospital. The staff was informed that at some time during the following 3 months, one or more pseudopatients would attempt to be admitted into the psychiatric hospital. Each staff member was asked to rate each patient who presented himself at admissions or on the ward according to the likelihood that the patient was a pseudopatient. A 10-point scale was used, with a 1 and 2 reflecting high confidence that the patient was a pseudopatient.

Judgments were obtained on 193 patients who were admitted for psychiatric treatment. All staff who had had sustained contact with or primary responsibility for the patient—attendants, nurses, psychiatrists, physicians, and psychologists—were asked to make judgments. Forty-one patients were alleged, with high confidence, to be pseudopatients by at least one member of the staff. Twenty-three were considered suspect by at least one psychiatrist. Nineteen were suspected by one psychiatrist *and* one other staff member. Actually, no genuine pseudopatient (at least from my group) presented himself during this period.

The experiment is instructive. It indicates that the tendency to designate sane people as insane can be reversed when the stakes (in this case, prestige and diagnostic acumen) are high. But what can be said of the 19 people who were suspected of being "sane" by one psychiatrist and another staff member? Were these people truly "sane," or was it rather the case that in the course of avoiding the type 2 error the staff tended to make more errors of the first sort—calling the crazy "sane"? There is no way of knowing. But one thing is certain: any diagnostic process that lends itself so readily to massive errors of this sort cannot be a very reliable one.

The Stickiness of Psychodiagnostic Labels

Beyond the tendency to call the healthy sick—a tendency that accounts better for diagnostic behavior on admission than it does for such behavior after a lengthy period of exposure—the data speak to the massive role of labeling in psychiatric assessment. Having once been labeled schizophrenic, there is nothing the pseudopatient can do to overcome the tag. The tag profoundly colors others' perceptions of him and his behavior.

From one viewpoint, these data are hardly surprising, for it has long been known that elements are given meaning by the context in which they occur. Gestalt psychology made this point vigorously, and Asch (13) demonstrated that there are "central" personality traits (such as "warm" versus "cold") which are so powerful that they markedly color the meaning of other information in forming an impression of a given personality (14). "Insane," "schizophrenic," "manic-depressive," and "crazy" are probably among the most powerful of such central traits. Once a person is designated abnormal, all of his other behaviors and characteristics are colored by that label. Indeed, that label is so powerful that many of the pseudopatients' normal behaviors were overlooked entirely or profoundly misinterpreted. Some examples may clarify this issue.

Earlier I indicated that there were no changes in the pseudopatient's personal history and cur-rent status beyond those of name, employment, and, where necessary, vocation. Otherwise, a veridical description of personal history and circumstances was offered. Those circumstances were not psychotic. How were they made consonant with the diagnosis of psychosis? Or were those diagnoses modified in such a way as to bring them into accord with the circumstances of the pseudopatient's life, as described by him?

As far as I can determine, diagnoses were in no way affected by the relative health of the circumstances of a pseudopatient's life. Rather, the reverse occurred: the perception of his circumstances was shaped entirely by the diagnosis. A clear example of such translation is found in the case of a pseudopatient who had had a close relationship with his mother but was rather remote from his father during his early childhood. During adolescence and beyond, however, his father became a close friend, while his relationship with his mother cooled. His present relationship with his wife was characteristically close and warm. Apart from occasional angry exchanges, friction was minimal. The children had rarely been spanked. Surely there is nothing especially pathological about such a history. Indeed, many readers may see a similar pattern in their own experiences, with no markedly deleterious consequences. Observe, however, how such a history was translated in the psychopathological context, this from the case summary prepared after the patient was discharged.

> This white 39-year-old male . . . manifests a long history of considerable ambivalence in close relationships, which begins in early childhood. A warm relationship with his mother cools during his adolescence. A distant relationship to his father is described as becoming very intense. Affective stability is absent. His attempts to control emotionality with his wife and children are punctuated by angry outbursts and, in the case of the children, spankings. And while he says that he has several good friends, one senses considerable ambivalence embedded in those relationships also. . . .

The facts of the case were unintentionally distorted by the staff to achieve consistency with a popular theory of the dynamics of a schizophrenic reaction (15). Nothing of an ambivalent nature had been described in relations with parents, spouse, or friends. To the extent that ambivalence could be inferred, it was probably not greater than is found in all human relationships. It is true the pseudopatient's relationships with his parents changed over time, but in the ordinary context that would hardly be remarkable—indeed, it might very well be expected. Clearly, the meaning ascribed to his verbalizations (that is, ambivalence, affective instability) was determined by the diagnosis: schizophrenia. An entirely different meaning would have been ascribed if it were known that the man was "normal."

All pseudopatients took extensive notes publicly. Under ordinary circumstances, such behavior would have raised questions in the minds of observers, as, in fact, it did among patients. Indeed, it seemed so certain that the notes would elicit suspicion that elaborate precautions were taken to remove them from the ward each day. But the precautions proved needless. The closest any staff member came to questioning these notes occurred when one pseudopatient asked his physician what kind of medication he was receiving and began to write down the response. "You needn't write it," he was told gently. "If you have trouble remembering, just ask me again."

If no questions were asked of the pseudopatients, how was their writing interpreted? Nursing records for three patients indicate that the writing was seen as an aspect of their pathological behavior. "Patient engages in writing behavior" was the daily nursing comment on one of the pseudopatients who was never questioned about his writing. Given that the patient is in the hospital, he must be psychologically disturbed. And given that he is disturbed, continuous writing must be a behavioral manifestation of that disturbance, perhaps a subset of the compulsive behaviors that are sometimes correlated with schizophrenia.

One tacit characteristic of psychiatric diagnosis is that it locates the sources of aberration within the individual and only rarely within the complex of stimuli that surrounds him. Consequently, behaviors that are stimulated by the environment are commonly misattributed to the patient's disorder. For example, one kindly nurse found a pseudopatient pacing the long hospital corridors. "Nervous, Mr. X?" she asked. "No, bored," he said.

The notes kept by pseudopatients are full of patient behaviors that were misinterpreted by well-intentioned staff. Often enough, a patient would go "berserk" because he had, wittingly or unwittingly, been mistreated by, say, an attendant. A nurse coming upon the scene would rarely inquire even cursorily into the environmental stimuli of the patient's behavior. Rather, she assumed that his upset derived from his pathology, not from his present interactions with other staff members. Occasionally, the staff might assume that the patient's family (especially when they had recently visited) or other patients had stimulated the outburst. But never were the staff found to assume that one of themselves or the structure of the hospital had anything to do with a patient's behavior. One psychiatrist pointed to a group of patients who were sitting outside the cafeteria entrance half an hour before lunchtime. To a group of young residents he indicated that such behavior was characteristic of the oral-acquisitive nature of the syndrome. It seemed not to occur to him that there were very few things to anticipate in a psychiatric hospital besides eating.

A psychiatric label has a life and an influence of its own. Once the impression has been formed that the patient is schizophrenic, the expectation is that he will continue to be schizophrenic. When a sufficient amount of time has passed, during which the patient has done nothing bizarre, he is considered to be in remission and available for discharge. But the label endures beyond discharge, with the unconfirmed expectation that he will behave as a schizophrenic again. Such labels, conferred by mental health professionals, are as influential on the patient as they are on his relatives and friends, and it should not surprise anyone that the diagnosis acts on all of them as a self-fulfilling prophecy. Eventually, the patient himself accepts the diagnosis, with all of its surplus meanings and expectations, and behaves accordingly (5).

The inferences to be made from these matters are quite simple. Much as Zigler and Phillips have demonstrated that there is enormous overlap in the symptoms presented by patients who have been variously diagnosed (16), so there is enormous overlap in the behaviors of the sane and the insane. The sane are not "sane" all of the time. We lose our tempers "for no good reason." We are occasionally depressed or anxious, again for no good reason. And we may find it difficult to get along with one or another person—again for no reason that we can specify. Similarly, the insane are not always insane. Indeed, it was the impression of the pseudopatients while living with them that they were sane for long periods of time—that the bizarre behaviors upon which their diagnoses were allegedly predicated constituted only a small fraction of their total behavior. If it makes no sense to label ourselves permanently depressed on the basis of an occasional depression, then it takes better evidence than is presently available to label all patients insane or schizophrenic on the basis of bizarre behaviors or cognitions. It seems more useful, as Mischel (17) has pointed out, to limit our discussions to *behaviors*, the stimuli that provoke them, and their correlates.

It is not known why powerful impressions of personality traits, such as "crazy" or "insane," arise. Conceivably, when the origins of and stimuli that give rise to a behavior are remote or unknown, or when the behavior strikes us as immutable, trait labels regarding the *behaver* arise. When, on the other hand, the origins and stimuli are known and available, discourse is limited to the behavior itself. Thus, I may hallucinate because I am sleeping, or I may hallucinate because I have ingested a peculiar drug. These are termed sleep-induced hallucinations, or dreams, and drug-induced hallucinations, respectively. But when the stimuli to my hallucinations are unknown, that is called craziness, or schizophrenia—as if that inference were somehow as illuminating as the others.

The Experience of Psychiatric Hospitalization

The term "mental illness" is of recent origin. It was coined by people who were humane in their inclinations and who wanted very much to raise the station of (and the public's sympathies toward) the psychologically disturbed from that of witches and "crazies" to one that was akin to the physically ill. And they were at least partially successful, for the treatment of the mentally ill *has* improved considerably over the years. But while treatment has improved, it is doubtful that people really regard the mentally ill in the same way that they view the physically ill. A broken leg is something

one recovers from, but mental illness allegedly endures forever (18). A broken leg does not threaten the observer, but a crazy schizophrenic? There is by now a host of evidence that attitudes toward the mentally ill are characterized by fear, hostility, aloofness, suspicion, and dread (19). The mentally ill are society's lepers.

That such attitudes infect the general population is perhaps not surprising, only upsetting. But that they affect the professionals—attendants, nurses, physicians, psychologists, and social workers—who treat and deal with the mentally ill is more disconcerting, both because such attitudes are self-evidently pernicious and because they are unwitting. Most mental health professionals would insist that they are sympathetic toward the mentally ill, that they are neither avoidant nor hostile. But it is more likely that an exquisite ambivalence characterizes their relations with psychiatric patients, such that their avowed impulses are only part of their entire attitude. Negative attitudes are there too and can easily be detected. Such attitudes should not surprise us. They are the natural offspring of the labels patients wear and the places in which they are found.

Consider the structure of the typical psychiatric hospital. Staff and patients are strictly segregated. Staff have their own living space, including their dining facilities, bathrooms, and assembly places. The glass quarters that contain the professional staff, which the pseudopatients came to call "the cage," sit out on every dayroom. The staff emerge primarily for caretaking purposes—to give medication, to conduct a therapy or group meeting, to instruct or reprimand a patient. Otherwise, staff keep to themselves, almost as if the disorder that afflicts their charges is somehow catching.

So much is patient-staff segregation the rule that, for four public hospitals in which an attempt was made to measure the degree to which staff and patients mingle, it was necessary to use "time out of the staff cage" as the operational measure. While it was not the case that all time spent out of the cage was spent mingling with patients (attendants, for example, would occasionally emerge to watch television in the dayroom), it was the only way in which one could gather reliable data on time for measuring.

The average amount of time spent by attendants outside of the cage was 11.3 percent (range, 3 to 52 percent). This figure does not represent only time spent mingling with patients, but also includes time spent on such chores as folding laundry, supervising patients while they shave, directing ward clean-up, and sending patients to off-ward activities. It was the relatively rare attendant who spent time talking with patients or playing games with them. It proved impossible to obtain a "percent mingling time" for nurses, since the amount of time they spent out of the cage was too brief. Rather, we counted instances of emergence from the cage. On the average, daytime nurses emerged from the cage 11.5 times per shift, including instances when they left the ward entirely (range, 4 to 39 times). Late afternoon and night nurses were even less available, emerging on the average 9.4 times per shift (range, 4 to 41 times). Data on early morning nurses, who arrived usually after midnight and departed at 8 a.m., are not available because patients were asleep during most of this period.

Physicians, especially psychiatrists, were even less available. They were rarely seen on the wards. Quite commonly, they would be seen only when they arrived and departed, with the remaining time being spent in their offices or in the cage. On the average, physicians emerged on the ward 6.7 times per day (range, 1 to 17 times). It proved difficult to make an accurate estimate in this regard, since physicians often maintained hours that allowed them to come and go at different times.

The hierarchical organization of the psychiatric hospital has been commented on before (20), but the latent meaning of that kind of organization is worth noting again. Those with the most power have least to do with patients, and those with the least power are most involved with them. Recall, however, that the acquisition of role-appropriate behaviors occurs mainly through the observation of others, with the most powerful having the most influence. Consequently, it is understandable that attendants not only spend more time with patients than do any other members of the staff—that is required by their station in the hierarchy—but also, insofar as they learn from their superiors' behavior, spend as little time with patients as they can. Attendants are seen mainly in the cage, which is where the models, the action, and the power are.

I turn now to a different set of studies, these dealing with staff response to patient-initiated contact. It has long been known that the amount of time a person spends with you can be an index of your significance to him. If he initiates and maintains eye contact, there is reason to believe that he is considering your requests and needs. If he pauses to chat or actually stops and talks, there is added reason to infer that he is individuating you. In four hospitals, the pseudopatient approached the staff member with a request which took the following form: "Pardon me, Mr. [or Dr. or Mrs.] X, could you tell me when I will be eligible for

Table 1. Self-initiated Contact by Pseudopatients with Psychiatrists and Nurses and Attendants, Compared to Contact with Other Groups

| | | | | University Medical Center | | |
| | | University Campus (Nonmedical) | | Physicians | | |
Contact	(1) Psychiatrists	(2) Nurses and Attendants	(3) Faculty	(4) "Looking for a Psychiatrist"	(5) "Looking for an Internist"	(6) No Additional Comment
Responses						
Moves On, Head Averted (%)	71	88	0	0	0	0
Makes Eye Contact (%)	23	10	0	11	0	0
Pauses and Chats (%)	2	2	0	11	0	10
Stops and Talks (%)	4	0.5	100	78	100	90
Mean Number of Questions						
Answered (out of 6)	*	*	6	3.8	4.8	4.5
Respondents (no.)	13	47	14	18	15	10
Attempts (no.)	185	1283	14	18	15	10

*Not applicable.

grounds privileges?" (or ". . . when I will be presented at the staff meeting?" or ". . . when I am likely to be discharged?"). While the content of the question varied according to the appropriateness of the target and the pseudopatient's (apparent) current needs the form was always a courteous and relevant request for information. Care was taken never to approach a particular member of the staff more than once a day, lest the staff member become suspicious or irritated. In examining these data, remember that the behavior of the pseudopatients was neither bizarre nor disruptive. One could indeed engage in good conversation with them.

The data for these experiments are shown in Table 1, separately for physicians (column 1) and for nurses and attendants (column 2). Minor differences between these four institutions were overwhelmed by the degree to which staff avoided continuing contacts that patients had initiated. By far, their most common response consisted of either a brief response to the question, offered while they were "on the move" and with head averted, or no response at all.

The encounter frequently took the following bizarre form: (pseudopatient) "Pardon me, Dr. X. Could you tell me when I am eligible for grounds privileges?" (physician) "Good morning, Dave. How are you today?" (Moves off without waiting for a response.)

It is instructive to compare these data with data recently obtained at Stanford University. It has been alleged that large and eminent universities are characterized by faculty who are so busy that they have no time for students. For this comparison, a young lady approached individual faculty members who seemed to be walking purposefully to some meeting or teaching engagement and asked them the following six questions.

1. "Pardon me, could you direct me to Encina Hall?" (at the medical school: ". . . to the Clinical Research Center?").

2. "Do you know where Fish Annex is?" (there is no Fish Annex at Stanford).

3. "Do you teach here?"

4. "How does one apply for admission to the college?" (at the medical school: ". . . to the medical school?")

5. "Is it difficult to get in?"

6. "Is there financial aid?"

Without exception, as can be seen in Table 1 (column 3), all of the questions were answered. No matter how rushed they were, all respondents not only maintained eye contact, but stopped to talk. Indeed, many of the respondents went out of their way to direct or take the questioner to the office she was seeking, to try to locate "Fish Annex," or to discuss with her the possibilities of being admitted to the university.

Similar data, also shown in Table 1 (columns 4, 5, and 6), were obtained in the hospital. Here too, the young lady came prepared with six questions. After the first question, however, she remarked to 18 of her respondents (column 4), "I'm looking for a psychiatrist," and to 15 others (column 5), "I'm looking for an internist." Ten other respondents received no inserted comment (column 6). The general degree of cooperative responses is considerably higher for these university groups than it was for pseudopatients in psychiatric hospitals. Even so, differences are apparent within the medical school setting. Once having indicated that she was looking for a psychiatrist, the degree of cooperation elicited was less than when she sought an internist.

Powerlessness and Depersonalization

Eye contact and verbal contact reflect concern and individuation; their absence, avoidance and depersonalization. The data I have presented do not do justice to the rich daily encounters that grew up around matters of depersonalization and avoidance. I have records of patients who were beaten by staff for the sin of having initiated verbal contact. During my own experience, for example, one patient was beaten in the presence of other patients for having approached an attendant and told him, "I like you." Occasionally, punishment meted out to patients for misdemeanors seemed so excessive that it could not be justified by the most radical interpretations of psychiatric canon. Nevertheless, they appeared to go unquestioned. Tempers were often short. A patient who had not heard a call for medication would be roundly excoriated, and the morning attendants would often wake patients with, "Come on, you m_____ f_____s, out of bed!"

Neither anecdotal nor "hard" data can convey

the overwhelming sense of powerlessness which invades the individual as he is continually exposed to the depersonalization of the psychiatric hospital. It hardly matters *which* psychiatric hospital— the excellent public ones and the very plush private hospital were better than the rural and shabby ones in this regard, but, again, the features that psychiatric hospitals had in common overwhelmed by far their apparent differences.

Powerlessness was evident everywhere. The patient is deprived of many of his legal rights by dint of his psychiatric commitment (21). He is shorn of credibility by virtue of his psychiatric label. His freedom of movement is restricted. He cannot initiate contact with the staff, but may only respond to such overtures as they make. Personal privacy is minimal. Patient quarters and possessions can be entered and examined by any staff member, for whatever reason. His personal history and anguish is available to any staff member (often including the "grey lady" and "candy striper" volunteer) who chooses to read his folder, regardless of their therapeutic relationship to him. His personal hygiene and waste evacuation are often monitored. The water closets may have no doors.

At times, depersonalization reached such proportions that pseudopatients had the sense that they were invisible, or at least unworthy of account. Upon being admitted, I and other pseudopatients took the initial physical examinations in a semipublic room, where staff members went about their own business as if we were not there.

On the ward, attendants delivered verbal and occasionally serious physical abuse to patients in the presence of other observing patients, some of whom (the pseudopatients) were writing it all down. Abusive behavior, on the other hand, terminated quite abruptly when other staff members were known to be coming. Staff are credible witnesses. Patients are not.

A nurse unbuttoned her uniform to adjust her brassiere in the presence of an entire ward of viewing men. One did not have the sense that she was being seductive. Rather, she didn't notice us. A group of staff persons might point to a patient in the dayroom and discuss him animatedly, as if he were not there.

One illuminating instance of depersonaliza-tion and invisibility occurred with regard to medications. All told, the pseudopatients were administered nearly 2100 pills, including Elavil, Stelazine, Compazine, and Thorazine, to name but a few. (That such a variety of medications should have been administered to patients presenting identical symptoms is itself worthy of note.) Only two were swallowed. The rest were either pocketed or deposited in the toilet. The pseudopatients were not alone in this. Although I have no precise records on how many patients rejected their medications, the pseudopatients frequently found the medications of other patients in the toilet before they deposited their own. As long as they were cooperative, their behavior and the pseudopatients' own in this matter, as in other important matters, went unnoticed throughout.

Reactions to such depersonalization among pseudopatients were intense. Although they had come to the hospital as participant observers and were fully aware that they did not "belong," they nevertheless found themselves caught up in and fighting the process of depersonalization. Some examples: a graduate student in psychology asked his wife to bring his textbooks to the hospital so he could "catch up on his homework"—this despite the elaborate precautions taken to conceal his professional association. The same student, who had trained for quite some time to get into the hospital, and who had looked forward to the experience, "remembered" some drag races that he had wanted to see on the weekend and insisted that he be discharged by that time. Another pseudopatient attempted a romance with a nurse. Subsequently, he informed the staff that he was applying for admission to graduate school in psychology and was very likely to be admitted, since a graduate professor was one of his regular hospital visitors. The same person began to engage in psychotherapy with other patients—all of this as a way of becoming a person in an impersonal environment.

The Sources of Depersonalization

What are the origins of depersonalization? I have already mentioned two. First are attitudes held by all of us toward the mentally ill—including those who treat them—attitudes characterized by fear, distrust, and horrible expectations on the one hand, and benevolent intentions on the other. Our ambivalence leads, in this instance as in others, to avoidance.

Second, and not entirely separate, the hierarchical structure of the psychiatric hospital facilitates depersonalization. Those who are at the top have least to do with patients, and their behavior inspires the rest of the staff. Average daily contact with psychiatrists, psychologists, residents, and physicians combined ranged from 3.9 to 25.1 minutes, with an overall mean of 6.8 (six pseudopatients over a total of 129 days of hospitalization). Included in this average are time spent in the admissions interview, ward meetings in the presence of a senior staff member, group and individual psychotherapy contacts, case presentation conferences, and discharge meetings. Clearly, patients do not spend much time in interpersonal contact with doctoral staff. And doctoral staff serve as models for nurses and attendants.

There are probably other sources. Psychiatric installations are presently in serious financial straits. Staff shortages are pervasive, staff time at a premium. Something has to give, and that something is patient contact. Yet, while financial stresses are realities, too much can be made of them. I have the impression that the psychological forces that result in depersonalization are much stronger than the fiscal ones and that the addition of more staff would not correspondingly improve patient care in this regard. The incidence of staff meetings and the enormous amount of record-keeping on patients, for example, have not been as substantially reduced as has patient contact. Priorities exist, even during hard times. Patient contact is not a significant priority in the traditional psychiatric hospital, and fiscal pressures do not account for this. Avoidance and depersonalization may.

Heavy reliance upon psychotropic medication tacitly contributes to depersonalization by convincing staff that treatment is indeed being conducted and that further patient contact may not be necessary. Even here, however, caution needs to be exercised in understanding the role of psychotropic drugs. If patients were powerful rather than powerless, if they were viewed as interesting individuals rather than diagnostic entities, if they were socially significant rather than social lepers, if their anguish truly and wholly compelled our sympathies and concerns, would we not *seek* contact with them, despite the availability of medications? Perhaps for the pleasure of it all?

The Consequences of Labeling and Depersonalization

Whenever the ratio of what is known to what needs to be known approaches zero, we tend to invent "knowledge" and assume that we understand more than we actually do. We seem unable to acknowledge that we simply don't know. The needs for diagnosis and remediation of behavioral and emotional problems are enormous. But rather than acknowledge that we are just embarking on understanding, we continue to label patients "schizophrenic," "manic-depressive," and "insane," as if in those words we had captured the essence of understanding. The facts of the matter are that we have known for a long time that diagnoses are often not useful or reliable, but we have nevertheless continued to use them. We now know that we cannot distinguish insanity from sanity. It is depressing to consider how that information will be used.

Not merely depressing, but frightening. How many people, one wonders, are sane but not recognized as such in our psychiatric institutions? How many have been needlessly stripped of their privileges of citizenship, from the right to vote and drive to that of handling their own accounts? How many have feigned insanity in order to avoid the criminal consequences of their behavior, and, conversely, how many would rather stand trial than live interminably in a psychiatric hospital—but are wrongly thought to be mentally ill? How many have been stigmatized by well-intentioned, but nevertheless erroneous, diagnoses? On the last point, recall again that a "type 2 error" in psychiatric diagnosis does not have the same consequences it does in medical diagnosis. A diagnosis of cancer that has been found to be in error is cause for celebration. But psychiatric diagnoses are rarely found to be in error. The label sticks, a mark of inadequacy forever.

Finally, how many patients might be "sane" outside the psychiatric hospital but seem insane in it—not because craziness resides in them, as it were, but because they are responding to a bizarre setting, one that may be unique to institutions which harbor nether people? Goffman (4) calls the process of socialization to such institutions "mortification"—an apt metaphor that includes the processes of depersonalization that have been described here. And while it is impossible to know whether the pseudopatients' responses to these processes are characteristic of all inmates—they were, after all, not real patients—it is difficult to believe that these processes of socialization to a psychiatric hospital provide useful attitudes or habits of response for living in the "real world."

Summary and Conclusions

It is clear that we cannot distinguish the sane from the insane in psychiatric hospitals. The hospital itself imposes a special environment in which the meanings of behavior can easily be misunderstood. The consequences to patients hospitalized in such an environment—the powerlessness, depersonalization, segregation, mortification, and self-labeling—seem undoubtedly countertherapeutic.

I do not, even now, understand this problem well enough to perceive solutions. But two matters seem to have some promise. The first concerns the proliferation of community mental health facilities, of crisis intervention centers, of the human potential movement, and of behavior therapies that, for all of their own problems, tend to avoid psychiatric labels, to focus on specific problems and behaviors, and to retain the individual in a relatively non-pejorative environment. Clearly, to the extent that we refrain from sending the distressed to insane places, our impressions of them are less likely to be distorted. (The risk of distorted perceptions, it seems to me, is always present, since we are much more sensitive to an individual's behaviors and verbalizations than we are to the subtle contextual stimuli that often promote them. At issue here is a matter of magnitude. And, as I have shown, the magnitude of distortion is exceedingly high in the extreme context that is a psychiatric hospital.)

The second matter that might prove promising speaks to the need to increase the sensitivity of mental health workers and researchers to the *Catch 22* position of psychiatric patients. Simply reading materials in this area will be of help to some such workers and researchers. For others, directly experiencing the impact of psychiatric hospitalization will be of enormous use. Clearly, further research into the social psychology of such total institutions will both facilitate treatment and deepen understanding.

I and the other pseudopatients in the psychiatric setting had distinctly negative reactions. We do not pretend to describe the subjective experiences of true patients. Theirs may be different from ours, particularly with the passage of time and the necessary process of adaptation to one's environment. But we can and do speak to the relatively more objective indices of treatment within the hospital. It could be a mistake, and a very unfortunate one, to consider that what happened to us derived from malice or stupidity on the part of the staff. Quite the contrary, our overwhelming impression of them was of people who really cared, who were committed and who were uncommonly intelligent. Where they failed, as they sometimes did painfully, it would be more accurate to attribute those failures to the environment in which they, too, found themselves than to personal callousness. Their perceptions and behavior were controlled by the situation, rather than being motivated by a malicious disposition. In a more benign environment, one that was less attached to global diagnosis, their behaviors and judgments might have been more benign and effective.

Notes

1. P. Ash, *J. Abnorm. Soc. Psychol.* 44, 272 (1949); A. T. Beck, *Amer. J. Psychiat.* 119, 210 (1962); A. T. Boisen, *Psychiatry* 2, 233 (1938); N. Kreitman, *J. Ment. Sci.* 107, 876 (1961); N. Kreitman, P. Sainsbury, J. Morrisey, J. Towers, J. Scrivener, *ibid.*, p. 887; H. O. Schmitt and C. P. Fonda, *J. Abnorm. Soc. Psychol.* 52, 262 (1956); W. Seeman, *J. Nerv. Ment. Dis.* 118, 541 (1953). For an analysis of these artifacts and summaries of the disputes, see J. Zubin, *Annu. Rev. Psychol.* 18, 373 (1967); L. Phillips and J. G. Draguns, *ibid.* 22, 447 (1971).

2. R. Benedict, *J. Gen. Psychol.* 10, 59 (1934).

3. See in this regard H. Becker, *Outsiders: Studies in the Sociology of Deviance* (Free Press, New York, 1963); B. M. Braginsky, D. D. Braginsky, K. Ring, *Methods of Madness: The Mental Hospital as a Last Resort* (Holt, Rinehart & Winston, New York, 1969); G. M. Crocetti and P. V. Lemkau, *Amer. Sociol. Rev.* 30, 577 (1965); E. Goffman, *Behavior in Public Places* (Free

Press, New York, 1964); R. D. Laing, *The Divided Self: A Study of Sanity and Madness* (Quadrangle, Chicago, 1960); D. L. Phillips, *Amer. Social. Rev.* 28, 963 (1963); T. R. Sarbin, *Psychol. Today* 6, 18 (1972); E. Schur, *Amer. J. Sociol.* 75, 309 (1969); T. Szasz, *Law, Liberty and Psychiatry* (Macmillan, New York, 1963); *The Myth of Mental Illness: Foundations of a Theory of Mental Illness* (Hoeber-Harper, New York, 1963). For a critique of some of these views, see W. R. Gove, *Amer. Sociol. Rev.* 35, 873 (1970).

4. E. Goffman, *Asylums* (Doubleday, Garden City, N.Y., 1961).

5. T. J. Scheff, *Being Mentally Ill: A Sociological Theory* (Aldine, Chicago, 1966).

6. Data from a ninth pseudopatient are not incorporated in this respect because, although his sanity went undetected, he falsified aspects of his personal history, including his marital status and parental relationships. His experimental behaviors therefore were not identical to those of the other pseudopatients.

7. A. Barry, *Bellevue Is a State of Mind* (Harcourt Brace Jovanovich, New York, 1971); I. Belknap, *Human Problems of a State Mental Hospital* (McGraw-Hill, New York, 1956); W. Caudill, F. C. Redlich, H. R. Gilmore, E. B. Brody, *Amer. J. Orthopsychiat.* 22, 314 (1952); A. R. Goldman, R. H. Bohr, T. A. Steinberg, *Prof. Psychol.* 1, 427 (1970); unauthored, *Roche Report* 1 (No. 13), 8 (1971).

8. Beyond the personal difficulties that the pseudopatient is likely to experience in the hospital, there are legal and social ones that, combined, require considerable attention before entry. For example, once admitted to a psychiatric institution, it is difficult, if not impossible, to be discharged on short notice, state law to the contrary notwithstanding. I was not sensitive to these difficulties at the outset of the project, nor to the personal and situational emergencies that can arise, but later a writ of habeas corpus was prepared for each of the entering pseudopatients and an attorney was kept "on call" during every hospitalization. I am grateful to John Kaplan and Robert Bartels for legal advice and assistance in these matters.

9. However distasteful such concealment is, it was a necessary first step to examining these questions. Without concealment, there would have been no way to know how valid these experiences were; nor was there any way of knowing whether whatever detections occurred were a tribute to the diagnostic acumen of the staff or to the hospital's rumor network. Obviously, since my concerns are general ones that cut across individual hospitals and staffs, I have respected their anonymity and have eliminated clues that might lead to their identification.

10. Interestingly, of the 12 admissions, 11 were diagnosed as schizophrenic and one, with the identical symptomatology, as manic-depressive psychosis. This diagnosis has a more favorable prognosis, and it was given by the only private hospital in our sample. On the relations between social class and psychiatric diagnosis, see A. deB. Hollingshead and F. C. Redlich, *Social Class and Mental Illness: A Community Study* (Wiley, New York, 1958).

11. It is possible, of course, that patients have quite broad latitudes in diagnosis and therefore are inclined to call many people sane, even those whose behavior is patently aberrant. However, although we have no hard data on this matter, it was our distinct impression that this was not the case. In many instances, patients not only singled us out for attention, but came to imitate our behaviors and styles.

12. J. Cumming and E. Cumming, *Community Ment. Health* 1, 135 (1965); A. Farina and K. Ring, *J. Abnorm. Psychol.* 70, 47 (1965); H. E. Freeman and O. G. Simmons, *The Mental Patient Comes Home* (Wiley, New York, 1963); W. J. Johannsen, *Ment. Hygiene* 53, 218 (1969); A. S. Linksy, *Soc. Psychiat.* 5, 166 (1970).

S. E. Asch, *J. Abnorm. Soc. Psychol.* 41, 258 (1946); *Social Psychology* (Prentice-Hall, New York, 1952).

14. See also I. N. Mensh and J. Wishner, *J. Personality* 16, 188 (1947); J. Wishner, *Psychol. Rev.* 67, 96 (1960); J. S. Bruner and R. Tagiuri, in *Handbook of Social Psychology*, G. Lindzey. Ed. (Addison-Wesley, Cambridge, Mass., 1954), vol. 2, pp. 634–654; J. S. Bruner, D. Shapiro, R. Tagiuri, in *Person Perception and Interpersonal Behavior*. R. Tagiuri and L. Petrullo, Eds. (Stanford Univ. Press, Stanford, Calif., 1958), pp. 277–288.

15. For an example of a similar self-fulfilling prophecy, in this instance dealing with the "central" trait of intelligence, see R. Rosenthal and L. Jacobson, *Pygmalion in the Classroom* (Holt, Rinehart & Winston, New York, 1968).

16. E. Zigler and L. Phillips, *J. Abnorm. Soc. Psychol.* 63, 69 (1961). See also R. K. Freudenberg and J. P. Robertson, *A.M.A. Arch. Neurol. Psychiatr.* 76, 14 (1956).

17. W. Mischel, *Personality and Assessment* (Wiley, New York, 1968).

18. The most recent and unfortunate instance of this tenet is that of Senator Thomas Eagleton.

19. T. R. Sarbin and J. C. Mancuso, *J. Clin. Consult.*

Psychol. 35, 159 (1970); T. R. Sarbin, *ibid.* 31, 447 (1967); J. C. Nunnally, Jr., *Popular Conceptions of Mental Health* (Holt, Rinehart & Winston, New York, 1961).

20. A. H. Stanton and M. S. Schwartz, *The Mental Hospital: A Study of Institutional Participation in* *Psychiatric Illness and Treatment* (Basic, New York, 1954).

21. D. B. Wexler and S. E. Scoville, *Ariz. Law Rev.* 13, 1 (1971).

22. I thank W. Mischel, E. Orne, and M. S. Rosenhan for comments on an earlier draft of this manuscript.

For Further Reading

Ennis, B. J. *Prisoners of Psychiatry.* New York: Avon, 1972.

Ennis B. J., and Emery, R. D. *The Rights of Mental Patients.* American Civil Liberties Union Handbook. New York: Avon, 1978.

Ethical Standards of Psychologists: 1977 Revision. Washington, DC: American Psychological Association, 1977.

Finkel, M. *Madness and Civilization: Its Legacy, Tensions, and Changes.* New York: Macmillan, 1976.

Gaylin, W., et al. *Doing Good: The Limits of Benevolence.* New York: Pantheon, 1978.

Laska, E. and Bank, R. *Safeguarding Psychiatric Privacy: Computer Systems and Their Uses.* New York: Wiley, 1975.

London, P. *Behavior Control.* New York: Meridian Books, 1977.

Martin, R. *Behavior Modification: Human Rights and Legal Responsibilities.* Champaign, Ill.: Research Press, 1974.

Price, R. H., and Denner, B., eds. *The Making of a Mental Patient.* New York: Holt, Rinehart & Winston, 1973.

Rappaport, J. *Community Psychology: Values, Research and Action.* New York: Holt, Rinehart & Winston, 1977.

Standards for Providers of Psychological Services. Washington, DC: American Psychological Association, 1977.

Szasz, T. S. *The Myth of Mental Illness: Foundations of a Theory of Personal Conduct.* New York: Hoeber-Harper, 1961.

Szasz, T. S. *Law, Liberty and Psychiatry.* New York: Macmillan, 1963.

Szasz, T. S. *Ideology and Insanity: Essay on the Psychiatric Dehumanization of Man.* Garden City, N.Y.: Doubleday, Anchor Books, 1970.

Szasz, T. S. *The Age of Madness: The History of Involuntary Mental Hospitalization.* Garden City, N.Y.: Doubleday, Anchor Books, 1973.

Part IV

Special Bioethical Issues

11
ABORTION

Miriam Tunney knew when she took the job as a nurse practitioner at the neighborhood health center that one day she would have to confront the issue of abortion. So far she had succeeded in avoiding the matter. But now, as mother and daughter went at each other in her office, Miriam wished she were better prepared to intervene.

Mrs. Fredericks insisted that her daughter have an abortion. Celia, who was just ending her first trimester, wanted no part of it. "I have a right to have my own baby if I want to," she told her worried mother defiantly.

"You're just a child yourself!" Mrs. Fredericks fired back.

"I'm eighteen."

"But unmarried!"

"Jack and I may get married before the baby's born. We haven't decided yet."

"Decided yet?" Mrs. Fredericks laughed scornfully. "You told me he has no intention of marrying you."

"He could change his mind," Celia said, suddenly humiliated, vulnerable.

"And if he doesn't—where will you be then?"

The girl pulled herself together, almost visibly.

"I'll quit school and get a job."

"Doing what for heaven's sake? Waiting on tables and mopping floors? Haven't you seen what that's done to me?" Obviously, the mother wasn't just being vicious—she was hurting. "And don't think I'm going to raise your kid, because I'm not. God knows I've had enough trouble putting food on the table for my own brood."

"I'm not expecting you to," Celia said sharply. "I'll take care of my own child."

The heated exchange continued for some minutes—Celia telling her mother that she was always trying to run her life, and Mrs. Fredericks insisting that she was thinking only of the welfare of her daughter and the baby. Finally, Mrs. Fredericks threw up her hands and turned to Miriam Tunney. "You try to talk some sense into this girl, will you?"

Miriam felt torn. She did not object to abortion to protect a pregnant woman's health or to prevent the birth of a seriously deformed child, but she was not sure she approved of it as a way of birth control, which is how it would function in Celia's case. On the other hand, Miriam had to agree with Mrs. Fredericks: Celia certainly was not ready to be a mother, and the child undoubtedly would have two strikes against it. Was it fair to the *baby to be brought into the world under such inauspicious circumstances, perhaps being born for no better reason than a confused girl's defiance of her mother? Miriam also wondered whether it was fair that the taxpayers ultimately might have to assume responsibility for the child, should Celia be unable or unwilling to do so. Now, as Mrs. Fredericks was asking Miriam to reason with her daughter, the young nurse practitioner wondered what she should do.*

Most HCPs will probably never find themselves in Miriam's situation, but many will. Of those who do not, a large number will still have to deal with issues related to abortion, which is the intentional termination of a pregnancy by inducing the loss of the fetus. Sometimes HCPs actually get involved in abortion procedures; other times they are expected to give counsel to those seeking abortions; still other times, as health care experts, they are called on to give thoughtful, well-informed views on various aspects of abortion. Clearly, abortion represents a biomedical issue of special interest to a cross section of HCPs. It is also an issue of intense social, moral, and legal interest.

If we assume that Celia Fredericks has the rights of an adult in this situation, then technically she is permitted under current law to have an abortion. This right was acknowledged nearly a decade ago in a momentous ruling involving a young woman who, to safeguard anonymity, was named Jane Roe. Jane Roe was unmarried, pregnant, and wished to have an abortion. Unfortunately for Ms. Roe, she lived in Texas, where the statutes forbade abortion except to save the life of the mother. So the young woman went to court to prove that the laws were unconstitutional.

The three-judge district court ruled that Jane Roe had reason to sue, that the Texas criminal abortion statutes were void on their face, and—most important—that the right to choose whether to have children was protected by the Ninth through the Fourteenth Amendments. Since the district court had denied a number of other aspects of the suit, however, the case went to the United States Supreme Court. On January 22, 1973, in the now-famous *Roe* v. *Wade* decision, the Supreme Court affirmed the district court's judgment.[1]

Expressing the views of seven members of the Court, Justice Blackmun pointed out that the right to privacy applies to a woman's decision of whether to terminate her pregnancy but that her right to terminate is not absolute. It may be limited by the state's legitimate interests in safeguarding the woman's health, in maintaining proper medical standards, and in protecting human life. Blackmun further indicated that the unborn are not included in the definition of *person* as used in the Fourteenth Amendment. Most important, he said that, prior to the end of the first trimester of pregnancy, the state may not interfere with or regulate an attending physician's decision, reached in consultation with the patient, that the patient's pregnancy should be terminated. After the first trimester and until the fetus is viable, the state may regulate the abortion procedure only for the health of the mother. After the fetus becomes viable, the state may prohibit all abortions except those to preserve the health or life of the mother.

In dissenting, Justices White and Rehnquist said that nothing in the lan-

---

1. U.S. Supreme Court Reports, *October Term 1972, lawyer's ed.* (Rochester, N.Y.: Lawyers' Cooperative Publishing, 1974), p. 147.

guage or history of the Constitution supported the Court's judgment, that the Court had simply manufactured a new constitutional right for pregnant women. The abortion issue, they said, should have been left with the people and with the political processes they had devised to govern their affairs.

Thus, for the time being at least, the abortion question was resolved legally. But the issue was hardly dead. Since then, a number of antiabortion movements have surfaced, which not only indicates that some people think abortion should be illegal but that many believe it is wrong. Whether legal or not, abortion remains an intensely personal moral concern for those who must confront it, patient and HCP alike. Obviously its legality provides options that would not have been present before, but these can make the moral dilemma that much thornier. In the past one could always rationalize away the possibility of an abortion on the basis of its illegality.

Some say an abortion is right if (1) it is therapeutic—that is, necessary to preserve the physical or mental health of the woman, (2) it prevents the birth of a severely handicapped child, or (3) it ends a pregnancy resulting from some criminal act of sexual intercourse. Others say that even therapeutic abortions are immoral. Still others argue that any restrictive abortion legislation is wrong and must be liberalized to allow a woman to have an abortion on demand, at the request of her and her physician regardless of the reasons. In order to evaluate such claims, it is important first to place abortion in a biological context. It is also important for HCPs to understand their profession's traditional views on abortion and the psychosocial dynamics operating today that can produce moral constraints for both HCP and patient. So before taking up the various prolife (antiabortion) and prochoice (proabortion) arguments, we will consider these matters.

Values and Priorities

Before starting to consider the morality of abortion, respond to the following inquiries, which are intended to flush out your values and priorities in this most important bioethical area.

1. Do you approve of abortion as a birth control measure?

2. How should Miriam respond?

3. Do you think Miriam should set aside her own personal views and present both sides as objectively as possible?

4. Do you think Miriam should excuse herself from this case and turn it over to someone else?

5. Assuming Celia is considered a minor, would it be right to force her to abort?

6. Make a list of the various parties who have a legitimate claim on Miriam in this situation; then rank the list in order of importance.

7. Is abortion ever morally justifiable?

8. Does a woman have a right to an abortion under any circumstances?

9. If you were personally opposed to abortion, would you ever help someone who wanted it to obtain one?

10. If you were personally opposed to abortion, would you ever assist in the care of an abortion patient?

11. Should HCPs set aside their personal views when counseling a pregnant woman who is contemplating an abortion?

12. Under what circumstances, if any, would you consider an abortion morally permissible?

13. Under which of the following conditions, if any, would you regard abortion as permissible? In cases of rape; incest; physical threat to the mother; psychological threat to the mother; sexual relations between unmarried minors, as in Celia's case; the patient's personal convenience, such as a woman's not wanting to disrupt a career.

14. Do you think that the marital status of the woman has any bearing on the permissibility of abortion?

15. Do you think that the medical and nursing professions' views on abortion are more or less conservative than the view expressed in *Roe* v. *Wade*, or are they about the same?

16. Would you consider the unborn a person at the moment of conception, after a heartbeat is detectable, at birth, or at some other stage?

17. What do you think is the strongest prolife argument? The strongest pro-choice argument?

18. How, if at all, has the women's rights movement affected the abortion issue?

19. Do you think that dealing with abortion and abortion patients is easier or more difficult for nurses than for physicians, or is it about the same?

20. Should the state pay for abortions for the poor?

Biological Background

Since most of the controversy that surrounds abortion concerns the question of precisely when during a pregnancy a human life is considered to begin, it is important to have some background information about the development of the human fetus and some familiarity with the terms that designate the various developmental stages.[2] Conception, or fertilization, occurs when a female germ cell (an ovum) is penetrated by a male germ cell (a spermatozoon). The result is a single cell called a zygote, which contains a full genetic code of forty-six chromosomes. The zygote journeys down the fallopian tube, which carries ova from the ovary to the uterus. This passage generally takes two or three days. During the journey, the zygote begins a process of cellular division that increases its size. Occasionally, the zygote ends its journey inside the fallopian tube, where it

2. *R. Munson, Intervention and Reflection (Belmont, Calif.: Wadsworth, 1979), pp. 41–42.*

continues to develop. Because the tube is so narrow, this kind of pregnancy must generally be terminated by surgery.

When the multicell zygote reaches the uterus, it floats freely in the intrauterine fluid and develops into what is termed a blastocyst, a ball of cells surrounding a fluid-filled cavity. By the end of the second week, the blastocyst implants itself in the uterine wall. From the end of the second week until the end of the eighth week, this entity is termed an embryo. During the embryonic stage, organ systems begin to develop, and the embryo takes on distinctly human external features.

The eighth week is important because it is then that the brain activity generally becomes detectable. From this point until birth, the embryo is termed a fetus. (In common parlance, *fetus* is used to designate the unborn entity at whatever stage.)

Two other terms sometimes arise in abortion discussions. One is *quickening*, which refers to the point at which the mother begins to feel the movements of the fetus. This occurs somewhere between the sixteenth and eighteenth weeks. The other is *viability*—the point at which the fetus is capable of surviving outside the womb. The fetus generally reaches viability around the twenty-fourth week. This, then, is the unfolding of events during pregnancy:

Zygote—day 1 through day 3

Blastocyst—day 4 through week 2

Embryo—week 3 through week 8

Fetus—week 9 until birth

Quickening—between weeks 16 and 18

Viability—around week 24

Should the unborn entity be terminated at any point, an abortion is said to occur. Thus *abortion* is simply the termination of a pregnancy.

Abortions can occur for a number of reasons. Sometimes the abortion occurs spontaneously, because of internal biochemical factors or because of injury to the woman. Such spontaneous abortions are ordinarily termed miscarriages.

Abortions can also result directly from human intervention, which can occur in a variety of ways. Sometimes it happens very early, as when a woman takes a drug such as the "morning-after pill" in order to prevent the blastocyst from implanting itself in the uterine wall. Subsequent intervention during the first trimester (through the twelfth week) usually takes one of two forms. In uterine or vacuum aspiration the narrow opening of the uterus, the cervix, is dilated. A small tube is then inserted into the uterus, and its contents are emptied by suction. In dilatation and curettage, the cervix is also dilated, but this time its contents are scraped out with a spoon-shaped surgical instrument called a curet. These procedures can sometimes be carried out through the sixteenth week, but after that the fetus is generally too large to make the procedure practical.

The most common abortion technique after the sixteenth week is by saline injection. The amniotic fluid (the fluid in the amnion, the membrane sac surrounding the fetus) is drawn through a hollow needle and replaced by a solution of salt and water. This leads to a miscarriage.

Another, far rarer method after the sixteenth week is the hysterotomy, or cesarean section. In this surgical procedure, the fetus is removed from the uterus through an incision.

To summarize, these are the various abortion possibilities:

Internally induced

> Spontaneous abortion—anytime

Externally induced

> "Morning-after pill"—immediately following intercourse
>
> Uterine or vacuum aspiration—through week 12
>
> Dilatation and currettage—through week 12
>
> Saline injection—after week 16
>
> Hysterotomy—after week 16 (extremely rare)

With this biological background in mind, let us now consider abortion in the context of the health professions.

Abortion in the Context of the Professions

Professional interactions with abortion patients and their families are largely determined by HCP attitudes toward abortion. These attitudes are shaped by numerous social, cultural, religious, and personal influences. It is useful, therefore, to recall the discussion in Chapter 2 about the various sources of values. Apart from these, a most influential determinant of the HCP attitude toward abortion is the profession itself, including its traditional outlooks on and approaches to the issue. Therefore, we should consider the professional context in which HCP attitudes are formed, subsequent decisions are made, and action is taken. Since physicians and nurses play principal roles in abortion interventions, we will focus on the context of professional medicine and nursing.

The Medical Profession

Even a cursory look at the modern history of professional medicine in the United States indicates an unmistakable antiabortion attitude over the last century. At various times between 1850 and 1900, the AMA adopted resolutions against the unwarranted destruction of human life and even solicited the clergy to impress upon the public the serious religious and moral implications of abortion.[3] The AMA continued to issue similar condemnations throughout the first half of the present century.

---

3. 22 Transcript (*Chicago: American Medical Association, 1871*), p. 258.

A number of events in the 1960s raised the abortion issue to the level of public debate and ultimately forced the AMA to crystallize its position. A truly catalyzing episode involved an Arizona resident named Sherri Finkbine, who in 1962 used the tranquilizer thalidomide during her pregnancy. Mrs. Finkbine subsequently read of seriously deformed babies being born to European women who had used thalidomide. She expressed her concern to her physician, who recommended termination of the pregnancy since the chances were very great that her baby would also be deformed. Under Arizona law, Finkbine had to present her case to a three-member medical board. She did, and the board approved the abortion.

Before the abortion took place, Mrs. Finkbine felt moraly obligated to inform other pregnant women in America who unknowingly might also have taken thalidomide. So she called a local newspaper and told her story to the editor. The ensuing article on the dangers of the drug was picked up by wire services and given national exposure. As a result, Mrs. Finkbine became the object of a great outpouring of antiabortion feeling. Given all the adverse publicity her case was attracting, the medical board reconsidered its position and withdrew approval of her abortion. Its reversal seemed on firm ground, since Arizona statutes permitted abortion only to save the life of the mother.

Convinced that the odds were great that she would bear a seriously deformed baby, Mrs. Finkbine sought to have the abortion in some other state but without success. Eventually she went to Sweden, where the abortion was performed in a hospital.

Understandably, the Finkbine case ignited an intensive public examination of the morality and legality of abortion. As it happened, the national debate was fueled by other events. An especially combustible one occurred in 1964, when 20,000 babies were born in the United States with serious abnormalities resulting from the German measles their mothers had contracted during pregnancy. Many of these women had requested abortions but could not legally obtain them.[4] The nation wondered whether these women—Sherri Finkbine and others in similar straits—should have been compelled to bear children who stood a high chance of serious deformity.

The question took on greater urgency after 1965, when amniocentesis moved out of the experimental stage and became an integral part of genetic counseling and diagnosis. Amniocentesis is a surgical procedure whereby amniotic fluid is withdrawn from a pregnant woman for diagnostic purposes. Done properly, the procedure can provide a great deal of information about the fetus, especially the presence or absence of such serious abnormalities as Tay-Sachs disease or Down's syndrome. The former is a fatal degenerative disease; the latter manifests itself in mental retardation and various physical abnormalities.

Against this backdrop, in 1967 the AMA's Committee on Human Reproduction urged adoption of a policy that would oppose induced abortions except when (1) incontrovertible medical evidence demonstrated a threat to the mother's health; (2) the child was likely to be born with an incapacitating physical or mental deformity; or (3) when the mother's mental health was threatened by a

---

4.   J. Hole and E. Levine, Rebirth of Feminism (New York: Quadrangle Books, 1971), pp. 283–284.

pregnancy that resulted from forcible rape or incest. In addition, the committee recommended two other guidelines: that the original medical recommendation for abortion be confirmed by two additional physician opinions and that the abortion be performed in a hospital accredited by the Joint Commission on Accreditation of Hospitals. The AMA House Delegates adopted the committee's resolution.[5] In 1970 the House Delegates reaffirmed the 1967 position, allowing that "no party to the procedure should be requested to violate personally held moral principles."[6]

Perhaps the most important aspect of this historical scenario is the medical profession's position on abortion, which clearly is more restrictive than the U.S. Supreme Court's, as articulated in 1973 in *Roe* v. *Wade*. Nowhere in the AMA guidelines is there a recognition of the woman's right to an abortion in the first trimester regardless of reason. The significance of this is that, although "the law of the land" currently recognizes such a right, the philosophical thrust of medical training, as shaped by the AMA's official position, apparently would inculcate a more conservative attitude. If it does, then we can wonder whether the physician who confers with a pregnant woman who is contemplating an abortion can be altogether objective. This issue is not merely speculative, for research indicates that at least some obstetrician-gynecologists consider abortion in the first trimester acceptable as long as women are not suspected of using it as a form of birth control.[7]

Physicians seem especially torn between the obstetrical objective of delivering a viable infant and the performance of second-trimester abortions. For example, here are the recorded feelings of some residents:

> I don't like doing abortions but I probably will because of the economics of it. . . . Midtrimester is very distasteful to me. I really don't get a kick out of doing them, I don't know too many guys who enjoy doing them.
>
> I cannot see myself fighting on the one hand to save a baby and on the other hand expediting its expulsion. I think if a woman has gone by the first trimester and is thinking of an abortion she should place [the baby] up for adoption. There are a lot of people that can't have children.
>
> I have no strong feelings against first-trimester abortions. They don't disturb me at all. Midtrimester abortions I find esthetically unpleasant. I don't think it is from a religious perspective necessarily; I just find them unpleasant. I guess it is difficult for me to excuse a woman for waiting until she is fourteen or sixteen weeks pregnant to make up her mind she doesn't want it, but aside from that it just goes against everything we have been training for.[8]

Part of the study from which these quotations are taken suggests that even residents who indicated that women should be able to obtain abortions said that they would only perform them if forced by their own economic necessity.

Such reports, although in themselves insufficient to warrant hard conclu-

---

5. Proceedings of the House Delegates *(Chicago: American Medical Association, June 1967), pp. 40–51.*

6. Proceedings of the House Delegates *(Chicago: American Medical Association, June 1970), p. 221.*

7. D. Scully, Men Who Control Women's Health: The Miseducation of Obstetrician-Gynecologists. *(Boston: Houghton Mifflin, 1980), pp. 89–90.*

8. Scully, p. 90.

sions, do raise a question about the relative weight physicians give to the patient's right to autonomy. Given their training and the attitudes it fosters, obstetrician-gynecologists may in fact be profoundly constrained in making the woman's and family's autonomy a major goal of their specialty. But if they cannot do so, then moral concerns arise about their capacity to provide a pregnant woman with adequate health care. Taken together with the obstetrician's monopoly on the abortion market, this raises the question of whether women might not be better served by a group that is psychologically better prepared for the task of abortion.

The Nursing Profession

In general, nurses share with physicians roughly similar attitudes toward abortion. Indeed, before entering the field, nurses in training seem, if anything, to be more conservative. More nursing students and their faculty oppose abortion on demand than other health professionals and the general population of comparable educational level. Similarly, fewer nursing students and faculty seem willing to help patients obtain an abortion than other HCPs.[9] In contrast, for example, social workers appear to have far more favorable attitudes toward abortion, probably because they work in a different social structure from nurses.[10]

Once in the field, nurses' attitudes apparently grow more permissive. For example, one study reported that nearly a quarter of the nurses surveyed favored unrestricted abortion; half or more favored abortion in cases of rape, defective fetuses, physical or mental impairment to the woman, or grave economic hardship; and three-fourths indicated that they would treat the abortion patient with as much understanding as any other patient.[11] In part, this less restrictive attitude can be attributed to such occupational determinants as the nurse's involvement in the dilemmas of unwanted pregnancies, the organization and administration of abortion services, and the social environment, including the attitudes of society and peers.[12]

These attitudinal generalities aside, it is important to distinguish the physician's practical role in abortion from the nurse's. The major difference is that the physician's involvement is ordinarily perfunctory, whereas the nurse's can be profoundly intimate. Specifically, the physician's involvement is generally confined to the clinic and operating room. In contrast, nurses often get involved in counseling and transmitting information to the patient. Furthermore, because they lack the professional autonomy of the physician, nurses are far more likely to get caught up in the crosscurrents of conflicting personal values and professional responsibilities, which can be exacerbated if the nurse happens to be a

---

9.   R. A. H. Rosen, H. H. Werley, J. W. Ayer, and F. B. Shea, "Some Organizational Correlates to Nursing Student Attitudes towards Abortion," Nursing Research, May–June 1974, pp. 253–259.

10.   G. E. Hendershot and J. W. Grim, "Abortion Attitudes among Nurses and Social Workers," American Journal of Public Health, May 1974, pp. 438–441.

11.   "What Nurses Think about Abortion," RN, June 1970, pp. 40–43.

12.   "Survey Finds Determinants of Attitudes toward Abortion," American Journal of Nursing, October 1971, p. 1900.

woman. Small wonder, then, that nurses sometimes suffer considerable stress in attempting to resolve their ambivalence about participating in abortion procedures, as well as anxiety and depression. As might be expected, it is not uncommon for nurses to feel anger toward abortion patients.[13]

Several moral concerns emerge from this profile of a nurse's role in and attitudes toward abortion. Most important, if nurses object to abortion on grounds of personal conscience, then it is questionable whether they can adequately care for abortion patients. Of course, various professional guidelines acknowledge this possibility and excuse participation that would violate personal values. Indeed, some states have passed laws that recognize this right and that protect HCPs from negligence and malpractice suits that might result from refusing to perform or participate in abortion. At the same time, it is likely that some nurses will feel conflict between exercising this right and performing what they think is a professional duty. What they fail to realize is that they risk subtle but serious injury to patients should they force themselves to participate in a procedure that they object to.

Supervisors have an obligation to inform subordinates of their rights and, just as important, to insist that those acting on those rights not be penalized. Similarly, patients would seem to have a right not to be nursed by someone who opposes abortion. If this is so, then the duty would seem to fall to supervisors to inform patients of their right and to ensure that patients who act on it (for example, by requesting that a particular nurse not be assigned to their case because she shows hostility toward them) are protected from administrative runarounds or vindictive action on the part of other nurses.

Far more difficult morally are cases of nurses who oppose abortions under certain circumstances but not under others; or who, like many physicians, believe abortions should be restricted to the first trimester. For example, some nurses may be able to participate in dilatation and curettage procedures, which occur during the first trimester, but find it objectionable to assist at a saline procedure, which occurs after the first trimester.[14] Such cases always call for a careful sorting of one's individual rights of conscience and one's professional obligations. Again, it seems that head nurses and supervising nurses can be especially helpful by creating a climate in which staff nurses feel free to air their conflicts and solutions can be effected. At the very least, nurses should indicate their conflict to supervisors and should not care for abortion patients while harboring unresolved feelings about the propriety of the procedure. Obviously, this means that nurses must clarify their own position on the various aspects of abortion—something they cannot do until fully apprised of its moral and philosophical dimensions.

Before turning to these moral concerns, we should examine the issue of abortion in the context of its psychosocial dynamics, with special reference to

13. F. J. Kane, M. Feldman, and S. Jain, "Abortion Service Personnel," Archives of General Psychiatry, March 1972, pp. 409–411.

14. A. J. Davis and M. A. Aroskar, Ethical Dilemmas and Nursing Practice (New York: Appleton-Century-Crofts, 1978), p. 106.

women's rights. A consideration of women's rights is relevant, because women who support the movement are often enlisted to support prochoice arguments. Furthermore, many pregnant women today can easily feel torn between asserting what they think is their right to abortion and abrogating the maternal role they were raised to play. Similarly, in dealing with the abortion decisions of other women, female HCPs can get caught at the intersection of a social movement calculated to advance their own personal and professional interests and a set of professional responsibilities, attitudes, and values that may be retarding those interests.

Psychosocial Dynamics: The Rights of Women

Throughout this book, we have seen how key psychosocial factors affect HCP-patient interactions and, consequently, the delivery of health care. Whether children see themselves as having any environmental control or whether the elderly view themselves as having social worth will affect these patients' reactions to illness and their dealings with health professionals and institutions. Similarly, HCP attitudes toward death and dying will affect their interactions with patients who are terminally ill. In all areas, psychological and social factors influence the interplay between patient and HCP, thus affecting the delivery of health care.

When the issue is abortion and the patients are pregnant women contemplating abortion, the psychosocial dimensions can be most powerful. This is especially noteworthy when one considers that the majority of women, particularly the poor, continue to derive most of their lifetime rewards from the family and therefore tend to support it. Women in this society are generally raised to embrace the maternal role wholeheartedly and to permit abortion only when their own lives are threatened by the unborn. Generally speaking, women as a group are not acculturated to accept nontherapeutic abortions.

Another aspect of the psychosocial dimension of abortion deals with the movement to secure basic rights for women. In recent years we have witnessed attempts by feminists to gain equal rights for women in education, politics, and employment. Part of this effort has been an attempt to gain control over their own reproductive systems, to secure the exclusive right to determine whether to carry a fetus to term. In fact, some feminists have argued that, without the capacity to limit reproduction, all of women's other rights and freedoms are illusory.[15] Others have gone so far as to suggest that the real reason many moralists in religion, politics, and the medical profession continue to oppose abortion on demand is that they recognize the revolutionary impact of fertility control. By this account, were women free to end unwanted pregnancies, they might choose to have no children at all or remain unmarried.[16] The merit of these positions does not concern us here. What does concern us is that in the minds of

---

15. L. Cisler, "Unfinished Business: Birth Control and Women's Liberation," in Sisterhood Is Powerful, ed. R. Morgan (New York: Vintage Books, 1970), pp. 248–249.

16. E. Frankfort, Vaginal Politics (New York: Quadrangle Books, 1972), xxiii–xxv.

some feminists there is a close connection between the right to abortion and the movement to secure equal rights for women. What is the moral significance of this link?

First, identifying the right to abortion as a fundamental female right at the very least provokes discussion about the patient's basic right to autonomy. Do pregnant women, as patients, have a right to autonomy to the extent that they may get an abortion on demand? If they do not, what argues against this right? Precisely how the right to autonomy applies to pregnant women has entered dramatically into the discussions about the morality of abortion, as we will see presently.

A second point of moral significance relates to the impact of the feminist movement. It is no secret that, largely as a result of the feminist movement, many women today feel intense pressure to assert their rights. This pressure has had an unsettling effect on some women. With their feminist consciousness heightened, many feel that they must now act on their new perceptions of sex and self, often in a way that clashes with how they have been socialized to behave. When the issue involves abortion and the suggestion is that an abortion decision is exclusively the woman's right, the clash can produce an inner struggle of epic proportions. The result for pregnant women can be confusion, intense anxiety, and severe moral constraints. And the impact on female HCPs is potentially no less severe than on pregnant women contemplating an abortion.

An awareness of these psychosocial factors will enable HCPs to understand an aspect of the pregnant woman's struggle that goes largely unstated and that the women themselves often cannot or will not articulate, even though they may feel it. Being aware of this dimension of the abortion issue gives female HCPs insight into their own states of mind, into the silent struggle that they themselves may be experiencing in sorting out their thoughts on abortion and their feelings about patients contemplating abortion. Conversely, an ignorance of the psychosocial dynamics of abortion decisions can easily subvert the delivery of health care by (1) leading to harsh and insensitive moral judgments about patients, (2) denying patients an opportunity to express their inner conflicts, and (3) precluding the help patients need in clarifying their own values and concept of rights and giving them priority. With these preliminary issues in mind, we can now turn to the key moral problem that abortion raises.

The Moral Problem

The key moral problem is this: Under what conditions, if any, is abortion morally justifiable? Three positions can be broadly identified.

First is the so-called conservative view, which holds that abortion is never morally justifiable or, at most, justifiable only when necessary to save the mother's life. This view is commonly associated with Roman Catholics, although they are far from the only persons who espouse it. At the opposite end of the spectrum is the so-called liberal view, which holds that abortion freely elected by the woman is always morally justifiable, regardless of the reasons or the point in fetal development. Most recently this view has been advanced by women's rights advocates, who focus on the woman's right to make all decisions that will affect her body. But again, this position has adherents outside the sphere of female liberation. Finally, there are the so-called intermediate or moderate views, which consider abortion morally acceptable up to a certain point in fetal development or hold that some reasons provide a sufficient justification for abortion but that others do not.

There is no consensus on the moral acceptability of abortion, but there is agreement that any answer to the question depends on one's view of what sort of entities fetuses are and whether such entities have rights. These two crucial problems are generally referred to as the ontological status and moral status of fetuses.

The Ontological Status of the Fetus

In philosophy, the term *ontology* refers to theory about the nature of being and existence. When we speak of the ontological status of the fetus, we mean the kind of entity the fetus is. Determining ontological status bears directly on the issue of fetal rights and, subsequently, on permissible treatment of the fetus.

The problem of ontological status embraces a number of questions: Is the fetus an individual organism? Is the fetus biologically a human being? Is the fetus psychologically a human being? Is the fetus a person?[17] Presumably, affirming that the fetus is biologically human attributes more significant status to it than affirming that it is an individual organism; and affirming that it is psychologically human assigns even greater status. To affirm that the fetus is a person probably assigns the most significant status to it. All these presumptions, however, depend on the precise meaning of the concepts involved.

Complicating the question of the fetus's ontological status is the meaning of the expression *human life*. The concept of human life can be used in at least two different ways. First, it can refer to biological human life, a set of physical characteristics that distinguish the human species from other nonhuman species. In this sense, human life may be coextensive with *individual organism*. Second, *human life* may refer to psychological human life, life characterized by the properties that are distinctly human. Among these might be the abilities to use symbols, think, and imagine. Abortion discussions can easily founder if these distinctions are not made. For example, many who would agree that abortion involves taking human life in the biological sense would deny that it involves taking human life in the psychological sense. What is more, they might see nothing immoral about taking life exclusively in the biological sense, although they would consider taking life in the psychological sense morally unacceptable.

Intertwined with the concept of human life is the concept of personhood. The definition of *personhood* may or may not differ from either the biological or psychological sense of human life. Some would argue that to be a person means simply to have the requisite biological or psychological properties or both. Others would propose additional conditions for personhood, such as conscious-

---

17.   *T. L. Beauchamp and L. R. Walters,* Contemporary Issues in Bioethics *(Belmont, Calif.: Dickenson, 1978), p. 188.*

ness, self-consciousness, rationality, even the capacities for communication and moral judgment. In this view, an entity must satisfy some or all these criteria, even additional ones, to be a person. Still others would extend the concept of personhood to include properties bestowed by human evaluation. For example, it has been argued that an entity must bear legal rights and social responsibilities and must be capable of being assigned the moral responsibility of being praised or blamed.

Clearly the conditions that we believe are necessary for "person" status directly affect our view of the ontological status of the fetus. For example, if the condition of personhood were only of an elementary biological nature, then the fetus could more easily qualify as a person than if the conditions included a list of psychological properties. And if personhood must be analyzed in terms of properties bestowed by human evaluation, then it becomes infinitely more difficult for the fetus to qualify as a person.

Further complicating the problem is the question of when the fetus gains full ontological status. Whether one claims that the fetus is an individual organism, a biological human being, a psychological human being, or a full-fledged person, one must specify at what point in its biological development the fetus attains this status. A judgment about when the fetus acquires its status bears as directly on abortion views as does a judgment of the status itself.

Again, a number of positions on when status is attained can be identified. An extreme conservative position would argue that the fetus has full ontological status from the time of conception. In direct contrast to this view is the extreme liberal position, which holds that the fetus never achieves full ontological status until birth. A cluster of moderate views falls between these polar positions. In every instance, full ontological status is achieved somewhere between conception and birth. Some would draw the line when brain activity is first present; others draw it at quickening; still others draw the line at viability.

In the last analysis, the ontological status of fetuses remains an open issue. But one's viewpoint about this status underpins one's position on the morality of abortion. Whether conservative, liberal, or moderate, ultimately one must be prepared to defend a view of the ontological status of fetuses.

The Moral Status

The issue of the fetus's moral status generally, but not always, is discussed in terms of the unborn's rights. What rights, if any, does the fetus have? Any seriously argued position on abortion must at some point address this question.

Various views currently circulate, each associated with one of the views on the fetus's ontological status. For example, the extreme conservative view holds that the fetus has full ontological status as well as full moral status at conception. Therefore, whether zygote, blastocyst, embryo, or fetus, the unborn entity enjoys the same rights that we attribute to any adult human. In this view, abortion would deny the unborn its right to life. Thus abortion is never to be undertaken without reasons sufficient to override the unborn's claim to life. In other words, only conditions that would justify the killing of an adult human, such as self-defense, would morally justify an abortion.

Liberals similarly derive their view of moral status from their position on ontological status. The extreme liberal view would deny the unborn any moral status. By this view, abortion need not be considered at all comparable to killing an adult person. Indeed, abortion may be viewed as little more than removing a mass of organic material, not unlike an appendectomy. Its removal raises no serious moral problems. A less extreme liberal view would grant the fetus ontological status as biologically human but deny that it is human in any significant moral sense and thus assign the fetus no significant rights.

Moderates would also assign moral status to the unborn at the point when it attains full ontological status. If brain activity was taken as the point of ontological status, then abortions conducted before that time would not raise serious moral questions; abortions conducted after that point would. Currently, viability seems to be an especially popular point at which to assign ontological status. Thus, many moderate theorists today insist that abortion raises significant moral questions only after the fetus has attained viability. This view is reflected in some of the opinions delivered in *Roe* v. *Wade*.

It is important to note that granting a fetus moral status does not at all deny moral status to the woman. Indeed, the question of whose rights should take precedence when a conflict develops raises thorny questions, especially for conservatives and moderates. For example, although granting the fetus full moral status, some conservatives nonetheless approve of therapeutic abortions, abortions performed to save the woman's life or to correct some life-threatening condition. These abortions are often viewed as cases of self-defense or justifiable homicide. Since self-defense and justifiable homicide are commonly considered grounds for killing an adult person, they are taken as moral justification for killing the unborn. But other conservatives disapprove of even therapeutic abortions.

Moderates grant moral status at some point in fetal development, but they too must arbitrate cases of conflicting rights. They must determine just what conditions are sufficient for allowing a woman's rights to override the fetus's right to life. Here a whole gamut of conditions that could surround a pregnancy must be evaluated, including rape, incest, fetal deformity, and of course physical or psychological harm to the woman.

With this philosophical background, we may now turn to the specific prolife (antiabortion) and prochoice (proabortion) arguments. These are presented through a dialectical method, in which a point is made to advance a position and a counterpoint is made to rebut it.

Prolife Arguments (against Abortion)

1. *Abortion is murder.*

POINT: "You don't have to be a lawyer to know what murder means: the intentional killing of an innocent human being. Who can imagine an uglier and more outrageous act? Well, I can think of at least one: abortion. At least when you kill adults, they usually have a chance to defend themselves. But what chance does a fetus have? None. Try as you will, you can't escape the inevitable fact that abortion is murder."

COUNTERPOINT: "You make it sound as though fetuses can go out and buy ice cream cones whenever they want, as well as think, imagine, wonder, hope, dream, and create. Why, a fetus isn't any more a human being than a cake mix is a cake. Certainly murder is a terrible thing—but murder of full-fledged walking, talking human beings, not of a glob of protoplasm."

2. *Abortion sets a dangerous precedent.*

POINT: "Everybody would agree that any action that leads to a casual attitude toward life is wrong. But that's just what abortion does. And that includes therapeutic abortion. As for abortions committed in the case of severely deformed fetuses, they're no more than the first step toward putting away the severely handicapped, the dysfunctional, the senile, and the incurably ill in our society. No, abortion leads to a shabby attitude toward life; it opens a Pandora's box of unspeakable affronts to human dignity and worth. Remember: The Holocaust under Hitler began with the legalization of abortion. Every abortion that's demanded or performed takes us one step closer to a systematic recognition of abortion and everything that implies."

COUNTERPOINT: "Why do you assume that the practice of abortion will inevitably lead to disrespect for life and usher in an age of nightmarish inhumanity? You and I both know that issues dealing with the severely handicapped or incurably ill are separate and distinct questions from abortion and have to be examined on their own merits. As for your Hitler reference, that has nothing to do with the question of whether abortion's right or wrong. It's outrageously irrelevant. Legal abortion didn't produce the Holocaust; Hitler's madness and those who cooperated with it did. Even if your point had any merit, it really argues against allowing abortions, not that they are immoral."

3. *Abortion involves psychological risks for the woman.*

POINT: "Let's face it, a woman and the child she's bearing are about as close as any two humans can ever get. And I don't mean just biologically but emotionally and psychologically, too. Ask any mother; she'll tell you. For a woman intentionally to harm her unborn violates the deepest levels of her unconscious needs, impulses, and desires. She's bound to pay a big price for this psychologically. Plenty of women already have; ask the psychologists. No, there's no question about it: The woman who has an abortion not only kills her unborn, she seriously damages herself."

COUNTERPOINT: "If there was an annual award for sweeping generalizations, you'd certainly get it. Why, you talk of women as if they were all the same. Sure, lots of women, maybe even most, want to carry their unborn to term. But 'lots' and 'most' don't mean all. Who knows what lurks in the so-called unconscious mind? Maybe a woman who outwardly wants to bear children secretly doesn't. Ask the psychologists, you say. All right, ask them. And while you're at it, ask about child abuse, postnatal and neonatal trauma, and nervous breakdowns. What makes you think every woman is suited to be a mother? Some women discover only after giving birth that they genuinely *don't* want the child. They assumed they would want it largely because of environmental conditioning, which as you well illustrated, has been scrupulous about heaping hot coals of guilt on the poor woman who might honestly admit that she doesn't want to carry, bear, and rear children. No, it's time to put aside these stereotypes about women for good. And while we're at it, let's not confuse what women actually want to do with what we believe they *should* want to do."

4.   *Alternatives to abortion are available.*

POINT: "All this talk about unwanted pregnancies and bringing undesired and unloved children into the world just won't cut it. Who are we kidding? There are countless individuals and couples who are dying to have kids but can't. If a woman doesn't want to carry a child to term, that's no reason to abort it. Just put it up for adoption at birth. As for the tragically deformed infant that no one may want to adopt, there are plenty of institutions and agencies set up for those pitiful souls."

COUNTERPOINT: "Sure there are plenty of people willing and able to adopt. But even if the child will be adopted, the woman still has to carry it for nine months and give birth. And she may be either unwilling or unable to do that. What's more, giving up your child for adoption after going through all that is no emotional picnic, you know. And often it's no easier on the child. Plenty of adoptees anguish over unresolved feelings of parental rejection. As for the deformed, why make society foot the bill for someone else's responsibility? I for one don't want my hard-earned tax dollars being spent caring for someone else's problems that could easily have been prevented in the first place."

5.   *The woman must be responsible for her sexual activity.*

POINT: "Nobody asked a woman to get pregnant. You don't have to be the 'happy hooker' to know that there are plenty of contraceptives readily available. When a woman doesn't practice some form of birth control, she should take responsibility for the consequences. For a woman to sacrifice an innocent human life just because she's been careless, ignorant, or indiscreet is the height of sexual and moral irresponsibility."

COUNTERPOINT: "Come now, everybody knows that there's no sure-fire birth control method, except abstinence. Any woman who has sex stands the chance of getting pregnant, as much as a driver stands a chance of having an accident. With drivers, even if they drive recklessly and have a serious accident, we allow them to repair the damage to themselves and others, don't we? In fact, we help them do it! To make a woman bear a child out of some warped sense of 'personal responsibility' is like turning our back on a torn and bloody motorist because he got what he deserved. Such an attitude reflects vindictiveness and punishment more than it does commitment to the principle of personal responsibility. No matter why a woman got pregnant, she still has the right to dispose of her pregnancy as she sees fit. Sure, she may have been irresponsible, but that's a separate issue from what she then chooses to do about the result of her indiscretion."

Prochoice Arguments (for Abortion)

1. *Pregnancies are dangerous for the woman.*

POINT: "There are very few adults, if any, who haven't had a direct or indirect experience that shows how dangerous pregnancies can be to the life and health of women. Let those without such knowledge look at the record. They'll see that in some cases the woman's life is actually on the line. In other cases, the woman must endure heroic suffering and hardship during the term of her pregnancy. In still other cases, the woman must endure lifelong health problems that result directly from the pregnancy. Given these possibilities, women should have the say over whether they want to run the risk."

COUNTERPOINT: "Your argument would be very persuasive if this were the nineteenth century. But it's not. Where have you been for the past thirty years? Don't you realize that modern medicine has, in effect, wiped out the dangers connected with pregnancy? Sure, there are still some cases where the woman's life is in jeopardy. But even in those cases, why assume that her life must be saved at all costs? There's another human life at stake, you know. As for your 'long-term' ill effects of pregnancy, what did you have in mind—varicose veins, hemorrhoids, and an aching back? When such 'horrors' become justification for abortion, we're really in trouble. What will be acceptable justification next—a ski trip to Aspen that, unluckily for the fetus, pops up in the seventh month?"

2. *Many unborn are unwanted or deformed.*

POINT: "Everybody knows that society has serious population, pollution, and poverty problems—not to mention crime, disease, and world hunger. Social problems are already straining our collective financial capacity. One way we can begin to deal with these problems is to make sure that every child that's brought into the world is wanted and healthy. By ignoring this responsibility, we just worsen the world's problems. Fortunately, we've got lots of ways to ensure that unwanted and unhealthy children are never born. Contraceptives are one way. But for one reason or another, some people don't play it safe, and it's usually those who can least afford to maintain children physically and emotionally. Such people need an alternative form of birth control: abortion. Abortion's especially necessary in the case of monstrously deformed fetuses. If a woman is willing and able to bear and care for the deformed or unintended child, fine. But if she isn't willing or can't, then it's unfair to make the rest of us pay the price for the problem that she's created and the responsibility she's shirked."

COUNTERPOINT: "Your thinking is really mind-boggling. You make it sound like the unloved and deformed are responsible for every social and global problem we have; or as though, if we can just keep these 'undesirables' from being born, then all our headaches will vanish. Sure, we've got massive problems, but lots of things account for them, including human mismanagement of resources, inadequate or misguided technology, nearsighted and unimaginative leadership, as well as the age-old conditions of human greed, prejudice, and downright stupidity. To think that abortions can help improve these conditions is exactly the kind of mindless proposal that contributes to our problems. You say it's unfair to make society pay the price for maintaining the unwanted or deformed child. What's so unfair about it? I always thought that part of society's function was to provide for those least able to fend for themselves. What would you suggest we do with our mental health services and facilities? Discontinue them? How about simply exterminating the mentally defective? That would really save us a bundle. And how about just executing criminals instead of incarcerating them? You shouldn't find that objectionable. After all, what could be more unfair by your standard than to make society pay the price for maintaining those who have flouted its conventions?"

3. *Some pregnancies result from rape or incest.*

POINT: "There's no more heinous a crime that can be committed against a woman than rape or incest. To oblige a woman to bear the child of such an outrage goes beyond adding insult to injury. It's barbaric. Nothing short of the woman's free choice would morally justify going through with pregnancy resulting from rape or incest.

COUNTERPOINT: "Agreed, rape and incest are ugly. But pregnancies resulting from them are rare. You're so concerned with not punishing the woman who has already been sexually abused. Well, by the same token, why punish the unborn by making them give up their lives? If forcing a woman to carry to term a fetus that's resulted from rape or incest is making the innocent suffer, then isn't terminating the life of the fetus an even more flagrant case? Besides, nobody is asking the woman to raise a child she doesn't want or can't maintain. There are plenty of individuals and institutions who'll do that if she's unwilling or unable."

4. *Women have rights over their own bodies.*

POINT: "Women bear the full and exclusive burden of carrying a fetus to term. They must endure the physical and emotional risks, the discomfort, the disruptions of career and routine living. Given this burden, they have the right to decide whether or not they want to go through with a pregnancy. But beyond this, the fetus is a part of the woman's body, and she should have absolute say over whether or not it's going to remain inside her and be allowed to grow and be born. To deny her that right is about as crude and basic a violation of free choice as imaginable."

COUNTERPOINT: "Of course women have certain rights over their own bodies. But having basic rights doesn't mean that those rights are absolute and take precedence over the rights of all others. My right to free speech doesn't justify my yelling 'Fire!' in a crowded theater when there is no fire. Nor does the woman's right over her own body mean she's justified in taking the life of the unborn. As soon as a woman gets pregnant, a relationship exists between her and the fetus. And like any other relationship, this one must involve the careful defining and assigning of rights and responsibilities and the establishing of priorities. This is especially important in pregnancy, because the unborn are in no position to argue for or defend their own rights. Let the woman who's so concerned about her own body make sure she doesn't get pregnant in the first place. But once she is pregnant, her right takes a back seat to the unborn's right to life, certainly as long as its life doesn't threaten the woman's."

It is important to realize that the preceding positions address the question of whether an abortion is ever morally justifiable. Whether women should have a *legal right* to an abortion is a different issue, although certainly related. To pinpoint the difference, as well as to raise additional moral concerns for HCPs that derive from the legalization of abortion, we should take a brief look at the legal issues.

Legal Issues

Undoubtedly, the morality of abortion bears on the morality of restrictive abortion legislation. People who believe that abortion is morally objectionable are likely to object to nonrestrictive legislation; those who reject the notion that abortion is inherently repugnant are likely to support nonrestrictive legislation. But it is perfectly possible and consistent to object to abortion while opposing restrictive legislation, for instance, because such laws would abridge the individual right to free choice. Conversely, it is consistent and commonplace to argue that individual acts of abortion may be moral, even obligatory, while opposing a loosening of abortion laws—for fear that such a systematic policy would lead to abuses, for example. Thus, one could be opposed to abortion but at the same time be opposed to restrictive abortion legislation, or one could condone individual acts of abortion but at the same time favor restrictive legislation.

At the same time, any of the prolife arguments can be marshaled in support of restrictive legislation, and any of the prochoice arguments can be enlisted on behalf of nonrestrictive legislation. Thus it could be argued that, since abortion is killing, the state has a moral obligation to protect its citizens by preventing their murder; or since abortion presents risks to women, the state has an obligation to protect its citizens (women in this case) from such risks. On the other side, it could be argued that the state has an obligation to relieve society's burden of caring for the unwanted or deformed by making it possible for women to have an abortion; or that the state has an obligation to minimize the profound hardship caused by a pregnancy resulting from rape or incest by allowing the woman to have an abortion.

Especially important to the legislation question are the probable consequences of restrictive legislation. These can perhaps be anticipated best by considering some statistics that characterized the situation prior to the Roe decision. For example, by some estimates, in the early 1960s a million women obtained abortions; most of these procedures were illegal.[18] Whatever the exact figure, prior to 1973 most abortions were obtained clandestinely. Given the secrecy that surrounded abortion, poor women seemed to have been particularly vulnerable to the threatening conditions that often accompanied clandestine abortions, particularly those that were self-induced or performed by unqualified persons. Not surprisingly, in 1960 alone 50 percent of the deaths in New York City associated with pregnancy and births resulted from illegal abortions.[19] It is safe to speculate, then, that if the legal status of abortion is returned to its pre-1973 status, there will probably be an increase in clandestine abortions, with the poor bearing the brunt of the procedures that most imperil health. In addition, physicians who believe in the morality of nontherapeutic abortions in the first trimester will presumably be forced to choose between obeying the law or following their consciences. Furthermore, restrictive legislation may create a fee system that borders on price gouging. These concerns should at least be part of an analysis of the morality of proposed abortion legislation.

All the same, a snarl of moral problems has developed since the passage of nonrestrictive legislation. For one thing, women today are often subjected to humiliating encounters with medical personnel. For example, sometimes they are asked to sign fetal "death certificates." Other times machines are set next to them that record fetal heartbeats; or abortions are performed in the same wards where other women are giving birth. Still other times they are given bags with a picture of a fetus that they are expected to return to the hospital after aborting at home.[20] At the very least, such procedures seem callous and raise doubt about how well some HCPs and institutions are meeting their obligations to provide adequate care for the pregnant woman and to respect her autonomy.

In addition, legalization of abortion has not necessarily meant an end to profiteering, inadequate facilities, or poorly trained personnel.[21] Nor does abortion reform appear to have significantly altered the plight of the poor. On the contrary, some evidence indicates that first- and second-trimester abortions are sometimes not available to institutional patients, whereas in the very same institution abortions are being performed on private patients.[22] More important, according to most recent federal legislation, Medicaid payments are permitted only for medical care following termination of a pregnancy by miscarriage or by an abortion to save the life of the mother; Medicaid payments are prohibited for all abortions except those performed to save the life of the mother.[23]

---

18.   N. Lee, The Search for an Abortionist (Chicago: University of Chicago Press, 1969), pp. 5–6.

19.   A. F. Guttmacher, "Abortion—Yesterday, Today, and Tomorrow," in The Case for Legalized Abortion, ed. A. F. Guttmacher (Berkeley, Calif.: Diablo Press, 1967), pp. 8–9.

20.   Frankfort, p. 40.

21.   S. Burt Ruzek, The Women's Health Movement: Feminist Alternatives to Medical Control (New York: Praeger, 1979), pp. 26–27.

22.   Scully, pp. 229–230.

23.   The Child Health Assurance Program (CHAP) was passed by the House of Representatives on December 11, 1979, HR 4962. For a synopsis of the bill see Health Policy: The Legislative Agenda (Washington, DC: Congressional Quarterly, Inc., 1980) p. 90.

Undoubtedly, then, any proposed abortion legislation will be fraught with serious moral questions of adequate care, autonomy, suffering, and social justice. Any discussion of the morality of the legislation that fails to consider these issues falls far short of being a thorough and responsible analysis.

Summary

This chapter considered a most important bioethical issue: abortion. After providing some biological background, placing abortion in the context of the medical and nursing professions, and relating it to the women's movement, the chapter examined its morality. The central moral problem of abortion concerns the conditions, if any, under which abortion is morally justifiable. Any answer to this question must be founded on a concept of the ontological and moral status of the fetus, which undergirds the various prolife and prochoice arguments. Finally, we looked at the legal side of abortion, directing special attention to the moral impact of abortion legislation on both patient and HCP.

Ethical Theories

In applying the theories to the issue of abortion, we will restrict ourselves to two questions: (1) What is the moral or ontological status of the fetus, and (2) is abortion ever morally permissible?

Egoism

Egoists would not necessarily be committed to any specific position on the ontological or moral status of the fetus. Conservative, liberal, and moderate theories could all be consistent with an egoistic view.

For egoists, the moral acceptability of any act of abortion ultimately depends on an evaluation of the long-term consequences for self. The egoistic pregnant woman, therefore, would determine whether an abortion is justified by carefully assessing the likely long-term advantages and disadvantages of having the child. Among the considerations would be the physical, psychological, and financial effects of bearing and raising the child. If she thought, after this calculation, that it was in her best long-term interests to have the child, then she would be morally obliged to go through with it. But if the calculation indicated greater disadvantage than advantage for her, then she would morally be obligated to terminate the pregnancy.

At no point in the evaluation would the egoist be concerned with such prolife arguments as abortion's being murder or the woman's having to assume responsibility for her sexual activity, except insofar as these might have an impact on the psychology of the woman and thus could affect her emotionally. By the same token, egoists would take no heed of the nonconsequential prochoice claim that women have a right to govern their own bodies. In the last analysis, egoism may serve as a theoretical basis for having an abortion or for not having one or for supporting or opposing restrictive abortion legislation.

Act Utilitarianism

Like egoists, act utilitarians would not necessarily be committed to any particular theory of the ontological and moral status of the fetus. Nevertheless, several empirical issues would matter to the utilitarian in determining a position on fetal status. One would be whether the fetus is capable of experiencing pleasure or pain. If it could be established that the fetus experiences pain in undergoing a saline abortion, for example, then this would be an evil that would need to be offset by the good achieved for the mother. Again, if it could be established that the onset of brain activity signals the onset of the experience of pleasure, then this would be a good that would be lost and, therefore, would also have to be balanced against the expected gains of the abortion. Such considerations become complex indeed when one introduces long-term goods and evils into these calculations, because then the expected pleasures and pains of the baby as it grows must be balanced against those of the mother.

In any event, it would be perfectly consistent for an act utilitarian to espouse a conservative view while endorsing a particular act of abortion. An act utilitarian might decide that treating the unborn as a full person from the time of conception produces more total good than any other position. At the same time, the utilitarian might consider an act of abortion morally justifiable because it is likely to produce more total good in a particular situation than having the child. So even if the fetus is considered a person, the principle of utility may still justify an abortion, because killing a person is not considered necessarily wrong. The morality of an act of abortion, then, depends on an evaluation of the likely consequences for all who will be affected. This would include a consideration of the consequences for the unborn—particularly important in the case of deformed fetuses whose birth might usher in profound suffering and hardship for the child, family, and society generally. As for abortion legislation, the act utilitarian would evaluate it on the basis of its social effects. If particular abortion laws would be likely to advance the general happiness of society, then they would be justifiable. Otherwise, they would not.

Rule Utilitarianism

All that has just been said about act utilitarianism also applies to rule utilitarianism, with the exception that the rule utilitarian would be concerned with the consequences of the rule under which an act of abortion falls and not with the consequences of the act itself. The rule utilitarian would probably view an unqualified prohibition of abortion as a rule whose consequences would be less socially beneficial than a less restrictive rule that would allow therapeutic abortions. However, the rule utilitarian might also argue that a rule allowing elective abortion or abortion on demand could produce more total unhappiness than a rule forbidding it. The same categories of consequences that concern the act utilitarian would figure in the rule utilitarian's analysis of the particular abortion rule. But whereas the act utilitarian might condone, say, one act of elective abortion but disapprove another, the rule utilitarian could not allow such latitude. Once forbidding elective abortion is calculated to produce more happiness than permitting it, the rule would have to hold consistently. The rule utilitarian would apply the same logic to abortion legislation. Ultimately, the morality of such laws would be determined by appeal to the utility principle.

The Categorical Imperative

The key issue in applying Kantian ethics to abortion is the status of the unborn, and Kant's thought is open to interpretation on this. If the fetus is a person, then it has the same inherent worth that any person has, and it must be treated with the same consideration and respect. Only the reasons that could also justify taking the life of an adult person could justify taking the life of the unborn. Self-defense certainly would be such a reason. Indeed, Kant would argue that a woman was not only morally justified but actually obliged to have an abortion if the fetus directly threatened her life. To act otherwise would be self-defeating. Similarly, destroying a seriously deformed fetus might be a way of recognizing the unborn's inherent dignity. By sparing it a life of suffering and indignity, we are treating it as a rational being would want to be treated. Again, assuming that the fetus is a person, Kant would argue for laws that recognize and protect its rights, because they would ensure that the fetus, like any other person, would be treated with the dignity and respect that it deserves.

But as long as the status of the fetus is in doubt, the legitimacy of protective laws is uncertain, because they would simply function to restrict the autonomous, self-directed nature of rational beings: They would deny the woman the right to choose. Again, if the fetus's status is in doubt, then the woman's right over her own body would be the overriding Kantian consideration. Clearly, then, the status of the fetus is the most important determinant in an application of Kantian ethics to abortion.

Prima Facie Duties

Abortion is no less complicated an issue for Ross. Again, the morality of abortion would depend on the status of the fetus. If the fetus is considered a person, then it has rights that must be acknowledged. Certainly the basic right would be its right to life. Thus although therapeutic abortions might be justified, it seems that nontherapeutic abortions would never be, because they would subordinate the prima facie duty of noninjury to lesser duties, such as self-improvement. Like Kant, Ross would endorse laws that acknowledge and protect the rights of the fetus. But if the status of the fetus is in doubt, then the issue of injuring a "person" dissolves. Not recognizing the "person" status of the fetus would not necessarily justify all or even most abortions, because a variety of situational prima facie duties would still have to be considered before determining one's actual duty. But where the fetus's status is in doubt, the most significant moral objection to abortion would have dissolved. Furthermore, restrictive abortion legislation might be considered unjust. Again, like Kant's ethics, Ross's nonconsequential view leaves the status of the fetus open to interpretation, and thus its position on abortion is unclear.

The Maximin Principle

Trying to apply Rawls's maximin principle raises many of the same issues apparent in Kant and Ross. Again, the big question involves the status of the fetus. If the fetus is a person, it has the right to life. Certainly no one in the original position would agree to a policy that would allow a life to be taken for matters of personal convenience or happiness. Thus nontherapeutic abortions would be unjustified. By the same token, rational, self-interested creatures behind the veil of ignorance surely would agree to be bound to a self-defense principle. Thus therapeutic abortions would be justified. Furthermore, Rawls could use his principle of paternalism to justify legislation intended to protect the rights of the unborn, who are in no position to protect themselves. But again, as with the other nonconsequential views, the issue grows complicated when the status of the fetus is in doubt. Then nontherapeutic abortions could be moral. And restrictive abortion legislation could be a violation of the equal liberty principle, since in denying a woman the right to choose, it would not be allowing the greatest amount of freedom consistent with a like freedom for all.

Roman Catholicism's Version of Natural-Law Ethics

Of central importance in applying the Roman Catholic position to abortion is its traditional view of the fetus as having full ontological and moral status from conception. Catholic theologians and moralists have consistently considered the fetus a person from conception to birth. By this account, the fetus is an innocent person; even where a pregnancy is due to rape or incest, the fetus may not be held accountable and made to suffer through its death. Accordingly, to Roman Catholics, direct abortion is never morally justifiable, and there should be laws that acknowledge and protect the rights of fetuses. Laws permitting direct abortion would be immoral.

Notice the word *direct* in the preceding sketch. Although Roman Catholic moral theology holds that the fetus may never deliberately be killed, it may be allowed to die as a consequence of an action that is intended to save the life of the mother. For example, suppose that a pregnant woman is found to have a malignant uterus. In order to save the woman's life, the uterus is removed, and as a result, the fetus is lost. Such an operation would be morally permissible by the principle of double effect. Thus, the action taken to save the mother's life is good (condition 1 in the principle of double effect). The bad effect, the death of the fetus, is not the means whereby the mother's life is preserved. After all, even if the fetus survived the successful operation, the mother would be saved (condition 2). Certainly, the intention is to save the mother, not to harm the fetus (condition 3). Finally, the mother's life surely is equivalent in importance to the fetus's (condition 4).

CASE PRESENTATION
Setting Priorities

Gail Belle was a dental hygienist; her husband, Mike, was a chemical engineer. Both were in their twenties. They had been married four years. Early in their marriage, Gail and Mike had decided that their careers and the opportunities for travel and personal enrichment that two salaries would afford were more important to them than having a family. They couldn't have been happier with the way things were turning out in general.

Then Gail got pregnant. The Belles assumed that it occurred when Gail was off the pill and using a diaphragm. But how it occurred was immaterial: Gail and Mike had to decide what to do.

Gail had decided she wanted an abortion immediately on learning of her pregnancy. Mike agreed. But it wasn't long before doubts set in.

"Maybe a child wouldn't be such a bad idea," Gail thought at one point.

Mike didn't dispute that, but he did catalog all the things that would change if they were to have a child. Gail didn't like that part of it at all.

The upshot was that about four and a half months passed before the Belles made up their minds to have an abortion.

"I can and will do it," Gail's doctor told her. "But I want you to be interviewed first by a counselor."

"Don't you think I know my own mind?" Gail challenged.

"That's not the reason," the doctor assured her. "It's just routine in abortion cases."

The interview wasn't at all bad. In fact, it helped the Belles crystallize the reasons they were having the abortion. For one thing, they came to grips with the fact that they wanted an abortion primarily for reasons of personal convenience. They also realized that whether or not the fetus was a person wasn't very important to them. Beyond that, they recognized that, in their view, one and only one condition would ever justify their having a child: if they both wanted one. As Gail put it at the end of the sessions, "I'm convinced that an unwanted baby should never be born." Mike agreed.

Following that, preparations were made. Shortly thereafter Gail had the abortion and the Belles returned to what they considered normal living.

Questions for Analysis

1. *Do you approve of the Belles' decision?*

2. *Do you think abortions for reasons of personal convenience are morally justified?*

3. *What is your conception of the role that a counselor should play in such situations?*

4. *Suppose you were the counselor who interviewed the Belles. What points would you introduce that would be relevant to the Belles' situation?*

5. *Do you think counselors are morally required to set aside their own moral judgments in counseling pregnant women who are contemplating abortion? Explain.*

6. *Do you think doctors have a moral obligation to set aside their own views on abortion and honor the wishes of their patients?*

7. *Do you think that the age of the fetus in any way affected the morality of this abortion?*

Exercise 1

Christian moralist Joseph Fletcher has written: "No unwanted and unintended baby should ever be born."[24] Would you agree?

Exercise 2

Reevaluate your responses under Values and Priorities.

24. J. Fletcher, Situation Ethics: The New Morality (*Philadelphia: Westminster Press, 1966*), p. 39.

Abortion in Adolescence: The Ethical Dimension

Tomas Silber

Tomas Silber is a pediatrician. He also holds an advanced degree in bioethics. In the following article, Silber merges medicine and ethics to present a bioethical approach to the abortion dilemma in adolescence. He addresses HCPs involved in the medical care of adolescents and those involved in adolescent counseling.

Silver begins by providing an overview of abortion in the United States: its medical significance and some historical-anthropological legal data about adolescent abortion. This is followed by a practical description of ethical theory, which Silber divides into consequential and nonconsequential (deontological) categories. But the bulk of the author's essay consists of a review of the most important contributions to the bioethical study of abortion.

Most important for our purposes, Silber locates the issue of adolescent abortion squarely in the context of basic patient rights, especially autonomy. He describes the levels of moral development in adolescence and provides an analysis of moral aspects of the treatment of adolescents. The article concludes with some personal comments and propositions intended to stimulate discussion. Silber's article is of particular value to HCPs who must care for adolescents, but it is also of general value to any HCPs who face abortion dilemmas.

From Adolescence 15 (Summer 1980): 461–474. Reprinted by permission of Libra Publishers, Inc. and the author.

A Personal Note

As a pediatrician involved in the practice of Adolescent Medicine, I have been faced, time and again, with the pregnant adolescent who considers the possibility of an abortion (15). Being socialized in a triple tradition of aiding life, providing help to the suffering and being loyal to my patient, I continue to find myself distressed by this situation. I have also sensed the issue to be extremely painful for my patients, not only from the viewpoint of what it actually does to their lives, but also from the viewpoint of their moral dilemma.

On examining my reaction I noticed two contradictory feelings. On one hand I felt an inherent repugnance to fetal destruction, stemming from my years of working with the fetuses "closest relative," the newborn premature baby. Having participated in bitter fights for the life of many "little ones," I can see an awful paradox in the termination of fetal development. In other words I have a particular relationship with the future human being. I recognize him/her as part of my species. More specifically, in my identity as a pediatrician I experience my loyalty to the fetus emotionally. On the other hand, on reviewing fifteen years of medical practice, I recall hundreds of unwanted pregnancies by teenage mothers. In one case conference after another I recognize their children in child abuse rounds, in D.O.A. cases, in the victims of lead poisoning, in learning disability clinics, in psychiatric clinics, and in juvenile delinquent programs. Too often I have seen what life has done to the teenage mother who enters the "failure syndrome." I also sense deeply an obligation toward my adolescent patients and their development as well as that of their future "possible children."

In addition to my practice, the issue reappears constantly in my dealings with and teaching of medical students, residents and fellows. In reviewing my own training in medical school, I have to admit that nothing was offered to help me reflect upon and resolve this moral dilemma. In 1976 I approached the faculty of a Special Studies Program at George Washington University with a proposal that I obtain a Master degree in Bioethics. This article is one by-product of two years of interdisciplinary study in this field. I offer this background information because it helps to explain my desire to share the material I have found helpful—material which is dispersed in the nonmedical literature, and also to focus attention on the problems of the ethical dimension, with the hope of stimulating further contributions.

Introduction

Epidemiological Data

It should be noted that although adolescent pregnancy has reached such magnitude that it has been termed a "global epidemic" (17), the sociology of this phenomenon is quite different throughout the world. Over 70% of females between the ages of 15 and 20 are already married in India, Bangladesh, Pakistan, and Tanzania. An average of 40% of adolescent girls in Africa and 30% in Asia are married, compared with only 15% in the Americas and 7% in Europe (9). In general, the pattern in western, industrialized nations has been one of delayed marriage, whereas in developing countries, early marriage is quite prevalent. Compounding this problem, earlier and earlier puberty is combined with later and later entry into the labor force. The societal implications of this trend are too complex to include in this article, but suffice it to say that they justify considering the abortion issue not only at an individual level but at the societal level.

Data on adolescent abortions throughout the world are limited. This is due to poor statistics and/or the problem of illegal (under-reported) abortions. For instance, in Egypt in the early seventies, 40% of hospital admissions for deliveries and pregnancy complications were really for abortions and their complications (9). Along the same line, it has been estimated that 50% of pregnancies in Latin America are currently terminated by illegal abortions. In the United States, in the seventies, one million pregnancies occurred in teenagers each year. Of those, approximately half a million resulted in live births, three hundred thousand were ended by therapeutic abortion, and the rest underwent spontaneous abortion (10,13).

Medical Significance of Abortion in Adolescence

Because the medical literature usually does not discriminate between abortion in adolescents and abortion in mature women, much of the current knowledge has been obtained by simply extrapolating knowledge of abortion in general to adolescents in particular.

Abortion is generally recognized as a safe procedure, but risk varies according to the gestation period during which the procedure is performed (16). This is so because of the difference between the types of procedure. While currettage and suction aspiration have a low complication rate, intrauterine instillation has a rather high rate. Since selection of the abortion technique is largely governed by the length of the gestation period, it is easily understood that the longer the pregnancy the more dangerous the abortion.

Adolescents are at a higher risk because as a group they are disproportionately represented among women receiving late abortions. At least one study has been reported of long-term consequences describing obstetrical and gynecological complications of early abortions later in life (14). This has been attributed to the fact that the cervix is small and tight in teenagers and therefore subject to a higher risk of trauma. Studies of short- and long-term psychological sequelae to abortion show transitory effects but few permanent adverse consequences (1). The difficulty is, however, that the findings cannot be evaluated in comparison with the outcome had the abortion not taken place. It does hold true that if the abortion takes place at a late date, the feelings of guilt and depression become more intense.

In summary, there is no question that abortion in adolescence may entail ill effects in the obstetric-gynecologic-psychological area. At the same time, not enough can be said about the ill effects of adolescent pregnancy carried to term!

Adoption has been hoped to be a "way out" of this situation. This may be so, and indeed, it is a topic that deserves study on its own merit. Here only two comments will be made. (1) At present, statistically, adoption plays a very minor role in adolescent pregnancy; (2) clinical experience has shown very serious problems with both the "giving up for adoption" mother and the adopted child.

Historical-Anthropological Legal Data

In antiquity, the people of the East opposed abortion with intensity. Buddhists punished offenders severely: in India the law books of the Aryas condemned abortion as homicide; among the Parsees, the induction of abortion was strictly forbidden by the Avesta religion; in the Assyrian code, women who procured abortions were to be impaled, and it was not permitted to bury their bodies.

In the West, on the other hand, especially in Greek and Roman society, abortion was practiced widely. There are no traces of any enactment against abortion before 200 B.C. It is extremely interesting that the Bible does not legislate on abortion. This seems to be related to the fact that in biblical times abortion was generally unknown among the Jews (7). With the advent of Christianity, a religious condemnation of abortion made its appearance in the West. In later times, the idea of "duty to the state" developed—an extreme case of which was Nazi Germany where abortion figured prominently in the criminal code. All the influences mentioned expressed themselves in laws relating to abortion. These laws followed the temper of their times, oscillating between criminal prosecution of abortion and abortion on demand in the first trimester.

The particular case of adolescent abortion has also received little attention. An exception might be found in early rabbinic literature which considered abortion legitimate for minors because of their condition as minors.

At present in the United States there is a strong legal trend (6) that: (1) the general rule (abortion in the first trimester as a matter of privacy), holds true also for adolescents (mature minor doctrine); (2) a minor may refuse to have an abortion requested by her parents; (3) a minor may consent to abortion against her parents' opposition.

Ethical Theory

The term "ethics" is derived from the Greek word "ethos," which originally meant "customs." The corresponding Latin term is "mores." Hence the origin of the term "morality." The attempt to give an account of morality is called an Ethical (11).

There is no easy answer to the question of the morality of abortion in moral philosophy. But here is where Ethical Theory can be of help—not so much to provide a concrete answer to a specific question, but to serve as a guide in looking for an answer. An Ethical Theory espouses criteria as to what is good. Thus, by focusing on relevant factors it explains morality, illuminates decisions, and tests for consistency. Basically there are two sets of ethical theories: (a) those that focus on what will be the consequences of a behavior and (b) those that focus on the kind of act. The followers of the former are known as partisans of a consequential ethical theory (e.g., hedonists, utilitarians). The followers of the latter are known as partisans of deontological ethics, from the Greek word for "duty," (e.g., Kantians, natural law).

So, one can see how the dilemma of abortion in adolescence can be approached differently from the point of view of the diverse Ethical Theories. A deontologist would concentrate on the *kind* of action involved in abortion. This doesn't mean that all deontologists would come to the same conclusion, but merely that they will coincide as to the angle from which the problem will be analyzed. A consequentialist, on the other hand, would place heavy emphasis on the sequelae to the act. With this consideration of moral philosophy relating to the abortion dilemma, we enter the field of applied ethics.

Bioethical Reflection on Abortion

Ethical reflection on abortion has been acquiring particular sophistication through the ages. However, I shall limit myself to the review of nine philosophical contributions of the seventies. These have been selected because they are representative of diverse, contemporary currents of thought, because they are clear, because they are well recognized in their field and because they have not found their way into the medical and psychosocial literature. I recognize the limitation of reducing a review to concepts developed in only the last few years, but I justify this by the fact that their language is more familiar to our "modern" mind. Together these works incorporate the substantial elements of previous centuries of reflection on abortion. They are presented in order of publication:

John T. Noonan (12) considers the morality of abortion from the historical religious tradition. He sees as a crucial point the determination of the humanity of a being. If the fetus is human, it deserves love. Since the religious perspective emphasizes the love of humanity, and since its destruction is the most unloving act,

abortion is wrong. Noonan insists on the question of deciding at what point an organism become human and is therefore entitled to love. He is concerned with finding some objectively identifiable criterion from which a distinction can be made between human and non-human life. Noonan states his conclusion in biological terms: whatever is conceived from human parents is human from the moment of conception and therefore entitled to human rights, foremost among which is the right to life. Noonan's philosophy may be thus synthetized: a fetus has full moral status.

Judith Jarvis Thomson (18) attempts to settle the question of whether abortion is immoral *apart* from the question of whether or not the fetus has full moral status (personhood). She emphasizes that the rights of the pregnant woman have been completely left out in the considerations of the abortion dilemma. She insists that determining what rights a fetus possesses is but one step in determining the moral status of abortion. The final step would be to find a just solution to the conflict between the rights of the fetus and the rights of the woman who is unwillingly pregnant. Thomson states that a woman has a right to an abortion regardless of what rights the fetus has. She illustrates this with the now classical "analogy of the violinist." (Picture yourself waking up one day in bed with a famous violinist. Imagine that your bloodstream has been hooked up to that of the violinist, who happens to have a condition which is certainly going to kill him, unless he is permitted to share your body for a period of nine months. No one else can do it. He will be unconscious all the time and you will have to stay with him, but after the nine months are over he may be unplugged, completely cured.) The question the analogy raises is: what are one's obligations under these circumstances? Her answer is that it would be commendable to save the violinist, but it would be incorrect to say that refusal to do so would be murder. The violinist's right to life "does not obligate you to keep him alive."

Roger Wertheimer (21) takes the position that the problem of the moral status of abortion is insoluble. This is so because the dispute centers around the status of the fetus and this is not a question of fact at all, but only a question of how one responds to facts.

Michael Tooley (19) also looks for a criterion to determine the right to life. He concludes that although it is true that the fetus is human, this biological feature has no moral significance. Being "a person," rather than being genetically human, is what determines the right to life. He argues that a right to life appears only when one has the capacity to desire to continue existing "as a subject of experience and other mental states." According to this interpretation, a human being may be eliminated until some time after birth when self-awareness develops. Tooley's philosophy may be thus synthetized: a fetus (and/or newborn) has no moral status.

Mary Ann Warren (20) also studies the topic of the ontological status of the fetus and she proposes a list of conditions for person-hood—self-consciousness, capacity to communicate, etc. She, too, reaches the conclusion that since the fetus does not have these characteristics, he/she has no moral status. Warren concludes that abortion would not be reprehensible and could simply be justified like any other surgical procedure. She doesn't stop there but introduces the issues of "potentiality"—that the fetus has some prima facie right to life because potentially he/she is a person. She then proposes that the fetus has "limited" rights, that is, they cannot "outweigh the right of a woman to obtain an abortion since the rights of any actual person invariably outweigh those of any potential person whenever the two conflict." Warren also points out, "there may well be something immoral, and not just imprudent, about wantonly destroying potential people, when doing so isn't necessary to protect anyone's rights." Thus she differs from Tooley in that she condemns infanticide.

H. Tristram Engelhardt (4) analyzes person-hood in a way reminiscent of Tooley. But he justifies this by constructing a theory which accounts for the development of varying levels of significance associated with human life. The determining value, he claims,

is that of social role. He sees the acquisition of rights as occurring by virtue of our membership in a social context. Infants do relate to persons and evoke responses, therefore, they are entitled to protection. Fetuses cannot play a social role and therefore abortion is permissible until the moment of viability.

Baruch Brody (3) rejects the idea that fetal humanity can be resolved on the basis of human definitions or decision. He believes that this decision can be made based on objective data. Brody does not accept that the crucial objective factors have to do with the genetic code or probability of development. He states that such an objective factor has to do with the "essence of humanity." By this Brody refers to what is lost when a human no longer exists: "After all when we die, when we stop being a living human being, we (but not our body and perhaps not our soul) cease to exist." Brody considers that the criterion for absent life is the non-functioning central nervous system and he argues next ". . . that the fetus becomes a living human being when it acquires that characteristic which is such that its loss entails that a living human being no longer exists." Thus he claims that the fetus is a human being from about 6 weeks, the time at which fetal brain activity starts to be noticeable. Brody's philosophy is that the fetus has no significant moral status during the initial stages of growth but *does* have significant moral status beginning at some later stage.

June English (5) says ". . . our concept of a person is not sharp enough or decisive enough to bear the weight of a solution to the abortion controversy." She proposes further that "we get the fallacious argument that since a fetus is something both alive and human it is a human being." But she recognizes that near the time of birth, a fetus does come *closer* to the diverse concepts of personhood, and she concludes "this could provide reason for making distinctions among the different stages of pregnancy as the U.S. Supreme Court has done." English brings up the issue of abortion as murder: "not all killings of humans are murders. Most notably, self-defense may justify even the killing of an innocent person." At the same time, she says that even if the fetus were not considered as a person, "that does not imply that you can do to it anything you wish. Animals, for example, are not persons, yet to kill or torture them for no reason at all is wrong."

John Badertscher (2) interprets the conflicting positions on abortion as grounded in two different moral orientations, which he describes as liberal and conservative. The liberal orientation understands freedom as autonomy and therefore claims that giving birth, like other human acts, should be freely chosen. The conservative orientation is based on a different concept of freedom: the freedom of the fetus has to be considered in any decision on abortion. It becomes easy to understand the mutual criticism. The liberal views the conservative as fatalistic and insensitive to the pregnant woman, and the conservative sees the liberal as exalting the power of the individual and excluding the defenseless fetus from any protection. Thus, Badertscher insists that the abortion debate has a religious dimension (orientation) and that therefore it is necessary for both positions to "begin to incorporate the deepest concerns of the other side into their own position." He concludes that as a result, the liberal could be led to consider whether the cause of freedom couldn't be advanced by protection of fetal life while the conservative could be pressed to take account of the real problems of pregnant women.

Having presented the varied and insightful views that ethicists have developed, it may now be useful to look at how the quality of moral reflection develops in the adolescent from its early and primitive manifestations to the above described mature reasoning showing ethical theories in action.

Moral Development in Adolescence

In adolescence, there is an intensified emotionality that differs qualitatively from that of the child in that the emotions are experienced as a result of states of the self rather than as direct correlates of external events. Associated with this heightening of subjective feelings comes the discovery of conflicts of feeling. This happens around the same time that the capacity for formal operational reasoning emerges.

To understand the adolescent's ethical thinking, however, we need to be aware not only of the stages of cognition but also of the particular stages in moral thought. This has been done by social psychologists (8) who have stratified moral development in several stages. For the sake of conceptual clarity, they may be divided into three major categories: "preconventional," "conventional," and "postconventional."

The "preconventional" adolescent is frequently very deprived or a moderately retarded youngster. These adolescents are often "well behaved" and are responsive to their culture's labels of "good" and "bad." They interpret "good" or "bad" in terms of physical consequences (reward-punishment) or in terms of the physical power of those that enunciate the rules, i.e., "My pregnancy is bad. My father told me that if I become pregnant he is going to kick me out of the house."

The "conventional stage" is most common in early adolescence. Maintaining the expectations and rules of the family, group or nation is perceived as valuable in its own right. There is concern not only to comfort but also to maintain and justify the social order, i.e., "I have decided against an abortion. I'm a Catholic and that would go against the teaching of my church."

The postconventional level is sometimes noticed around mid-adolescence but it usually appears (if it does at all) in late adolescence. It is characterized by a strong impulse toward autonomous moral principles which carry their weight apart from the authority of the group or persons that follow them, i.e., "I have decided to have this abortion because I'm not ready yet to be a mother. I think it is my duty to be well myself before I take care of a baby."

From the above it may be understood that the types of moral development are called stages because they represent a developmental sequence. Cognitive maturity is a necessary but not a sufficient condition to reach moral judgment maturation. A final comment to this section: a mature moral judgment can be expressed in different (and at times opposing) principles!

Bioethical Reflection on Abortion in Adolescence

Characteristics of Adolescents

There are some peculiar characteristics of adolescents which must be taken into consideration in determining the meaning of abortion in adolescence:

1. Adolescents are in the midst of a cognitive and moral developmental stage.

2. The adolescent aims to exercise autonomy.

3. The adolescent shares with the child a potential for a wide range of outcome.

4. The adolescent also shares with the child, but to a degree, a certain limitation in the understanding of the possible consequences of her actions. (Though this limitation is not of such a magnitude as to render obsolete the "mature minor" doctrine, it is sufficient to question the degree of responsibility.)

5. The adolescent shares with the adult a right to privacy.

Next, I would like to stress a peculiarity of adolescent pregnancy. In spite of its adverse consequences, it still remains a fact that pregnancy is a landmark in the girl's life which (especially if carried to term) differentiates her from her non-pregnant peers. It means that she has crossed the threshold into biological womanhood. No matter how horrible the circumstances surrounding this experience, it remains phenomenologically a significant (though often thwarted) event.

Having presented these basic thoughts, I suggest that in the case of an adolescent pregnancy, the moral dilemma of abortion is accentuated by the fact that we are dealing with *two* growing individuals claiming protection. I consider that we have obligations to *both minors*.

Another consideration is that the abortion dilemma confronts a *morally developing* being. This means that in addition to the personal decision of the moment, it will also have a considerable impact on the further adolescent moral development.

Moral Aspects of Treatment of Adolescents

The abortion issue highlights a situation that is often present in the doctor (counselor)-patient relationship: the need to address "ethically charged" material.

The usual questions then arise: What does it mean to be "non-judgmental"? What is the adolescent's responsibility? What are the doctor's obligations? What is the role of "advice"? To be "non-judgmental" does not mean to relinquish one's own values, or to be "blind" or indifferent to the ethical. Being "non-judgmental" means to do one's utmost to be objective. An essential part of objectivity is not to impose personal value judgments on the patient.

The reason for taking this position is both ethical and practical. "Imposing values" would: violate the patient's autonomy; lead to a power struggle; deprive the adolescent of an opportunity to work and exercise her own moral code. The adolescent should be encouraged to weigh decisions, to investigate alternatives and to choose her own way. By applying the Socratic method, the clinician will identify the stage of moral development of his/her patient. The questions asked will induce reflection, and reflection may promote a higher level of moral functioning. This doesn't mean that a higher level of functioning will produce a "right answer" or even one that is better than a lower level. It is always possible to do the right thing for the wrong reason. The point is that for many people it has been helpful to do moral reasoning, to compare it with their moral intuition and then to proceed.

Socrates, as well as his enemies, knew that the "Socratic method" required a framework within which reflection was to be induced. And, in the case in question, this is the awesome responsibility of the physician/counselor.

No matter what my conviction may be as to the advisability of abortion for a given patient, it is overruled by my adherence to the principle of autonomy. By this I mean that we should support the adolescent patient in her autonomous decision making. This, I insist, we can do actively, raising questions in an unbiased fashion. (If we can't do this, our obligation is to refer the patient to someone else.) This approach may increase the anxiety and suffering of the patient, but elevating the adolescent's anxiety by forcing her to think through her decision is necessary for psychological and ethical reasons. The increased anxiety, the working at the ethical decision may lead to further exploration, reading of material on abortion, talking to people who agree or disagree with her, etc. She will then make a decision with more understanding, one which is less likely to haunt her in later years. By guiding the adolescent not to avoid stressful questions, not to rely on denial, a counselor also is preparing her for a better future. The tragedy of the adolescent facing the abortion decision is that she has to choose between the "sin of aborting" and the "sin of harming one's life." In her dilemma the adolescent might have to "sin boldly."

As stated above, the adolescent and her decision merit all our respect and the decision should be supported even if it contradicts our own judgment. But this raises an issue. Do we have to conceal our judgment if it differs? Or does honesty require revealing our judgment while still encouraging autonomy? The answer is difficult. I will start by saying that it is hard to avoid preconceptions. Sometimes one cannot avoid presenting an ethical framework in the way one formulates one's questions. In addition, this age group is often an easy prey to undue coercion. Because of all these factors I proceed to a full disclosure of my viewpoint whenever I'm genuinely asked to do so by my patient. I also have learned humility. I tell all my patients that I have on occasion erred badly in my judgment and that at times I am so baffled by a situation that I do not know how to proceed. Perhaps that is why intuitionist theories in general, such as existentialism, argue so strongly against classical Ethical Theory and go out of their way to refuse to postulate criteria as to what is good or bad.

Ethical Propositions

The time has come to tie things together. Since the task is ethical reasoning, I shall speak first person singular in order to accept responsibility and in the hope of generating discussion and future contributions.

Like Noonan, I have come to think of abortion as killing. Arguments that the fetus has no moral status or that the fetus acquires this status at a particular stage of gestation violate biological reality. It is not the killing of a person but it is still a killing.

The reality of abortion as killing should not remove abortion as an alternative; it merely means that we should face up to the consequences of the decision to abort. My observations have shown

that the ethical question of abortion is the major source of conflict for the adolescent in deciding whether or not to abort. (Their intuitions are indeed of great significance!) Like Warren, I also think that in many circumstances, abortion is justified as a lesser evil. There are numerous circumstances, unfortunately, in which other serious considerations may override a prohibition of fetal killing.

Perhaps two cases from my medical practice will illustrate these points:

Jean, 17 years old, had been dating her boyfriend for a year. She had had strong convictions that she would remain a virgin and that he would "respect her." She was the only daughter of elderly parents, raised in a fundamentalist tradition. Only "bad girls" would anticipate sexual encounters and use contraceptives. When Jean saw me about her pregnancy she was suffering deeply over the idea that she had betrayed her parents' trust and that she was "no good." She also felt very close to her boyfriend, she had her maternal instinct awakened by the birth of a nephew, and felt a desire to "keep the baby." She was torn by the conflict between sacrificing her pregnancy for the sake of the happiness of her parents (and perhaps as an expiation) and bringing pregnancy to term "to have Jim's baby." In our counseling sessions she became aware of how she was considering different alternatives and contradictory values. Her first step was to discover what her genuine values were, as opposed to impulsive thoughts that were arising in order to obtain something (marriage, peace of mind, etc.). Ultimately she decided to continue her pregnancy, motivated by the conviction that she had *this* responsibility.

Joan, 15 years old, had been promiscuous and depressed for a long time. After a suicidal gesture she accepted my suggestion for a psychiatric consultation. In her 6th month of therapy, a week after deciding on a *sexual moratorium*, she found herself two months pregnant. Her mother, an alcoholic in a recovery phase, who was taking care of her three younger siblings, encouraged her to continue the pregnancy because "she didn't believe in abortion." Joan wasn't sure who the father-to-be was. She was torn between her perception of the fetus as a defenseless person in need of protection and love (like herself) and her new understanding that she was not ready for maternity; that in the present circumstance her baby's future was as bleak as her mother's and her own. In her second counseling session she decided to have an abortion. She cried bitterly, "I need to grow first!"

If the adolescent has chosen abortion, the question may be raised: What is the moral significance of the act? It has been variously seen as morally indifferent, as murder, as wrong-doing (killing), as the lesser evil. As expressed earlier, I tend to oscillate between the latter two, but in addition, I would like now to contribute two concepts to further illuminate moral aspects of abortion in adolescence: the concept of "incomplete responsibility" and the "sacrificial" viewpoint.

By "incomplete responsibility" I mean that in many instances the adolescent girl does not have a full understanding of the nature and long-range implications of sexual interactions. This recalls Thomson's analogy: the violinist has been attached without any good previous agreement. Inasmuch as responsibility relates to morality, there is indeed a difference in many instances between adolescent and adult abortion.

By "sacrificial" viewpoint I suggest that abortion in adolescence is not morally indifferent and certainly not murder. Instead, I view it as a painful offering made toward the avoidance of suffering and the promotion of a better future.

Relating to this, I do see the imperative for the adolescent girl not to repeat her mistake. The sacrifice of the fetus should be followed by a compromise (abstinence, contraception, responsible sex, family planning). From this viewpoint, a second abortion should be considered unethical when it implies betrayal of the assumed compromise. This view does *not* imply that 2nd abortions should not be allowed.

This brings us to the issue of compulsion. Everything written so far pertains to free ethical discourse. It is inadmissible to force an adolescent to have a compulsory abortion and, conversely, to stop her from having an abortion if she so desires. It is equally immoral to demand that doctors or nurses should participate in an abortion procedure against their will. Even if adolescents have a right to undergo an abortion, this does not imply a right to force others to perform such acts. Ethical conduct is here seen as independent of legal considerations, that is, the ethical may inform the legal though every unethical behavior should not necessarily be illegal. The legal should have little bearing on the morality of the subject matter. The ethical impact of abortion in adolescence on society in general is not to be ignored but it is beyond the scope of this article.

It is obvious that debate on abortion is only one component of a much larger struggle in moral philosophy. Our society stands at the crossroad of a number of different histories. Each history is the bearer of a highly particular kind of moral tradition. And of course, when these moral traditions encounter each other, they are to some large degree hurt and fragmented in the process. Therefore, it comes as no surprise that the confusions of pluralism are expressed at the level of moral argument in the form of a salad of conceptual fragments. Out of this "mish-mash" we have to recognize the diverse origins of the premises that underly our moral reasoning.

Summary and Conclusions

In the developing adolescent, abortion presents a special dilemma since it affects "two minors."

The ethical question is usually a major source of conflict for the adolescent in deciding whether or not to abort.

The principle of autonomy may be a primary guide for dealing with adolescent abortion. The issue of incomplete responsibility and the sacrificial viewpoint also deserve consideration.

While the adolescent and her decision merit our respect and support, neither individual adolescents nor society should view abortion as a policy to be encouraged at the expense of contraception and adoption.

Notes

1. Athanasiou, R. et al. "Psychiatric Sequelae to Term Birth and Induced Early and Late Abortion: A Longitudinal Study," *Family Planning Perspectives*, 1973, 5(4), pp. 227–231.

2. Badertscher, J. "Religious Dimensions of the Abortion Debate," *Studies in Religion*, 1976–77, 6(2), pp. 177–183.

3. Brody, B. "On the Humanity of the Fetus," in R. L. Perkins (Ed.), *Abortion: Pro and Con.* Cambridge, Mass.: Schenkman Pub. Co., 1974.

4. Engelhardt, H. T. "The Ontology of Abortion," *Ethics,* University of Chicago Press, April 1974, 84, pp. 217–34.

5. English, J. "Abortion and the Concept of a Person," *Canadian Journal of Philosophy,* Oct. 1975, 5.

6 Holder, A. R. *Legal Issues in Pediatrics and Adolescent Medicine.* New York: John Wiley and Sons, 1977.

7. Jakobowitz, I. *Jewish Medical Ethics,* 2nd Edition, New York: Bloch Publishing Company, 1975, pp. 170–191.

8. Kohlberg, L. and Gilligan, C. "The Adolescent as a Philosopher," *Daedalus,* Fall 1971.

9. Lee, L. and Poxman, J. M. "Pregnancy and Abortion in Adolescence: A Comparative Legal Survey and Proposals for Reform," *Columbia Human Rights Law Review,* 1974–75, 6(2), pp. 307–55.

10. McAnarney, E. "Adolescent Pregnancy—A National Priority," *Am. J. Dis. Child,* Feb. 1978, 132, pp. 125–26.

11. Melden, A. *Ethical Theories.* Englewood Cliff, N.J.: Prentice-Hall, Inc. 1967.

12. Noonan, J. T., Jr. "An almost Absolute Value in History," in J. T. Noonan, Jr. (Ed.), *The Morality of Abortion—Legal and Historical Perspectives.* Cambridge, Mass: Harvard University Press, 1970.

13. Population Reference Bureau, Inc., "Adolescent Pregnancy and Childbearing: Growing Concern for Americans," *Population Bulletin,* September, 1976, 31(2).

14. Russell, J. K. "Sexual Activity and Its Consequences in the Teenager," *Clinics in Obstetrics and Gynecology,* 1974, 1(3), pp. 68–698.

15. Silber, T. "Attitudes of Inner City Adolescents Towards Abortion," Presentation at the XVIII Congress of the Ambulatory Pediatric Association, New York, June 1978.

16. Sources. VS Center for Disease Control, Abortion Surveillance, 1974 (April 1976).

17. Stokes, B. "Teenager Pregnancy: Global Epidemic," *Washington Post,* 5/6/78, Worldwatch Institute.

18. Thomson, J. J. "In Defense of Abortion," *Philosophy and Public Affairs,* 1971, 1(1), pp. 47–66.

19. Tooley, M. "Abortion and Infanticide," *Philosophy and Public Affairs,* 1972, 1(1), pp. 37–65.

20. Warren, M. A. "On the Moral and Legal Status of Abortion," *The Monist,* Jan. 1973, 57(1).

21. Wertheimer, R. "Understanding the Abortion Argument," *Philosophy and Public Affairs,* 1971, 67(1).

For Further Reading

Brody, B. *Abortion and the Sanctity of Human Life.* Cambridge, Mass.: Harvard University Press, 1975.

Callahan, D. J. *Abortion: Law, Choice, and Morality.* New York: Macmillan, 1970.

Dickens, B. H. *Abortion and the Law.* Bristol, England: Mac-Gibbon & Kee, 1966.

Ehrlich, P. *The Population Bomb.* New York: Ballantine, 1968.

Feinberg, J. *The Problem of Abortion.* Belmont, Calif.: Wadsworth, 1973.

Finnish, J., et al. *The Rights and Wrongs of Abortion.* Princeton, N.J.: Princeton University Press, 1974.

Gebhard, P. H., Pomeroy, W. B., Martin, C. E., and Christenson, C. V. *Pregnancy, Birth and Abortion.* New York: Harper & Row, 1958.

Granfield, D. *The Abortion Decision.* Garden City, N.Y.: Doubleday, 1971.

Group for the Advancement of Psychiatry, Committee on Psychiatry and Law. *The Right to Abortion: A Psychiatric View.* New York: Charles Scribner's Sons, 1970.

Hall, R. E. *Abortion in a Changing World,* vols. 1 and 2. New York: Columbia University Press, 1970.

Huser, R. J. *The Crime of Abortion in Common Law.* Canon Law Studies no. 162. Washington, DC: Catholic University Press, 1942.

Lader, L. *Abortion.* Indianapolis, Ind.: Bobbs-Merrill, 1966.

Noonan, J. T. *The Morality of Abortion: Legal and Historical Perspectives.* Cambridge, Mass.: Harvard University Press, 1970.

St. John-Stevas, N. *The Right to Life.* New York: Holt, Rinehart & Winston, 1964.

12
EUTHANASIA

"I'm afraid I have bad news for you," Dr. Channing told the Stockards. "Your baby was born with certain birth defects."

The blood drained from the young couple's faces. For months they had joyfully been anticipating the birth of their first child. Now the doctor had turned their moment of joy into tragedy.

It seems that the Stockard baby had been born with Down's syndrome and duodenal atresia. Dr. Channing explained: "Normally humans have twenty-three pairs of chromosomes. But sometimes nature fools us. It provides one extra chromosome. The result is Down's syndrome."

Dr. Channing pointed out that somehow this extra chromosome alters the normal process of development. As a result, the child is born with retardation and various physical abnormalities. "Typically," he said, "these abnormalities are relatively minor: a broad skull, a large tongue, and upward slant of the eyelids."

"Mongolism," Leo Stockard said.

Dr. Channing nodded. "Some call it that." He told the Stockards that there was no known cure, no way to compensate for the abnormalities. "Those with the condition," he said, "generally have an IQ of between 50 and 80. They can be taught simple tasks and, as far as we know, are happy individuals, but they do require care."

"And the other problem?" Irene Stockard asked. "What is that?"

"Duodenal atresia," Dr. Channing said. He explained that in duodenal atresia the upper part of the small intestine, the duodenum, is closed off. Food cannot pass through and be digested. "Unlike Down's syndrome, we can successfully treat duodenal atresia through surgery." Then Dr. Channing paused before saying in muted tones, "Of course, you may not want to."

"But that would mean. . . ." Leo Stockard's voice trailed off. He couldn't bring himself to say the obvious: Without the surgery, the child would starve to death.

There is no more difficult decision to make in health care than the one that faces the Stockards and Dr. Channing. It involves a determination of whether an individual will be kept alive or permitted to die and raises the torturous issue of euthanasia.

The case of Karen Ann Quinlan in 1976 probably did most in recent years to rivet public attention on the legal and moral aspects of *euthanasia*, the act of painlessly putting to death a person suffering from an incurable disease. Karen Ann Quinlan was a young woman who lapsed into a coma of mysterious origin and remained unconscious for months, kept alive by a respirator. Eventually her father won a New Jersey Supreme Court decision that allowed him to authorize withdrawal of the respirator if he so chose. Joseph Quinlan did, but Karen Ann did not die. As of this writing, she continues to lie comatose in a New Jersey convalescent home. As publicized as the Quinlan case has been, it is far from being an isolated instance of the question of euthanasia.

Improvements in biomedical technology have increased euthanasia's relevance to individuals, institutions, and society as a whole. Respirators, artificial kidneys, intravenous feeding, new drugs, sophisticated surgery—all have made it possible to sustain an individual's life artificially long after he or she has lost the capacity to sustain life independently. In cases like Quinlan's, individuals have fallen into a state of irreversible coma, what some term a vegetative state. In other instances, as for example after severe accidents or with congenital brain disease, the individual's consciousness has been so dulled and the personality has so deteriorated that the person lacks the capacity for satisfaction. In still other cases, such as terminal cancer, individuals vacillate between agonizing pain and a drug-induced stupor, with no possibility of ever again enjoying life. Not too long ago, nature would have taken its course, and such individuals would have died. Today we have the technological capacity to keep them alive artificially. But should we? Or are we justified, even obliged, at least in some instances, to let them die? Would the Stockards be right in refusing surgery for their baby?

These are questions that beset patients, relatives, HCPs, and society generally. Surrounding discussions of euthanasia are a number of conceptual issues that largely determine one's view of the morality of euthanasia: personhood, death, extraordinary versus ordinary treatment, killing versus allowing to die, euthanasia, and voluntary versus nonvoluntary decisions. Only after airing these conceptual issues can we confront the central moral question of whether euthanasia is ever morally permissible.

Death decisions always raise problems for HCPs, but the problems can be most agonizing for pediatric professionals caring for defective newborns like the Stockard baby. So after unearthing the conceptual issues and various positions on the morality of death decisions, we will examine the special case of seriously defective newborns. In addition, we will look at some relevant intrainstitutional and interprofessional problems that confront HCPs. Finally, as with abortion, we will learn to distinguish between questions about the morality of euthanasia and those about the morality of euthanasia legislation.

Values and Priorities

Before tackling any of these issues, take stock of your own values and attitudes by responding to the following questions:

1. Do you think the Stockards should allow their baby to starve to death?

2. Do you think it would be right to allow the baby to starve to death but wrong to inject it with a lethal medication?

3. What part should Dr. Channing play in determining what to do?

4. Suppose you were a nurse who, at the Stockards' request, was instructed by Dr. Channing to allow the infant to starve to death. What would you do?

5. In a case like Karen Ann Quinlan's, should positive measures be permitted to cause the patient's death in the event that death did not result from withdrawal of life-support systems?

6. Under what conditions would you say a person is no longer alive?

7. Is it possible for an individual to be alive technically but no longer to qualify as a person?

8. What role, if any, should the consent of the patient play in death decisions?

9. Do you think it is ever morally permissible to allow patients to die in the absence of their explicit consent?

10. Who should make a death decision—physician, patient, family, or a board?

11. Should HCPs ever try to influence a person's death decision?

12. Do you think patients who are terminally ill and near death should be allowed to take their own lives?

13. Do you think physicians should ever help terminally ill patients who are near death commit suicide if the patients ardently request it?

14. Is it ever wrong to keep a patient alive?

15. Do you think it would ever be immoral to maintain the life of a seriously deformed newborn?

16. Do you think financial considerations should ever be a factor in a death decision?

17. What part, if any, should the quality of life play in a death decision?

18. Do you believe in maintaining life at all costs?

19. Should HCPs take an active role in social policy debates concerning euthanasia?

Personhood

Just as the question of personhood influences abortion questions, so does it affect euthanasia decisions. What conditions should be used as the criteria of personhood? Can an entity be considered a person merely because it possesses certain biological and psychological properties? Or should other factors be introduced, such as consciousness, self-consciousness, rationality, and the capacities for communication and moral judgment? If personhood is solely an elementary biological matter, then patients like Karen Ann Quinlan and the Stockard baby could more easily qualify as persons than if personhood depends on a complex list of psychosocial factors.

The significance of the personhood issue lies in the assignment of basic patient rights: Once the criteria for personhood are established, those qualifying presumably enjoy the same general rights as any other patient. Conversely, for those who do not qualify and have no reasonable chance of every qualifying, the rights issue is far less problematic. This does not mean that a death decision necessarily follows when an entity is determined to be a nonperson. But it does mean that whatever may inherently be objectionable about allowing or causing a *person* to die dissolves, because the entity is no longer a person. So the concept of personhood bears directly on euthanasia decisions.

Death

Another conceptual issue, related to personhood, concerns death. Chapter 9 indicated three distinct categories of factors that can be identified in any discussion of death: philosophical, physiological, and methodical. The first two are particularly relevant to euthanasia decisions.

The philosophical category refers to one's basic concept of death, which emanates from a view of what it means to be a person. The absence of the properties of personhood would not necessarily define death, but it would incline one toward such a definition.

The physiological category refers to standards that define the functioning of specific body systems or organs. The traditional physiological standard for recognizing death has been irreversible loss of circulatory and respiratory functions. More recently, however, physiological standards defined by the Harvard Ad Hoc Committee (1968) have focused on the central nervous system. By this account the physiological standards for determining death relate to (1) reflex activity mediated through the brain or spinal cord, (2) electrical activity in the cerebral neocortex, and (3) cerebral blood flow. The significance of the physiological category in death decisions is that a patient who might be declared alive by one set of criteria might be ruled dead by another. If a patient is dead, then euthanasia becomes academic, since one cannot kill or allow someone to die who already is dead. On the other hand, if the person is considered alive, then euthanasia emerges as a central concern.

Ordinary versus Extraordinary Treatment

A third conceptual issue involves the concepts of ordinary as opposed to extraordinary treatment, terms used to differentiate two broad categories of medical intervention. The terms and concepts are often applied glibly, but their meanings elude hard-and-fast definition.

Moralist Paul Ramsey has applied *ordinary* to all medicines, treatments, and surgical procedures that offer a reasonable hope of benefit to the patient but do not involve excessive pain, expense, or other inconveniences. In contrast, he has identified *extraordinary* as measures that are unusual, extremely difficult, dangerous, or inordinately expensive or that offer no reasonable hope of benefit to the patients.[1]

Such descriptions are useful and probably find widespread acceptance, but they raise questions. An obvious one concerns the concepts used to define *ordinary* and *extraordinary*. What can be considered "reasonable hope" or "hope of benefit to the patient"? What measures qualify as unusual? Ramsey mentions cost, but some would claim that cost has no place in a moral calculus. And then there is always the question of whether these criteria should be used individually or in combination; if in combination, what is the proper mix? Furthermore, patient idiosyncrasies inevitably influence a determination of ordinary and extraordinary in a particular case. The use of antibiotics for a pneumonia patient undoubtedly qualifies as ordinary treatment. But does it remain ordinary treatment when the patient with pneumonia happens to have terminal cancer with metastasis to the brain and liver? The institutional setting can also affect evaluations of what constitutes ordinary and extraordinary. For example, what is extraordinary treatment in a small community hospital could very well be ordinary in a large teaching hospital.[2]

The significance of trying to define these two concepts is that euthanasia arguments often rely on them to distinguish the permissible from the impermissible act of euthanasia. Some ethicists argue that HCPs should provide ordinary treatment for the moribund but not extraordinary treatment; extraordinary measures may be withheld or never started. In contrast, other ethicists insist that HCPs must initiate extraordinary measures. Indeed, as we will see, the medical profession makes similar operational distinctions in making death decisions.

Killing versus Allowing to Die

A fourth conceptual issue that we should try to clarify is the difference between killing a person and allowing a person to die. Presumably, killing a person is a definite action taken to end someone's life, as in the case of the physician who, out of mercy, injects a terminally ill patient with a lethal dose of medication. Killing is an act of commission. In contrast, allowing someone to die is presumably an act of omission, whereby the steps needed to preserve someone's life simply are not taken. For example, a doctor, again out of mercy, may fail to give an injection of antibiotics to a terminally ill patient who happens to contract penumonia. As a result of this omission, the patient dies.

Those making this distinction between killing and allowing to die, such as

---

1. P. *Ramsey*, The Patient as Person (*New Haven, Conn.: Yale University Press, 1970), pp. 122–123.*

2. A. J. *Davis and M. A. Aroskar*, Ethical Dilemmas and Nursing Practice (*New York: Appleton-Century-Crofts, 1978), p. 117.*

the AMA, say it is reasonable to do so because in ordinary language and everyday life we distinguish between actively causing someone harm and permitting the harm to occur. If the distinction is not made in cases of euthanasia, then we lose by default the important distinction between causing someone harm or permitting harm to occur.

It is also claimed that the distinction between killing and allowing to die acknowledges cases in which additional curative treatment serves no purpose and in fact interferes with a person's natural death. It recognizes that medical science will not initiate or sustain extraordinary means to preserve the life of a dying patient when it is abundantly clear that such action serves no useful purpose for the patient or the patient's family.

Finally, it is argued that the distinction is important in identifying the cause of death and ultimately in pinpointing responsibility. In instances where the patient dies following nontreatment, the cause of the death is the patient's disease—not the treatment or the person who did not provide it. In failing to distinguish between killing and allowing to die, we blur this reality. If allowing to die is subsumed under the category of euthanasia, then the nontreatment is the cause of death, not the disease.

But not everyone agrees that the distinction between killing and allowing to die is a valid one. Some argue that withholding extraordinary treatment or suspending heroic measures in terminal cases is tantamount to the intentional termination of life, that it is an act of killing. Thus they claim that no logical distinction can be made between killing and allowing to die.

Whether the distinction between the two can be sustained logically is only one question that the debate raises. Another is the moral relevancy of such a distinction. Even if the distinction is logical, does it have any bearing on the rightness or wrongness of acts commonly termed euthanasia?

For those who do make the distinction, allowing a patient to die under carefully circumscribed conditions could be moral. On the other hand, they would probably regard killing a patient, even out of mercy, an immoral act. But those who oppose the distinction between killing and letting die would not accept killing a dying patient as an immoral act. Yes, killing may be wrong. But in some cases it may be the right thing to do. What determines the morality of killing a patient, what is of moral relevance and importance, is not the manner of causing the death but the circumstances in which the death is caused.

In sum, those who distinguish between killing and allowing to die claim that the distinction is logically and morally relevant. Generally, they would condemn any act of killing a patient while recognizing that some acts of allowing a patient to die (for example, in cases where life is being preserved heroically and death is imminent) may be moral. On the other hand, those who hold that the distinction between killing and letting die is not logical, that allowing to die is in effect killing, claim that killing a patient may be morally justifiable depending on the circumstances, *not* the manner in which the death is caused. Thus the debate that surrounds the question about killing versus allowing to die is basic to the very meaning of the term *euthanasia*.

Interpretations of Euthanasia

A fifth conceptual issue deals with the meaning of *euthanasia*. Construing euthanasia narrowly, some philosophers have taken it to be the equivalent of *killing*. Since allowing someone to die does not involve killing, it is not properly speaking an act of euthanasia at all. By this account, then, there are acts of allowing to die, which may be moral, and acts of euthanasia, which are always wrong.

Other philosophers interpret *euthanasia* more broadly. For them, euthanasia includes not only acts of killing but also acts of allowing to die. In other words, euthanasia can take an active or passive form. Active (sometimes called positive) euthanasia refers to the act of painlessly putting to death persons suffering from incurable conditions or diseases. Injecting a terminally ill patient with a lethal dose of a medication would constitute active euthanasia. Passive euthanasia, on the other hand, refers to any act of allowing a patient to die. Not providing a terminally ill patient with the antibiotics needed to survive pneumonia would be an example of passive euthanasia.

It is tempting to view the debate about the interpretations of *euthanasia* largely in terms of semantics. Although the meaning of *euthanasia* certainly is a factor in the disagreement, there is more to the issue than merely word definition and concept clarification.

One side, the narrow interpretation, considers killing a patient always morally wrong. Since euthanasia, by definition, is killing a patient, it is always morally wrong. But allowing a patient to die does not involve killing. Thus, allowing a patient to die does not fall under the same moral prohibition as euthanasia; allowing a patient to die may therefore be morally right.

The other side, the broad interpretation, considers acts of allowing patients to die as acts of euthanasia as well, albeit passive euthanasia. They argue that, if euthanasia is wrong, then allowing patients to die is wrong, too, since it is a form of euthanasia. But if allowing patients to die is not wrong, then euthanasia is not always wrong. Generally speaking, those favoring the broad interpretation claim that allowing patients to die is not always wrong and that euthanasia may therefore be morally justifiable. Having established the possible moral justifiability of euthanasia, it is conceivable that acts of active euthanasia, as well as passive, may be moral. What determines their morality are the conditions under which the death was caused, not the manner in which it was caused.

Voluntary versus Nonvoluntary Decisions

There is one final conceptual issue that arises in discussions of euthanasia. It concerns the difference between voluntary and nonvoluntary decisions about death.

Voluntary decisions about death are those in which a competent adult patient requests or gives informed consent to a particular course of medical treatment or nontreatment. Voluntary decisions include cases in which persons take their own lives either directly or by refusing treatment. Voluntary decisions also include cases where patients deputize others to act in their behalf. For example, a woman who is terminally ill may instruct her husband and family not to permit antibiotic treatment should she contract pneumonia or not to use artificial support systems should she lapse into a coma and be unable to speak for herself. Similarly, a man may request that he be given a lethal injection after an industrial explosion has left him with third-degree burns over most of his body and no real hope of recovering. For a decision about death to be voluntary, the individual must give explicit consent.

A nonvoluntary decision about death is one in which the decision is not made by the person who is to die. Such cases would include situations where, because of age, mental impairment, or unconsciousness, patients are not competent to give informed consent and where others make the decisions for them. Suppose that as a result of an automobile accident a woman suffers massive and irreparable brain damage, falls into unconsciousness, and can be maintained only by artificial means. Should she recover, she is likely to be little more than a vegetable. Given this prognosis, the woman's family, in consultation with her physicians, decides to suspend artificial life-sustaining means and allow her to die.

In addition to voluntary and nonvoluntary decisions, a third kind of decision is at least theoretically possible. An involuntary decision would be a decision to die that is contrary to the expressed wishes of the individual. It is important to distinguish this kind of decision from the nonvoluntary one, for the involuntary decision is clearly the most objectionable and indefensible on moral grounds. Failing to distinguish between nonvoluntary and involuntary decisions, one might confuse cases like the preceding nonvoluntary examples with those that involve a violation of the expressed wishes of the patient. But morally the two are quite different.

In actual situations, the difference between voluntary and nonvoluntary decisions is often troublesome. For example, take the case of a man who has heard his mother say that she would never want to be kept alive with "machines and pumps and tubes." Now that she is in fact being kept alive that way and is unable to express a life-or-death decision, the man is not sure whether his mother would actually choose to be allowed to die. Similarly, a doctor might not be certain that the tormented cries of her stomach-cancer patient to be "put out of my misery" are an expression of informed consent or of excruciating pain and momentary despair.

The distinction between voluntary and nonvoluntary decisions is relevant to both the narrow and broad interpretations of the meaning of *euthanasia*. Each distinguishes four kinds of death decisions, in which the voluntary or nonvoluntary aspect plays a part. Thus the narrow interpretation recognizes cases of

1. Voluntary euthanasia

2. Nonvoluntary euthanasia

3. Voluntary allowing to die

4. Nonvoluntary allowing to die

In this view, the first two are generally considered immoral; instances of the second two may be moral under carefully circumscribed conditions.

The broad interpretation, recognizing no logical or morally relevant distinction between euthanasia and allowing to die, yields four forms of euthanasia:

1. Voluntary active euthanasia

2. Nonvoluntary active euthanasia

3. Voluntary passive euthanasia

4. Nonvoluntary passive euthanasia

In this view, any of these subtypes of euthanasia may be morally justifiable under carefully circumscribed conditions.

In conclusion, then, a decision to allow someone to die—passive euthanasia—may be morally justifiable. But the narrow and broad interpretations of *euthanasia* differ sharply in their moral judgment of *deliberate* acts taken to end or shorten a patient's life—that is, acts that the narrow interpretation terms voluntary or nonvoluntary euthanasia and that the broad interpretation terms voluntary or nonvoluntary *active* euthanasia. Generally speaking, the narrow interpretation considers such acts morally repugnant; the broad interpretation views them as being morally justifiable only under carefully circumscribed conditions.

With these six complex conceptual issues as a basis, we can now turn to the arguments for and against death decisions. We will focus on the most conservative of death decisions: cases of the voluntary decision to be allowed to die. Having raised relevant arguments pro and con, we will then see how they are applied to other forms of death decisions. These arguments will use the same dialectical style adopted for the abortion debate.

Arguments for Voluntary Allowing to Die

1. *Individuals have the right to decide about their own lives and deaths.*

POINT: "What more basic right is there than to decide whether you're going to live? There is none. A person under a death sentence who's being kept alive through so-called heroic measures certainly has a fundamental right to say, 'Enough's enough. The treatment's worse than the disease. Leave me alone. Let me die!' Ironically, those who would deny the terminally ill this right do so out of a sense of high morality. Don't they realize that, in denying the gravely ill and suffering the right to release themselves from pain, they commit the greatest villainy?"

COUNTERPOINT: "The way you talk you'd swear people have absolute rights over their bodies and lives. You know as well as I that it just isn't true. No individual has absolute freedom. Even the Patient's Bill of Rights, which was drawn up by the American Hospital Association, recognizes this. Although it acknowledges that patients have the right to refuse treatment, the document also recognizes that they have this right and freedom only to the extent permitted by law. Maybe people should be allowed to die if they want to. But if so, it's not because they have an absolute right to dispose of themselves if they want to."

2.   *The period of suffering can be shortened.*

POINT: "Have you ever been in a terminal cancer ward? It's grim but enlightening. Anyone who's been there knows how much people can suffer before they die. And not just physically. The emotional, even spiritual, agony is often worse. Today our medical hardware is so sophisticated that the period of suffering can be extended beyond the limit of human endurance. What's the point of allowing someone a few more months or days or hours of so-called life when death is inevitable? There's no point. In fact, it's downright inhumane. When someone under such conditions asks to be allowed to die, it's far more humane to honor that request than to deny it."

COUNTERPOINT: "Only a fool would minimize the agony that many terminally ill patients endure. And there's no question that by letting them die on request we shorten their period of suffering. But we also shorten their lives. Can you seriously argue that the saving of pain is a greater good than the saving of life? Or that the presence of pain is a greater evil than the loss of life? I don't think so. Of course, nobody likes to see a creature suffer, especially when the creature has requested a halt to the suffering. But we have to keep our priorities straight. In the last analysis, life is a greater value than freedom from pain; death is a worse evil than suffering."

3.   *People have a right to die with dignity.*

POINT: "Nobody wants to end up plugged into machines and wired to tubes. Who wants to spend their last days lying in a hospital bed wasting away to something that's hardly recognizable as a human being, let alone his or her former self? Nobody. The very prospect insults the whole concept of what it means to be human. People are entitled to dignity, in life *and* in death. Just as we respect people's right to live with dignity, so we must respect their right to die with dignity. In the case of the terminally ill, that means people have the right to refuse life-sustaining treatment when it's obvious to them that all the treatment is doing is eroding their dignity, destroying their self-concept and self-respect, and reducing them to some subhuman level of biological life."

COUNTERPOINT: "You make it sound as though the superhuman efforts made to keep people alive are not worthy of human beings. What could be more dignified, more respectful of human life, than to maintain life against all odds, against all hope? In situations like those, humans live their finest hours. And that includes many patients. All of life is a struggle and a gamble. At the gaming table of life, nobody ever knows what the outcome will be. But on we bash—dogged in our determination to see things through to a resolution. Indeed, humans are noblest when they persist in the face of the inevitable. Look at our literature. Reflect on our heroes. They are not those who have capitulated but those who have endured. No, there's nothing undignified about being hollowed out by a catastrophic disease, about writhing in pain, about wishing it would end. The indignity lies in capitulation."

Arguments against Voluntary Allowing to Die

1. *We shouldn't play God.*

POINT: "Our culture traditionally has recognized that only God gives life, and only God should take life away. When humans take it upon themselves to shorten their lives or to have others do it for them by withdrawing life-sustaining apparatus, they play God. They usurp the divine function; they interfere with the divine plan."

COUNTERPOINT: "Well, I'm impressed! It's not everyday that I meet somebody who knows the mind of God. How you can be so sure about the 'divine plan' is beyond me. Did it ever occur to you that the intervention of modern medicine might be interfering with God's will by keeping people alive who otherwise would have long since died? If there is a God and a divine plan, it seems pretty clear that God didn't intend that His creatures should live forever. Judging from history, God meant for all humans to die. Don't you think that modern medicine has interfered with this plan? What more gross a case of 'playing God' can you imagine than to keep people alive artificially? Let's face it: Nobody knows what God's plan is for allowing people to die."

Natural

2. *We can't be sure consent is voluntary.*

POINT: "Many of those opposed to nonvoluntary death decisions are quick to approve of voluntary ones. What they overlook is that we can't ever be sure consent is voluntary. In fact, the circumstances that surround most terminal cases make voluntary consent impossible. Take the case of the terminal patient who's built up a tolerance to drugs and, as a result, is tortured by pain. Just when are we supposed to get this person's consent? If we get it when they're drugged, then they're not clearheaded enough for the consent to be voluntary. If we withdraw the drugs, then they'll probably be so crazed with pain that their free consent still will be in question. Since consent in such cases can't be voluntary, it can only be presumed. And to allow a death decision on the basis of presumed consent is wrong."

COUNTERPOINT: "Agreed, in the situations you set up, rational free choice is in question. But you've overlooked cases where people facing a death due to a dreadful disease make a death request *before* they're suffering pain. Maybe you'll reply that the consent of people in such situations is uninformed and anticipatory and that patients can't bind themselves to be killed in the future. Okay, but what about cases where patients not under pain indicate a desire for ultimate euthanasia and reaffirm that request when under pain? Surely, by any realistic criteria, this would constitute voluntary consent."

3. *Diagnoses may be mistaken.*

POINT: "Doctors aren't infallible, that's for sure. The AMA readily admits that far too many medical procedures and operations aren't even necessary. This doesn't mean that physicians are malicious, only that they're human. They make mistakes in their diagnoses. Any instance of electing death runs the tragic risk of terminating a life unnecessarily, and for that reason it's wrong."

COUNTERPOINT: "Sure, physicians make mistakes, but not so often as you imply. In fact, in terminal cases, mistaken diagnoses are as rare as the kiwi bird. But that fact probably won't satisfy you, because you claim that any risk makes a death decision wrong. Well, by the same token every diagnosis, from the simplest to the most complex, carries a chance of error and with it the possibility of needless treatment. Sometimes the treatment involves operations, and operations always carry some jeopardy for the patient. So are we to say that it's wrong for people to opt for procedures that expert medical opinion says they need, because there's always a chance of mistaken diagnoses? Of course not. The correctness of a diagnosis is a separate issue from the individual's right to request and receive treatment. This applies with equal force to death decisions."

4.   *There's always a chance of a cure or of some new relief from pain.*

POINT: "People easily forget how final death is. That may sound obvious, but it's pertinent to death decisions. After all, people can never be recalled from the grave to benefit from a cure for the diseases that ravaged them or to benefit from a new drug to relieve the pain they suffered. Instances of such 'wonder' drugs, even of spontaneous remissions of disease, are common enough to make a death decision precipitate and therefore wrong."

COUNTERPOINT: "First of all, it's highly unlikely that some kind of cure will benefit those who are already ravaged by a disease. Ask physicians. They'll tell you that a cure is most likely in the early stages of a disease and most unlikely in its final stages. But the issue of death decisions pertains precisely to people in the final, torturous stages of a disease. So even if a cure is discovered, the likelihood of its helping these patients is virtually nonexistent. Another thing. You fail to distinguish cases where a cure is imminent from those where it's remote from those that fall somewhere in between. To treat all terminal diseases as if they had an equal probability of being cured is unrealistic. Generally speaking, there's a time lapse between a medical discovery and the general availability of a drug. I might agree that it would be 'precipitate' and even wrong to elect death in that interim between discovery and general availability of the drug. But I don't see why it would be precipitate and wrong to elect death if you're suffering from a terminal disease for which no cure or relief has even been discovered."

5.   *Allowing death decisions will lead to abuses.*

POINT: "Of course it's easy to think of death decisions as isolated instances of individuals electing to die. Looked at this way, those who make them appear to be above moral reproach. But when individuals choose to die and are subsequently allowed to, their actions open the door for all sorts of abuses. The chief abuse is *nonvoluntary* death decisions. A terminal patient who elects to die at the very least brings individuals and society closer to accepting nonvoluntary killing, as in cases of defective infants, the old and senile, and the hopelessly insane. Indeed, such private decisions are likely to set off a chain reaction that will lead to horrible abuses."

COUNTERPOINT: "The treatment of defective infants, the old and senile, and the hopelessly insane is a distinctly different issue from the issue of permitting death decisions. There is no connection between them; allowing death decisions does not inevitably lead to the abuses you fear. But even if there were a connection, that wouldn't of itself demonstrate that voluntary death decisions are immoral—only that they shouldn't be legalized."

These, then, are the arguments generally marshaled for and against the voluntary decision to be allowed to die or voluntary passive euthanasia. These basic arguments are commonly enlisted in discussions of other forms of death decisions as well.

For example, those arguing for voluntary active euthanasia often stress the inherent freedom of individuals to do as they choose, as long as their actions do not hurt anyone else. They also contend that it is cruel and inhumane to make people suffer when they have requested to have their lives ended. In contrast, those opposing voluntary active euthanasia cite the sanctity of human life, arguing that the intentional termination of an innocent human life is always immoral. They also express concern about mistaken diagnoses, the possibility of cure and relief, and especially the potentially dangerous consequences of eroding respect for human life.

Similarly, those supporting nonvoluntary active or passive euthanasia generally appeal to principles of humaneness and human dignity; those opposing it use all the arguments that have already been cataloged while stressing the moral objections to allowing people to die or killing people without voluntary consent.

Of course, various positions are possible and likely within this broad outline. For example, some who favor letting people choose to die (voluntary passive euthanasia) might well oppose nonvoluntary allowing to die. And many who may support nonvoluntary allowing to die (nonvoluntary passive euthanasia) might object to nonvoluntary (active) euthanasia. In other words, for some moralists the key issue in determining the morality of a death decision is whether or not it is voluntary. Only if a death decision is voluntary is it moral. Others concern themselves primarily with the distinction between active and passive, not voluntary and nonvoluntary. In their view, the focus should not be on the person who makes the death decision but on whether there is a morally significant difference between active and passive euthanasia. For them, even if passive euthanasia is acceptable, active euthanasia is not.

Defective Newborns

Much of the preceding discussion applies to the issue of babies born with serious birth defects, such as Down's syndrome and duodenal atresia, as in the opening anecdote of this chapter. There are, however, some aspects of treating defective newborns that deserve highlighting.

It is important to note that, in the case of defective newborns, allowing to die includes withholding ordinary and not just extraordinary treatment. The Stockard situation is a case in point. In that instance, ordinary nourishment may be withheld, resulting in the infant's death. Is this moral? In answering this question, it is necessary to come to grips with the central moral issue involving seriously defective infants, which involves determining the conditions, if any, under which it is morally permissible to allow defective newborns to die.

It is possible to identify three broad positions on the moral acceptability of allowing severely defective newborns to die.[3] Each is supported by controversial value assumptions. The first argues that it is permissible only if there is no significant potential for a meaningful human existence. Clearly implied here is a quality-of-life judgment, which, of course, elicits debate.

The second position agrees that allowing seriously defective infants to die is morally permissible when there is no significant potential for a meaningful human existence. But it also sanctions such an act when the emotional or financial hardship of caring for the infant would place a grave burden on the family. By introducing emotional or cost factors by way of justification, adherents of this view distinguish themselves from those espousing the first position. They often argue their case on the grounds that the seriously defective newborn does not have person status. Again, however, this position is marbled with questionable assumptions.

The third position claims that it is *never* morally permissible to allow a defective newborn to die. Stated more cautiously, it is never moral to withhold from a defective newborn any treatment that would be provided a normal one. By this account, allowing the Stockard baby to starve to death would be immoral, because we would never treat a normal infant that way. The clear implication here is that the defective infant has full personhood and must be treated accordingly. Just as clearly, this view rejects any quality-of-life or cost factors in determining the acceptability of allowing a defective infant to die. Like the other positions, however, this one is embedded with several debatable value judgments. It also ignores the fact that a normal infant would not require corrective surgery in order to digest food.

It is quite apparent, then, that whether cases involve defective newborns or adults who are terminally ill, the central moral question concerns the acceptability of a death decision and subsequent action. This question must receive the most deliberate analysis by HCPs and others. But there are additional moral problems relating to death decisions that HCPs must confront in their interactions with the gravely ill and with other professionals. These are well worth considering.

Special Professional Problems Related to Death Decisions

A basic problem that faces HCPs and others concerns the criteria for a death decision. There are several aspects to this issue. One deals with who decides.

---

3. *T. A. Mappes and J. S. Zembaty,* Biomedical Ethics *(New York: McGraw-Hill, 1981), p. 346.*

Should it be the physician, the patient, the family, a member of the clergy, or a board? When patients are competent, their rights to autonomy and adequate care would presumably require their participation in the decision. When they are not competent, more difficult questions arise. For example, some ethicists believe that death decisions should never be made for patients who cannot themselves ask to die. Thus it would be wrong for HCPs or anyone else to permit or cause the death of defective newborns and incompetent older patients being kept alive artificially. Others disagree, arguing that in such cases the decision rightfully shifts to those most knowledgeable about the patient: parents, guardians, HCPs, or perhaps others. Because of the complexities involved in any death decision, it is understandable that many HCPs, especially nurses, prefer a board decision.[4] Indeed, a collective decision (as opposed to a decision made exclusively by the physician) seems most likely to ensure that the legitimate interests of all parties will be considered and that undue personal bias will not warp the decision.

But who decides is only one issue. Another concerns the decision criteria. Under what conditions, if any, should death decisions be made? The answer to this question will inevitably reflect perspectives about such concepts as personhood, death, and ordinary versus extraordinary treatment, as well as general views on allowing to die and euthanasia and the importance placed on consent. The question of who should formulate the criteria is related to this issue. Again, some kind of collective body might be in order—one that represents the various medical, ethical, religious, and legal aspects of death decisions. This seems a most appropriate way to ensure that patient interests and rights are guarded, that issues receive comprehensive coverage, and that undue bias is eliminated.

In addition to problems concerning decision criteria, there are others that relate to implementing a decision after it has been made. One such problem involves vague physician orders. Nurses who must work with such ambiguity can get caught between their legal obligation to follow medical orders, which may imply termination of treatment, and their professional obligation to provide life-preserving services. Such a dilemma can be torture for personnel in old-age institutions, whose nurses may be the patient's and the family's only daily resource. Problems for nurses can grow even more complex when the patient's family desires one course of action and the rarely-present physician orders something different. It would seem, then, that physicians are under a special obligation to clarify their orders. By the same token, nurses are obliged to keep physicians informed of family wishes and of changes in the patient's condition that may warrant a change in standing orders.

Related to this last problem is the physician's obligation to draw on the nurse's perceptions of the dying patient. Nurses, because of their daily contact with patients, often know more about the patient's condition than physicians do. This observation is especially relevant to institutions with guidelines for orders not to resuscitate. Although physicians are responsible for recording the

---

4. N. K. Brown, J. T. Donovan, R. J. Bulger, and E. H. Laws, "How Do Nurses Feel about Euthanasia and Abortion?" American Journal of Nursing, July 1971, pp. 1415–1416.

decision, they risk considerable harm to patients if they do not consult on a continuing basis with those nurses who may be best informed about the patients' conditions. Clearly, the treatment of the terminally ill, perhaps more than any other group of patients, calls for the closest cooperation between physicians and nurses. Anything less jeopardizes these patients' right to adequate care.

Although professional staff deal with dying, death, and death decisions more directly than administrative staff do, the bureaucratic line of authority does get involved in these issues, thereby raising additional matters. For one thing, administrators have an obligation to ensure a free flow of communication among HCPs involved in implementing a death decision. For another, administrative and professional staff must make sure that their organizational clashes do not jeopardize the care of the dying. Chapter 5 noted that a health care setting generally has two lines of authority, the bureaucratic and the professional. These parallel chains of command have divergent values that can conflict, leading to clashes between professional and administrative staff. Death decisions are a case in point.

Physicians, nurses, and other HCPs tend to identify with the interests of patients. In contrast, administrators tend to view patients as managerial problems, relating specifically to the allocation of scarce hospital resources. When a patient happens to be terminally ill, these incompatible perspectives can cause problems for everyone, especially the dying patient. HCPs generally seek to provide treatment tailored to the individual's needs and requests, but they can still be indifferent to when and under what conditions patients should choose death. On the other hand, administrators, given their preoccupation with efficiency and equality of services for all patients, often ignore individual needs.[5]

So although determining the moral acceptability of a death decision stands as the central moral issue, there are other moral questions that HCPs must face. These pertain to decision criteria and to policies for implementing death decisions in an institutional setting. In addition, there remains the issue of the moral acceptability of social policies related to death decisions. Because HCPs often get entangled in these social policy debates by virtue of their expertise or role as patient advocates, we should examine some of the legal aspects of euthanasia.

Legal Considerations

Even if we decide that some forms of death decisions are morally acceptable, another question arises: Should individuals have a legal right to make a death decision? Should people be permitted under law to have their lives terminated?

Currently it is illegal deliberately to cause the death of another person. It is generally recognized, however, that people have a right to refuse life-saving treatment. In recent years numerous attempts have been made to legislate the individual's right to refuse such treatment—that is, passive euthanasia. By and large, these efforts have been rebuffed. Some object to the proposed legislation

---

5. *Davis and Aroskar, p. 127.*

because of inherent difficulties in trying to define such phrases as *death with dignity, natural death, extraordinary means,* and *heroic measures.* Others observe that there is no need for such formal legislation, since there is already a widely recognized informal right to reject life-sustaining treatment. Still others express concern that legalizing passive euthanasia will lead to the legalization of active euthanasia.

In the absence of specific legislation, various documents and directives have been developed to allow people to inform others about the nature and extent of the treatment they wish to receive should they become seriously ill. Such documents usually are termed *living wills.* Living wills specify the person's wishes and relieve others of having to make momentous life-or-death decisions, but generally speaking they are not legally binding. Thus there is no guarantee that the person's wishes will be respected.

In 1977 the State of California, as part of the Natural Death Act, created a version of a living will called the Directive to Physicians (see Appendix C). What makes this document unique is that it has the same legal power as an estate will. Of course, as with estate wills, the Directive to Physicians can be contested. But the document takes a giant step toward according legal status to a living will.

It is important to note that the morality of legalizing euthanasia is a separate issue from the morality of euthanasia itself. Although many of the arguments for and against the legalization of various death decisions capitalize on the general arguments for and against their morality, it is entirely possible that one could approve of individual death decisions but at the same time object to any systematic social policy permitting them. For example, legalization of voluntary euthanasia might be opposed on grounds that it will lead to abuses by physicians, families, and others or that it will lead to the legalization of nonvoluntary euthanasia.

Summary

This chapter inspected a biomedical issue of foremost concern: euthanasia. We saw that it was crucial to clarify such concepts as personhood, death, extraordinary versus ordinary treatment, killing versus allowing to die, euthanasia, and voluntary versus nonvoluntary consent before grappling with the morality of death decisions. Numerous arguments were offered in this chapter for and against the most conservative death decision: allowing to die. These arguments can be applied to other forms of death decisions as well. Particularly agonizing are cases involving seriously defective newborns. The morality of withholding ordinary treatment emerges as a central concern. In addition to determining the morality of death decisions, HCPs must also face moral questions about decision criteria and about implementing policies within an institutional context. Finally, although related to the morality of death decisions, the morality of euthanasia legislation is a separate and distinct issue, which HCPs must sometimes engage in their roles as expert resources or as patient advocates.

Ethical Theories

The central issues to be considered here concern the morality and legality of death decisions: Are death decisions ever morally permissible? Should individuals have the right to make death decisions?

Egoism

As on all moral issues, egoists would approach the morality of death decisions from the viewpoint of self-interest. Such conceptual issues as the meaning of euthanasia and the distinctions between killing and allowing to die, voluntary and nonvoluntary decisions, and active and passive means would not figure significantly in the egoist's calculation. Rather, the central question is: What would be the likely consequences for self? For example, if a death decision would produce a net saving of pain, then a dying patient tortured by pain would be morally justified in making such a decision. Whether it is proper for an outside party to allow or help such a patient die is a separate question whose answer depends on an application of the self-interest yardstick. If a doctor sees his or her best long-term self-interests served by allowing or helping such a patient to die, then the doctor should assist the patient. The key thing is that egoists would not be overly concerned with the difference between voluntary and nonvoluntary or active and passive.

The egoist's position on legalizing some form of euthanasia would depend on whether the legislation would be likely to advance one's best long-term interests. Figuring prominently in the evaluation would be the possibility of ultimately facing a death decision oneself. Lacking specific legislation to acknowledge an individual's right to make a death decision and have it implemented, one might not be able to die with dignity. Egoists would consider other things, such as the social impact of such a law, only insofar as they would affect the egoist's personal happiness.

Act Utilitarianism

Act utilitarians would examine each proposed act of euthanasia. If the act would be likely to produce the greatest total happiness, then it would be morally justifiable. In this view, no form of a death decision would be in and of itself objectionable. However, the act utilitarian would be sensitive to the possible negative long-range social effects of any act of euthanasia. This is particularly relevant with nonvoluntary euthanasia, whose potentially sinister consequences for individuals and society need no cataloging. Nevertheless, if in the final analysis any death decision—including cases of nonvoluntary euthanasia—would probably produce the most total good, then it would be morally justifiable. (We should note that either the act or rule utilitarian could argue that, since life is a necessary condition for happiness, euthanasia would be wrong. The destruction of life would preclude the possibility of all future happiness.)

As for legalizing euthanasia, if such an act would be likely to produce the greatest social good, it would be justifiable. If not, it would be wrong. Act utilitarians who would agree that some acts of euthanasia are morally justifiable could object to any systematic social policy permitting even voluntary euthanasia if, for example, they foresaw abuses, needless deaths, or other undesirable consequences.

Rule Utilitarianism

Rule utilitarians would apply the utility principle to the rule under which the specific death decision falls. For example, applying the various forms of euthanasia to terminal cases who have no relief from torturous pain, they might come up with a variety of rules: (1) Allowing a human life to end is permissible when suffering is intense and the condition of the person permits no legitimate hope. (2) Even if the death decision is not voluntary, allowing a human life to end is permissible when suffering is intense and the condition of the person permits no legitimate hope. (3) When a person requests it, the taking of human life is permissible when suffering is intense and the condition of the patient permits no legitimate hope. (4) Even when a person has not requested it, the taking of human life is permissible when suffering is intense and the condition of the person permits no legitimate hope.

In theory, the rule utilitarian could endorse any of these rules, if following the rule generally produced more total good than not following it. In evaluating the consequences of the proposed rule, rule utilitarians would be alert to the abuse to which a rule might be open. This is especially noteworthy with rules 3 and 4. But the caution would apply to all the rules.

Regarding the legalization of euthanasia, the rule utilitarian would evaluate the morality of any proposed law on the basis of its probable social consequences. If such a law appeared likely to produce the most total happiness, then it would be justifiable. Rule utilitarians would consider very carefully the implications for abuse of having a systematic social policy that acknowledged and vouchsafed the right to a death decision.

The Categorical Imperative

A quick reading of Kant might lead one to conclude that he would oppose all forms of death decisions. After all, Kant claims that autonomous, rational beings have a duty to preserve their lives. Refusing needed medical care would be inconsistent with this obligation and therefore immoral.

Yet Kant asserts that it is by the very fact that we are rational and autonomous that we have inherent worth and dignity. But what happens when our status as rational and autonomous creatures is severly damaged or impaired, as it is when people are disoriented, comatose, or irrational due to severe and prolonged injury or illness? Since their status is damaged, even destroyed, it is no longer clear that there is a moral imperative to sustain them. Even if they do have the status of rational, autonomous creatures, our obligation to accord one another dignity may make it acceptable for loved ones or HCPs to permit patients to die or even to kill them. In short, voluntary death decisions could be justified, even obligatory, under Kantian ethics.

Given the ambiguity of the Kantian position, it is impossible to nail down Kant's reaction to any proposed legislation about death decisions. If voluntary death decisions are potentially moral, then it would seem that people should have a legal right to make them. To deny them that right or not expressly to grant them the right would appear to be incompatible with the freedom that rational, autonomous creatures are entitled to.

Prima Facie Duties

In applying Ross's prima facie duties to death decisions, the duty of noninjury would receive paramount consideration. Ross holds that we have a prima facie duty not to kill any human being except in justifiable self-defense, *unless we have an even stronger prima facie duty to do something that can only be accomplished by killing.* What might such a "strong prima facie duty" be? Not to injure another person. It could be argued, therefore, that there may be times when we can meet an overriding prima facie obligation not to cause injury to another person only by killing. But this smacks of a paradox, for isn't killing someone causing the person injury? Perhaps not. If we treat comatose or irrational terminal patients in a manner that they would choose if they had their faculties, then we might not be causing them injury at all. In fact, perhaps people in such a state are beyond injury. In any event, if we can make a fair and honest presumption that such a person would want to be allowed to die or even to be killed, then we cannot be said to be causing the person injury in implementing the implied request. Where patients have left specific instructions to that effect—for example, in a living will—then again we would not be injuring them in meeting their requests. In fact, it seems that we would even have a prima facie obligation to do so. On the other hand, where it is clear that the person wants to be maintained at all costs, then despite our own feelings we would have a prima facie obligation to honor that preference. Not to do so would be to injure the person. Apparently, then, the issue of voluntary consent could be crucial for Ross in determining whether the death decision involves injustice. When there is expressed or implied consent, no injury seems to be involved and the death decision would be morally justifiable.

Furthermore, people would have a legal right to make death decisions. Lacking appropriate legislation, patients could easily be prevented from embarking on a perfectly moral course of action in a matter of the greatest urgency.

The Maximin Principle

Many of the observations about Ross apply equally to Rawls. Assuming the original position, behind the veil of ignorance, people probably would not want to be bound by a rule that would prohibit them from ending the sort of intense suffering under discussion. Thus Rawls's ethics would appear to sanction at least voluntary death decisions. What is more, Rawls's paternalism would allow, perhaps encourage, death decisions based on the presumption of consent, providing of course the evidence for presumption was overwhelming. In no instance, however, would it approve of killing patients who wished to be maintained at all costs. And in all probability, it would condemn a death decision where there was not enough evidence for presuming consent. Of course, what constitutes enough evidence is problematic. Again, as with Ross, the key issue seems to be whether the consent is voluntary.

By this account, Rawls, like Ross, could support legislation that would permit voluntary death decisions. Such laws would maximize individual liberty in a way that is compatible with a like liberty for all.

Roman Catholicism's Version of Natural-Law Ethics

Roman Catholic theorists would disapprove of any deliberate action taken to terminate life. Whether the action is voluntary or nonvoluntary is irrelevant. The key factor is whether the action is positive. A positive, deliberate act to terminate a life is the killing of an innocent life, which violates the will and law of God.

However, Roman Catholic moralists do distinguish between euthanasia and allowing to die. The former is killing and is never permissible, its voluntary nature notwithstanding. But allowing to die is not killing and may be moral under carefully circumscribed conditions. It follows that the Roman Catholic view could support legislation that permits "allowing to die" decisions but would oppose legislation that permits euthanasia.

Roman Catholicism does not recognize any moral obligation to maintain hopeless cases through extraordinary means. Thus, by Roman Catholic moral theology, it would be justifiable for someone to refuse extraordinary treatment and for HCPs to withhold such treatment, thereby letting the will of God determine the outcome. This is precisely what occurred in the Quinlan case. The Quinlans, who happened to be Roman Catholics, requested that life-sustaining treatment be discontinued, an action that was perfectly compatible with the moral tenets of their faith. Curiously, Karen Ann did not die when the extraordinary means were removed. Theologically speaking, God evidently did not want her to die. Had the Quinlans then initiated some direct action (active euthanasia) to terminate Karen Ann's life, they would have acted immorally according to Roman Catholicism. They would have interfered with the will of God by deliberately taking her life. Had Karen Ann been conscious and tortured by pain, under Roman Catholicism it would have been permissible to administer drugs for the purpose of relieving the pain, even if the drugs indirectly happened to shorten her life. As long as relief of pain is intended, and not the patient's death, such an action is not considered killing.

If these applications are sound, then each of the theories provides a basis for at least allowing a patient to die. As always, these tentative applications are merely intended as conceptual frames within which HCPs and others may begin the challenge of moral clarification and analysis.

CASE PRESENTATION
The Bitter End

A lot of people around here—doctors and nurses alike—think I should have acted differently. But some have come forth privately and confessed that they think I did the right thing. A few even said they think it took guts. I don't know about that. If anybody showed guts that day, it was Esther Minturn, not I.

Who would have thought such a gentle woman, who hardly ever spoke above a whisper, would have the stuff to make a decision like that? But she bit the bullet all right; nobody can deny that. Maybe she'd already bitten it, when she and her husband first found out that he'd have to undergo cancer surgery.

And a nasty bit of business it was, a radical neck dissection that left a gaping wound in Sam's throat, with his right carotid artery vulnerable. But even in the long period in the intensive care unit that followed, Sam remained cheerful and optimistic. And Esther almost never left his bedside. The surgeon had taught her how to care for Sam's wound, and so she became his private duty nurse.

At first Sam made good progress. Every day he got stronger and more self-sufficient. But then abscess set in. We didn't know it at the time, but it foreshadowed the end. Grafts were started, but infection thwarted them. I guess it was sometime during those two weeks, when Sam went to the operating room about six times, that we realized he wouldn't make it. Even the attending physicians admitted that his chances were slim.

From then, on, it was all downhill. Infection raged; treatment failed. Sam lost his will to live. He talked less and less. After awhile he spoke only to request more Demerol. When he became disoriented, he had to be restrained.

We all felt for him. A nicer guy you'd never want to meet. And look how he was going out—not in the warmth and security of his own bed but in a hospital, drugged and restrained like some sort of dangerous beast.

Two days before the end, I went into Sam's room to check his vital signs and discovered blood spurting upward from a spontaneous carotid rupture. I thought about just turning on my heels and returning later to take his blood pressure. If the surgeons hadn't been nearby, I probably would have. But instead I summoned the surgical resident, who patched him up and then admitted he couldn't do it again.

The next day Sam had to be put on a respirator. We told Esther. In a quiet, determined voice, she simply said, "I won't allow it." We followed her request.

From then on Esther never left Sam's side. She'd see it through to the bitter end. She asked nothing for herself, not even a cup of coffee. All she wanted was for Sam not to suffer, that it be over.

The hours hung like weeks. Then things happened fast. They always do at the end. Out in the hall the intern informed me that Esther had asked the resident to stop the I.V. The resident had refused. The intern said I could slow it just to keep it open.

When I returned to Sam's room, I found Esther removing the I.V. I didn't know what to say. Every human instinct told me she had a right to. But the nurse in me spoke, "I can't condone what you're doing. I have my orders."

"Your orders be damned!" Esther said sharply. "He belongs to me, not to your hospital!"

With that she did what she had to do. I didn't stop her.

Questions for Analysis

1. *Was Esther morally justified in withdrawing Sam's I.V.?*

2. *Would you regard Esther's action as active or passive?*

3. *Did the narrator have a moral obligation to stop Esther?*

4. *Was the intern morally justified in ordering that the I.V. be slowed merely to keep it open?*

5. *Apply the various ethical theories to Esther's action and to the narrator's inaction.*

Exercise 1

Neurosurgeon M. S. Heifetz has argued that under certain circumstances those who are near death from a terminal illness and who wish to commit suicide should be assisted by their physicians. Indeed, laws exist in several countries (for example, Uruguay, Switzerland, Peru, Germany, and Japan) for physician assistance. Do you think such laws should exist in the United States? Would they eliminate moral problems for physicians faced with such situations?[6]

Exercise 2

Reevaluate your responses under Values and Priorities.

---

6. M. S. Heifetz, The Right to Die *(New York: Putnam, 1975).*

Death by Decision

Jerry B. Wilson

In the following selection, professor of religious studies Jerry B. Wilson attempts to place death decisions within the context of the requirements of Christian love. Wilson begins with some general guidelines for medical care of the dying, including defective infants and the aged. Following the lead of moralists like Joseph Fletcher, he then considers death decisions within the framework of *theocentric (literally, "God-centered") love.* Theocentric love, Wilson points out, takes root in the concrete needs of patients as persons. It is not static but dynamic; it takes into account the needs of individuals in a changing and increasingly complex society. On the basis of situational nuances, then, theocentric love formulates moral norms that are relevant to the problems of contemporary medical practice, specifically to death decisions.*

Wilson admits that sometimes the requirements of responsible love conflict with the medical and legal standards of practice. In those cases, he says, responsible love needs to reform these moral norms. In order to do this, it is necessary to locate a basis for agreement on the level of ethical values and principles. Wilson thinks that the "sanctity of life" principle provides a basis for a moral consensus from which to respond to moral standards. It is at this point that Wilson embarks on the most controversial aspects of his presentation. Not only does he think that sanctity of life can and should be used to update the concept of death and permit passive euthanasia, he thinks it should be invoked to allow active euthanasia.

Clearly, Wilson's essay speaks directly to the medical and legal aspects of the death decisions that HCPs face. Beyond this, it also illustrates how a moral principle (sanctity of life) that finds vast acceptance can be interpreted in a way that many would find objectionable, even many Christians and certainly traditional Catholics.

There is little question concerning the requirements of responsible medical care when a physician can offer the hope of recovery or can prolong life without undue suffering. Under these circumstances, there are several general responsibilities that seldom conflict. In the first place, there is a duty to recognize the autonomy of each patient. So long as a person is competent, decisions should not be made and treatment rendered without his informed consent. In the second place, every available means should be offered to restore his health or to preserve his life as long as possible. In the third place, the doctor should alleviate suffering and secure the most rewarding life for each patient.

In numerous cases, these duties and expectations conflict. Efforts to prolong the patient's life frequently result in the prolongation of his suffering and dying. Measures to provide the necessary relief often precipitate his death. This moral dilemma, however, can be resolved when those who are responsible for the patient's care are attentive to his needs and desires. If he is able, the patient should be allowed to determine the course of treatment to be followed. Otherwise, someone else must be prepared to make this decision in his behalf. In either case, the physician has a respon-

sibility to counsel both the patient and his family in order to help them cope with the problem.

1. Sharing the Burden of Responsibility

Patients who are suffering and dying usually cannot make responsible medical decisions on their own. In the majority of cases, the responsibility for their care rests entirely on their families or their doctors. In a complex and mobile society, however, familial relationships are often tenuous and undependable. Furthermore, because the practice of medicine is highly specialized, the doctor-patient relationship tends to be technical and impersonal. These are factors that must be taken into account in guidelines for terminal medical decisions.

In the place of the conventional pattern of private practice, a social systems model provides a more inclusive perspective from which to interpret the structure of medical practice. From this point of view, the moral and the legal responsibilities of the physician can be defined. The doctor-patient relationship is a unit or a subsystem that is part and parcel of a larger system of interrelated collectivities. On the one hand, this dyadic relationship is a minimal-level system. It constitutes an adequate basis for medical practice in emergency situ-

From Death by Decision: The Medical, Moral, and Legal Dilemmas of Euthanasia, *by Jerry B. Wilson, pp. 173–195, 207–208. Copyright ©1975 The Westminster Press. Reprinted by permission.*

**J. Fletcher,* Situation Ethics: The New Morality *(Philadelphia: Westminster Press, 1966).*

ations and in routine cases when the bearing of other related systems is insignificant insofar as medical decisions are concerned. On the other hand, the doctor and the patient are members of a larger social system. This frame of reference is important in cases of serious injury or illness and when medical resources are expensive or limited. In such cases, a plurality of needs and interests impinge on the therapeutic relationship.

This analysis exposes a cross section of the relationships in which both doctors and patients are involved in terminal medical care. These relationships provide the basis on which they can share the burden of responsibility for the decisions that must be made in this context. Teams of doctors and clergy, psychiatrists and counselors, along with members of other professions, including nurses, social workers, and lawyers, can contribute insight and can assure that terminal decisions are made at a high level of integrity. Through discussions and counseling, the physician can enable the patient and his family to understand the diagnosis and the prognosis. Counselors and psychiatrists can help the dying patient preserve his identity and dignity as a unique individual, despite the disease. They can also attend the living and help them to overcome feelings of anxiety, resentment, and guilt and to adjust to their loss. Chaplains and clergymen often can foster communication, assist the doctor in understanding the patient's beliefs and desires, and help both physicians and patients to find meaning and values that transcend suffering and death.

The primary purpose of interdisciplinary cooperation and understanding is to assure that every patient is offered the opportunity to live as long and as fully as possible and is given an appropriate emotional world in which to die. It is essential for cases in which the only relevant medical resources are limited, or experimental, or in which they promise little hope and cause undue suffering or expense. Such teamwork is particularly important when a patient's condition is irremediable and it becomes necessary to redirect the skills of medicine in order to preserve his dignity, to relieve suffering, and to make possible an easy death. This procedure is fitting in a wide variety of circumstances, especially when disease or the degenerative processes of age progress too far or when injury or deformity is too radical to permit even a limited but choice-worthy existence. In such cases, three alternatives remain within which specific courses of action must be determined:

a. *Prlongation of life* involves ameliorative and emergency treatment in order to sustain life as long as possible. Although supportive efforts are given priority, palliative measures are applied to make the limited remaining life as comfortable as possible. All remedial treatment that causes distress is omitted.

b. *Passive euthanasia* includes palliative therapy and limited ameliorative treatment in order to relieve suffering without prolonging life. Emergency and remedial procedures are omitted and supportive measures are reduced to allow the patient to die.

c. *Active euthanasia* entails the application of palliative treatment and deliberate measures to cause death. In extreme cases, the borderline between passive and active euthanasia is often very subtle, both in intention and in action. Direct action to end life may involve administering lethal pain-killers or terminating artificial supportive measures, for example, turning off a respirator or a heart-lung machine.

2. General Guidelines for Medical Decisions

When it becomes apparent that medical care ought to be restricted to one of these levels, every aspect of the case should be examined thoroughly, and each decision should be endorsed by a committee of medical and nonmedical specialists. This committee would be analogous to panels that are created to approve abortions and to select patients for treatment when facilities are limited. The authority of such committees should be strictly limited, but they should have the following responsibilities:

a. To guarantee each patient's right to live and his right to die.

b. To verify the doctor's diagnosis and to protect him from false and dangerous criticism.

c. To offer advice and understanding to patient and family.

d. To assure that hope for recovery and efforts to prolong life are not given up prematurely.

e. To document each case as carefully as possible in order to prevent irresponsible decisions.

Committees that are responsible for evaluating terminal medical care should have at least five members. They should include no fewer than two physicians in fields related to the patient's condition. These doctors should not be related to him in any other way, directly or indirectly, as members of his family, as his personal physician, or as doctors connected with cases for which he might be regarded as a donor of vital organs. A representative of the hospital administration should be on these committees to assure that the policies of the hospital are followed. Furthermore, such committees should include a counselor or psychiatrist and a chaplain or the patient's rabbi, priest, or minister. Any other person requested by the patient or his family should be allowed to participate in the deliberations and the decisions of the committee.

Responsibility to a patient as a person requires that he be permitted to make his own medical decisions. Thus, so long as he is conscious and competent, medical care should neither be applied nor denied without his knowledge and consent. The mere fact that he is terminally ill or fatally injured and suffering does not undermine the value of his life or relieve the obligation of family and physician to care for him. In many cases, a desperate hope for recovery, or deep personal relationships, or important goals yet to be fulfilled add meaning and purpose to the brief time that remains. Therefore involuntary euthanasia should never be permitted.

The fact that a patient places himself and remains under the care of a doctor does not indicate implied consent to any and every therapeutic measure. Furthermore, the fact that he wants to die does not in itself indicate a mental condition that renders him incompetent to make his own medical decisions. Those who are responsible for his care must make certain that he is fully capable of understanding his condition and that his request to die is an expression of his genuine and consistent desire. Nevertheless, when a patient wishes to die and his suffering cannot be relieved adequately or his condition renders his life hopelessly intolerable, he should be permitted to refuse treatment to prolong his life. In extreme cases, he should even be allowed to choose measures to end his life directly. In short, voluntary euthanasia, both active and passive, ought to be sanctioned in response to the needs and claims of the dying.

For the patient who is unconscious or incompetent, the family or guardian usually assumes a much greater role in determining the course of medical care. In such cases, a committee is especially important to assure that they fully understand the patient's condition, that the treatment that doctors can offer is explained, and that their decisions are made in the patient's best interest. The fact that someone else had to decide in his behalf in no way undermines the value of his life, and so long as recovery is possible, even if it is only limited or temporary recovery, his right to live should be preserved. This does not imply that a person's life should be prolonged at all costs simply because he is not able to make his own decisions, but it underscores the seriousness of interrupting therapy aimed at recovery or prolongation of life.

In the course of prescribing care for an incompetent patient, the problem of suffering should be taken into consideration. It is irresponsible to require a person to endure useless emotional or physical distress simply because he is not able to give informed consent either to the withdrawal of efforts to prolong his life or to measures to cause his death. Consequently, the patient who loses consciousness or experiences cardiac arrest often should not be revived simply to undergo hopeless suffering. Through consultation with doctors and counselors, his family should be informed concerning his condition. Their questions and uncertainties should be resolved, and they should be prepared to choose or to share the responsibility of deciding whether or not to continue to prolong his life.

In making terminal medical decisions, the extent to which vital systems have been destroyed by accident or by disease should be evaluated thoroughly. It is not always possible to determine this with absolute certainty; nevertheless, there are clinical tests which are indispensable. According to Dr. Hannibal Hamlin, "the EEG [electroencephalogram] can signal a point of no return, although the cardiovascular system continues to respond to supportive therapy that produces a respectable ECG [electrocardiogram]."[1] Dr. Denton Cooley has reported that there is a general agreement among the surgeons who have performed heart transplants concerning three criteria for declaring a patient dead in addition to a flat EEG: the patient should no longer have natural heartbeat, respiration, or reflex.[2] These clinical signs of brain death have also been recommended by a committee of the Harvard Medical School under the chairmanship of Dr. Henry K. Beecher.[3]

The way in which these signs are interpreted reflects the way in which death is defined within the medical profession. Increasingly, "brain death" is replacing the traditional definition of death as the permanent absence of respiration and circulation. The same criteria are applied, however, because respiratory and to some degree cardiovascular activities are functions of the brain stem. In addition to conventional tests for death, the electroencephalograph provides confirmatory data.[4] When these tests reveal the absence of spontaneous signs of life, all of the so-called life-sustaining measures should be terminated.

In most cases involving permanently comatose patients, brain damage is not serious enough to warrant declaring the patient dead as the basis for discontinuing medical treatment. Terminal decisions in these cases tend to be especially difficult. Although it may be pointless to sustain the life of a patient in this condition, relatives and physicians are usually reluctant to do otherwise. At the same time, the irretrievably unconscious patient is totally unable to respond to their care. In fact, he lacks even the potential for personal relationships which make life human. In such cases, as Ramsey has observed, "the duty always to keep caring for the dying is suspended by their inaccessibilty to any form of care and comfort."[5] Furthermore, it makes no difference, insofar as the patient is concerned, whether his death is brought about by passive or active means.

3. Guidelines concerning Defective Infants

The ideal solution to the problem of birth defects is, of course, to prevent them altogether, but our knowledge of their genetic sources is insufficient to reduce seriously the frequency of malformations. The discovery of abnormalities after conception has been limited because of the dangerous level of radiation exposure required for X-ray examinations;[6] nevertheless, other diagnostic procedures, such as prenatal chromosome studies, offer good prospects of early detection. In this event and when there is evidence of radical deformity as the result of an infection, such as

rubella, or following the consumption of a teratogenic agent, such as thalidomide, abortion may be indicated. At the present time, however, this approach is applicable in only a relatively few cases, for most abnormalities remain unsuspected until after birth and are not caused by disease or by toxic drugs.[7]

As a general rule in responsible medical practice, it is the child which matters and not the parents when the application of a particular treatment is in question.[8] This rule is also relevant for the care of most defective children, for an increasing majority are able to lead useful and rewarding lives. It is especially applicable for cases in which it is not possible to determine early in life the extent or consequences of a recognized defect. When the child involved is severely defective, however, the wishes and desires of the parents should be taken into account. The difficult decisions concerning the proper course to follow must still give priority to the welfare of the child. They cannot be made on the basis of the parents' interests alone. Furthermore, the question should not be resolved by appeal to expedience, economic necessity, or the good of society at large. In short, responsible medical care cannot condone euthanasia as a eugenic measure.

A second general rule of medical practice requires that the abnormal patient be given the same quality of care as the normal patient.[9] This also should apply to "normal" children with specific defects of intellectual function and for those of normal intelligence with physical abnormalities, such as talipes, polydactylism, cleft lip, cleft palate, and remediable cardiac malformations. These conditions may have relatively little effect on the quality of life, and in the course of these lives much more can be done to ameliorate the consequences of such defects.[10] Obviously, the desirability of continued existence under these circumstances cannot be seriously questioned. With regard to deformities caused by thalidomide, for example, Drs. Charles H. Franz and George T. Aiken insist that we would be defeatists to consider euthanasia for phocomelic children. "The proponents of this procedure," they argue, "fail to realize that many of these children are of normal and high normal intelligence, indicating an excellent possibility of emotional and social habilitation."[11]

In contrast with these, for whom doctors and

parents should assume the responsibility of securing the maximum possible quality of life, there are infrequent cases of gross physical deformities (such as major cardiac abnormalities) and severe defects of the central nervous system (for example, anencephalus) for which there can be little or no hope of survival. Between these two extremes is a range of conditions that are severely defective—mentally or physically, or both. These often raise difficult questions concerning the desirability and the effectiveness of medical intervention. The following three kinds of abnormalities illustrate the medical dimensions of this dilemma:

a. Spina bifida is a relatively common defect which poses this problem more acutely, perhaps, than any other condition. The majority of infants with this condition die without treatment or in spite of it, but in a few cases modern medical and surgical procedures can offer a chance of reasonably normal life. On the other hand, the risk is great that the child will remain paralyzed for the rest of his life.[12]

b. Mongolism is a second syndrome that often poses difficult medical and moral problems. Once again, early mortality is high, usually as a result of cardiac and other serious malformations, but it is not uncommon for mongols to survive to adult life and even old age.[13] Although physically deformed and mentally subnormal, many are able to enjoy life, and so long as their condition does not deprive them of this value, life should not be denied them simply because they are unable to conceptualize their desires. Nevertheless, they should not be required to endure needless suffering; and when this occurs, or when radical corrective treatment is required, decisions must be rendered in their behalf by those entrusted with their care.

c. Hydrocephalus is a third common type of serious abnormality which requires appropriate surgical measures in order to increase the chance of survival and minimize disability. Without treatment approximately half of the hydrocephalic infants that are live-born die within five years,[14] and those that survive usually suffer severe deformities, such as an enormously unsightly head, mental and neurologic damage, and progressive optic atrophy.[15] While surgical intervention often limits and occasionally prevents damage, it also greatly increases the possibility that the patient whose life is saved will be radically disabled.[16]

For any of these conditions, it is not responsible practice simply to postpone decisions and allow the patient to die rather than to face the alternatives and consequences and to choose an appropriate response to the needs of the patient and those intimately related to him. The doctor should not be expected to shoulder the full responsibility for determining the proper course to follow. As in cases in which the patient is unconscious or otherwise incapable of giving informed consent to medical decisions, the physician can provide the necessary information concerning the condition of the defective child and recommend one or several alternatives. A committee of specialists in several fields can offer insight and guidance. In the last analysis, however, the primary responsibility for decisions belongs to the parents.

4. Guidelines concerning the Aged

Age *per se* has nothing at all to do with the quality and value of life, and the experience, the wisdom, and the relationships that most people develop throughout their lives often make the advancing years most rewarding. Improved environmental conditions and health care enable many to remain well and self-sufficient beyond the Biblical threescore and ten years. Even though the incidence of chronic disease is high among persons over sixty-five, much can be done to ameliorate their conditions, and with patience and proper treatment, the majority of elderly incapacitated persons can eventually be rehabilitated. Therefore, necessary medical and surgical care should not be denied merely because of advanced age.

The term "aged" is for all practical purposes a physiological rather than a chronological concept. In this sense, age is often an important criterion for determining how long to prolong life and when to apply heroic measures. Younger patients usually have a much greater chance of withstanding serious illness and recovering from critical injury. In cases of cerebral accidents, for example, they are often able to make satisfactory recovery even after a long period of unconsciousness, because the brain has a remarkable capacity to compensate for injury. For older patients, however, this possibility is very slight, especially when their condition stems from cerebrovascular diseases.

When a cure could offer the prospects of a reasonably full life, the rapid pace of medical research and the hope for new life-saving measures usually warrant efforts to prolong the lives of patients who are presently incurable. For the very elderly who are afflicted with advanced cardiovascular, cerebrovascular, or malignant diseases this support may not be justifiable, especially when suffering is involved. Furthermore, if they must endure severe, intractable pain from rheumatoid arthritis, for instance, it might on occasion be more merciful to terminate treatment for intercurrent infections such as pneumonia, which Dr. William Osler used to call "the old man's friend."[17]

So long as the patient is capable, he should always be allowed to determine the goal toward which care should be directed and the measures that should be applied. Dr. Walter C. Alvarez, emeritus consultant at the Mayo Clinic, points out that

> with these old persons who have suffered long, the physician is usually safe in discussing death and dying. With such persons he need fear no embarrassment about mentioning these things. The patient will show little fear, and all he may ask of his physician is that he prevent suffering at the end.[18]

When the elderly patient cannot make this decision, Dr. Alvarez suggests

> that the relatives should be asked if they wish the physician to keep carrying out efforts at resuscitation with much oxygen and endless injections of stimulants, or if they would prefer to let the loved one pass peacefully when his or her time has come. Sometimes, as in the case of a patient suffering from a brain tumor, the physician should point out that even if through some miracle he could prolong life for a few weeks, the person would be so badly crippled in mind and body that he or she would be utterly miserable and perhaps in constant pain. Often the family knows that the patient would be better off dead, but they do not have the courage to say this, fearing that they would be criticized by someone.[19]

In every case, a team of competent physicians and counselors should assist the patient, his family, and his doctor in making the difficult decisions required by suffering and death.

Medical Ethics in a Pluralistic Society

Theocentric love does not begin with general or abstract principles from which to deduce specific moral norms for medical care. Instead, it begins with the concrete needs of patients as persons. It seeks to understand and to respond to the needs and claims of each patient in the context of a society that is increasingly complex. In specific cases, responsible love takes into account the long-range results, as well as the immediate consequences, of its decisions and actions. On the basis of its insight and experience, it formulates moral norms that are relevant to the problems of contemporary medical practice.

When the requirements of love conflict with medical and legal standards of practice, responsible love seeks to reform these moral norms. In order to change accepted moral rules in a pluralistic society, it is necessary to locate a basis for agreement on the level of ethical values and principles. Daniel Callahan suggests that the principle of the sanctity of life provides the basis for a moral consensus from which to evaluate and to affirm or to amend moral standards having to do with human control over life and death.[20] It is important to recognize, however, that the logic of moral discourse is not strictly deductive science. Furthermore, such ethical concepts are indeterminate principles which convey broad ranges of meaning rather than specific, determinate meanings.[21] In fact, the "sanctity of life" is interpreted in a number of different ways in our society. Thus an appeal to this generic principle does not *entail* specific moral conclusions.

This principle is often understood to imply that life *per se* is of ultimate value. Professional medical ethics tends to give priority to the value of biological life and to order other values and obligations accordingly. Because they are usually compatible, these values make possible standards of practice that are relatively consistent and applicable in most cases. These standards require that doctors secure the maximum longevity possible. So long as a patient can be cured or at least offered a reasonably full life, the doctor's responsibility to work toward this end is seldom challenged. When this goal becomes unrealistic, however, his duty to prolong life as long as possible can be called into question. Nevertheless, traditional legal and professional medical standards strictly forbid him to practice euthanasia.

For many people, including both doctors and laymen, the sanctity of life often seems to be destroyed by desperate efforts to prolong life in the face of hopeless suffering. They interpret this principle to refer to more than mere biological existence. From their perspective, a terminal patient's request to die is not always considered to be irresponsible. Furthermore, a doctor's decision to omit resuscitative procedures, to suspend supportive measures, or to prescribe lethal pain-killers is not necessarily regarded as a violation of the sanctity of life. In some cases, such omissions and actions appear to reflect a reappraisal of the meaning of this principle in the light of new medical options and human needs and as a reassessment of the implicit moral norms for terminal medical decisions.

The sanctity of life is often judged in relation to other ethical principles. When it is contingent upon the quality of life, the doctor is not required to prolong life indefinitely after the quality of life has been undermined by suffering, age, deformity, injury, or illness. If he cannot cure, he may be called upon to ease suffering by every possible means, including euthanasia. When the sanctity of life is based upon its worth to society, a doctor may not be expected to "save" patients who have no social value. If there are heavy demands on limited medical personnel and facilities, he may choose to allow the "hopeless" and the "useless" to die prematurely in order to make space and treatment available for other patients. In the last analysis, medical care is provided only for "life that is worth living" when either hedonistic or utilitarian principles constitute the primary basis for the moral norms of medical practice.

Within Christian medical ethics, the sanctity of life is a basic ethical principle. Underlying this principle, however, is a fundamental commitment to God as the ultimate source of being and value. On the basis of this commitment, theocentric faith affirms and orders ethical values and moral norms. It also rejects those systems which absolutize finite centers of value. Consequently, this faith challenges the vitalism implicit in legal and professional standards of medical practice. Furthermore, it opposes hedonism and utilitarianism as bases for medical ethics.

Unlike these secular faiths, theocentric faith interprets the principle of the sanctity of life in relation to God as Creator, Lord, and Redeemer. It emphasizes the importance of the physical-biological dimension of a person's existence and the significance of the sociocultural matrix of his life. Nevertheless, these factors are not the source of life or of its sacredness. As a basic principle of Christian ethics, the sanctity of life affirms a person's right to live and safeguards this right against the conflicting values and claims of society. At the same time, this principle does not translate this right into a necessity. From the perspective of Christian faith, the sanctity of life is not destroyed by death, for death is understood as a process of life as it is created and sustained by God.

Respect for the sanctity of life constitutes the basis of many of the professional and legal standards of medical practice. The way in which this principle is interpreted from the point of view of theocentric faith has important implications for medical care. It also provides a basis for transforming the moral norms governing the practice of medicine. Because there is a general consensus within our society concerning this principle, Christian medical ethics finds areas of basic agreement with regard to its meaning in terms of responsible medical care. Where there are differences of opinion, as is the case with the question of euthanasia, Christian ethics seeks to foster a more profound consensus on the basis of its understanding of the meaning of the sanctity of life. Agreement on this level would make it possible to transform traditional standards of terminal medical care in keeping with the requirements of responsible medical practice.

Guidelines for Legal Norms

Innovations in medical science and technology often create complex moral dilemmas. This in turn makes it especially difficult to change laws that have to do with the practice of medicine. Questions that arise in the context of terminal medical care cannot be answered solely on the basis of the rules of the past or on the basis of prevailing medical and legal practices. Although these factors should be taken into account, new standards with which to resolve these problems must be derived from the fundamental ethical values underlying our moral and legal norms. Thus it is necessary to reexamine the meaning of these principles in the light of the new options that are available in medical practice.

1. Updating the Concept of Death

Death is a social as well as a private event and a legal as well as a medical question. Nevertheless, legal and medical definitions of death no longer coincide. The legal concept of death is quite general and does not take into account the changes that have occurred in the practice of medicine. Courts, for example, usually interpret death simply as a state that is the antithesis of life and as an event that takes place at a specific point in time when vital functions cease and can no longer be revived.[22] From this perspective, a patient is considered to be alive as long as any heartbeat and respiration can be perceived with or without instruments, regardless of how these signs of life are maintained.[23]

The medical concept of death is much more complex, for doctors understand death as a dynamic process rather than as a single event. Their distinction between *organismic* or *clinical* death (the loss of vital functions, which is sometimes reversible) and *organic* or *medical* death (the death of all systems, which is final) should be taken seriously in efforts to update legal definitions of death. This is especially important because of the possibility of prolonging the signs of life long after the loss of vital functions is permanent. The ability to save lives by means of organ transplants also makes it necessary to have a legal definition of death that distinguishes between organismic and organic death.

Irreversible cerebral failure has been suggested as an indication of death. In fact, courts and legislatures are increasingly taking this medical concept of death into account. Several states—for example, Kansas, Maryland, and California—have already enacted legislative changes to include brain death in addition to permanent cessation of respiration and circulation in legal definitions of death. This should not be interpreted as an expression of a sophisticated vitalism. When the principle of the sanctity of life is understood in human terms rather than with reference to biological existence alone, brain death seems to be an appropriate basis for a legal as well as a medical definition of death. As Dr. Hamlin points out, "the sanctity of life is not generated by cardiac signs of its presence or absence when the brain has already died."[24]

The life of the brain is an indispensable dimension of human existence. Respect for a patient's life and responsibility to him as a person require that every reasonable means to prolong life be offered as long as his brain survives. To amend conventional legal concepts of death to include neurological death would allow doctors to introduce artificial life-sustaining measures without the obligation to continue their use indefinitely. Clinical tests that determine when irreversible brain death has occurred would provide criteria for declaring a patient dead. This would also enable doctors to "save" essential organs from donors in order to prolong the lives of other patients. In the last analysis, however, the greatest care should be taken and safeguards should be established to assure that death is declared because a patient is dead and not because he is a donor.

2. Sanctions for Passive Euthanasia

The law seeks to protect each person's right to live. Furthermore, it does not recognize mercy as a legitimate excuse or justification for homicide. Thus there are no sanctions for active or passive euthanasia against a patient's will (involuntary euthanasia). Theoretically, in the absence of mitigating circumstances, involuntary euthanasia is a felony. When it is committed intentionally, as in the case of a deliberate overdose of a lethal narcotic, it reflects a willful disregard for the victim's right to life. Because of the sanctity of each patient as a person, this should be defined as murder and punished accordingly. When involuntary euthanasia is performed unintentionally, as in the case of an accidental overdose of a lethal narcotic, it should be treated as involuntary manslaughter or negligent homicide. The fact that the patient's life is not taken deliberately provides the rationale for the lesser charge. At the same time, however, this provision would serve to protect patients from negligence.

In addition to the right to life, the law guarantees personal autonomy so long as an individual's exercise of his freedom does not interfere with the rights or the freedom of others. The right of each person to make his own medical decisions, for example, is protected even when his life is at stake. Except in emergencies, the doctor's right to act in order to help or to save is strictly limited by the patient's right to refuse his services. This, however, enables the doctor to practice passive euthanasia, for he must refrain from efforts to prolong life when a patient refuses to consent. At the same time, he may yield to the patient's request and direct his attention fully toward relieving suffering and preparing both the patient and the family for death.

When it is understood in personal terms, respect for the sanctity of life involves a recognition of the freedom of each individual. In the practice of medicine, this respect requires that the patient be permitted to determine the course of the medical treatment that he receives. There are circumstances, however, in which a person's wishes cannot be known or must be disregarded. When, for example, a patient is not competent to give or to withhold consent, the doctor is not bound by his refusal of necessary treatment. In such cases, medical decisions are usually made by the doctor in conjunction with the patient's closest relative or his guardian. Nevertheless, the principle that each person is the master of his own body is important and should be preserved in the law. On this basis, there should be legislation to sanction passive euthanasia.

3. Justification for Active Euthanasia

A good case can also be made for sanctioning active euthanasia. There are occasions in which the patient's desire to live is undermined by suffering and disease even though his death is not imminent. Out of regard for his freedom over his own life and death, he should be allowed to choose to die by active means rather than to endure hopeless suffering. Furthermore, those who assist him should be freed from liability for his death. The moral distinction between killing out of mercy and killing from malice constitutes a basis for distinguishing euthanasia from murder. The extreme leniency that prosecutors, judges, and jurors have accorded those who have killed out of mercy indicates their tacit approval, or at least their acceptance, of active euthanasia in direct opposition to existing laws.

The dangers of errors and abuse are often cited in arguments against legalizing active euthanasia. In the absence of adequate safeguards, such problems would inevitably occur if either a patient's doctor or his family were permitted to practice mercy killing at his request. In some cases, there would be mistakes because of their emotional involvement in the patient's suffering. In other cases, unscrupulous doctors and relatives would take advantage of the law by disguising homicide motivated by greed or enmity as an act of mercy. Because of these dangers, the right to die and the freedom to terminate a patient's life at his request must be limited by conditions that are necessary to secure the right of others to life.

For the good of "society," that is, to protect every patient's right to live, the law should treat active euthanasia as a felony except under legally sanctioned circumstances. Because of the inviolability of each patient, involuntary euthanasia should be defined as murder in the absence of mitigating circumstances. The moral distinction between mercy and malice as motives for homicide provides a basis for declaring unauthorized mercy killing a second-degree felony when it is not in opposition to a patient's desire to live. Although voluntary euthanasia *per se* is not morally reprehensible, it should be defined as manslaughter when it is not performed in accordance with legally prescribed procedures which are designed to preserve both the right to live and the right to die.

Finally, in order to be humane and relevant to the problems that arise in terminal medical care, the law should prescribe conditions that would justify active measures to terminate life and would provide the maximum safety for each patient. Because of the tragic consequences of radical birth defects, advanced degenerative diseases, and hopeless suffering in terminal illness at any age, the right to die should not be denied. Sometimes it is impossible to distinguish between allowing and causing a patient to die on the basis of an objective analysis of the doctor's action. In either case, both the end and the means are the same. After emergency and ameliorative treatment has begun, the doctor must often perform overt acts in order to allow the patient to die. There also may be occasions in which he should be permitted to take direct action to terminate the life of a patient who is not yet on the verge of death.

Legislation establishing legal justifications for

voluntary active euthanasia must provide ample safeguards against errors and abuse. Nevertheless, its requirements should not be so complex and restrictive that it would increase the suffering and anxiety of the dying patient and his family. For their safety and for the protection of the doctor, these provisions should be included:

a. Euthanasia may be justifiable when it is performed at the request of a competent patient whose condition is terminal. When he is permanently incapable of making a request or giving consent to die, his nearest relative or legal guardian may be allowed to act in his behalf. Euthanasia is never justifiable against a person's wishes.

b. A committee composed of at least five members should evaluate his condition, verify his desire to die, and decide whether active or passive euthanasia is warranted. The committee should include at least two physicians in fields related to the patient's condition, a representative of the hospital administration, a counselor or psychiatrist, and a chaplain, minister, priest, or rabbi. Under no condition should a member of the committee be connected with any case for which the patient might be considered as a donor of organs for transplants.

c. The patient's condition, his request for euthanasia (or that of his nearest relative when the patient is incompetent to give informed consent), and the decision of the committee should be indicated on appropriate legal forms, which should be properly attested and recorded.

d. Euthanasia in accordance with these conditions should be defined as a legitimate cause of death and should in no way affect insurance benefits or survivorship rights.

Legislation to permit voluntary active euthanasia in accordance with these regulations would serve to protect each patient's right to die and to preserve his right to live. In the absence of such safeguards, he is dependent upon the moral sensitivity and integrity of those who are responsible for his care. Until there are legal justifications and procedures to allow doctors to practice euthanasia, those who do act out of mercy must in turn rely on the mercy of others who judge their actions. The same moral insight and concern is required in order to reform professional and legal standards of medical practice to make them more responsive to patients who are suffering and dying. This challenge must be met by men and women of faith and goodwill in order to exercise wisely and humanely the power over life and death created by contemporary science and technology.

Editor's Notes

1. Hannibal Hamlin, "Life or Death by EEG," *Journal of the American Medical Association*, Vol. CXC (Oct. 12, 1964), p. 113.
2. Denton Cooley, M.D., "Summit for the Heart," *Time*, Vol. XCII (July 26, 1968), p. 49.
3. Henry K. Beecher, "A Definition of Irreversible Coma," *Journal of the American Medical Association*, Vol. CCV (Aug. 5, 1968), pp. 85–88.
4. *Ibid.*, pp. 85–88.
5. Paul Ramsey, *The Patient as Person* (New Haven, Conn.: Yale University Press), 1970, p. 161.
6. Thomas McKeown, "The Community's Responsibilities to the Malformed Child," Symposium on the Cost of Life, *Proceedings of the Royal Society of Medicine*, Vol. LX (Nov. 1967), p. 220.
7. *Ibid.*, p. 220.
8. R. S. Illingsworth and Cynthia M. Illingsworth, "Thou Shalt Not Kill, Should Thou Strive to Keep Alive?" *Clinical Pediatrics*, Vol. IV (May 1965), p. 307.
9. *Ibid.*, p. 308.
10. Denis Hill, "Economic and Ethical Considerations Arising from the Care of the Defective Child and the Very Old," Symposium on the Cost of Life, *Proceedings of the Royal Society of Medicine*, Vol. LX (November, 1967), p. 1233.
11. "Seven Authorities Discuss the Thalidomide Tragedy," *Illinois Medical Journal*, Vol. CXXII (September, 1962), p. 265.
12. McKeown, "The Community's Responsibilities to the Malformed Child," *loc. cit.*, pp. 1222–1223.
13. *Ibid.*, p. 1221.
14. *Ibid.*, p. 1222.
15. Illingworth and Illingworth, "Thou Shalt Not Kill . . . ," *loc. cit.*, p. 306.
16. *Ibid.*
17. Walter C. Alvarez, "Care of the Dying," *Journal of the American Medical Association*, Vol. CL (Sept. 13, 1952), p. 88.
18. *Ibid.*
19. *Ibid.*, p. 91.

20. Daniel Callahan, "The Sanctity of Life," in Donald R. Cutler (ed.), *The Religious Situation: 1969* (Beacon Press, 1969), pp. 300–301.

21. *Ibid.*, p. 314.

22. Marshall Houts and Irwin H. Houts (eds.), *Court Room Medicine,* Vol. III (Matthew Bender & Company, Inc., 1967), p. 1–14.

23. M. Martin Halley and William F. Harvey, "Medical vs. Legal Definitions of Death," *Journal of the American Medical Association,* Vol. CCIV (May 6, 1968), p. 424.

24. Hamlin, "Life or Death by EEG," *loc. cit.,* pp. 112–113.

For Further Reading

Behnke, J. A., and Bok, S. *The Dilemmas of Euthanasia.* New York: Doubleday, Anchor Books, 1975.

Brim, O., et al., eds. *The Dying Patient.* New York: Russel Sage Foundation, 1970.

Caughill, R. E., ed. *The Dying Patient: A Supportive Approach.* Boston: Little, Brown, 1976.

Choron, J. *Death and Western Thought.* New York: Collier Books, 1968.

Cooper, I. S. *Hard to Leave When the Music's Playing.* New York: Norton, 1977.

Cutler, D. R., ed. *Updating Life and Death.* Boston: Beacon Press, 1968.

Downing, A. B., ed. *Euthanasia and the Right to Die.* New York: Humanities Press, 1969.

Gould, J., and Craigmyle, L., eds. *Your Death Warrant?* New York: Arlington House, 1971.

Kohl, M. ed. *Beneficent Euthanasia.* Buffalo, N.Y.: Prometheus Books, 1975.

Kübler-Ross, E. *On Death and Dying.* New York: Macmillan, 1969.

Kübler-Ross, E. *Questions and Answers on Death and Dying.* New York: Macmillan, 1974.

Russel, O. R. *Freedom to Die: Moral and Legal Aspects of Euthanasia.* New York: Human Sciences Press, 1975; Dell, 1976.

Williams, G. *The Sanctity of Life and the Criminal Law.* New York: Alfred A. Knopf, 1957.

Williams, R. H., ed. *To Live and to Die: When, Why and How.* New York: Springer, 1974.

Winter, A., ed. *The Moment of Death: A Symposium.* Springfield, Ill.: Charles C Thomas, 1969.

13
HUMAN EXPERIMENTATION

The object of the experiment was clear enough: to find out which, if either, of two antibiotics crossed the placental barrier into the fetal tissues and which did so more effectively. Such knowledge would prove extremely useful in determining how to provide treatment and protection for a fetus still in its mother's womb and at risk of infections because the mother was allergic to penicillin. As their subjects, researchers chose a number of living fetuses scheduled for abortions. Permission was obtained from the pregnant women, and the experiments were conducted. In due time the findings were written up in a leading medical journal. But although the experiment had been concluded, the episode had only just begun.

Not long after the findings were published, five physicians involved in the research were arrested under a centuries-old grave-robbing statute and charged with "violation of sepulcher." The medical community was furious. "Horrendous!" was how one medical newspaper portrayed the turn of events. It described the "attack upon physicians openly engaged in medical research under protocols subject to the approval of a committee of peers" as a "violation of humanity." The editorial further charged that the act would have a "chilling effect on research utilizing fetal tissue," without which medical researchers could not have developed, for example, polio vaccine.[1]

Justice officials claimed that, if researchers had obtained maternal consent to perform the legal equivalent of an autopsy on the dead, aborted fetuses, there would have been no problem. For their part, researchers insisted that they had indeed received consent. Naturally, a question arose concerning precisely what the women had consented to: If they had agreed only to take the antibiotics and allow physicians to take blood samples, then clearly they had not consented to the whole experiment, which included studying the postabortion fetal tissue. But if they had consented to the entire procedure, then the case against the researchers seemed to dissolve. Thus the entire case turned on the nature of consent.

A case similar to this one actually occurred in Boston in the mid-1970s.[2] It points up an area of increasing bioethical concern: research carried out on human subjects.

---

1. *All quotes are taken from an editorial titled, "It's a Nightmare,"* Medical Tribune, *June 5, 1974.*
2. *J. A. Robertson, "Medical Ethics in the Courtroom,"* Hastings Center Report 3 *(September 1973): 1–3.*

In a sense, medicine has always experimented, but the concept of a medical experiment is a modern phenomenon. As bioethicist Robert Veatch points out: "If the term is taken to mean a procedure systematically designed and controlled for the purpose of gaining information instead of or in addition to curing a particular patient . . . [then] the clearly differentiated concept of the medical experiment did not emerge until the nineteenth century."[3] In any event, there is no specific mention of the ethical problems of the researcher in any of the classical codes of conduct formulated prior to the nineteenth century.

Today we recognize that many complex issues cluster around the subject of human experimentation. In recent years the subject's right to informed consent has captured most interest, but there are other issues about human experimentation that are of equal importance. For example, there is the question about the conditions under which human experimentation is acceptable, which has implications in the area of adequate health care. Also there are very serious questions involving certain groups of subjects, such as children, prisoners, ward patients, and—as we have just seen—fetuses. Inevitably, concerns about autonomy and informed consent arise when any member of these groups is a research subject. There are also legitimate moral inquiries to be made about the design of the experiment, concerning fairness and injury. In addition, there is the matter of controlling the actual conduct of the experiment, with particular regard to protecting human subjects. In recent years, this responsibility has fallen to peer review committees. Whether or not the use of such committees is the best way to ensure subject rights, and what unique problems peer committes raise for HCPs in research fields, are important aspects of human experimentation. And as if these issues were not enough, there are additional concerns that gather around the duties to obtain consent, to continue an experiment, and to repair the damage that results from an experiment's undesirable side effects. Clearly, then, the issues surrounding human experimentation are many, serious, and complex.

Values and Priorities

Before getting into the issue of human experimentation, try responding to the following questions, which are intended to help you pinpoint your own values.

1. Do you think the researchers acted immorally in the preceding case?

2. Do you think the parents were wrong in volunteering their fetuses for any kind of medical experiment?

3. Would you say that the reactions of the medical community were justified?

4. Do you think the state should oversee such experiments? If so, to what extent?

---

3. *R. M. Veatch,* Case Studies in Medical Ethics *(Cambridge, Mass.: Harvard University Press, 1977), p. 266.*

5. Who do you think is the best judge of the worth of a medical experiment and of the risks involved?

6. As an HCP, would you report what you thought was an unethical medical procedure?

7. As a subordinate involved in a research project, where do you think your primary loyalties should lie—with the research team, the human subjects, the institution, or society in general?

8. What would you say is the best source of human subjects for experiments?

9. Do you think people have an obligation to participate in medical experiments?

10. Would you participate in a drug experiment designed so that neither you nor your physician would know for sure that you were receiving the drug being tested as opposed to a placebo?

11. Would you say that medical research rarely involves HCPs other than researchers?

12. Do you think humans should ever be used as subjects in medical experiments?

13. Under what conditions, if any, do you think it would be permissible to use humans who had not consented as experimental subjects?

14. Do you think it is ever right to use prisoners as experimental subjects?

15. Do you think you would ever be justified in volunteering your child for experimentation in research that did not promise to benefit him or her directly?

16. Which of the following economic groups do you think is used the most in high-risk experiments: the poor, the middle class, or the affluent? Or do you think they are used equally?

17. Do you believe it is right for a pregnant woman (or both parents) to volunteer an intended-for-abortion fetus for medical experimentation?

18. In your view, is it moral to keep a fetus intended for abortion alive in order to conduct a medical experiment?

19. Who do you think should oversee the conduct of medical experimentation?

The Climate of Research and Involvement of Subordinates

It is tempting for health care students and for those in the field whose occupational goals, interests, and duties are not directly related to research to regard human experimentation as an issue more of theoretical than practical concern. This attitude is altogether understandable, since far more HCPs are involved in primary care than in research. Nevertheless, the issues raised in this chapter should be of more than passing interest, because professionals at any level can easily get caught up in them. So, before turning to any specific research issues, let's briefly examine how the moral aspects of human experimentation can ensnare nonresearch HCPs.

At the outset, it is helpful to understand the complex psychosocial climate in which research is often begun and conducted. Generally speaking, research involves considerable rewards in prestige, position, and satisfactions for investigators. Indeed, professional reputations often depend primarily on the publication of research results. In addition, investigators, like other HCPs, are attempting to meet their own role expectations and to satisfy career interests. Often there is a wall of obstacles that investigators must surmount in order to achieve these personal and professional goals. One result is that researchers operate under a certain degree of moral constraint. Thus they too can feel acute pressure, which initiates the flurry of activity that is necessary to do the research.

Part of the activity is locating a source of funds: No money, no project. Another part is bringing various kinds of expertise to bear on different aspects of the research. Sometimes collaboration is necessary. But even if it is not, access to facilities must be gained. For example, hospital beds for research purposes may be allocated by a special peer review committee. Thus it is necessary to secure approval of the committee. Moreover, in actually conducting the research, investigators may have to coordinate the activities of a group of assistants and technicians. And of course, investigators must always locate subjects. Therefore, when investigators do research, they inevitably cast a net upon the professional waters, which ensnares a variety of HCPs in moral issues, even though their involvement in the research may never be more than peripheral.

In addition, sometimes HCPs inadvertently come across medical procedures that raise ethical concerns. Take, for example, the case of student Jane Struthers, who works in a clinical chemistry laboratory during the summer. One muggy August afternoon, while running the glucose machine, Jane notices that the lab has received half a dozen samples of amniotic fluid—the fluid surrounding a fetus—for glucose tolerance tests. Jane is puzzled because the lab usually receives about one sample a week from the clinic. She mentions this to her supervisor, Jack Fuller. Concerned by the irregularity, Jack promises to speak to the clinic director about it.

The following week Jane notices that the lab has received no samples of amniotic fluid at all from the clinic. She mentions this to Jack, who informs her that Dr. Dreyfus had again sent over half a dozen samples but that the clinic director sent them back and told the physician to stop. It seems that Dr. Dreyfus had a theory about the possible relation of glucose tolerance to some aspect of fetal health and thought she would try out her idea on a small scale before designing a formal experiment.[4]

This little episode raises several questions about the ethics of human experimentation, including the question of when an experiment becomes an experiment and who the subjects of an experiment actually are. It also shows how

---

4.   *Veatch, p. 273.*

subordinates at various levels can get involved in research. Specifically, it raises a question about the role of the medical student, lab technician, nurse, and others in subordinate positions when they discover medical procedures that raise ethical concerns.

There are other ways that HCPs in subordinate positions get enmeshed in human experimentation issues. In some research projects, a great many subordinate personnel are involved. But any morally controversial decision is usually made at higher levels, by researchers and committees. Subordinates are then expected to carry these out. Suppose, for example, that an experimental project is set up to try to rehabilitate persons who are comatose as a result of severe brain injury. Such therapeutic research clearly offers great potential benefit to patient-subjects. Naturally enough, questions arise about which patients will be admitted and how determinations will be made. The therapists who must ultimately work with the patient-subjects will be carrying out the decisions of others, but to a degree they will share the moral responsibility for those decisions. Just as important, therapists can easily feel conflict when they believe that some patients have been excluded from participation on questionable grounds. What are subordinates to do in such a situation? Where does their primary responsibility lie?

The point is that human experimentation inevitably involves more people than the investigators alone. More often than not, many other HCPs get involved. As a result, they get caught up in the moral dilemmas that the research may pose. Lacking familiarity with the moral aspects of human experimentation, these subordinates are in no position to detect the moral problems or, having detected them, to sort out various rights and responsibilities. Furthermore, without sufficient background, they are not qualified to function as patient advocates. This is a most important consideration, because (1) research is often conducted with little visibility outside the research environment and (2) those most involved, the investigators, sometimes cannot objectively evaluate the moral implications of their research or calculate the risk-benefit ratio. For these reasons, then, the moral aspects of human experimentation should be of concern to all HCPs.

Bearing this in mind, we can now turn to specific issues. To begin, let's consider some conceptual matters.

Conceptual Considerations

Before turning to the more prominent issues in human experimentation, we should consider some preliminary remarks of a conceptual nature. Specifically, it is important to understand the concepts of therapy, experimentation, and human experimentation and to distinguish between therapeutic and nontherapeutic experimentation.[5]

In biomedicine, *therapy* ordinarily refers to a set of activities intended primar-

---

5.   *T. A. Mappes and J. S. Zembaty,* Biomedical Ethics *(New York: McGraw-Hill, 1981), pp. 138–140.*

ily to relieve suffering and to restore or maintain health. Therapy takes a variety of forms, ranging from diagnosis to treatment to some preventive measures. Whatever its form, therapy is *always* intended to help a particular patient.

In contrast, *experimentation* or *research* refers to a set of scientific activities intended primarily to contribute general knowledge about the processes involved in human functioning. Notice that the primary aim is not to help a particular patient but to acquire information about human chemical, physiological, or psychological processes. Naturally, the hope is that such knowledge will ultimately prove helpful in delivering health care, but the primary aim of experimentation is scientific rather than therapeutic. When experimentation or research involves humans as subjects, it is termed *human experimentation.*

Drawing on the concepts of therapy and experimentation, researchers distinguish two kinds of projects: therapeutic and nontherapeutic, both of which are forms of human experimentation. Like all research, therapeutic research or experimentation aims to acquire general knowledge. But in *therapeutic research,* patient-subjects are also expected to benefit medically from the projects that they are participating in. For example, the first beneficiaries of the polio vaccine were the human subjects who participated in the research; the same can be said of the experiments with kidney dialysis machines, coronary bypass surgery, and organ transplants. In contrast, *nontherapeutic research* is intended solely to provide information required by researchers; patient-subjects in such projects do not stand to receive any therapeutic benefit from participating.

A good illustration of a nontherapeutic project can be seen in some of the LSD experiments that were conducted in the 1960s. In one experiment, researchers who were studying the human personality recruited over a hundred subjects through a newspaper advertisement for experimental subjects to be paid $2 per hour. Those responding to the ad were carefully screened, and a group was selected for the experiment. Before receiving the LSD, each subject was given a battery of tests, including projective tests and tests of anxiety, attitudes and values, and esthetic sensitivity and creativity. After other preparatory measures, subjects took the LSD and experienced various reactions, which were dutifully recorded and subsequently reported by researchers. Obviously such a test was nontherapeutic. In other words, no plausible argument could be made that it was conducted for the benefit of the subjects.

Although the LSD experiment was clearly nontherapeutic, the line between therapeutic and nontherapeutic is not always so easy to draw. For one thing, therapeutic experiments are never conducted solely for the patient's benefit, since the purpose of all research is to provide general knowledge. For another, therapeutic research often has a nontherapeutic component; subjects sometimes are expected to submit to tests that are strictly nontherapeutic and that may even entail risk to the patient. Similarly, nontherapeutic experimentation may indirectly benefit patient-subjects.

A good example of the latter possibility is the so-called Willowbrook experiments, conducted between 1956 and 1970 at Willowbrook State Hospital, an institution for the care of the mentally retarded. The research concerned hepatitis, which was very prevalent at Willowbrook, and other diseases such as measles, shigellosis, and parasitic and respiratory infections. In an attempt to gain a better understanding of hepatitis and possibly to develop methods of immunizing against it, researchers injected infected serum to produce hepatitis in the patient-subjects in their research unit. This experiment was decidedly nontherapeutic, but it did provide a couple of indirect benefits for the participants. One was that the patients were housed in special units that were well equipped and well staffed, where they were isolated from exposure to other infectious diseases prevalent in the institution. As a result, they were exposed to less risk while in the hepatitis unit than they would be in the normal institutional environment. A second benefit was that patient-subjects were likely to develop a subclinical infection, after which they would be immune to that particular hepatitis virus. So in this case, a nontherapeutic project had a medical payoff for patient-subjects.[6]

Although it is not always easy to distinguish between therapeutic and nontherapeutic research, the distinction is still worth making to help set guidelines for the conduct of human experimentation. From the viewpoint of ethics, the distinction also helps us focus on a basic moral question concerning nontherapeutic projects. On what grounds, if any, can we justify experimentation that puts human subjects at risk? Some would say there are none. Many more would argue that the benefits to all outweigh the risks to some. Since we will focus on this question at the end of the chapter in applying the ethical theories, we need not air it now. Suffice it here to underscore the need for clear conceptualizations in order to understand and deal with the complex moral issues involved in human experimentation.

When Human Experimentation is Acceptable

One basic moral issue in human research deals with the conditions under which it is acceptable. When is it morally permissible to use humans as subjects in experimentation?

Since World War II, at least thirty-three different guidelines and codes have been formulated in an attempt to establish the conditions that make the involvement of humans as experimental subjects morally justifiable.[7] Perhaps the best-known of these documents are the Nuremberg Code and the Declaration of Helsinki. The former, developed by the Allies after World War II, provided the standards for judging the practices of Nazis involved in human experimentation. The Declaration of Helsinki was adopted in 1964 by the Eighteenth World Medical Assembly, meeting in Finland, to be used as a guide by physicians engaged in biomedical research involving human subjects. This code was subsequently revised by the Twenty-ninth World Medical Assembly, held in Tokyo in 1975. From a consideration of these and other codes, such as the AMA's in 1966,

---

6. R. Ward et al., "Infectious Hepatitis: Studies of Its Natural History and Prevention," New England Journal of Medicine 258 (February 27, 1958): 407–416; and S. Krugman et al., "Infectious Hepatitis: Detection," New England Journal of Medicine 261 (October 8, 1959).

7. H. K. Beecher, Research and the Individual (Boston: Little, Brown, 1970), appendix A.

and the Public Health Service (PHS) requirements in 1966, five basic principles concerning human experimentation have received widespread acceptance:

1. Research subjects must have volunteered after receiving all the information necessary for their decision to be an informed one.

2. Research subjects should be allowed to withdraw at any point in the research.

3. All unnecessary risks should be eliminated in the design of the research and through prior animal experimentation.

4. Benefits either to the subjects or to society should outweigh the risks to the subjects.

5. Experiments should be conducted only by individuals qualified to conduct them.

Inasmuch as these principles speak directly to such basic patient rights as adequate care, information and truth, and self-determination, they deserve moral commendation. And yet these principles, like the formulation of the rights themselves, are so general as to leave some questions unanswered and raise others.

For example, while acknowledging subjects' right to informed consent, the principles say nothing about how informed consent applies to special groups of patient-subjects, such as children, prisoners, and the poor. Are members of these groups capable of giving consent? If so, under what conditions? Similarly, the principles do not address the unique problems involving fetal research. Is it to be assumed that the consent of the mother (or both parents) is sufficient? Even if it is, under what conditions may fetal research be conducted? What limits, if any, should govern fetal research?

The risk-benefit issue points up another inadequacy in these principles. Although they all agree that the benefits to the subject or society should outweigh the risks to the subject, they do not acknowledge that this principle can easily conflict with the principle that ensures full disclosure. As we have seen in discussions of paternalism, HCPs have a prevailing, but perhaps unexamined, prerogative to intervene on their patients' behalf without full disclosure whenever it is supposed that such intervention is in the best interests of the patients. Indeed, current policies of the U.S. Department of Health and Human Services (formerly the Department of Health, Education, and Welfare) in effect sanction incomplete disclosure:

> Where an activity involves therapy, diagnosis, or management, and a professional-patient relationship exists, it is necessary to recognize that each patient's mental and emotional condition is important . . . and that in discussing the element of risk, a certain amount of discretion must be employed consistent with full disclosure of facts necessary to any informed consent.[8]

---

8. The Institutional Guide to DHEW Policy on Protection of Human Subjects, *Department of Health, Education, and Welfare Publication No. NIH, December 1, 1972, p. 8.*

Similarly, although risk-benefit concerns in nontherapeutic research rarely raise such disclosure conflicts, they do raise troublesome questions about the weight that should be given various benefits and risks, which in part will determine subjects' consent. How much weight, for example, should be given to the benefits of scientific knowledge as opposed to personal risks? The LSD experiments mentioned earlier raise this issue. In the view of researchers, the scientific benefits resulting from the experiment likely outweighed the risks to subjects. But clearly this is a controversial value judgment that requires airing.

In short, the five principles derived from the various codes are commendable, but they nonetheless leave many questions unanswered and many issues unresolved. So a closer look at some of these is warranted, foremost among them those related to informed consent and special groups of subjects.

Informed Consent

There is little question that consent is central to the ethical conduct of experimentation with human subjects. Since the concept of informed consent has been discussed elsewhere, we need not retrace that ground. We need only recognize that although the meaning of informed consent is simple (that is, subjects have a right to decide whether they want to participate in research), applying the concept can be extremely difficult. Problems arise because, first of all, patient-subjects must be *competent*; then they must be *informed*; and finally they must *voluntarily* make the decision. Not only are these terms difficult to define in themselves, but determining whether the conditions are met in a particular case can be troublesome.

For example, competency can be next to impossible to determine when certain patient-subjects are involved. Are the mentally retarded competent to give informed consent? Are the dying and the terminally ill? What about children and the mentally ill? Even if members of these groups are capable in theory of giving informed consent, questions still remain about situational conditions that would preclude competency.

Similar concerns surround the concept of voluntariness. For example, sometimes patients agree to participate out of fear or through intimidation. This is especially likely when the researchers also happen to be the patients' primary care providers. Needing subjects for an experiment in which they are involved, HCPs may coax their patients to participate. Patients who actually have or feel themselves to have little power in their HCP relationships may agree for fear of jeopardizing future treatment should they refuse. The voluntary nature of a decision is just as questionable when patients who are desperate to be cured or relieved of suffering allow unrealistic expectations to color their decision. Of course, such a likelihood always exists when the research involves potential cures for terminal diseases. When false hope blinds objective analysis, the question arises whether consent is really voluntary. In addition, the circumstances under which consent is obtained can raise questions about its voluntary nature. For example, it is doubtful that the consent of a woman already in labor for participation in an experiment concerning a labor-inducing drug is voluntary. Had the consent been obtained months before, that would be a different matter; but obtained under conditions of stress, when little time has been given for evaluation, the consent can hardly be called voluntary.

Another set of cases deals with the "informed" part of informed consent. For example, there are prospective subjects who simply do not have the background to understand the nature and implications of an experiment as explained by researchers. This is especially true of people with little or no education, many of whom end up as experimental subjects, as we will see. In such cases, HCPs bear special obligations to go to extreme lengths to ensure understanding. Again, researchers can preclude full understanding by being biased in their disclosure. Investigators and subordinates involved in nontherapeutic research are particularly prone to biased explanations, since the subjects must convince people to participate even though they stand to gain no medical benefit from their participation. Hospital house staff can also provide something less than full disclosure while functioning in their role as agents of researchers.

Whenever informed consent is violated, serious moral questions arise. First, by definition, violations of informed consent are violations of the basic patient right to information. Second, without informed consent, patients cannot make the significant medical decisions that affect their lives with open eyes and thus cannot exercise their right to autonomy. Third, denial of informed consent can preclude participation by those who, had they been fully informed, might have volunteered. As a result, not only does the needed research suffer, but individuals lose an opportunity to participate in what might be an uplifting and rewarding experience. Fourth, insults to the principle of informed consent damage the trust that is needed in the HCP-patient relationship. Finally, breaches of the principle of informed consent ultimately pollute the atmosphere in which research is done. They place researchers, their subordinates, other HCPs, and all medical science in the untenable and hypocritical position of having to defend their commitment to the betterment of humankind through medical research, while at the same time violating a most basic human right.

Another moral issue concerning the human subjects of experimentation deals with the procedures used to select them. This issue is especially troublesome when the research involves a possible cure for a terminal disease. In such cases, participants stand to benefit enormously; by the same token, nonparticipants have much to lose. How will the selection decisions be made? Since these decisions resemble those involved in allocating scarce medical resources, such as organs for transplant, we will leave this subject for Chapter 14.

Vulnerable Subjects

Another aspect of subject selection concerns the sources of the subjects. Where will they come from? Some argue that the first call for volunteers in human experimentation should go out to the scientific community, because scientists and allied personnel understand best what is at stake in an experiment. Thus they are most qualified to give informed consent.

Maybe so, but subjects are not usually obtained this way. For one thing, there probably would not be enough scientific personnel to provide a sufficient pool of subjects. For another, the best-informed personnel probably are those who must oversee the experiment, a task made no easier by their participation in it. In any event, the scientific community rarely provides the human subjects. Subjects more often come from economically and socially disadvantaged groups, especially from groups that happen to reside in public institutions and are therefore relatively defenseless: geriatric patients, the mentally retarded, patients in public hospital wards, and so on.

The participation of such vulnerable subjects raises special moral concerns about possible exploitation. We cannot examine all the participant groups in this chapter, but we can examine some that typify the moral problems involved. We will focus on children, prisoners, and ward patients; then we will consider the problems unique to fetal research.

Children

Chapter 7 explained that in some states individuals are considered children until they reach the age of twenty-one; in others the age of majority is eighteen. In this discussion *children* will refer either to those who lack completely the capacity to consent (for example, neonates) or have, for one reason or another, only a limited capacity to understand and consent (for example, a four-year-old). The central moral issue here is whether it is morally permissible to use in human experimentation children who have either no capacity to consent or a limited capacity.

At the outset it is useful to recall the distinction between therapeutic and nontherapeutic research. The former stands to benefit the participant, the latter does not. The moral problems of involving children in therapeutic research are far less complex than when the research is nontherapeutic. Indeed, it has been argued that parents are under a special obligation to help provide their child with adequate health care, and so they are not only justified but obliged to give proxy consent for their child's participation in therapeutic research. It should be noted, however, that some would question this analysis on the grounds that therapeutic research, in contrast to therapy, is not directed solely to the benefit of the child-subjects. This aside, the more serious moral matters surround nontherapeutic research, for there parents are expected to provide proxy consent for participation in research that does not stand to benefit the child directly and may even harm it.

The general justification for including children in any form of research begins with the recognition of children as unique human beings. They are not merely little adults whose medical treatment can be extrapolated from experimentation on adults. We saw earlier that pediatric care presents special problems. HCPs cannot administer intravenous fluids to children on the basis of adult requirements; nor can they use adult standards to determine the proper administration of drugs to children. Indeed, pediatric HPCs cannot even assume that a drug will produce the same results in a newborn infant as it does in a two-year-old. Thus, the right that children have to adequate health care gives rise to a concomitant professional obligation to provide it. But the only way for HCPs to meet this obligation is to have available as much pertinent pediatric data as possible. In many instances, this information can only emanate from research that involves child-subjects.

Notice that this analysis has both consequential and nonconsequential aspects. The consequential deal with the positive benefits for children; the nonconsequential deal with the professional obligation to provide adequate pediatric health care. A second justification for using children in human experimentations is strictly nonconsequential. By this account, all members of society, including children, have minimal moral duties. One of these is to participate in nontherapeutic research. Thus, when the risks are minimal and the possible benefits are great, one has a duty to volunteer. It is concluded that parents can therefore feel morally justified in providing proxy consent for their children, and researchers can feel justified in accepting it as legitimate informed consent.

Against these positions are those claiming that proxy consent is never permissible. Moralist Paul Ramsey, for one, has argued that no human subjects should be used in nontherapeutic research without their informed consent. Since children are incapable of providing informed consent, they should never be used.[9]

It is clear, then, that several serious moral problems are embedded in this issue that researchers, and HCPs generally, must deal with. First, there is the conflict between providing adequate health care and respecting informed consent. Second, there is an equal conflict between consequential and nonconsequential perspectives, between considerations for the welfare of the group (all children) and considerations for the rights of the individual (the child-subject). Third, there is the conflict medical researchers and HCPs can feel between their role as advocates for the interests of all children and their more generalized and historical duty to safeguard the rights and interests of the individual child-subject.

Prisoners

Prisoners have long been used as subjects for human experimentation. For example, in the early part of this century, a group of prisoners in the Philippines was infected with plague bacilli. The prisoners were not asked their consent. For their participation, they received cigars and cigarettes. Such episodes were not at all uncommon in the late nineteenth and early twentieth centuries. Today, only prisoners who volunteer may participate as subjects in human experimentation. Nevertheless, the issue of informed consent remains of central concern when the subjects are prisoners.

One key aspect of informed consent as it applies to prisoners is voluntariness. Can prisoners make a voluntary decision to participate in research? If they cannot, then they cannot give informed consent and so their participation becomes morally suspect; if they can, then they are at least capable of providing

9.   P. Ramsey, The Patient as Person (New Haven, Conn.: Yale University Press, 1970), pp. 11–17.

informed consent, which means that any opposition to prisoner participation must develop along lines other than the incapacity of prisoners to consent.

Some, like author Jessica Mitford, have argued that the climate of prisons precludes voluntary consent. They argue that crowded living conditions, inadequate medical care, limited opportunities to make money, and indeterminate sentences all make voluntary consent impossible.[10] The clear implication here is that prison life is such that a decision to participate in research can only be made under duress and consequently cannot be considered voluntary. Whether that is a fact is debatable.

Others have argued that it is not so much the prison atmosphere that makes voluntary consent impossible but the very nature of total institutions. Prisons, like old-age or mental institutions, are inherently coercive. They put pressures on inmates to do what is desired and expected of them. Some prisoners may "volunteer" out of boredom, some out of fear, and some with an eye on impressing a parole board. The point is that the nature of total institutions is such that their inmates can never be free of undue influence and duress, which is necessary if they are to make a voluntary decision.[11]

Still others have raised serious questions of social justice in their objection to using prisoners as subjects in nontherapeutic research. They claim that prisoners carry a disproportionately large share of the burdens and risks involved in nontherapeutic experiments when compared with the population as a whole. In this view, society as a whole benefits by the drug testing that, according to law, must be conducted before new drugs are made available to the public. The first phase of such testing requires their use by normal subjects—individuals who do not have the medical problem related to the drug being tested. Since World War II, most such tests have been conducted in prisons on prisoners. This means that prisoners as a group carry a disproportionate share of the risks of such tests compared with any other social group. It is concluded, therefore, that this is unjust and indeed raises broader questions about the distribution of risks and benefits in society.[12]

Against these positions are those favoring the use of prisoners in research. One pro argument begins by observing that all individuals have a right to autonomy, which means in part that they should be allowed to volunteer as experimental subjects in nontherapeutic research. To deny prisoners the opportunity to volunteer, then, undercuts their basic right to autonomy. Opponents are quick to point out that in their view the right to autonomy belongs only to free citizens. Having been sentenced for crimes, prisoners have lost the right to volunteer. This was essentially the position taken by the House Delegates of the AMA in 1952 when they expressed disapproval of using people convicted of murder, rape, arson, kidnapping, treason, and other heinous crimes as research subjects.

---

10.   J. Mitford, Kind and Usual Punishment: The Prison Business *(New York: Knopf, 1972)*.

11.   C. Cohen, *"Medical Experimentation on Prisoners,"* Perspectives in Biology and Medicine 21 *(Spring 1978): 357–372.*

12.   *Mappes and Zembaty, p. 144.*

Of course, the most common argument for using prisoners is the utilitarian one that points out the potential benefits to society. This view holds that, since more is to be gained by using prisoners than by not using them, it is right to employ them as research subjects.

Ward Patients

One of the most plentiful and frequently used sources of research subjects are the clinics and wards of our hospitals.[13] What makes this fact of special significance is that ward patients are generally the poor, the powerless, and the uneducated. They are at an additional disadvantage because their physicians on the house staff generally are constrained to some extent to act as agents for medical school faculty in locating suitable research subjects.[14] (Recall from Chapter 5 the conflicting obligations that arise from the incompatible roles that HCPs in institutions are required to play.) Thus the potential for the abuse of these most vulnerable of patients is a clear and present moral concern in nontherapeutic research.

The fact is that studies involving the highest level of risks relative to benefits are likely to use ward patients.[15] This unequal distribution of risks and benefits cannot be explained by convenience or accessibility. Indeed, students of this problem, such as Henry Beecher and especially Bradford Gray, make a strong case for the claim that a more informed group would not agree to the high-risk experiments in which so many ward patients end up.[16] In other words, research with unfavorable risk-benefit ratios seem most likely to involve unknowing subjects, a large number of whom are ward patients—that is, the poor and uneducated.

Assuming the accuracy of these accounts, two intertwined moral issues emerge. One involves the equity of the distribution of risks and benefits in society; the other involves informed consent. Apparently they cannot be separated in the use of ward patients as research subjects. If it is true that the poor and least educated groups in society are being used in high-risk research, how can this be justified, especially when these same groups are least likely to receive good medical care—that is, to receive the benefits of past research? As Gray points out, responsibility for this risk-benefit pattern is difficult to pinpoint, because it is the result of many investigators' selection of subjects. More important, he observes that, to the extent that the pattern results from the organization and financing of health care in this country, procedural requirements are not a useful way of approaching the problem. Rather, since the pattern seems to be

---

13. B. Barber, J. J. Lally, J. Loughlin Makarushka, and D. Sullivan, Research on Human Subjects: Problems of Social Control in Medical Experimentation (New York: Russell Sage Foundation, 1973), pp. 53–57.

14. B. Gray, Human Subjects in Medical Experimentation (Melbourne, Fla.: Krieger Publishing, 1981), p. 242; S. J. Miller, Prescription for Leadership: Training for the Medical Elite (Chicago: Aldine, 1950), pp. 145–154; and Barber, Lally, Loughlin Makarushka, and Sullivan, pp. 104–107.

15. Barber, Lally, Loughlin Makarushka, and Sullivan, p. 54.

16. H. Beecher, "Ethics and Clinical Research," New England Journal of Medicine 274 (1966): 1354–1360; Gray, p. 242.

based on the ease of evading informed consent requirements with subjects who are poor and uneducated, informed consent procedures must be made more effective.[17]

Given the correctness of this analysis, HCPs in public institutions would seem to be under special obligation to advocate procedures that are geared to protect their patients' right to informed consent. This is not to minimize or ignore conflicting obligations HCPs may have that arise from other professional relationships—for example, as members of a research team or as employees of a hospital with affiliations to a research center. Nevertheless, at the very least HCPs should surely scrutinize all their obligations with an eye to priorities, being especially mindful of their obligations to their most vulnerable patients.

Fetuses

In the discussion that follows, a fetus is a biologically human organism in the uterus, from the moment of fertilization to the time of physical separation from the mother, or a human organism outside the uterus from the moment of fertilization to the time of clear viability.[18] In discussing experiments on fetuses, it is important to distinguish between fetuses that are scheduled to go to term and those that are scheduled for abortion or have already been aborted. From the viewpoint of ethics, experimentation on term fetuses would raise essentially the same concerns as experimentation on children. In the case of aborted fetuses, these concerns grow more complex, and additional ones intrude. Thus this section will confine its observations to live human fetuses intended for abortion.

There are three main ways of involving live human fetuses in research. In one kind of research, the subject is the fetus *in utero*, most frequently a fetus scheduled for abortion. In another, the subject is the still-living but unviable product of a spontaneous or induced abortion, after it is disconnected from the placenta. A third kind of research falls between these two. In cases of abortion by hysterotomy (a procedure similar to cesarean section), the fetus may be removed from the uterus but the placenta left in place; the fetus is subjected to experiments before the umbilical cord is cut.[19]

Each of these techniques serves important, even unique, functions in medical research. For example, research on the fetus *in utero* may consist of tests to determine whether substances pass over the placental barrier to the fetus. As the opening case indicated, such research is vital for ensuring the protection and health of the fetus. Similarly, research using the still-living but unviable fetus separated from its mother is essential in developing ways of saving immature fetuses or improving incubators for immature or premature neonates. Finally, experiments on the fetus still connected with the mother can show whether a

---

17. *Gray, pp. 253–254.*

18. *T. L. Beauchamp and L. R. Walters,* Contemporary Issues in Bioethics *(Belmont, Calif.: Dickenson, 1978), p. 450.*

19. *P. Ramsey,* The Ethics of Fetal Research *(New Haven, Conn.: Yale University Press, 1975), pp. ix–xxii.*

substance passes from the maternal into the fetal circulation. From the scientific point of view, compelling arguments can be made for using fetuses as subjects in research. Nevertheless, moral concerns are present.

One moral issue involves consent. As suggested earlier, the ethics of proxy consent for nontherapeutic research involving children always raises legitimate moral questions. These questions are no less pressing when the subjects are fetuses, especially ones scheduled for abortion. In such cases, the tenuous parental relationship with the organism casts serious doubt on the moral legitimacy of proxy consent. After all, the parents are assuming no long-term responsibility for the fetus or the possible child. What happens, for example, when an experiment begun on a fetus *in utero* causes serious damage but not death? Who is responsible for any subsequent decisions that must be made concerning disposition of the fetus? Such a possibility provokes an inquiry into the grounds on which biological parents can give proxy consent for the use of the fetus as a research subject. When term fetuses are involved, the issue of proxy consent does not emerge in the same way, because the parents presumably assume extended responsibility for the fetus or child. Of course, there remains the question of whether the parents are justified in subjecting the fetus to any risk.

Experiments on fetuses scheduled for abortion raise another concern. Suppose the parents volunteer their fetus as a research subject. Shortly after experiments are begun, the parents change their mind about the abortion; they now want to carry the fetus to term. Should they be permitted to do this—even if the fetus has been irreparably damaged? And if a severely damaged child is born, who bears the responsibility for trying to repair the damage? Paramount here are issues of parental rights to autonomy and of potential injury to the child, as well as issues of subsequent care of the child.

In addition to these matters, discussions of fetal research inevitably raise questions about the disposition of a fetus that is either deliberately or spontaneously aborted. Such concerns may better be discussed along with matters dealing with the disposition of corpses, but they can be mentioned here because they relate directly to experimentation. For example, it is possible that fetal tissue could be commercially grown and distributed to biological supply companies, much the way that a variety of animal tissues are.[20] Should this be permitted? Should there be any moral restraints on the disposition of fetal remains? Additional disposition questions concern the determination of precisely when the fetus may be considered dead, the use of placental tissue, and the use of fetal organs for transplant.

Of course, underlying all these issues are the ones concerning the ontological status of the fetus, a topic addressed in Chapter 11. If the fetus is given full person status from the moment of conception, then presumably it has all the rights to protection in matters of human experimentation that any person does. If the fetus is not a person, then questions of fetal experimentation are not so urgent. Nevertheless, if the fetus is to be considered a potential person, then issues reemerge about the dignity and respect that it deserves.

---

20. R. Munson, Intervention and Reflection *(Belmont, Calif.: Wadsworth, 1979), p. 232.*

Even from these sketchy remarks, it should be apparent that the area of fetal research presents a tangle of moral questions. Again, these are issues that involve not only researchers but many other HCPs as well, especially those in obstetrics, who are nearest the sources of the experimental subjects. It is not uncommon for pressure to be put on these HCPs to acquire the raw material for proposed research or for these HCPs to be instructed to help conduct procedures and dispose of fetal tissue in ways that they find morally objectionable. At such times a close inspection of one's conflicting obligations is called for, which would include an examination of the broader issues involved in fetal research.

Additional Concerns

Before concluding this brief overview of the moral aspects of human experimentation, we should consider a couple of additional areas of concern: experimental design and control of experimental conduct.

Experimental Design

Whenever research is done, some procedural blueprint or design is formulated within which the research is to be conducted. Obviously the design should be valid and reliable, otherwise it squanders resources and breeds unfounded conclusions, which may ultimately victimize those who act on them. Also, the experiment or research should be so designed as to minimize subject risk. Apart from these general moral concerns, there are more specific ones that deal with the experimental methods selected. One method of particular moral significance is the double-blind experiment, which is often used in testing drugs.

In double-blind methodology, neither investigators nor subjects know which drug is being administered. Thus the chances of biased results are greatly minimized. At the same time, a double-blind setup raises moral concerns. For one thing, some of the subjects do not receive the drug that is being tested. Ordinarily this is irrelevant in the first phase of testing, but it is significant in later phases, in which subjects are used who might benefit from the new drug. The fact that some of these subjects will be receiving a placebo or inert drug means that their role in the experiment is essentially nontherapeutic, as opposed to the therapeutic roles of those actually receiving the drug. Although those receiving placebos do not run the risks that might accompany the drug, they do risk being denied what may prove to be a life-saving treatment.

A good example of the potential injury to subjects in a double-blind test can be seen in the development of polio vaccines in 1960. Initial research indicated that the vaccine was extremely effective in preventing polio. To confirm their expectations, researchers injected 30,000 children with a substance known to be useless in the prevention of polio, a placebo injection. Based on statistics, investigators knew that a certain number of these children would contract polio and die from it. In the meantime, those injected with the real vaccine were being protected from the ravages of the disease.

The double-blind methodology, then, raises a couple of important moral questions. One concerns the potential injury to some subjects. The injury can result from being denied the benefit of the drug being tested or from receiving medication with intended undesirable side effects, which can occur when researchers want to simulate the effects of the real drug in order to maintain the test's integrity. Double-blind testing also raises a question about informed consent. Although the idea behind double-blind testing seems understandable, researchers and their subordinates frequently fail to explain it to subjects in terms that the subjects can fully understand. In such cases, the consent that subjects give cannot be considered informed.

In addition, moral problems stemming from a conflict in professional roles arise for physician-researchers. As physicians, they are obliged to provide their patients with adequate care. But as researchers conducting a double-blind experiment, they are administering a useless, perhaps even a harmful, treatment to their patients.

The dilemma facing researchers in double-blind experimentation has another facet. Researchers are expected to uphold the canons of scientific experimentation to the highest degree, in order to ensure the most reliable experimental results, thereby precluding the introduction of harmful therapies into medical practice. But they are also expected to obtain informed consent, which normally would entail telling subjects exactly what protocol they would be subjected to. How can experimenters meet both these requirements?

There are other concerns that are indirectly related to design. One deals with the termination of the experiment. Specifically, do researchers have a special obligation to continue treating a subject once the treatment appears successful? Or does the obligation cease at the experiment's predetermined point of conclusion?

Another indirect concern involves responsibility for the harmful consequences of an experiment. Who should bear the responsibility of caring for a patient-subject injured in an experiment—an insurance company, the research director, the local physician, the drug company? If the subject happens to be paid for participating, does the locus of responsibility shift?

Such concerns and all the others mentioned in this chapter point inexorably to investigators' overriding obligation to exercise extreme care in experiments involving humans. They should aim at effecting the proper conditions for the inclusion of human subjects, obtaining informed consent, protecting the rights of special patient groups, and ensuring proper experimental design. Indeed, these obligations should not fall exclusively to investigators or health care personnel but to society as well, since profound questions of social justice are often involved. Given the correctness of these suppositions, a most important consideration emerges: how best to exercise control over experiments involving human subjects.

Control for the Conduct of Experimentation

Without effective control of the conduct of experimentation, the chances of ensuring the rights of subjects and of others are slim. In other words, in the absence of effective control, the potential for abuse to the subject, the scientific community, and society loom large. Because important questions of injury and justice are at stake, control of the conduct of experimentation is a legitimate moral issue. Indeed, it is a fundamental concern.

In addressing the issue of control, the U.S. Public Health Service (PHS) in 1966 passed a resolution that requires medical research institutions to establish peer review committees to oversee the conduct of human experimentation. These committees are to ensure that (1) the potential benefit of any human experiment outweighs the risks and (2) the subjects are fully informed of the risks they run in participating. Initially, the resolution applied only to PHS-funded research, but since the mid-1960s it has been so broadened that now peer committees commonly review any research at their institutions that involves human subjects, regardless of funding. Indeed, in 1971 the U.S. Department of Health, Education, and Welfare made peer review committees an essential part of all research that it underwrites, including social science research. (Of course, HCPs will know that the peer review concept has also increased in importance outside the research context. Perhaps the best example is legislation requiring the establishment of Professional Standards of Review Organizations to review the delivery of health service for which the government acts as a third party.)

Assuming that peer review committees are currently the most common way of controlling the conduct of experimentation, we must wonder how effective they are. How well do they ensure that the potential benefit of any human experiment outweighs the risks and that the subjects are fully informed of the risks they run in participating? Undoubtedly some committees are quite effective. By the same token, some committees certainly fall far short of their goals. Most probably fall somewhere in between. Actually, it is difficult to evaluate committee performance except on an individual, case-by-case basis. It is possible, however, and useful, to make some general observations about the composition and practice of these committees, on the basis of which relevant moral questions can be asked.

First, peer review committees are, by and large, not very active. Second, most committees seem to approve research projects without modification.[21] Third, even when proposals are modified, there is no guarantee that the actual conduct of the research will be altered. Fourth, peer committees continue to be dominated by medical researchers. As Gray points out, a committee composed primarily of clinical investigators is qualified for certain aspects of the review function, such as making recommendations for reducing risks and for judging whether a consent form covers all relevant aspects of a project. But is it equipped to judge whether a consent form is written in language understandable to subjects?[22] Perhaps this was one concern that led the Department of Health, Education, and Welfare to state that review committees should include persons capable of assessing proposed research from the standpoint of "applicable law" and "community standards."[23] Certainly the regulation implies that risk-benefit

21. Barber, Lally, Loughlin Makarushka, and Sullivan, p. 55.

22. Gray, p. 53.

23. U.S. Department of Health, Education, and Welfare, "Protection of Human Subjects: Proposed Policy," Federal Register, 38, 184, October 1973, pp. 27882–27885.

decisions need to be shared with the larger community, which may not share the committee's biases in favor of research.

A fifth relevant observation is that peer committees inevitably serve multiple functions, which may end up competing for priority. For example, the committee's primary function is supposedly to protect the subject. In practice, however, the committee serves to legitimize the research. It is possible that the legitimization function could take priority over the protective function. The upshot may be worse for subject protection than if no committee existed. As Gray puts it, "From the standpoint of protecting subjects, we would be in worse shape if we believed in a bogus solution than if we had no solution at all; if there is a group present whose task it is to deal with the 'ethical problem,' then such problems can be given less attention by the rest of the members of the institution."[24] These five observations alone would lead one to question whether peer committees as they currently function do or can ensure subject rights.

But apart from these concerns, there is the problem of how peer review committees function in an institutional context. As noted in Chapter 5, HCPs often get caught in a web of bureaucratic relationships that entail conflicting responsibilities. This is no less true of members of peer committees. Indeed, when committee members take their jobs seriously, they often get embroiled in destructive in-house squabbles that pit the committee against research faculty. Such a turn of events can only poison professional relationships and threaten patient care.

The issue of controlling the conduct of human experimentation is therefore marbled with important moral concerns of justice and injury. At stake are patient-subject rights to informed consent and safety, as well as a cluster of professional rights and responsibilities. In the final analysis, the moral challenge shapes up as striking a balance between two pressing needs: the need to conduct research and the need to protect human subjects.

Summary

This chapter focused on the moral aspects of using humans as subjects in medical experimentation and research. After examining the climate in which research is done and the way in which research can involve various HCPs, it clarified such important concepts as research, therapy, and therapeutic as distinct from nontherapeutic experimentation. It then examined the conditions under which human participation is acceptable and inspected the central issue in research: informed consent. Of particular note were issues that surround the use of special subjects in research: children, prisoners, ward patients, and fetuses. Finally, it considered the moral aspects of experimental design, giving special consideration to double-blind methodology. It concluded by emphasizing the importance of controlling the conduct of human experimentation. In that regard, it raised important considerations about the function of peer review committees as control mechanisms.

---

24. *Gray, p. 53.*

Ethical Theories

The ethical theories will now be applied to the following question: What grounds, if any, can justify experimentation that puts human subjects at risk—that is, uses humans in nontherapeutic research? In all that follows, informed consent and an appropriate risk-benefit ratio will be assumed.

Egoism

Egoists could justify using humans as subjects of research by appealing to best long-term self-interests. From the viewpoint of researchers whose professional growth depends on conducting research, much can be argued egoistically to defend experimenting with humans. Similarly, from the view of participants who stand to benefit, a comparably compelling argument could be made. In terms of a general policy that allowed or disallowed the use of humans in research, egoists would consider the likely long-term impact on self. Assuming that the subject's interests are respected, which would preclude the victimization of self-interests, there seems little for the egoist to object to about human experimentation. Allowing such research promises great benefit to self.

Act Utilitarianism

Act utilitarians would evaluate each case of human experimentation on the basis of total benefit. Certainly if the experiment promised the greatest ratio of good, the act utilitarian would approve it. Act utilitarians would consider any policy on human experimentation within the socio-temporal context in which it was proposed, in effect converting the policy into an individual act. At a given time and place, if such a policy would be likely to yield the greatest common good, act utilitarians would support it. The fact is that proponents of human experimentation inevitably erect their case on a utilitarian foundation. Thus it is commonly argued that human experimentation permits the discovery of new diagnostic and therapeutic techniques and that such experimentation is therefore needed for sound medical procedure. In this way, act utilitarians conclude that human experimentation is not only justifiable but obligatory.

Rule Utilitarianism

Much the same can be said of the rulist view. Would a rule stating that human experimentation should not be part of medical research be likely to produce more total good than its contradiction, which would allow human experimentation at least in some cases? For reasons previously mentioned, the rulist would probably conclude that a prohibition against all human experimentation would not yield as much total happiness as permitting the use of humans in research.

The Categorical Imperative

It is tempting to focus on Kant's prohibition against using people as a means to an end and thereby to conclude that Kant would not permit the use of humans in research. But it is most important to remember that Kant's recognition of the inherent dignity and worth of human beings entails respect for their autonomy or self-determination. To prohibit all human participation in research would flout this autonomy by denying people the right to perform what is potentially a selfless, humanitarian act of the highest caliber. In addition, it would preclude an individual's acting out of recognition of a duty to participate. Furthermore, Kant could argue that each of us is the beneficiary of the advances made by past biomedical research, which would have been impossible without human participation. Since we have benefited by the participation of past research subjects, we have an obligation, out of justice, to see that the research continues. An unqualified prohibition against the use of humans in research would disallow someone's exercising autonomy based on a recognition of this duty to justice. This is not to imply that we necessarily have a duty to participate, for the duty would likely be an imperfect one. Rather, by this analysis, if individuals recognize an obligation to meet this imperfect duty, they would not be permitted to act on it if participation were prohibited. It is safe to conclude, then, that Kant's ethics could be used to justify human participation in medical research.

Prima Facie Duties

Ross's ethics could employ numerous prima facie duties to justify, even require, participation in human experimentation. One operative duty is justice. Since we have benefited by past human participation, we have an obligation out of fairness to ensure the continuance of such research. Similarly, out of gratitude, we owe it to those who have enhanced our welfare by participating in past experiments to return their generosity when we can. The duties of beneficence and self-improvement could also be used to justify participation. Accordingly, since we are in a position to enhance the physical well-being of others and ourselves—at the same time enhancing our own moral development through altruistic acts—we have a duty to participate when we can. Given that Ross's ethics could be used to justify, even oblige, participation, it follows that a public policy prohibiting participation would prevent a person from acting morally. Therefore, participation should be allowed.

The Maximin Principle

Rawls's natural duties prescribe the same moral course, inasmuch as they are identical to Ross's prima facie duties. In addition, assuming the original position, rational creatures would hardly agree to bind themselves to a rule that prohibited human participation, because they could easily be victimized by such a rule. On the other hand, allowing human participation promises great benefit for the individual.

Roman Catholicism's Version of Natural-Law Ethics

Roman Catholic ethics could also be used to justify the use of humans in medical research. Arguments similar to those used in applying Kant's, Ross's, and Rawls's ethics could be marshaled. In addition, by the principle of double effect, participants would be willing a major good while only risking harm to themselves. Assuming that the risk is minimal, the intended good would out-weigh it, and the act would be permitted. Furthermore, it could be argued that, as a member of the human family, individuals have a duty to help improve the lot of humankind when they can do so with minimal risk to themselves. Given this analysis, a policy forbidding human participation would deny individuals the right to perform acts that are morally justifiable, even obligatory. Thus the use of humans as research subjects would be permissible.

If the preceding analyses are correct, each of the theories could be used to justify the use of humans as subjects in biomedical research. Indeed, under carefully circumscribed conditions, they might even argue for an obligation to participate.

CASE PRESENTATION
No Authorization

"In my opinion, Dean Sundren hasn't gone far enough," Dr. Royce Bellinger, chairperson of the university's Human Subject Protection Committee, told committee members. "Yes, he has warned Dr. Steinman that further violations will be dealt with harshly, but I think the time for action is now."

Committee members sat silently, doubtless replaying in their minds the incidents of the past several weeks, which had resulted in a confrontation between the committee and the university's research faculty.

It had all begun when the committee learned that a university medical research team, headed by Dr. Steinman and including several other investi-gators of international reputation in medical science, was performing ex-periments on dying patients without the required committee approval. The experiments were taking place within Dr. Steinman's bone marrow transplant unit. Evidently bone marrow transplants were being performed on leukemia patients. In some cases, patients were being treated experimentally with high doses of anticancer drugs along with bone marrow transplants. But the drugs being used had not received committee approval. In other cases, patients with one form of cancer were treated with combinations of chemotherapy and bone marrow transplants that had been approved by the committee—but only for other varieties of cancer. Upon learning of these experiments, the committee set up a special investigative panel of physicians to look into the matter.

Steinman told the committee that the procedures being used were not regarded as research at all but as the best available therapy for terminal patients. "Please understand," Dr. Steinman said passionately, "that patients and their families often plead for unorthodox therapy when all else fails—even when they know that the chances for success are remote." He assured the panel that such was the case in every instance of experimentation under investigation.

But some research team members had other views. One even told the panel that, once diagnosed as terminally ill, a patient is no longer considered a human being but becomes a research subject with a disease fitted to protocol.

In the end, the committee decided that the violations were real and deliberate. They reported their findings to the dean of the medical school and recommended disciplinary action.

Dean Sundren agreed that committee regulations had been violated. So he reprimanded Steinman and the others and told them that further violations would mean an end to federal and university support for their work. With that, he pronounced the matter closed.

But Dr. Bellinger thought otherwise. He characterized the dean's action as mere wrist-slapping, whereupon Dean Sundren informed him that the authority for disciplining faculty lay not with the committee but with the administration. Indeed, the dean went so far as to imply that should the committee insist on usurping this authority, he would press for Dr. Bellinger's removal as chairperson in order to prevent further deterioration in relations between the research faculty and the committee.

Dr. Bellinger wouldn't cave in. In fact, as he now made his case before the committee, he felt more convinced than ever that the committee had to force the issue if it was ever to have more than a cosmetic function. Some on the committee agreed with Bellinger. Others weren't sure. But each knew that the moment of truth was at hand; a decision had to be made.[25]

Questions for Analysis

1. Indicate how a conceptual issue has led Dr. Steinman into this situation.

2. Do you think that peer review committees can operate effectively without punitive powers?

3. Indicate the conflicting obligations that Dr. Steinman feels as a physician and a researcher.

4. Do you think any rights of the patients were violated?

5. Which parties in this situation have a claim on Dean Sundren, and what is the nature of the claim? Where do you think Sundren's primary duty lies?

6. Do you think dying patients have a right to experimental therapy and that their physicians have an obligation to provide it?

7. Do you think the rights of dying patients must be viewed differently from those of other patient groups?

8. Suppose that Steinman had been reasonably sure the committee would not have approved the experiment. What bearing, if any, would that have on the morality of his action?

9. Evaluate Steinman's behavior in the context of each of the ethical theories.

10. Assume you were a committee member. Make a list of all the things you would consider in deciding how to vote; then order the items by priority. What would be your recommendation?

---

25. *Based on an actual case reported in P. Jacobs, "Furor over UCLA Tests on Dying," Sacramento Bee, December 27, 1980, p. 1.*

Exercise 1

Although informed consent continues to be a central issue in medical research using human subjects, there is growing evidence that concern with patient benefit is being subordinated to claims of society's needs and interests. For example, the World Medical Association, which authored the International Code of Medical Ethics in 1949, emphasized in its "Principles for Those in Research and Experimentation" (1954) that *healthy* subjects should be "fully informed," presumably implying a different standard for "sick" subjects. In 1964 in the Declaration of Helsinki, the World Medical Association made a positive case for the necessity of research, including research that does not benefit the subject directly. It asserted the importance of laboratory experiments for furthering scientific knowledge and social benefit. The recommendations included a set of guidelines for nontherapeutic clinical research. What do you think of this trend to emphasize society's needs and interests over the patient's benefit?

Exercise 2

Reconsider your responses under Values and Priorities. Would you change or qualify any?

Departmental Dynamics and Development of Technology

Diana Crane

Today, medical technology is clearly being developed and used for intervention in special health care situations. The best-known examples are in situations involving heart transplants, kidney transplants, and cancer chemotherapy. What ethical issues do such experimental interventions raise? This question concerns sociologist Diana Crane in the following selection. But the title of the essay suggests an additional important concern: how the organizational setting of the experimentation raises ethical issues.

Specifically, Crane examines medical decision making on two cancer chemotherapy wards. Her objective is to find out if and how organizational variables affect patient care in the context of this kind of medical experimentation. In extracting the ethical issues involved in cancer chemotherapy research, Crane begins by describing the nature of the experimental procedures and their effects on patients. She then considers these issues in relation to two different settings, one associated with a prestigious university and the other with a federal hospital. By Crane's account, the social organization of these two cancer chemotherapy wards do indeed affect the quality of patient care and the extent to which patient needs and rights are observed. She concludes by urging stronger external controls on experimentation through making peer review committees responsible to a central organization that includes non-HCPs.

Crane's article is important for HCPs not only because it raises moral questions about experimentation on cancer patients but also because it shows how organizational structure affects professional behavior and raises moral concerns. In so doing, this selection underscores earlier observations (in Chapters 4 and 5) about the interplay between organizational structure and HCP conduct.

In order to understand the ethical problems which arise in the course of cancer chemotherapy, it is necessary first to describe briefly the nature of the experimental procedures and their effects upon the patients. Ethical problems are likely to occur when patients in advanced stages of cancer are given a series of drug treatments which include high doses of exceedingly toxic drugs, alone or in combination, over a period of several days or weeks. The goal of the treatment is to bring about a remission in the disease process. Remissions when they occur are generally brief in duration although occasionally patients go into remission for a year or more. A number of drugs are tried in succession. If none of them works, the whole series may be repeated.

The immediate effects of the drugs upon the patients are generally negative. Nausea, vomiting, diarrhea, loss of hair, hematuria, and gastrointestinal upset are some of the debilitating symptoms which are likely to occur. Since patients become very susceptible to infection, it is sometimes necessary to isolate them in order to avoid infection. In this case, they are allowed to see family members and staff members only when they are wearing masks. The diagnostic procedures such as bone marrows which are necessary to evaluate their progress can be exceedingly painful. Psychological stress is not uncommon.

In response to increased concern in recent years for the subjects of medical experimentation,* a number of ethical codes have been developed. A primary difficulty with all such codes is that they are of necessity very general. It is not always clear how these general directives could be translated into medical practice. Sections of these codes which have the most relevance for experimenta-

tion with cancer chemotherapy are those concerning the patient's freedom to initiate the treatment process and to halt it while it is underway, the necessity of avoiding undue physical and mental suffering for the patient, and the stipulation that the risks to the patient should not outweigh the anticipated benefits to humanity if the medical problem is solved.

Voluntary consent by the patient to participation in a medical experiment is considered to be absolutely essential. In practice, it is an ideal which is difficult to realize particularly if the potential subjects are relatively uneducated and of lower social class status than the physician. The patient's freedom to bring the experiment to an end if he has reached the physical or mental state where continuation of the experiment seems unendurable to him is likely to be transgressed in practice. The physician is reluctant to terminate an experiment before it is finished since he thereby loses his total investment of time and money in the patient as a research subject. Since most experiments with cancer chemotherapy involve considerable discomfort for the patient, the point at which additional discomfort becomes "unnecessary" is extremely difficult to define in practice. Similarly, it is difficult to weigh potential risks to life against potential benefits to humanity in cases of patients who are terminally ill.

These issues were examined in two different settings. Ward I was associated with a prestigious university on the East Coast. Ward II was located in but administratively separate from a federal hospital in the same city. The two wards will be described separately. Comparisons between them will be made in a subsequent section.

Barber (1967, p. 96) reports that in addition to the Nuremberg Code which has been the model for many subsequent codes, codes have been written by the United Nations, the World Medical Association, the United States Public Health Service, the French National Academy of Medicine, and the American Medical Association.

Social Organization and Ethical Experimentation: Ward I

Ward I was a 22-bed research ward consisting of a combination of private rooms and a few four-bed rooms. The freshly painted white walls gave the area a clean but rather sterile appearance. The staff of the ward consisted of 10 senior researchers, three fellows (one of whom was directly in charge of the ward), and two residents. The occupants of the last two positions continually changed as residents rotated through the service for a few weeks at a time. In addition there were two or three medical students assigned to the ward for short periods. The senior researchers had their offices in a wing which was adjacent to the ward but separated from it by a short hallway.

The work was divided in such a way that the treatment and research roles were kept separate. Senior men did research; house staff were in charge of the day-to-day care of the patients. Research physicians took turns at monthly stints on the ward as attending physicians, countersigning the orders on the patients' charts. This is often described in the literature on medical experimentation as an ideal situation: the physician in charge of the research on a patient is not solely responsible for evaluating its benefits to the patient.

The senior members of the ward had developed a strong ideology which supported their activities. The most important aspect of the ideology consisted of a justification of the research role. In general, the senior men tended to justify what they were doing on the grounds that it was better for such patients to be treated than not to be treated. One of these physicians said:

> We are the only people who treat these patients. They are the outcasts of modern medicine.

The self-image of these physicians is suggested by the comment of one of them that he was the only physician in the city who was doing anything about leukemia.

The group's ideology also emphasized the amount of consensus which existed among the members. They saw themselves as a tightly-knit group who shared the same views toward the value of doing research on cancer patients. Much credit was given by members of the group to their director who had been on the service for many years. One of them said:

> We were selected by one man whose philosophy we probably had before we came here or we wouldn't be here. We are unanimous.

Many comments in the interviews referred to his influence and to their admiration for him as a physician. Along with this ideology went high morale among the senior men and a strong conviction that their ethical decisions were correct, as the following comment indicates:

> On Fridays at staff meetings we go over our morality which is quite high and which is controlled by each other.

The senior staff's ideology affected other members of the staff. The nursing staff appeared to be extremely committed to the goals of the ward. Some of the nurses had been working on the ward for long periods of time, a remarkable fact when one considers that their work involved daily contact with terminally ill patients undergoing extremely unpleasant courses of treatment. Members of the house staff who served for relatively short periods on the ward appeared to be rapidly socialized. A resident who was interviewed near the beginning of his rotation was somewhat critical of the activities of the ward but two weeks later the same resident was critical of the interviewer for raising questions about the functioning of the service.

The effects of this social system upon the quality of care which the patients received are difficult to evaluate precisely without a detailed examination of a series of patients including study of their records and interviews with physicians who had been associated with their course of treatment. This approach was not attempted during this study and there is some doubt as to whether it would have been possible to obtain sufficient cooperation in order to do so. Members of the senior staff tended to be defensive about their roles and to perceive the investigator as raising unnecessary issues. Persons occupying more peripheral roles in the setting such as medical students and psychiatrists were often more critical of the activities of the ward. Using their comments combined with an analysis of the kinds of controversies which were described by the ward's staff, it is possible to infer some of the effects of the ward's social structure upon patient care.

For example, it appeared that the technical

separation of the research and treatment roles was not entirely effective. The disparity in status between researchers and practitioners was too great to permit a true separation of these roles. The practitioners who were the residents were transients on the ward, junior physicians with little experience handling these difficult types of cases, and numerically a minority (two as compared to ten). They were dependent upon the research staff for references to advance their future careers. This point was stressed by the physician who was in charge of the ward. On the one hand, he said that residents were encouraged to make their own decisions concerning patients. On the other hand, he also pointed out:

> In a university hospital everyone is very career-minded. . . . One is working for a chief of service and one does what his policy dictates. People who are not prepared to tow the line are out in private practice. . . . One's motives are dictated by one's career opportunities.

A psychiatrist who was attached to the ward brought out this theme even more sharply:

> The two residents are supposedly the responsible physicians but only theoretically. . . . Patients are not sure who their doctor is. . . . The residents are in the middle. They don't have complete autonomy. They are under more pressure in a sense because of this.

One type of conflict between researcher and practitioner which probably occurred fairly frequently was over whether or not a patient's treatment should be continued or whether he should be sent home to die. During the period when the field research was being conducted on the ward, several controversies of this kind arose. One patient, for example, had received the full series of drugs without going into remission. The research physician wanted the patient to stay on the ward and to receive the series of drugs a second time. The resident who was in charge of his case felt that the patient had experienced sufficient discomfort already, that it was unlikely that he would benefit from a repeat series of drugs, and that it would be desirable for him to go home to die where he could be with his family. Although the residents were low status members of the ward's social structure, their close involvement with the patients made it

possible for them to manipulate the other members of the ward in order to obtain the kinds of decisions they wanted. For example, in presenting cases to the entire staff at the weekly conference, the resident could stress certain aspects of the case rather than others in order to win the group's approval of the course of action that he wanted rather than that wanted by the researcher who was associated with the case. A resident described this strategy very clearly:

> There is a tremendous amount of material on each patient. I obviously select certain things from it to emphasize. Today I stressed the social aspects of the case of Mr. Brown. I could have just glossed over these aspects of the case if I had wanted to. One of the staff wanted to use a very toxic drug on Mr. Brown. I don't want it used. I think most people would support me if they knew the case. Mr. Brown wants to go home. However, by myself, I can't say no. However, I used the rounds to get the group to say that they would rather let him go. So when this happened the staff member who wanted to continue the treatment had to agree.

Psychiatric observers claimed that the psychological problems of the patients were neglected. For example, the effects of being in isolation were at times very stressful for patients. One woman became hysterical under such conditions. The house staff were not prepared to handle the psychiatric aspects of their cases. As a result, they often abandoned patients by neglecting both their emotional and physical care. The nurses attempted to deal with the patient's psychological problems but this in turn meant that the patients discussed their problems less with the physicians and as a result obtained less information about their cases. The research staff were not interested in such problems and rarely attended a weekly psychiatric conference which was run by the ward psychiatrist. The staff's neglect of the psychiatric aspects of their cases was seen most clearly when they began to undertake a new type of operation, bone marrow transplants. When it was suggested to them that psychiatric screening of patients who were being selected for these operations would be desirable, the head of the program is reported to have said that he did not want any psychiatrists "messing around" with the project. The operation proved to be very traumatic for patients; some patients broke down emotionally. One patient proved to be a psychotic.

Members of the senior staff did not appear to recognize that there might be negative aspects to their organization's activities. One senior staff member commented:

> Recently a medical student who had spent a quarter on the ward made a note on a patient's chart over at _____ Hospital. This was a 60-year-old female with lymphosarcoma. She said the woman was probably a candidate for chemotherapy but that it would be unfortunate to subject her to such a traumatic experience. She should be allowed to die. This comment took us by surprise. We wondered what sort of impression we were making on the medical students and the house staff. We've noticed that many medical students and house staff are very depressed. They think that what we are doing is wrong.

Social Organization and Ethical Experimentation: Ward II

Observations and interviews were also conducted on a second ward which was devoted to cancer chemotherapy research. This was a 45-bed ward composed of large rooms holding a dozen or more beds and a few private rooms. The walls were painted drab green and were in need of a fresh coat. The staff consisted of four senior men and 13 clinical associates, five who were spending their second year on the ward and eight who were spending their first year there. The clinical associates performed both experimental and treatment roles. Senior men were engaged primarily in research. All physicians except the director of the service had their offices inside the treatment complex.

Unlike the other group of researchers, this one lacked a strong ideology to support their activities. This was reflected in the attitude expressed by 12 out of 15 members of the department that there was no policy in the department regarding the treatment of cases. In Ward I only two out of 11 physicians expressed such an attitude. The absence of policy was seen as a result of the absence of strong leadership on the ward. The following comments illustrate this point of view:

> This service has no policy-maker. If the head of a medical service feels one way or the other and makes his opinion heard and is looked up to, these questions become easier to deal with. You can defer your decision to the leader. There is nothing like that here. Each physician is pretty much encouraged to make his own decisions about his own patients.
>
> X and Y are terrible leaders. Y has great difficulty dealing with patients or with colleagues. X is afraid of dying patients. . . . There is a need for a philosophy but they don't provide it. I wanted to bring the new doctors together at the beginning of the year to prepare them a bit for what they were going to face. There isn't a philosophy from the top here. They are too mealy-mouthed.
>
> Dr. X told us the first day that in the terminal situation we were to make our own decisions. Heroic measures were not expected if we felt them to be unnecessary. But we should work as hard as possible to save lives. Be active but don't overdo it. But this is very loosely stated so that it is interpreted differently by different people. It is a relatively non-specific thing. Dr. X is often excessively non-specific. He could have ended a controversy at rounds this morning simply by stating what should be done but he doesn't do that. Some things should be set down since the people who are making the decisions are not the ones who have the most experience. The most experienced person ought to set the policy.

The absence of an ideology to support their activities affected every aspect of the functioning of the ward. Morale was so low that it verged upon a state of demoralization. Ten out of 15 of the staff members reported that they were very or somewhat depressed when working with terminal patients. On Ward I, no physicians reported that they were very depressed. Four out of ten reported being somewhat depressed.

One source of the low morale on Ward II was a lack of conviction about the value of what they were doing. One of the younger physicians said:

> We test drugs. That's what we're here for. I think that certain drugs and certain protocols are not worth the morbidity to the patient. They cause so much discomfort, it's not worth it. I haven't decided whether I believe there is any future in this.

The combination of the lack of an ideology and their low morale affected patient care in a number of ways. There was a considerable amount of controversy concerning the appropriate treatment for particular patients. Fourteen out of 17 physicians on Ward II reported such controversies compared to eight out of 13 on Ward I. Only five out of 14 responded positively to a question concerning whether other physicians in the ward had influenced their thinking concerning the treatment of these patients. On Ward I, the comparable figure was nine out of 11. Some physicians reported that the director of the ward failed to criticize sloppy work on the part of the clinical associates. One physician said:

> Dr. X found that one of the men had not done an adequate write-up on one patient and had done no write-up on another patient. He should have given an ultimatum to that guy to do the write-ups but he didn't. He has no guts. He is wishy-washy.

Lack of leadership affected the performance of the nursing staff which was criticized by the physicians. In turn the nurses were critical of the physicians. They claimed on the one hand that the physicians avoided patients, and on the other, that some patients were overtreated. The social worker who was assigned to the ward corroborated the accusation of avoidance. He claimed that the patient charts showed that some physicians neglected their patients. At the same time, some of the doctors developed close relationships with certain patients.† The interviews suggested a tendency to identify with male patients who were approximately the same age as the physicians. Such patients became very dependent upon the physicians. Close relationships of this kind may have produced a lack of objectivity on the part of the physician with consequent effects on patient care.

It seems clear that the social organization of these two cancer chemotherapy wards affected the quality of patient care and the extent to which the patient's needs and rights were observed. In the following section, we will attempt to specify the effects of ward structure upon the ethical aspects of patient care in the two wards.

Social Organization and Ethical Codes

According to the Nuremberg Code for medical research involving human subjects, "the voluntary consent of the human subject is absolutely essential" (this and subsequent principles quoted in Barber 1967, pp. 96–97). A number of the problems involved in observing this principle are summarized in the following quotation from a physician who administered the day-to-day activities of Ward I. He was asked whether the patient's attitude influenced his decisions to continue their treatment:

> One case in point was the case of Mr. Jones who is an intelligent, well-educated attorney. He was here last summer. He knew a lot about the drugs and he asked that his life not be resuscitated or prolonged in a semi-coma. We respected his views. He didn't want to hang around indefinitely, if the medical possibilities had been exhausted. On the other hand, I think that Mr. Brown has only a hazy idea that he is not doing well and that he is in a very serious condition. I don't think he could grasp what is going on. I don't think he is capable of it. He doesn't understand his situation. He wants to go home. But the final decision has been dictated by medical policy and not by his personal feelings.

This quotation suggests that differences between social classes affect the extent to which informed consent actually takes place. It implies that scientific considerations rather than the attitude of the patient influence decisions to continue treatment in some cases. The same physician expressed general skepticism that patients could understand and benefit from the information which was given them:

> Patients who are going to have bone marrow transplants have the operation explained to them and they decide whether they want to go ahead. To me, this is ludicrous. Do they really understand what they are doing? I think that 99 percent of what they are told is lost.

A senior physician on Ward I doubted that even other physicians could really understand the value of their research. He commented:

---

†Only one senior physician was reported to have close relationships with his patients on Ward I.

I can't even explain the value of a bone marrow transplant to physicians except after several hours of talk. If you asked me if other doctors at this hospital have difficulty understanding what we do, the answer is yes. We have serious conflicts with them.

The psychiatrist attached to Ward I doubted whether the patients were actually told what alternative treatments were available to them. He said:

Very few people have refused treatment on that ward. This is probably because the alternatives are not presented to them. They are told that you will be killing yourself if you don't take these drugs, but this may not be true.

On the other hand, a member of the senior staff presented a very different picture of this process:

The decision to prolong life is a two-way decision. The patient has to consent. It is very rare to treat a patient who is incapable of making the decision. Occasionally they are mentally confused and a member of the family makes the decision.

A case on Ward II illustrated another type of problem which is involved in obtaining the patient's consent. In this case, the patient stated an opinion which conformed to cultural expectations of her social role but which she evidently did not really believe. The patient was a housewife with three children who was carrying a fourth child when the diagnosis of leukemia was made. The physicians faced the dilemma of deciding whether to abort and treat her or to wait until the baby was born and treat her afterward. The patient expressed the desire to have the child but evidently suffered from a conflict between her desire for chemotherapy and her belief that abortion was immoral. At one point when she was five months pregnant she developed abdominal pains while on the ward. She interpreted these pains to the physician in charge as labor pains and expressed satisfaction that she would have the child soon and then be able to begin chemotherapy. Absence of psychiatric assessment of her case probably prevented the physicians from discovering her true feelings toward the child and toward treatment.

There was relatively little discussion of in-formed consent on Ward II. No direct questions were asked on this topic on the interview schedule but questions about prolongation of life for experimental purposes and the patient's attitude toward treatment which elicited such discussion on the other ward did not do so on this ward.

Item 9 of the Nuremberg Code states that "During the course of an experiment, the human subject should be at liberty to bring the experiment to an end if he has reached the physical or mental state where continuation of the experiment seems to him impossible." The case of Mr. Brown on Ward I which has been referred to earlier was perhaps the most clear-cut example of this situation during the period when observations were being made on that ward although there were other similar cases. Mr. Brown had received the full cycle of drugs without going into remission and some of the physicians wished to begin the treatment again. The physician who was in charge of the ward commented:

We really face the question of whether we should go ahead when there is very little hope of remission. However, we intend to carry on. The alternative is that he goes home, gets pneumonia, bleeds and dies.

One of the medical students said:

I sense that Mr. Brown wants to go home. They want to give him an extremely toxic agent. That will make him miserable and the chances of curing him are very poor.

The physicians's ambivalence toward this principle was clearly stated by an informant who had formerly served as a clinical associate on Ward II:

I think that you have to be more stringent about keeping people alive in that situation because your data depend on how long the patient lives.

There did not appear to be specific controversies of this kind on Ward II, although it was evident that some physicians felt that patients were overtreated. Physicians on both wards were asked the following question: "It has been claimed that lives of patients upon whom new drugs are being tested are sometimes prolonged until the experiment is finished." Two out of 11 physicians on Ward I replied affirmatively to this question compared to 11 out of 14 on Ward II. These questions evoked a great deal of resistance on Ward I but not on Ward II, perhaps due to the absence of an ideology or policy concerning these matters on that ward.

There were also allusions to another practice which reflects overtreatment, that of trying drugs "just to see what will happen." This type of approach was criticized by several physicians on Ward I on the grounds that its scientific value was nil. Another set of experiments which had been conducted on Ward II involved resuscitation of every patient on the ward regardless of his case in order to study the effects of resuscitation upon these kinds of patients. If the resuscitation succeeded in reviving the patient, he experienced considerable discomfort while the gain in useful life was minimal.

However, when asked how much the patient's attitude affected their decisions to continue treatment, the majority of physicians on both wards said that it did (seven out of 11 on Ward I; nine out of 14 on Ward II). But they also indicated that they were more responsive to the patient who wanted to be treated than to the patient who did not want to be treated. The family's attitude toward these matters had no influence whatsoever on Ward I (none of the physicians reported that they were influenced by the family's attitude) while half the physicians on Ward II said that they were influenced by the family's attitude. The physician who was in charge of Ward I said:

Most families don't understand. They're not in a position to assess what is going on. In general the family doesn't influence me here.

A senior physician on Ward II said:

You can't let the family influence you. But it is necessary to handle them diplomatically, so they will go along with your decisions. I steer them to my way of thinking. They don't want to make the decisions themselves.

The Nuremberg Code also states that "the experiment should be so constructed as to avoid all unnecessary physical and mental suffering and injury." As far as mental suffering is concerned, it appeared that the physicians on both wards tended to deny or ignore the psychiatric aspects of these cases. Avoidance of physical suffering was complicated by the fact that narcotics interfere with the effects of chemotherapy. As a result the physicians on Ward I tended to use placebos in place of narcotics to alleviate pain. The senior physicians denied that the patients experienced severe pain although the house staff reported that they did. Narcotics were apparently used more frequently on Ward II. However, a former member of that ward commented:

You have to hurt people to do the research. Knowing what the therapy is like, I wouldn't choose to have it. . . . You couldn't plan a death more horrible than intensive chemotherapy.

Finally, the Nuremberg Code states that "the degree of risk to be taken should never exceed that determined by the humanitarian importance of the problem to be solved by the experiment." This issue also becomes ambiguous in this type of setting. A young physician on Ward II described such ambiguities in a case which he was handling at the time:

Mrs. Smith is 76. Recently she had a temperature of 107 degrees and she went into shock. She was given antibiotics and she came through shock. This was remarkable. . . . The question now is what to treat her with. There are two ways of thinking about this. Is this a good time to treat her? She is so badly off that we might just make things worse by treating her. I used to think we shouldn't treat people in this situation but I have changed my mind. The chances of success are low but the outcome is so certain that she is going to die that I don't think treatment will really make her worse. If she gets a remission, then her white count will go up and her normal body responses to infection will be increased. Otherwise she will just get infection or maybe bleed to death. The other possibility is not to treat her and just transfuse her. But you can't do that for long because the bleeding will continue. You would just be putting the blood in one place and it will come out at another. Her ultimate chances of survival are zero. But you can try and see what happens with various drugs. However, you can also argue that you are prolonging her misery if you get a remission because she will relapse very soon after. I can't answer that question.

In conclusion, additional insight into the nature of this experimental situation can be obtained from the fact that informants commented that they would treat patients differently if they were in private practice. The Nuremberg Code and others like it imply that patients should not be treated differently when they are experimental subjects than when they are private patients.

Ethical Treatment of Unsalvageable Patients

The data which have been presented in this [article] are based on field studies so that it is impossible to generalize the findings to other settings. Since studies have been done in similar settings, it is appropriate to compare this study with others and to attempt to develop a model of the factors that affect the behavior of physicians in such settings.

It is necessary to consider not only situations in which experimentation on human beings is taking place but all situations in which heroic attempts are being made to prolong the lives of unsalvageable patients. This type of situation is contrary to the norms of the medical profession as revealed by the present study. The unsalvageable patient is typically not treated heroically. Attempts to treat him are exceedingly stressful for both patient and physician. As a result, physicians who decide to treat such patients actively must develop a set of norms and a special organizational structure to protect themselves from demoralization and to protect the patients from exploitation. From this point of view, the problems of artificial kidney units where experimentation is not generally conducted are similar to those of cancer chemotherapy units, to the kind of situation described by Fox (1959), and to those in which an innovative operation such as a heart or kidney transplant is used to treat the patient. On the other hand, a large category of medical experiments, those which are conducted upon salvageable patients, is excluded from consideration.

What are the factors that affect the ethical treatment of unsalvageable patients undergoing experimental or heroic treatments? Fox found that the physicians on the ward which she studied typically treated their patients like close friends and almost like colleagues in some respects. Since the physicians needed to obtain the cooperation of these patients over periods of several years, they explained many of the scientific aspects of their illness to them in order to motivate them to continue to participate in the activites of the ward.

On Ward I, the possibilities for such involvements were limited because of the transience of the house staff and the sharp separation between the roles of researcher and practitioner. In addition, the researchers' strong ideology provided them with all the emotional support which they needed for their work. In fact, on Ward I there was a tendency to dehumanize the patients and to think of them as being incapable of making decisions.[‡]

The demoralized junior physicians on Ward II also became friendly with their patients occasionally but for different reasons. The absence of a strong ideology to justify their work meant that colleague relationships were strained and that it was very difficult for them to maintain their own motivation to continue their work. Under these circumstances, they sometimes sought emotional satisfaction through friendships with patients.

Kuty (1973) who made field observations in four artificial kidney units in France and Belgium concluded that the patients' rights were more adequately protected in units which were democratically organized as compared to those which were hierarchically organized. He advocates a situation somewhat like the one Fox described, where the physicians treat the patients almost like colleagues. He suggests that the physician in these circumstances becomes somewhat deprofessionalized (Kuty 1973, p. 446). Under these conditions, the patient is aware of what is happening to him and can participate in the decision-making process.

However, this type of doctor-patient relationship is ineffective if the physicians are demoralized and deprofessionalized as a result of inadequate leadership. It would appear that both strong leadership, including a clearly defined set of norms and values and effective social control, is needed as well as the willingness to treat the patient as a kind of collaborator rather than as an object.

As we discussed above, ethical codes gener-

[‡]*The one physician on Ward I who did form close relationships with his patients was involved in developing techniques for bone marrow transplant surgery. The consequences of experimental surgery and of experimental treatments for relationships between physicians and patients will be discussed shortly.*

ally specify desirable consequences of physician behavior. However it appears that such codes are of limited usefulness. It is more important to specify the organizational variables which are conducive to ethical behavior. It appears that where medical experimentation and life-prolonging technologies in general are being used, the traditional relationships between physicians and between physician and patient are inappropriate. On the one hand, a more cohesive set of relationships among the physicians involved in such units including strong leadership is necessary.§ On the other hand, a less hierarchical, less authoritarian relationship between doctor and patient in which information can be freely exchanged is required. These conditions are most likely to occur when the social value of the patient to the physician and the social visibility of the unit are both high. In the absence of these two conditions, considerably greater effort and self-discipline on the part of the physician is required to create the appropriate organizational environment for ethical treatment of unsalvageable patients.

Fox and Swazey's studies of heart surgeons suggest that the patient upon whom experimental surgery is performed is the one whose rights are most likely to be protected for several reasons. First, the social value of the patient to the surgeon is very high due to three factors: (a) the number of patients who can be treated by each physician in this fashion is relatively small; (b) important results can be obtained from a single patient (by contrast, most forms of treatment and especially drug tests require trials on extensive series of patients); (c) the amount of time invested in each patient by the physician is high.[11] Second, the social visibility of the unit is likely to be high. Other physicians and the public are likely to be watching its activities and monitoring the results. If the mortality rate is too high, other physicians will exert social control, thus bringing the work at least temporarily to an end (Swazey and Fox 1970, Fox and Swazey 1974, Chapter 6).

The variables which tend to protect the surgical patient from exploitation by his physician are less likely to occur when the patient is the subject of medical research or treatment. The social value of each patient to the investigator is likely to be low because he deals with many patients on a short-term basis and the results obtained from each one are significant only as part of a series of trials. The social visibility of such units is apt to be low, since the research is not sufficiently dramatic or innovative to attract the attention of other physicians or of the public. It is not surprising that four out of the six case studies of unethical experiments which Barber and his associates (1973, Chapter 8) present are experiments involving drug tests.

On the other hand, when ethical standards are high in experimental units involving medical treatment, such as in the setting described by Fox (1959), the two variables which we have specified are likely to be positive. On Fox's Metabolic Ward, the social value of the patient to the investigators was high since they dealt with few patients, could obtain results from a single patient, and invested extensive amounts of time in each patient. Since the researchers were associated with a very prestigious medical school, the social visibility of the unit was probably higher than it would have been had it been connected with a less prestigious medical school.

Conclusion

As we have suggested, the use of any kind of treatment to prolong the life of an unsalvageable patient is stressful for patient and physician alike, whether experimental or not (Simmons and Simmons 1970 and 1971). The high suicide rates among patients in artificial kidney units is one indication of this (Abram et al. 1971). Under such conditions, the emotional well-being of a patient requires that he be more fully informed about the hazards and potentialities of his treatment than would be necessary under normal conditions.

On the other hand, the physician tends to reject these difficult patients. In order to handle them, he needs to be part of a cohesive group of physicians which provides both a set of norms to guide the behavior of its members and exercises social control when deviations occur.

---

§ *Barber et al. (1973) suggest that in some cases the absence of strong leadership by senior men in a research unit was associated with unethical behavior by their junior colleagues.*

11. *See Fox and Swazey 1974, Chapter 4, for case studies of kidney and heart transplant surgery teams which substantiate these observations.*

In addition to the organizational variables which we have discussed, stronger external controls are needed, probably in the form of modifications of the already existing peer review mechanism, local committees of physicians which currently review medical research before it is undertaken (Barber *et al.* 1973, Chapter 9). Making these committees responsible to a central organization, as Pappworth (1968) proposes, would have the effect of making all of these groups conform to minimum standards. In addition, as Barber *et al.* suggest, the inclusion of more non-medical members and the institutionalization of reviews of research in progress would provide important additional checks on the ethical behavior of researchers. Extension of the activites of these peer-review committees to cover all those who apply life-prolonging technologies to unsalvageable patients would be advisable.

Bibliography

Abram, H. S., *et al.* Suicidal behavior in chronic dialysis patients. *American Journal of Psychiatry*, 127 (1971), 1199–1204.

Barber, B. Experimenting with humans. *The Public Interest*, 6 (Winter, 1967), 91–102.

Barber, B., *et al. Research on Human Subjects: Problems of Social Control in Medical Experimentation*. New York: Russell Sage Foundation, 1973.

Fox, R. C. *Experiment Perilous*. Glencoe: Free Press, 1959.

Fox, R. C., and Swazey, J. P. *The Courage to Fail: A Social View of Organ Transplants and Dialysis*. Chicago: University of Chicago Press, 1974.

Kuty, O. *Le Pouvoir du Malade: Analyse sociologique des Unités de Rein artificiel*. Thèse de doctorat présentée à la Faculté des Lettres et des Sciences humaines de Paris-V (René Descartes), 1973.

Pappworth, M. H. *Human Guinea Pigs: Experimentation on Man*. Boston: Beacon Press, 1968.

Simmons, R. G., and Simmons, R. L. Sociological and psychological aspects of transplantation and dialysis as a special case. In J. Nazarian and R. G. Simmons (eds.), *Transplantation*. Philadelphia: Lea and Febiger, 1970.

Simmons, R. G., and Simmons, R. L. Organ transplantation: a societal problem. *Social Problems*, 19 (1971), 36–57.

Swazey, J. P., and Fox, R. C. The clinical moratorium: a case study of mitral valve surgery. In P. A. Freund (ed.), *Experimentation with Human Subjects*. New York: George Braziller, Inc., 1970, 315–357.

For Further Reading

Annas, G. J., Glantz, L. H., and Katz, B. F. *Informed Consent to Human Experimentation: The Subject's Dilemma*. Cambridge, Mass.: Ballinger, 1977.

Beecher, H. K. *Research and the Individual*. Boston: Little, Brown, 1970.

Fried, C. *Medical Experimentation: Personal Integrity and Social Policy*. New York: American Elsevier, 1974.

Hilton, B., and Callahan, D., eds. *Ethical Issues in Human Genetics: Genetic Counseling and the Use of Genetic Knowledge*. New York: Plenum Publishing, 1976.

Jonsen, A. R. *Biomedical Experiments on Prisoners*. San Francisco: University of California School of Medicine, 1975.

Regan, T., and Singer, P. *Animal Rights and Human Obligations*. Englewood Cliffs, N.J.: Prentice-Hall, 1976.

U.S., National Commission for the Protection of Human Subjects. *Research on the Fetus: Report and Recommendations and Appendix.* Washington, DC: U.S. Department of Health, Education, and Welfare, 1975.

U.S., National Commission for the Protection of Human Subjects. *Research Involving Prisoners: Report and Recommendations and Appendix.* Washington, DC: U.S. Department of Health, Education, and Welfare, 1977.

Visscher, M. B. *Ethical Constraints and Imperatives in Medical Research.* Springfield, Ill.: Charles C Thomas, 1975.

14
ALLOCATION OF RESOURCES

RATS EAT BABY. *Sensational as it was, the headline probably would have caught the eye of the most shockproof of readers. But to someone studying to be a public health nurse, it was positively riveting. Jane Sutton proceeded to read the back-page story with intense interest.*

Two huge gray rats chewed the hand off a newborn infant last night in the Rocklin district, an East Side ghetto area.

According to the child's mother, Millicent Washington, she and her other three small children were awakened in the middle of the night by the cries of the baby and found two huge gray rats on top of him.

"I used a broom handle to beat them off," the badly shaken and husbandless woman reported. She said that when she picked up the infant she saw that his hand was mangled and bloody. "I didn't know what to do," she told a *Times* reporter. "We have no doctor, you know, and no car or phone, and the hospital's over an hour away."

The emergency room of the county hospital is about 1½ hours by bus from the Washington residence, a dugout basement under a dilapidated row house.

A neighbor reported that shortly after 3 A.M. a panic-stricken Mrs. Washington came pounding on her door asking for some money for a taxi to take her child to the hospital. "He's bleeding to death!" Mrs. Washington was quoted as saying. The neighbor said she gave the distraught woman all she had—just change—then helped her solicit help from other neighbors. It took them nearly an hour to raise taxi fare.

Fearing for the safety of her other children, Mrs. Washington took them with her to the emergency room, where the entire family spent the night waiting for the baby to be admitted and treated.

The child is reported in critical condition, suffering from blood loss and shock. Medical officials were guarded about the child's prospect for a full recovery. "Our resources are limited here," a resident was reported as saying. "We did the best we could. But there's just so much we could do under the circumstances." He added that it would have helped had the baby been brought in immediately.

The story stunned Jane Sutton. For some minutes she thought about the deplorable conditions under which the Washingtons must be living. She wondered how many other people were living the same way. There must be hundreds in this city alone, she thought, thousands throughout the country. How futile it all seemed. Even if they can save the baby's hand, she thought, what's the point? He'll just end up back where it all happened, and who knows about the next time. . . .

Jane folded the paper and was about to discard it when another story, this one on the front page, caught her attention. She had read the story only minutes before and had been elated by it. It concerned a medical breakthrough, the development of a completely artificial heart that researchers had effectively tested on a calf and were now ready to try on humans. The device was described as a "great technological advance" and as "another stunning example of the wonders of modern medicine."

Looking at the story now, Jane didn't feel the mix of elation and pride that she had felt when she first read the article. Looking at the photograph of the contented cow this time, she felt depressed. I wonder why there was no picture of the baby the rats tried to eat? she thought. Somehow the paper made a mistake, she told herself. It really meant to put the story of babies and rats on the front page and the story of hearts and calves on the last. But then she decided it was no mistake, that the stories had been situated according to their perceived importance.

"Priorities," Jane Sutton mumbled to herself as she folded the paper and returned to her studies. "Our priorities are all screwed up."

It is a safe bet that Jane means more than journalistic priorities. She is probably referring to society's health priorities. She may even be thinking about the large sums of money that go into the biomedical research that yields technological wonders like the artificial heart and to the relatively small sums spent on eliminating or minimizing the conditions that lead to the need for such a device or to the rat attack on the Washington baby. Jane Sutton's reading of the daily paper has apparently led her to ponder an issue of increasing social importance: the allocation of scarce medical resources. Just how much of a nation's resources should be earmarked for health care? What is the proper distribution of available funds to medical research and clinical practice? Of what is available for research, how much should go to the production of expensive machines used in treatment facilities? How much to cancer research? How much to preventive medicine?

The scarcity of medical resources is clearly associated with the expansion of medical services. As modern medicine made its services available to more and more people, it was inevitable that scarcities would develop. These scarcities touch almost every facet of health care—from equipment to medicine, from highly specialized practitioners to organ donors. Understandably, a problem of allocation has arisen.

The issue of allocation can be viewed as having both an economic and an ethical dimension.[1] On the economic side, the issue concerns the most *efficient* distribution of resources; on the ethical side, it concerns the *fairest* distribution. One thing that makes dealing with either of these difficult is that neither can entirely be separated from the other.

---

1. *T. L. Beauchamp and L. R. Walters,* Contemporary Issues in Bioethics *(Belmont, Calif.: Dickenson, 1978), p. 347.*

In sorting out the economic and ethical issues, it helps to view the allocation problem as existing on two levels. The first level is *macroallocation*, which refers to the amount that a society should expend for medical resources and how it will be distributed. Jane Sutton is reacting to the newspaper accounts on this level. In her view, priorities are misplaced: More should be spent on prevention and primary care, less on crisis intervention. Others would disagree. They would say that the lion's share of medical funds should go to crisis intervention (for example, surgery, kidney dialysis, intensive care units, artificial hearts, organ transplants, and so on). These opposite views create the central debate in macroallocation discussions.

But no matter which side one takes, one can always ask who should have access to the available resources. This question deals with *microallocation*, which refers to decisions concerning who should obtain the resources that are available.

There are far too many allocation issues to deal with all of them in this chapter, so we will focus on those that seem especially urgent. Regarding macroallocations, we will consider whether allocations for preventive medicine should take priority over allocations for crisis intervention. This will require a look at what a preventive bias implies. One of the most important questions concerns the poor, who more than any other group stand to benefit from a preventive emphasis and to lose from a crisis approach. Thus we will take a close look at the position of the poor in the distribution of finances and resources. Finally, we will consider various microallocational issues, chief among them how HCPs and others should go about deciding who will get available resources when there are not enough for all.

Values and Priorities

Before we begin our exploration of allocation issues, ask yourself the following questions. They may help clarify your feelings about the allocation of scarce medical resources.

1. What are your feelings about the Washington case?

2. Would you evaluate the service provided the Washington baby as adequate or inadequate?

3. How do you feel about millions of dollars being spent for the development of such technology as the artificial heart?

4. Do you agree with Jane that our priorities are misplaced?

5. Do you think that society should allocate a greater or lesser share of the gross national product for health, or is it allocating just about the right amount?

6. Where you do think most of the money that society allocates for health care should go (for example, research, development of sophisticated technology, prevention, and so on)?

7. Is there any area of health care that you believe is underfunded?

8. Which area do you think should get the greater allocation—crisis or preventive medicine? Or should they get an equal share?

9. Do you think that society is spending too much money on crisis medicine, not enough, or just about the right amount?

10. Which economic group do you think benefits more from current health allocations—the poor or the more affluent? Or do you think both benefit equally?

11. How you you think scarce resources should be allocated—on the basis of need, equal treatment, ability to pay, or some other basis?

12. Do you think the poor have about the same access to medical resources as the more affluent?

13. Is there anything necessarily objectionable about one economic group having less access to health resources than another?

Do you agree or disagree with the following?

14. The poor have mostly themselves to blame for their health problems.

15. Society must spend less money on the development of technology and more on eliminating the causes of illness.

16. Kidney machines and organs for transplant should be allocated on the basis of the relative social value of the potential recipients.

17. Everybody who needs a medical resource should have an equal chance of obtaining it, regardless of other considerations.

18. The decisions about who will get scarce resources, such as organs for transplants, should be left to HCPs.

19. The best way to determine who gets a scarce resource is to allow physicians to follow their moral intuitions.

20. If the only way to save some patients is by denying others a scarce resource and thus sacrificing them, then no one should receive the needed available resource.

The Problem of Macroallocation

In order to understand the emergence of macroallocation decisions as central concerns in the distribution of health resources, it helps to know how much of our domestic resources are available for health care and where most of these resources are channeled. We can begin by noting that what we as a nation spend on health care depends on the relative values we assign to competing domestic and nondomestic budget items. This observation immediately raises a moral concern about how much value should be given to health in the overall budgeting. This question, laced as it is with considerations about social justice, is beyond the scope of this chapter. We will focus instead on the priorities that are evident in our current allocations of the medical share of the gross national product (GNP). From all evidence, our priority is clearly in the area of crisis medicine as opposed to preventive medicine.

Crisis Medicine

Medical care, as an industry, constitutes a huge part of the U.S. economy. Second only to the construction industry, it makes up about 8 percent of the GNP. A considerable portion of this share goes into medical research and specialized treatment, which are chiefly intended to support crisis intervention. Both of these high-cost items eventually turn up in the most expensive part of the medical enterprise: hospital care.

There is little question that hospital care today is based largely on advanced technology. For example, hospitals throughout the country have hundreds of coronary care units for intensive monitoring and hundreds of devices for computerized transaxial tomography, a sophisticated and costly type of x-ray for diagnosis without surgery and dangerous drugs. Physicians also have at their disposal an impressive array of other high-powered equipment, as well as a smorgasbord of drugs ranging from antibiotics for infection to tranquilizers for mental illness to synthetic hormones. In brief, we have spent mightily in the areas of biomedical research and technology, often reaping mind-boggling dividends.[2]

What if the trend continues? What if we continue to put the lion's share of medical allowances into crisis intervention? Obviously, we will see more of the same results. Thus there will be more impressive technology, further concentration of it in hospitals, and of course, increased costs to support the technological base for the delivery of health services. At the same time, improvements in crisis intervention hardware and techniques, such as organ transplants, cardiac surgery, and sophisticated diagnostics, will continue to benefit the relatively few. Virtually untouched by these innovations are the many who could benefit from preventive and primary health care but who get little of either because these are not priority areas.

Clearly, the obvious advantage of continued concentration on crisis medicine is that it addresses, often dramatically, the health needs of those whom the intervention is designed to reach. And indeed, to ignore or minimize in our allocations those in need of sophisticated surgery or organ transplants seems morally unjustifiable. On the debit side, crisis medicine is extraordinarily expensive in proportion to the number it benefits. Widespread prevention of disease may be a cheaper and more efficient way to raise health levels and save lives. What the health professions and society face, then, are decisions that must weigh the maximum cost efficiency of allocations against the needs of individuals, as well as decisions that must strike some kind of reasonable and justifiable balance between crisis and preventive medicine.

---

2. R. F. Greifinger and V. W. Sidel, "American Medicine," Environment, May 1976, pp. 6–18.

Preventive Medicine

Even if a decision has been reached to put more emphasis on prevention, questions of social policies still arise.[3] One such question is how much emphasis should be placed on reducing the incidence of health problems. Specifically, which facet of prevention should be stressed—primary or secondary? Primary prevention deals with reducing or eliminating the causes of health problems; secondary prevention focuses on early detection and treatment of diseases (for example, screening and multiphasic testing for various conditions). If some consideration should be given to primary prevention, then we must decide the relative weight to be allocated to socioeconomic and environmental issues.

For example, about 20 percent of all disabilities are caused on the job. Should we therefore allocate 20 percent of our primary prevention resources to occupational safety? Or should we first decide who should bear the costs of disability or hazard reduction: employer, employee, consumer, or society? We would and probably should ask similar questions about primary prevention in nonoccupational areas. How much should be allocated to increasing product safety? To altering possibly destructive human behaviors such as smoking, drinking, and overeating? To exterminating vermin and improving the atmosphere in which people like the Washingtons live?

Reduction of the number of health problems is only one social policy concern. Another deals with the severity of the health problems that occur. The individual's total health status, together with advances in physical rehabilitation, are important considerations in reducing the severity of specific health problems. But there are many areas in which broader social policies help determine whether a health problem results in job change and unemployment. In an illuminating work on poverty and health, Howard S. Luft identifies a number of these areas:

> Should vocational rehabilitation programs be expanded along present lines or changed substantially? What is the role of racial and sexual discrimination in the labor market, and would the reduction of such discrimination differentially affect the disabled? Can educational programs be established to provide people job skills and flexibility that can serve as implicit insurance in case they develop a health problem? Can employers be given effective incentives to hire the disabled; should they be required to do so? Do business cycle-swings differentially affect the disabled?[4]

Still another set of policy questions concerns prevention. These deal with reducing the impact of health problems, including social policies to alleviate the effects of disability on the individual and family. Prominent questions here are about health insurance, specifically the nature and extent of coverage. Do existing insurance programs adequately cover all those who need health care? If not, what changes should be made and who should pay for them? Should some form of comprehensive national health insurance be established? If so, what should it

---

3. *H. S. Luft,* Poverty and Health: Economics Causes and Consequences of Health Care *(Cambridge, Mass.: Ballinger, 1978).*

4. *Luft, pp. 10–11.*

include and how should it be financed? Most important, should it prescribe a wide range of preventive actions (for example, hazard reductions, diet improvements, and behavioral changes), so that ultimately it might develop its own dynamic toward improving the overall health and well-being of the population?

Any decision to allocate more resources for preventive medicine must ultimately engage social policy concerns, which themselves raise questions of values and justice. Allocating more money for prevention would probably mean that fewer people in need of crisis intervention would be served. On the other hand, preventive medicine might ultimately be more cost efficient because it would probably do more to raise the total health level of the nation. It would also be likely to address the health needs of the most desperate group in society, the poor.

For several reasons, the health needs of the poor are of important moral and economic concern in macroallocation decisions. First, their needs appear primarily in the area of preventive medicine. This means that they have much at stake in what approach to health care the nation takes; there are real concerns about whether the poor receive adequate health care. Second, the economics of prevention are as much tied to the costs of meeting the health needs of the poor as the economics of crisis medicine are wedded to the costs of meeting the needs of those with more exotic health problems. Third, the ultimate question concerning the most just basis for distributing resources, entails in part an analysis of need as a principle of distributive justice. But in the area of macroallocations, one cannot speak intelligently about need as a basis for distributing resources without understanding the needs of the most desperate. Given their central importance in macroallocation discussions and specifically in the crisis-prevention debate, the poor and their health plight deserve the consideration of HCPs and society alike.

Poverty and the Poor

There are several ways to determine poverty level. One way establishes an amount of income at a particular time—for example, $5000 at a specific economic state of the nation. Families falling below that level are considered poor. Another way, used by the Social Security Administration (SSA), establishes poverty criteria that take into account the different consumption needs of families, based on their size, composition, age, and location of residence. By these criteria, 40 million Americans, or 22 percent of the population, were living below the poverty level in 1959. In 1972 the figure was 24 million, or 12 percent of the population. Today, given the past decade's inflation, this figure should probably be adjusted upward. By SSA criteria, somewhere between 12 and 20 percent of the population is living below the poverty level.[5]

Poverty and the poor relate to health care in a number of ways. First, economic deprivation is unquestionably associated with health problems.

---

5. R. L. Kane, J. M. Kasteler, and R. M. Gray, The Health Gap (New York: Springer, 1976), p. 3.

Whether health problems are viewed in terms of pathology (the mobilization of the body's defenses and coping mechanisms), illness (recognition of symptoms), impairment (a physical, emotional, or psychological loss), or disability (a behavior pattern that evolves when functional limitations interfere with an individual's ordinary daily activities), the poor have a greater incidence of health problems than any other economic group in society.[6] Statistics generated largely through national health surveys over the past twenty years prove the point:

1. 60% of the children coming from families defined as poor have never seen a dentist.

2. 30% of their parents have one or more chronic diseases.

3. Incidence of all forms of cancer [is] inversely related to income, that is, the poorer the person the greater the likelihood of contracting cancer.

4. Heart disease and diabetes are more prevalent among the poor than among the non-poor.

5. Six times as many cases of hypertension, arthritis, and rheumatism; eight times as many cases of visual impairments; and far more psychiatric illnesses, especially schizophrenia, beset the poor than the more affluent.

6. The poor are troubled with liver and stomach problems at a rate of two to one over the more affluent.

7. The poor experience four times as many cases of emphysema [as] persons whose annual family income is $15,000 or more.

8. Infant mortality is inversely related to income.

9. The poor run a risk of dying before age twenty-five that is four times higher than the national average.[7]

When the health problem is disability, the picture for the poor is just as ugly:

1. Sixty-five percent of all poor families (composed of at least a husband and wife) will include a disabled adult.

2. At least 30% of the disabled who are currently poor are poor because of health problems; among white males, the percentage rises to 75%.

3. Among the non-aged (16 to 64 years old), disability is responsible for at least 9 to 18% of all poverty.

4. Twenty-three to 31% of all non-aged poor white males are poor because of disability.

5. The probability that an adult in a low income family (under $4,000) will become disabled in a given year is twice as high as one in a family with $8,000 or more, even after adjusting for age, sex, race, and income.[8]

Additional statistics underscore the sobering state of the minority poor. For example, vascular lesions of the central nervous system are twice as high among

---

6. Luft, pp. 18–19.

7. Kane, Kasteler, and Gray, p. 6.

8. Luft, p. 7–8.

poor blacks as among middle-class whites. Similarly, black men have twice as high a disability rate as white men, which can be explained solely in terms of income and education.

Such sobering statistics lead to the inescapable conclusion that the health status of the poor is well below that of other income groups. By missing this point, HCPs overlook an important dynamic in the interplay between income and health problems and run the risk of engaging in inappropriate, even irrelevant, health care interventions.

The Connection between Poverty and Health Problems

As a group, the poor do not enjoy the same level of health that other income groups do. Why? Does poverty cause health problems? The answer is important, because it helps determine the direction and amount of at least part of our society's allocation for health care. If poverty causes health problems and if prevention is a priority, then a certain amount should be spent to combat poverty. Society should also attack the financial barriers to adequate health care, and it would seem that HCPs should help.

But what if poverty is not the cause of health problems? What if other factors that commonly accompany poverty are part of the cause or are actually the true conditions for health problems? Then allocating resources to remove financial barriers would be misguided and would raise moral concerns on at least three levels. First, the tax monies allocated to remove financial barriers would be squandered. Second, the health status of the poor would not materially be improved. Third, and perhaps most important, society would probably think that much was being done when in fact little was. Furthermore, when it became apparent that the health plight of the poor had not been improved, society might even label the problem intractable and choose to allocate its resources elsewhere or blame the poor for their own abysmal health status. Thus there are serious moral questions of injury and justice embedded in assumptions about the connection between poverty and health problems.

Whether poverty causes health problems (or illness causes poverty) is perhaps a question best left to social scientists. But it is noteworthy for ethics that, although health care literature abounds with studies that associate low income with health problems, the data do not support the easy conclusion that poverty as such causes them. Considerable evidence points to other factors. One that has received much attention is the claim that the poor simply do not get an equal share of the available resources. Another is that the poor do not use what is available. Since both of these factors affect macroallocation decisions and the delivery of health care to the poor, we should consider them.

Inequities in Health Care Delivery

Almost everybody agrees that the poor receive less than the nonpoor in the way of actual medical care, both in quantity and quality. And yet, such generalizations about inequity should be made very cautiously, since they rest on definitions that not everyone will agree with. For example, what is to be considered "quality" care? In one sense, it may refer to medical specialists; in another, to the use of private practitioners as opposed to public clinics. The point is that considerable data must be amassed and scrutinized before generalizations are warranted.[9] This having been noted, some observations are in order about the poor person's access to health resources.

First, studies indicate that although the poor have more health problems than the nonpoor, they do not use a greater share of the health care facilities. In fact, there is not only a positive correlation between social class and frequency of using health care services but also between social class and the type of service used, with a greater percentage of care among the upper classes going for prevention. Precisely why this is so remains to be explained. Part of the explanation may lie in the pattern of services that has developed.

To understand the pattern of service that exists today, it is necessary to keep in mind that we are an urban society.[10] Most of the population is concentrated in metropolitan areas. This urbanization, which has resulted in a nearly 300 percent increase in the number of city dwellers in the past thirty years, is also seen in the disposition of physicians. The concentration of physicians in urban areas has increased faster than the concentration of all other occupational groups. Likewise, hospitals and medical schools have become primarily an urban phenomenon. The result is that a number of rural areas have fewer services available today than they did in 1950.

A second part of this pattern is the location of the health care setting, the chief forms of which are the hospital and the physician's office. About one-fifth of all settings are publicly run general hospitals, located primarily in the poor areas of our major cities. In contrast, anywhere from 50 to 80 percent of the hospitals in most metropolitan regions are voluntary nonprofit hospitals. These are generally located near medical schools or in the suburbs and primarily serve working-class, middle-class, and upper-class patients. The location of doctors' offices generally follows the pattern of voluntary nonprofit hospitals—that is, they are concentrated in the affluent downtown areas or in the suburbs. Finally, private, profit-making hospitals tend either to be a major element in the city's health care system, as in newer and faster-growing metropolises like Houston and Los Angeles, or to make up an insignificant portion of it, as in the East and Midwest.

Given this pattern, some conclusions seem warranted about the poor person's access to medical resources. Elliott Krause has said:

> Oversimplifying, but not excessively, we have, combining doctors and
> office and hospital settings, a two-class medical care system. On the one
> hand, few practitioners and a few public settings for the poor in either the
> ghetto or rural areas; on the other hand, many practitioners and voluntary
> nonprofit hospitals for the middle class and upper class in the suburbs.[11]

Krause's conclusion seems to be supported by the types of services available to the poor and nonpoor. For example, hospitals generally provide emergency

---

9.   M. J. Fefcowitz, "Poverty and Health: A Reexamination," |Inquiry 10 (March 1973): 3–13.

10.   E. A. Krause, Power and Illness: the Political Sociology of Health and Medical Care (New York: Elsevier, 1977), pp. 144–150.

11.   Krause, p. 146.

services for those in crisis and for those who have no alternatives. But the urban and rural poor, who lack private physicians and community clinics, have taken to using the public city hospital's emergency room as a kind of general practitioner for all their complaints, major and minor. As a result, emergency room staff and services are overtaxed, and the delivery of care to those in genuine emergency need is jeopardized. Furthermore, since the poor rarely have a private physician to safeguard their interests in an emergency situation, the way the more affluent do, chances are considerable that they will be subjected to various bureaucratic abuses.

Similarly, outpatient or ambulatory services are different for the poor than for the rich. Most of the facilities available to the chronically ill poor are run on an outpatient basis. They also seem to reflect the interest of medical specialists or the government in certain health problems more than concern for the health needs of the poor. Thus the large hospitals generally offer an array of outpatient clinics, each established for teaching the diagnosis and treatment of a particular disease.[12] There is nothing inherently objectionable about this phenomenon, but it does raise some concerns. What happens to the poor whose health needs cannot be met in a specialty clinic? How are the general health needs of such people to be met? Just as important are the severe constraints placed on the poor in selecting an HCP. Whereas the more affluent patients can pick and choose among HCPs, the poor must accept what is offered them.

A look at the pattern of health care services as it currently functions, then, suggests that a "two-class medical care system" may exist, in which the poor do not get an equal share of the available resources. If this is the case, then the central moral question involves equity. Specifically, is it fair that health care is less available to those most in need—the poor—than to those less in need? In part, the answer to this question involves an airing of the various ways that resources can be distributed and a determination about which is the fairest—which, of course, is the conceptual issue beneath all allocation decisions. Should people be provided health care exclusively on the basis of need? Or should they receive similar care for similar problems? Or should they be given care solely on the basis of their ability to pay? Since this fundamental issue is discussed thoroughly in the article that follows this chapter, we will not belabor it here. Suffice it to say that whatever the choice, it will affect not only the delivery of care to the poor but also the costs of providing health care in general and the direction that the nation will take in providing it.

Use of Health Services

Although there may be inequitable distribution of health care and although the health problems of the poor may result in part from this, the reasons for their receiving less-than-average care extend beyond simple availability. One important factor is underutilization. Why the poor do not make use of available resources is a complex issue. To see how and why, let us return to the case that opened this chapter.

---

12. L. S. Robertson et al., Changing the Medical Care System (New York: Praeger, 1974), p. 8.

Her curiosity piqued by the newspaper articles, Jane Sutton decided to use the Washington episode as the basis for a class report on what she perceived as inequities in the distribution of health care. Even from the little she had read in the newspaper, Jane suspected that the poor simply do not have access to the same health care services that the more affluent do. After all, had Mrs. Washington had her own physician, quicker access to emergency care, and the financial wherewithal, her baby surely would have received more skillful medical attention sooner. As a thoughtful, sensitive HCP in training who already showed a marked awareness of patient interests, Jane intended to argue in her report that HCPs must advocate for greater availability of health care for the poor. She was convinced that, if they failed to do this, they would not meet their obligation to provide the poor with adequate health care.

To establish her thesis, Jane visited the Washingtons. She was appalled, although not surprised, by the deplorable conditions she found. These she had anticipated. What caught her off guard were the attitudes of Mrs. Washington and her neighbors toward their health.

While talking with Mrs. Washington, Jane noticed that one of her other children had a hacking cough. Mrs. Washington admitted that the child had had the cough for some time, although she did not know exactly how long. When Jane asked her if she was concerned, Mrs. Washington said no. "Everybody around here has some kind of cough," she told Jane with a shrug. "Besides, I can't afford a doctor every time one of the kids coughs or sniffles." Jane reminded Mrs. Washington that she could have the child examined free at the inner-city clinic. But Mrs. Washington dismissed the idea with a wave of her hand. "I don't like it down there," she said. Jane pressed her for an explanation. At first Mrs. Washington was reticent, but then she said, "They're cold, if you know what I mean. They don't talk to you like a person. They just don't seem to understand. They just don't seem interested." She paused, screwed up her world-weary face, and said, "I guess I just don't understand all their fancy medical talk."

Before she left, Jane asked Mrs. Washington whether she was fearful for her family's welfare, given the attack on her baby. Of course she was, Mrs. Washington told her. But what could she do? "There's just so much you can do, isn't there? Some things you just have to live with—or die with."

Jane talked with several of the neighbors, who were as poor as the Washingtons. She found similar attitudes among them about their health.

All the way home, Jane was troubled that her thesis was inaccurate. She decided that it was not solely the unavailability of services that caused health problems for the poor. After all, Mrs. Washington could receive care for her child with the cough. What was stopping her was her definition of health and illness. Because everyone in Mrs. Washington's circle had "some kind of cough," she simply did not perceive her child's chronic cough as significant. The reason she was not using the services at the free clinic was in part that she felt uncomfortable around HCPs. No doubt her modes of thinking and talking were different from those of the HCPs, who probably thought and spoke more in abstract terms. As a result, Mrs. Washington found it difficult to understand their diagnoses and treatment procedures. So, rather than feel foolishly uncomfortable, she simply avoided the clinic and HCPs generally. Another reason that she did not use the available health services was a fatalistic outlook on life, one that is all too common among the poor.

The point of this story is that, although inequity in the distribution of services may partially explain the below-average care that the poor get, there are numerous other reasons. One of the most important is their failure to use what is available. This lack of use must be explained by reference to the social structure of poverty and the whole lifestyle of the poor. The ideals, values, and beliefs to which the poor adhere play a formidable role in their use of the resources that are available and thus, ultimately, in their health status. If HCPs and society fail to acknowledge and deal with this fact, they stand little chance of delivering adequate health care to the impoverished.

If progress is to be made in alleviating the health problems associated with poverty, it seems that a substantial part of the effort must be aimed at education. People like Mrs. Washington must not only learn to recognize symptoms and distinguish the innocuous from the serious but must also alter their concepts of health and illness, change their perceptions of HCPs and the health care system, learn to extract from the system the health services they need, and most important, develop attitudes and values that advance, not retard, their health status. Therefore, merely making services available to the poor does not appear sufficient to meet their right to adequate health care. In fact, failure to buttress increased distribution of services to the poor with a concerted effort to alter their lifestyles through education may itself be morally unjustifiable, for the following reasons: (1) Distribution alone does not alleviate the health plight of the poor; (2) it wastes precious resources; (3) it diverts resources that might be more effectively used elsewhere; (4) it gives the impression that something substantive is being done while the real problem is left to fester; and (5) it sets up society for a big letdown and a subsequent cruel and unjustifiable indictment of the poor's complicity in their own health status.

What are the implications of this analysis for HCPs? If it is true that any effort to help the poor must include an instructional component, then HCPs must function as much as educators as providers of health services. In one sense, HCPs are always instructing, but in delivering health care to the poor, the educational function may predominate. If this is so, then far greater attention must be paid to care than to cure, to prevention than to treatment. However, this emphasis requires the decision that our society should channel its resources in a different direction and requires HCPs to commit themselves to a vigorous and conscientious pursuit of these new goals.

We have now taken a look at two important issues in macroallocation decisions. One involves the kind of health care that our society wishes to emphasize: crisis or preventive care. Each has its advantages and disadvantages from both an economic and an ethical point of view. The other issue, related to the first, concerns the poor. Any macroallocation decision concerning crisis or preventive medicine must consider them, because they have the most at stake in terms of health care. There are additional macroallocation issues, but these two should be enough to provoke subsequent thought and inquiry. The remainder of the chapter examines some microallocation issues.

The Problem of Microallocation

Problems at the microallocational level involve decisions made by HCPs about who shall obtain whatever resources are available. There are more patients than medical supplies, so HCPs face agonizing choices about who will get the resources and who will be denied them. In some cases, as in the allocation of hemodialysis machines or heart and kidney transplants, a microallocation decision is tantamount to a decision about who shall live and who shall die. But whatever the nature of the resource, the paramount moral question always concerns the fairest way to decide who gets what.

Decision Making

Some still advocate leaving microallocational decisions to the intuition of the physician, but almost everyone who has addressed the subject agrees that reliance on HCP moral intuitions is insufficient. Therefore, in recent years several positions have emerged that attempt to establish ground rules for deciding who shall get the scarce resources that are available. Two of these systems—complex criteria systems and random selection systems—are especially noteworthy.[13]

COMPLEX CRITERIA SYSTEMS.   One position that has been advanced entails a cluster of criteria for allocation decisions. Among the criteria that have been suggested are the relative likelihood that the patient will benefit from the procedure without complicating ailments; the patient's ability to contribute either financially or experimentally as a research subject; the patient's age, family role, and life expectancy; the patient's past and potential social contributions; and the patient's status in the community served by the institution. The precise nature and number of criteria remain problematic, as does the relative weight given to each. But those who hold to such a system generally argue that the patient who satisfies most of the key criteria should be preferred. They buttress their position by pointing out that such criteria are both consequential (for example, future social contributions) and nonconsequential (for example, past contributions).

RANDOM SELECTION SYSTEMS.   Some ethicists have criticized complex criteria systems on several grounds. For example, some say that they are primarily consequential, specifically utilitarian, since individuals are viewed solely in terms of their social roles. Others claim that such a system jeopardizes the HCP-patient relationship insofar as patients realize that HCPs do not necessarily have their best health care interests at heart but rather are "taking their measure" in relation to other patients. Still others insist that a complex criteria system simply is not workable.

Some of these critics have proposed instead some version of a random selection system. One version is natural random selection, more commonly characterized as "first come, first served" Another version is formal random selection, as in a drawing or lottery. Either of these alternatives, claim their proponents, preserves an individual's personal and transcendental dignity, which they say is lost when one emphasizes a person's social role and function. Moreover, they argue that a random selection system precludes the intrution of biased criteria and minimizes the chances of unfair discrimination. In addition, it

---

13.   *Beauchamp and Walters, p. 350.*

preserves the trust between HCP and patient. Thus, although determining who will receive scarce resources is of overriding concern to HCPs who are faced with such decisions, other microallocation issues are also morally important.

Additional Concerns

One additional concern centers around who makes allocation decisions. Should it be the physician, a committee, or a board? If a group, who should be included? Should non-HCPs be part of it? Another issue concerns an extension of the patient's right to adequate health care. Possibly the patient should have an advocate, a representative who will argue that person's interests before the decision-making body. Still another issue deals with the macroallocation implications of microallocation decisions, perhaps best illustrated in the case of allocating hemodialysis machines and kidneys for transplant.

Those in need of hemodialysis or kidney transplant currently are eligible for Medicare payments. As a result, well over 25,000 patients are now receiving dialysis supported by Medicare. Within the next few years, the cost of the dialysis program could run over a billion dollars a year. That much less will be available for other health needs (for example, preventive health care); so the macroallocational implications of this situation are noteworthy.

The implications grow even more complex when one realizes that dialysis rarely solves all the problems of kidney patients. On the contrary, even when the procedures work effectively, they are time-consuming and typically result in medical and psychological problems, including neurological disorders, severe headaches, gastrointestinal bleeding, bone diseases, and depression.[14] Indeed, a study conducted about ten years ago indicated that 5 percent of all dialysis patients commit suicide. A causal connection cannot automatically be inferred, but this together with other alarming data must at least alert HCPs to the possibility that (1) considerable harm may await the patient who undergoes dialysis and (2) the debilitating side effects may argue against such heroic medical intervention. Such considerations might well unsettle HCPs, especially since the main thrust of their training is to preserve life at all costs. At the same time, the rather grim outlook for microallocation to such dialysis patients heightens the macroallocation issues of where best to channel limited health care funds.

Beyond these microallocation concerns with moral overtones, there is the issue of the precise nature of the HCP's role in resource acquisition. Just what should be the role and responsibility of HCPs and institutions in acquiring exotic resources needed for patient care? Never does this question take on a more dramatic form than when the resources happen to be organs.

Organ Acquisition

As indicated earlier, the demand for certain organs for transplant has outstripped supply. As a result, thousands of kidney patients across the country

---

14. R. Munson, Intervention and Reflection *(Belmont, Calif.: Wadsworth, 1979), p. 399.*

must wait several years for transplants. Similarly, the list of blind people waiting for corneas numbers in the thousands. The same grim statistical picture applies to other areas as well. Burn centers have only about 10 percent of the amount of skin they need to save the lives of seriously burned patients; only a small number of middle ear bones are available for the many who need them to hear; only a fraction of the needed number of pituitary glands are available for dwarfed children to help them grow. Where are these organs to come from? Barring the development and mass production of synthetic organs, they will come from donors. It is at this point that the HCP's role and responsibility begins to blur; it is here that a tangle of incompatible obligations threatens to ensnare the HCP.

One set of understated problems involves the sheer logistics of acquiring organs. Some states, such as California, now send "donor cards" to drivers with their driver's licenses. If drivers choose to, they can fill in the card before two witnesses and thereby authorize the use of their organs for transplantation in the event of their death. On paper, this sounds like an ingenious way of acquiring organs, and indeed millions of drivers have signed such cards. Nevertheless, only a small percentage of these organs are ever harvested; the organ silos remain unfilled. Why?

The main reason is that there is no system for using donor cards. The people who have the most access to the driver's licenses of accident victims—police, paramedics, nurses, coroners—are rarely trained to look for the cards or told precisely what to do if they find them. Although highway patrol personnel sometimes are told to inform medical authorities if they come upon a donor card, precisely who these "medical authorities" are remains vague. It could be an ambulance driver, a nurse, or any HCP for that matter. In addition, what these "medical authorities" are then supposed to do usually goes unspecified. Indeed, most hospitals have no set protocol for checking driver's licenses. More often than not, in the confusion that follows tragedy, the licenses are put in hospital files with personal effects or given to relatives. What is more, hospitals shy away from asking relatives whether a loved one has a donor card for fear of alarming the family; HCPs, fearing malpractice suits for "turning patients off too soon," rarely press the issue. In short, hospitals, HCPs, and other professionals who deal with accident victims have little if any knowledge of their logistical roles in organ acquisition.

This situation raises moral concerns about adequate health care, autonomy, and justice as they apply to patients, accident victims, families, and professionals. Potential recipients stand to have their right to health care undermined if they do not get the benefit of readily available organs. Also, the wishes of the donors, who freely and humanistically donated their organs, are violated. In addition, there is injury to the victim's family, inasmuch as research indicates that families often find the pain of death easier to bear if the "gift of life" offered by the victim is in fact received. Beyond these concerns, the absence of a system to implement the wishes of donors undercuts the duty of HCPs to provide adequate health care to potential recipients, intensifies the pressure of transplantation teams to secure organs, perhaps by morally questionable means, and invites clashes among various professional groups.

There are additional microallocation issues that could be raised, but this brief discussion should be enough to indicate their nature and seriousness. Together with macroallocation decisions, these make up a body of truly formidable moral decisions facing HCPs and society.

Summary

This chapter dealt with the allocation of scarce medical or health care services. One dimension is macroallocation decisions, which concern the amount to be expended for medical resources in society and how it will be distributed. One important macroallocation issue is whether allocations for preventive medicine should take priority over the traditional emphasis on crisis medicine. A crucial element in this debate is the poor, their health status, their access to services, and their capacity to use what is available. The second dimension of the allocation issue involves microallocations—decisions concerning who shall obtain the resources that are available. The crucial question in microallocation decisions is on what basis the decision will be made. This chapter examined two systems that attempt to provide this basis: the complex criteria system and the random selection system. Finally, it touched on other microallocation issues, such as who should make the decisions and whether patients should be represented, as well as the personal and social implications of some decisions and the role and responsibility of HCPs in resource acquisition.

Ethical Theories

In applying the ethical theories to the issue of allocating scarce medical resources, we may confine ourselves to the most crucial aspect of microallocation decisions, which is how specific decisions will be made. This chapter has discussed two popular systems, using complex criteria and random selection. A third system, although endorsed by only a small minority of ethicists and even fewer HCPs, is nonetheless influential in theoretic discussions. It is termed the "no-treatment" system.[15]

The no-treatment system begins by observing that the context of microallocation decisions is such that some patients must be selected to die and that others can save themselves only by allowing that other to die. Given this dilemma, no-treatment proponents conclude that we should not select anyone at all, that treatment should be given to none since no one should live when not all can.

The situation here is often compared to the plight of a group of people adrift in a lifeboat. Room and resources are limited; if the boat had fewer occupants, those who remained would have a better chance of survival. Given this outlook, should some be abandoned in order to maximize the survival chances of the others? No-treatment adherents would say no. Simply by virtue of each person's being a human, each occupant has equal worth. Therefore, any action that

---

15. *Beauchamp and Walters, p. 350.*

sacrifices any one member for the good of the others would be morally unjustifiable. The same kind of analysis can be applied to microallocation decisions.

Egoism

In determining how precious medical resources should be allocated, egoists would focus on their own best long-term interests. Thus they would endorse in any given situation the system that promised the most happiness for themselves. But which of the three competing systems would this be? The answer would depend on a variety of situational nuances. For example, the complex criteria system might be to an egoist's advantage, providing that the key criteria were ones that he or she could meet. Thus egoists could not endorse a criteria system without first examining the criteria. Failing to meet the egoist's interests, a criteria system might well be abandoned for a random selection approach, which at least would give the individual a chance. It is hard to see how egoists could endorse the no-treatment system, since it would preclude even the possibility of medical intervention on their behalf and it is based on nonconsequential priorities. In summary, egoism cannot offer a decisive position on which of the two most popular systems should be endorsed. The choice depends on the circumstances and on which of the two would be most likely to advance self-interests.

Act Utilitarianism

At the outset, we must recognize that act utilitarians would look askance at the no-treatment system, because of its nonconsequential bias. Of the other two systems, the complex criteria system would have considerable appeal because of its decidedly social considerations. This does not mean, however, that the random system does not have social import; it does. For one thing, such a system may be viewed as preserving the HCP-patient relationship; for another, it may preclude dissension among members of groups who cannot qualify under selected criteria. It is conceivable, then, that under just the right conditions a random system would produce the most social benefit, and thus the act utilitarian would endorse it. Indeed, it is not outside the realm of logical possibility that under carefully circumscribed conditions even the no-treatment system might be the most socially desirable. In any event, in theory act utilitarians would be open to any one of these three systems. In practice, they would probably favor a complex criteria approach because of its clear social consciousness. Of course, if they adopt a criteria approach, act utilitarians must then select the proper mix of criteria that would best ensure the highest social dividend.

Rule Utilitarianism

Much the same can be said of rule utilitarianism. In theory, any of the three systems is compatible with a rulist approach. But given the criteria system's preoccupation with social benefit, a selection rule formulated around complex criteria would probably yield the greatest ratio of total happiness. In addition, the criteria approach speaks directly to the utilitarian's concept of justice, which is rooted in efficiency and productivity. In contrast, a concept of justice based on fair play and equal treatment is embedded in both the random selection and no-treatment systems.

The Categorical Imperative

On first look, Kant's ethics seems to endorse a no-treatment approach. After all, this system recognizes the inherent worth of each human; it avoids the complex criteria system's penchant for making some individuals (nonrecipients) the means to socially desirable ends (the preservation of those who promise to benefit society); and it treats everyone equally. What is more, it precludes doing harm to people, which humans presumably have a perfect duty to honor.

However, it could be argued that a no-treatment system allows people to suffer and die through omission. Do people not have an obligation to prevent suffering and death where they can? They do, although the obligation is an imperfect one. As with all imperfect duties, the nature and extent of our obligation to prevent harm is unclear. But if we assume that we are bound to preserve the lives of those who are in need of available medical resources, then perhaps a random selection system is more in keeping with Kantian ethics. A formal random selection process, for example, would not only help meet this imperfect duty but would do it in a way that gives each patient an equal chance of being selected. Thus the dignity and worth of all would be respected. Furthermore, it could be observed that nonrecipients ultimately suffer, even die, not because others receive the resources but because they themselves did not receive them. Under a no-treatment system, they would not have received the resources anyway. In addition, a no-treatment system intentionally allows others to die as much as a random system does, although admittedly it does not allow a choice to live at another's expense. Thus, a strong case can apparently be made under Kantian ethics for some version of a random selection process.

Prima Facie Duties

Paramount in Ross's view would probably be the prima facie duty of justice. Since Ross, like all nonconsequentialists, bases his concept of justice on a recognition of the essential equality of all humans, he would be concerned that the system selected treat everyone equally. This suggests that Ross would favor a random selection or no-treatment system.

But other prima facie duties inevitably intrude in microallocation decisions. Most important would be the duty of noninjury. A random system can be viewed as indirectly injuring nonrecipients; a no-treatment system injures everyone. In addition, there are duties of beneficence to be considered. Given the availability of life-sustaining resources, HCPs and society are in a position to help at least some of those in need. Also embedded in the concept of beneficent duties would be the duties that patients have to one another. Thus it could be argued that, given a choice between a random and a no-treatment system, a patient has a duty of beneficence to select the random approach; a no-treatment approach patently ignores this obligation. Finally, there are fidelity duties that

HCPs have to patients to provide health care. Again, a no-treatment approach precludes any HCP's meeting this duty, whereas a random system would preclude some HCPs' meeting it. Such considerations lead to the tentative conclusion that Ross's ethics would endorse some version of a random selection system.

The Maximin Principle

Taking the view of the original position, individuals would be irrational to bind themselves to a rule that would disallow even the possibility of their receiving available resources that they need. Thus a no-treatment system seems incompatible with Rawls's ethics. Of the remaining systems, the random system seems more compatible with Rawls's difference principle, because it improves the lot of the least well-off—those who could not qualify under a criteria system. At the same time, the random system benefits everyone in the sense of providing all an equal chance of being selected. It is also consistent with the equality principle, because compared with the criteria system, it allows a greater amount of liberty (call it opportunity) consistent with a like liberty (opportunity) for all. In other words, under a random system everyone will have an opportunity to be selected, whereas under a criteria system most will have no opportunity. Given these observations and remembering that those operating behind a veil of ignorance do not know their relative social status, it seems that inhabitants of the original position would prefer a system that guarantees them a chance of being selected as opposed to a system that offers them only the possibility.

Roman Catholicism's Version of Natural-Law Ethics

Roman Catholics would probably begin by acknowledging the worth and dignity of all humans and their right to equal treatment. This view alone would presumably rule out a strictly utilitarian determination of who is to get what. A question still remains, however, about which of the two remaining systems Roman Catholicism would prefer.

First, a no-treatment system would guarantee that no one would benefit— that in effect everyone would suffer and possibly die. In contrast, a random system is intended to ensure that the suffering of at least some will be alleviated. It could be argued, then, that one has an obligation to prevent suffering whenever possible, and thus the random system should be preferred. But as mentioned earlier, implementing the random system means that some in the group will be sacrificed for others.

Introducing the principle of double effect at this point, the significant question becomes this: Is the suffering and possible death of nonrecipients the means by which the lives of the recipients are preserved? No, the use of medical resources is. Looked at another way, the nonrecipients could go on living, even enjoy spontaneous remission of their ailments, without at all affecting the well-being of the recipients. If this analysis is correct, injury to nonrecipients cannot be the means by which the recipients benefit. All of which suggests that a Roman Catholic version of natural-law ethics would likely endorse some version of a random system as the fairest way to determine who gets scarce medical resources.

As always, these analyses represent one person's attempt to throw light on a complex issue in health care by applying ethical theory. As such, they should be viewed as tentative conclusions intended to open, not close, debate.

CASE PRESENTATION
The Candidate

Professor Lundstrom was just leaving class when she learned of her appointment as chair of a select committee on health care. She couldn't believe it! But there it was, on Congressional stationery no less, in words testifying to her "substantial contributions over the past thirty years to the fields of nursing, nursing education, and sociology."

In the days ahead, the professor focused on the committee's charge, which was to draft a proposal to provide health services to the population. Having published extensively on the need for society to come to grips with deficiencies in the health care delivery system, Professor Lundstrom relished her task and believed that she had something unique to contribute.

Not that the committee's work would be a picnic. Indeed, Professor Lundstrom's appointment was only a few weeks old when she found herself besieged by individuals and groups who had their own ideas about what such a proposal should contain. Some urged that the bill ensure that everyone have access to whatever medical resources are available. Others were pushing for a limited package of health care to all citizens, with emphasis on protecting them against the ravages of catastrophic illness. Still others wanted to restrict the limited package to those who cannot afford to obtain health services on their own, such as the elderly, the unemployed, and the poor. There were even those who opposed any direct government intervention, urging instead that Congress allow the free-market system and charitable institutions to deal with the nation's health care problems. Small wonder, then, that as her plane was taking off for Washington, where she was to spend the next several months working on the proposal, Professor Lundstrom felt a profound responsibility to help formulate the fairest proposal possible.

At the very hour that the professor's plane was lifting off, Dr. Merriwether, head of a renal unit in a hospital not far from Dulles Airport in Washington, learned that the brain waves of a patient in the intensive care unit had just flattened out. "That gives us a kidney," he said to his associate, Dr. Feinberg.

"Too bad the other one's shot," Dr. Feinberg said. Dr. Merriwether agreed but reminded Dr. Feinberg that one kidney was better than none at all. "Sure, sure," Dr. Feinberg said halfheartedly, "some bank we have. Our total accounts now add up to one measly kidney."

"Well," said Dr. Merriwether, "I guess Clyde Magnusson is going to have a few more years to pollute his liver."

Dr. Merriwether was referring to the candidate for the kidney. Clyde Magnusson was a man in his early sixties who had spent most of the past thirty years knocking around various skid rows before finally settling in the Washington area two years earlier. Clyde had no family, as far as anyone knew, and no friends to speak of—at least none who visited him at the hospital, where he had been admitted the week before with hopelessly diseased kidneys. When the doctors informed Clyde of his good fortune, he reacted with practiced indifference, pausing only long enough in his television watching to hear them out.

"You know," said Dr. Feinberg to Dr. Merriwether when they had left Clyde's room, "it's too bad a kidney has to be wasted on a wastrel." Merriwether agreed with a laugh.

The doctors had no way of knowing that within a few hours the picture would change dramatically when Brigitte Lundstrom, RN and Ph.D., became a victim of a hit-and-run accident just outside Dulles Airport. The accident left her kidneys beyond repair. Other than that, and various bruises and contusions, her physical condition was good.

"I know this woman!" Dr. Feinberg said when he learned of Lundstrom's admission. Indeed he did. Over the years Dr. Feinberg had attended several symposiums that featured the professor and had even served on a panel with her. He had recently read of her appointment and decided to contact her when she arrived in Washington. Now this.

"Let's do a tissue check on her," Feinberg said to Merriwether.

"But we already have a candidate," Merriwether reminded him. But Feinberg was not to be put off. Reluctantly Merriwether agreed. When the match proved close enough for the transplant, Feinberg suggested that they give Lundstrom the kidney.

"But what do we tell Magnusson?" Merriwether asked him.

"Tell him it was a mistake, tell him anything!" Feinberg said.

Merriwether demurred. "I don't think it's right. If we hadn't told him, that would be one thing, but. . . ."

"Well I don't think it's right to give that rummy what a person of this quality needs to survive!" Feinberg said sharply.

"I know," said Dr. Merriwether, "I know. . . . Maybe we should mull this over a little longer."

Questions for Analysis

1. *Who do you think should get the kidney? Why?*

2. *Apply the ethical theories to the choice facing the physicians. What moral courses do the theories suggest?*

3. *Do you think the situation would be morally different if the doctors had not informed Magnusson?*

4. *Do you think that Magnusson and Lundstrum should be consulted?*

5. *Suppose you were on the committee that Professor Lundstrom had been appointed to chair and were asked your opinion of how to allocate health care resources. Which of the positions would you favor? On what moral grounds would you base your judgment?*

6. *If you selected either of the "limited package" approaches to providing care, how would you go about defining the boundaries of the services that would be made available?*

7. *Can you think of other public policy positions that offer a fairer solution to the distribution of health services than the ones presented to Lundstrom?*

Exercise 1

Some moral dilemmas outside the health care field are similar to microallocation decisions. Indeed, ethicists have tried to set up analogous nonmedical cases in the hope that they might help in the development of criteria and procedures for making microallocation decisions. One such case is found in *U.S.* v. *Holmes*.[16]

In 1841 an American ship, the *William Brown*, struck an iceberg near Newfoundland. The crew and half the passengers were able to escape in two available vessels. One of these vessels was overloaded and began to leak. After twenty-four hours, it foundered in the rough seas. Desperate to keep the vessel afloat, the crew threw fourteen men overboard. Two sisters of one of the men either jumped overboard to join their brother or instructed the crew to throw them out. As their criteria for determining who should live, the crew had used the rule "not to part man and wife, and not to throw over any woman." Several hours later, the vessel's occupants were rescued. Most of the crew then disappeared. But one, Holmes, who acted on orders from the mate, was indicted, tried, and convicted of "unlawful homicide." In handing down his decision, the judge said that lots should have been cast, for in such situations there was no procedure "so consonant both to humanity and to justice." Do you agree with the judge and the basis for his decision? Could the crew's action be defended on grounds of utility? Do you think that no one should have been thrown overboard, even if that meant all would die? Do you think a case like this is analogous to microallocation decisions and helps develop needed criteria and procedures?

Exercise 2

Reconsider your responses under Values and Priorities. Are there any you would change?

---

16. *U.S.* v. *Holmes, 26 Fed. Cas. 360, C.C.E.D. Pa. (1842); P. E. David, ed.,* Moral Duty and Legal Responsibility: A Philosophical-Legal Casebook *(New York: Irvington, 1966), pp. 102–118.*

Social Justice and Equal Access to Health Care

Gene Outka

On what basis should we decide who gets what when there is not enough for all? This question of distributive justice underlies allocation decisions in health care, and so it is one that HCPs and others should engage before embarking on any allocation program.

In the following article, professor of religious ethics Gene Outka addresses this fundamental issue of social justice. Outka begins by examining the more popular conceptual bases of social justice: merit, social contribution, social value, personal need, similar cases. In Outka's view, the only relevant criterion for distributing health care resources is the individual's need for care. This standard is most compatible with justice, which he feels demands equal treatment for individuals unless there is a relevant difference between them.

Outka concedes, however, that the limits of medical resources pose a problem for his goal of equal access. Given the scarcity of resources, it may be necessary to establish priorities among categories of illnesses, so that some kinds of care are not funded. In that event, all individuals who needed the nonfunded care would be denied it, thereby preserving the spirit of the equal access doctrine. Outka claims that such discrimination in patient care, although not the ideal situation, is more just than discrimination on any other basis—for example, on the basis of economic status.

Is it possible to understand and to justify morally a societal goal that increasing numbers of people, including Americans, accept as normative? The goal is: the assurance of comprehensive health services for every person irrespective of income or geographic location. Indeed, the goal now has almost the status of a platitude. Currently in the United States politicians in various camps give it at least verbal endorsement.[1] I do not propose to examine the possible sociological determinants in this emergent consensus. I hope to show that whatever these determinants are, one may offer a plausible case in defense of the goal on reasonable grounds. To demonstrate why appeals to the goal get so successfully under our skins, I shall have recourse to a set of conceptions of social justice. Some of the standard conceptions, found in a number of writings on justice, will do.[2] By reflecting on them it seems to me a prima facie case can be established, namely, that every person in the entire resident population should have equal access to health care delivery.

The case is prima facie only. I wish to set aside as far as possible a related question which comes readily enough to mind. In the world of "suboptimal alternatives," with the constraints, for example, that impinge on the government as it makes decisions about resource allocation, what is one to say? What criteria should be employed? Paul Ramsey in *The Patient as Person* thinks that the large question of how to choose between medical and other societal priorities is "almost, if not altogether, incorrigible to moral reasoning."[3] Whether it is or not is a matter that must be ignored for the present. One may simply observe in passing that choices are unavoidable nonetheless, as Ramsey acknowledges, even where the government allows them to be made by default, so that in some instances they are determined largely by which private pressure groups prove to be dominant. In any event, there is virtue in taking up one complicated question at a time, and we need to get the thrust of the case for equal access before us. It is enough to observe now that Americans attach an obviously high priority to organized health care. National health expenditures for the fiscal year 1972 were $83.4 billion.[4] Even if such an enormous sum is not entirely adequate, we may

Reprinted with permission of the publisher from Journal of Religious Ethics 2 *(Spring 1974): 11–32. This version taken from T.A. Mappes and J. S. Zembaty (eds.),* Biomedical Ethics *(New York: McGraw-Hill, 1981), pp. 523–531.*

still ask: How are we to justify spending whatever we do in accordance as far as possible with the goal of equal access? The answer I propose involves distinguishing various conceptions of social justice and trying to show which of these apply or fail to apply to health care considerations. . . .

Which then among the standard conceptions of social justice appear to be particularly relevant or irrelevant? Let us consider the following five:

1. To each according to his merit or desert.

2. To each according to his societal contribution.

3. To each according to his contribution in satisfying whatever is freely desired by others in the open marketplace of supply and demand.

4. To each according to his needs.

5. Similar treatment for similar cases.

In general I shall argue that the first three of these are less relevant because of certain distinctive features that health crises possess. I shall focus on crises here not because I think preventive care is unimportant (the opposite is true), but because the crisis situation shows most clearly the special significance we attach to medical treatment as an institutionalized activity or social practice, and the basic purpose we suppose it to have.

To Each According to His Merit or Desert

Meritarian conceptions, above all perhaps, are grading ones: advantages are allocated in accordance with amounts of energy expended or kinds of results achieved. What is judged is particular conduct that distinguishes persons from one another and not only the fact that all the parties are human beings. Sometimes a competitive aspect looms large.

In certain contexts it is illuminating to distinguish between efforts and achievements. In the case of efforts one characteristically focuses on the individual: rewards are based on the pains one takes. Some have supposed, for example, that entry into the kingdom of heaven is linked more directly to energy displayed and fidelity shown than to successful results attained.

To assess achievements is to weigh actual performance and productive contributions. The academic prize is awarded to the student with the highest grade-point average, regardless of the amount of midnight oil he or she burned in preparing for the examinations. Sometimes we may exclaim, "It's just not fair," when person X writes a brilliant paper with little effort while we are forced to devote more time with less impressive results. But then our complaint may be directed against differences in innate ability and talent which no expenditure of effort altogether removes.

After the difference between effort and achievement, and related distinctions, have been acknowledged, what should be stressed is the general importance of meritarian or desert criteria in the thinking of most people about justice. These criteria may serve to illuminate a number of disputes about the justice of various practices and institutional arrangements in our society. It may help to explain, for instance, the resentment among the working class against the welfare system. However wrongheaded or self-deceptive the resentment often is, particularly when directed toward those who want to work but for various reasons beyond their control cannot, at its better moments it involves in effect an appeal to desert considerations. "Something for nothing" is repudiated as unjust; benefits should be proportional (or at least related) to costs; those who can make an effort should do so, whatever the degree of their training or significance of their contribution to society; and so on. So, too, persons deserve to have what they have labored for; unless they infringe on the works of others their efforts and achievements are justly theirs. . . .

. . . I would simply hold now (1) that the idea of justice is not exhaustively characterized by the notion of desert, even if one agress that the latter plays an important role; and (2) that the notion of desert is especially ill suited to play an important role in the determination of policies that should govern a system of health care.

Why is it so ill suited? Here we encounter some of the distinctive features that health crises possess. Health crises seem nonmeritarian because they occur so often for reasons beyond our control or power to predict. They frequently fall without discrimination on the (according-to-merit) just and unjust, i.e., the virtuous and the wicked, the industrious and the slothful alike.

While we may believe that virtues and vices cannot depend upon natural contingencies, we are bound to admit, it seems, that many health crises do. It makes sense therefore to say that we are equal in being randomly susceptible to these crises. Even those who ascribe a prominent role to desert acknowledge that justice has also properly to do with pleas of "But I could not help it."[5] One seeks to distinguish such cases from those acknowledged to be praiseworthy or blameworthy. Then it seems unfair as well as unkind to discriminate among those who suffer health crises on the basis of their personal deserts. For it would be odd to maintain that a newborn child deserves his hemophilia or the tumor afflicting her spine.

These considerations help to explain why the following rough distinction is often made. Bernard Williams, for example, in his discussion of "equality in unequal circumstances," identifies two different sorts of inequality, inequality of merit and inequality of need, and two corresponding goods, those earned by effort and those demanded by need.[6] Medical treatment in the event of illness is located under the umbrella of need. He concludes: "Leaving aside preventive medicine, the proper ground of distribution of medical care is ill health: this is a necessary truth."[7] An irrational state of affairs is held to obtain if those whose needs are the same are treated unequally, when needs are the ground of the treatment. One might put the point this way. When people are equal in the relevant respects—in this case when their needs are the same and occur in a context of random, undeserved susceptibility—that by itself is a good reason for treating them equally.[8]

In many societies, however, a second necessary condition for the receipt of medical treatment exists de facto: the possession of money. This is not the place to consider the general question of when inequalities in wealth may be regarded as just. It is enough to note that one can plausibly appeal to all of the conceptions of justice we are embarked in sorting out. A person may be thought to be entitled to a higher income when he works more, contributes more, risks more, and not simply when he needs more. We may think it fair that the industrious should have more money than the slothful and the surgeon more than the tobacconist. The difficulty comes in the misfit between the reasons for differential incomes and the reasons for receiving medical treatment. The former may include a pluralistic set of claims in which different notions of justice must be meshed. The

latter are more monistically focused on needs, and the other notions not accorded a similar relevance. Yet money may nonetheless remain a casually necessary condition for receiving medical treatment. It may be the power to secure what one needs. The senses in which health crises are distinctive may then be insufficiently determinative for the policies which govern the actual availability of treatment. The nearly automatic links between income, prestige, and the receipt of comparatively higher quality medical treatment should then be subjected to critical scrutiny. For unequal treatment of the rich ill and the poor ill is unjust if, again, needs rather than differential income constitute the ground of such treatment.

Suppose one agrees that it is important to recognize the misfit between the reasons for differential incomes and the reasons for receiving medical treatment, and that therefore income as such should not govern the actual availability of treatment. One may still ask whether the case so far relies excessively on "pure" instances where desert considerations are admittedly out of place. That there are such pure instances, tumors afflicting the spine, hemophilia, and so on, is not denied. Yet it is an exaggeration if we go on and regard all health crises as utterly unconnected with desert. Note for example that Williams leaves aside preventive medicine. And if in a cool hour we examine the statistics, we find that a vast number of deaths occur each year due to causes not always beyond our control, e.g., automobile accidents, drugs, alcohol, tobacco, obesity, and so on. In some final reckoning it seems that many persons (though crucially, not all) have an effect on, and arguably a responsibility for, their own medical needs. Consider the following bidders for emergency care: (1) a person with a heart attack who is seriously overweight; (2) a football hero who has suffered a concussion; (3) a man with lung cancer who has smoked cigarettes for forty years; (4) a sixty-year-old man who has always taken excellent care of himself and is suddenly stricken with leukemia; (5) a three-year-old girl who has swallowed poison left out carelessly by her parents; (6) a fourteen-year-old boy who has been beaten without provocation by a gang and suffers brain damage and recurrent attacks of uncontrollable terror; (7) a college student who has slashed his wrists (and not for the first time) from a psychological need for attention; (8) a woman raised in the ghetto who is found unconscious due to an overdose of heroin.

These cases help to show why the whole subject of medical treatment is so crucial and so perplexing. They attest to some melancholy elements in human experience. People suffer in varying ratios the effects of their natural and undeserved vulnerabilities, the irresponsibility and brutality of others, and their own desires and weaknesses. In some final reckoning, then, desert considerations seem not irrelevant to many health crises. The practical applicability of this admission, however, in the instance of health care delivery, appears limited. We may agree that it underscores the importance of preventive health care by stressing the influence we sometimes have over our medical needs. But if we try to foster such care by increasing the penalties for neglect, we normally confine ourselves to calculations about incentives. At the risk of being denounced in some quarters as censorious and puritannical, perhaps we should for example levy far higher taxes on alcohol and tobacco and pump the dollars directly into health care programs rather than (say) into highway building. Yet these steps would by no means lead necessarily to a demand that we correlate in some strict way a demonstrated effort to be temperate with the receipt of privileged medical treatment as a reward. Would it be feasible to allocate the additional tax monies to the man with leukemia before the overweight man suffering a heart attack on the ground of a difference in desert? At the point of emergency care at least, it seems impracticable for the doctor to discriminate between these cases, to make meritarian judgments at the point of catastrophe. And the number of persons who are in need of medical treatment for reasons utterly beyond their control remains a datum with tenacious relevance. There are those who suffer the ravages of a tornado, are handicapped by a genetic defect, beaten without provocation, etc. A commitment to the basic purpose of medical care and to the institutions for achieving it involves the recognition of this persistent state of affairs.

To Each According to His Societal Contribution

This conception gives moral primacy to notions such as the public interest, the common good, the welfare of the community, or the greatest good of the greatest number. Here one judges the social consequences of particular conduct. The formula can be construed in at least two ways.[9] It may refer to the interest of the social group considered collectively, where the group has some independent life all its own. The group's welfare is the decisive criterion for determining what constitutes any member's proper share. Or the common good may refer only to an aggregation of distinct individuals and considered distributively.

Either version accords such a primacy to what is socially advantageous as to be unacceptable not only to defenders of need, but also, it would seem, of desert. For the criteria of effort and achievement are often conceived along rather individualistic lines. The pains an agent takes or the results he brings about deserve recompense, whether or not the public interest is directly served. No automatic harmony then is necessarily assumed between his just share as individually earned and his proper share from the vantage point of the common good. Moreover, the test of social advantage *simpliciter* obviously threatens the agapeic* concern with some minimal consideration due each person which is never to be disregarded for the sake of long-range social benefits. No one should be considered as *merely* a means or instrument.

The relevance of the canon of social productiveness to health crises may accordingly also be challenged. Indeed, such crises may cut against it in that they occur more frequently to those whose comparative contribution to the general welfare is less, e.g., the aged, the disabled, children.

Consider for example Paul Ramsey's persuasive critique of social and economic criteria for the allocation of a single scarce medical resource. He begins by recounting the imponderables that faced the widely discussed "public committee" at the Swedish Hospital in Seattle when it deliberated in the early 1960s. The sparse resource in this case was the kidney machine. The committee was

---

*Editor's note: Agapeic *is from the Greek word* agapé, *meaning "selfless love." In Christian morality,* agape *is used to mean "Christian love." For an analysis of agapeic ethics, see* G. Outka, Agapé: An Ethical Analysis *(New Haven, Conn.: Yale University Press, 1972).*

charged with the responsibility of selecting among patients suffering chronic renal failure those who were to receive dialysis. Its criteria were broadly social and economic. Considerations weighed included age, sex, marital status, number of dependents, income, net worth, educational background, occupation, past performance, and future potential. The application of such criteria proved to be exceedingly problematic. Should someone with six children always have priority over an artist or composer? Were those who arranged matters so that their families would not burden society to be penalized in effect for being provident? And so on. Two critics of the committee found "a disturbing picture of the bourgeoisie sparing the bourgeoisie," and observed that "the Pacific Northwest is no place for a Henry David Thoreau with bad kidneys."[10]

The mistake, Ramsey believes, is to introduce criteria of social worthiness in the first place. In those situations of choice where not all can be saved and yet all need not die, "the equal right of every human being to live, and not relative personal or social worth, should be the ruling principle."[11] The principle leads to a criterion of "random choice among equals" expressed by a lottery scheme or a practice of "first come, first served." Several reasons stand behind Ramsey's defense of the criterion of random choice. First, a religious belief in the equality of persons before God leads intelligibly to a refusal to choose between those who are dying in any way other than random patient selection. Otherwise their equal value as human beings is threatened. Second, a moral primacy is ascribed to survival over other (perhaps superior) interests persons may have, in that it is the condition of everything else. "Life is a value incommensurate with all others, and so not negotiable by bartering one man's worth against another's."[12] Third, the entire enterprise of estimating a person's social worth is viewed with final skepticism. "We have no way of knowing how really and truly to estimate a man's societal worth or his worth to others or to himself in unfocused social situations in the ordinary lives of men in their communities."[13] This statement, incidentally, appears to allow something other than randomness in *focused* social situations; when, say, a president or prime minister and the owner of the local bar rush for the last place in the bomb shelter, and the

knowledge of the former can save many lives. In any event, I have been concerned with a restricted point to which Ramsey's discussion brings illustrative support. The canon of social productiveness is notoriously difficult to apply as a workable criterion for distributing medical services to those who need them.

One can go further. A system of health care delivery that treats people on the basis of the medical care required may often go against (at least narrowly conceived) calculations of societal advantage. For example, the health care needs of people tend to rise during that period of their lives, signaled by retirement, when their incomes and social productivity are declining. More generally:

> Some 40 to 50 per cent of the American people — the aged, children, the dependent poor, and those with some significant chronic disability are in categories requiring relatively large amounts of medical care but with inadequate resources to purchase such care.[14]

If one agrees, for whatever reasons, with the agapeic judgment that each person should be regarded as irreducibly valuable, then one cannot succumb to a social productiveness criterion of human worth. Interests are to be equally considered even when people have ceased to be, or are not yet, or perhaps never will be, public assets.

To Each According to His Contribution in Satisfying Whatever Is Freely Desired by Others in the Open Marketplace of Supply and Demand

Here we have a test which, though similar to the preceding one, concentrates on what is desired de facto by certain segments of the community rather than the community as a whole, and on the relative scarcity of the service rendered. It is tantamount to the canon of supply and demand as espoused by vairous laissez-faire theoreticians.[15] Rewards should be given to those who by virtue of special skill, prescience, risk-taking, and the like discern what is desired and are able to take the requisite steps to bring satisfaction. A surgeon, it may be argued, contributes more than a nurse because of the greater training and skill required, burdens borne, and effective care provided, and should be compensated accordingly. So too, perhaps, a star quarterback on a pro football team should be remunerated even more highly because of the rare athletic prowess needed, hazards involved, and widespread demand to watch him play.

This formula does not then call for the weighing of the value of various contributions, and tends to conflate needs and wants under a notion of desires. It also assumes that a prominent part is assigned to consumer free choice. The consumer should be at liberty to express his preferences, and to select from a variety of competing goods and services. Those who resist many changes currently proposed in the organization and financing of health care delivery in the U.S.A.—such as national health insurance—often do so by appealing to some variant of this formula.

Yet it seems health crises are often of overriding importance when they occur. They appear therefore not satisfactorily accommodated to the context of a free marketplace where consumers may freely choose among alternative goods and services.

To clarify what is at stake in the above contention, let us examine an opposing case. Robert M. Sade, M.D., published an article in *The New England Journal of Medicine* entitled "Medical Care as a Right: A Refutation." He attacks programs of national health insurance in the name of a person's right to select one's own values, determine how they may be realized, and dispose of them if one chooses without coercion from other men. The values in question are construed as economic ones in the context of supply and demand. So we read:

> In a free society, man exercises his right to sustain his own life by producing economic values in the form of goods and services that he is, or should be, free to exchange with other men who are similarly free to trade with him or not. The economic values produced, however, are not given as gifts by nature, but exist only by virtue of the thought and effort of individual men. Goods and services are thus owned as a consequence of the right to sustain life by one's own physical and mental effort.[16]

Sade compares the situation of the physician to that of the baker. The one who produces a loaf of bread should as owner have the power to dispose of his own product. It is immoral simply to expropriate the bread without the baker's permission. Similarly, "medical care is neither a right nor

a privilege: it is a service that is provided by doctors and others to people who wish to purchase it."[17] Any coercive regulation of professional practices by the society at large is held to be analogous to taking the bread from the baker without his consent. Such regulation violates the freedom of the physician over his own services and will lead inevitably to provider apathy.

The analogy surely misleads. To assume that doctors autonomously produce goods and services in a fashion closely akin to a baker is grossly oversimplified. The baker may himself rely on the agricultural produce of others, yet there is a crucial difference in the degree of dependence. Modern physicians depend on the achievements of medical technology and the entire scientific base underlying it, all of which is made possible by a host of persons whose salaries are often notably less. Moreover, the amount of taxpayer support for medical research and education is too enormous to make any such unqualified case for provider autonomy plausible.

However conceptually clouded Sade's article may be, its stress on a free exchange of goods and services reflects one historically influential rationale for much American medical practice. And he applies it not only to physicians but also to patients or "consumers."

> The question is whether the decision of how to allocate the consumer's dollar should belong to the consumer or to the state. It has already been shown that the choice of how a doctor's services should be rendered belongs only to the doctor: in the same way the choice of whether to buy a doctor's service rather than some other commodity or service belongs to the consumer as a logical consequence of the right to his own life.[18]

This account is misguided, I think, because it ignores the overriding importance that is so often attached to health crises. When lumps appear on someone's neck, it usually makes little sense to talk of choosing whether to buy a doctor's service rather than a color television set. References to just tradeoffs suddenly seem out of place. No compensation suffices, since the penalties may differ so much.

There is even a further restriction on consumer choice. One's knowledge in these circumstances is comparatively so limited. The physician makes most of the decisions: about diagnosis, treatment, hospitalization, number of return visits, and so on. In brief:

> The consumer knows very little about the medical services he is buying — probably less than about any other service he purchases. . . . While [he] can still play a role in policing the market, that role is much more limited in the field of health care than in almost any other area of private economic activity.[19]

For much of the way, then, an appeal to supply and demand and consumer choice is not quite fitting. It neglects the issue of the value of various contributions. And it fails to allow for the recognition that medical treatments may be overridingly desired. In contexts of catastrophe, at any rate, when life itself is threatened, most persons (other than those who are apathetic or seek to escape from the terrifying prospects) cannot take medical care to be merely one option among others.

To Each According to His Needs

The concept of needs is sometimes taken to apply to an entire range of interests that concern a person's "psychophysical existence."[20] On this wide usage, to attribute a need to someone is to say that the person lacks what is thought to conduce to his or her "welfare"—understood in both a physiological sense (e.g., for food, drink, shelter, and health) and a psychological one (e.g., for continuous human affection and support).

Yet even in the case of such a wide usage, what the person lacks is typically assumed to be basic. Attention is restricted to recurrent considerations rather than to every possible individual whim or frivolous pursuit. So one is not surprised to meet with the contention that a preferable rendering of this formula would be: "to each according to his essential needs."[21] This contention seems to me well taken. It implies, for one thing, that basic needs are distinguishable from felt needs or wants. For the latter may encompass expressions of personal preference unrelated to considerations of survival or subsistence, and sometimes artificially generated by circumstances of rising affluence in the society at large.

Essential needs are also typically assumed to be given rather than acquired. They are not constituted by any action for which the person is responsible by virtue of his or her distinctively greater effort. It is almost as if the designation "innocent" may be linked illuminatingly to need, as retribution, punishment, and so on, are to desert, and in complex ways, to freedom. Thus essential needs are likewise distinguishable from deserts. Where needs are unequal, one thinks of them as fortuitously distributed; as part, perhaps, of a kind of "natural lottery."[22] So very often the advantages of health and the burdens of illness, for example, strike one as arbitrary effects of the lottery. It seems wrong to say that a newborn child deserves as a reward all of his faculties when he has done nothing in particular that distinguishes him from another newborn who comes into the world deprived of one or more of them. Similarly, though crudely, many religious believers do not look on natural events as personal deserts. They are not inclined to pronounce sentences such as, "That evil person with incurable cancer got what he deserved." They are disposed instead to search for some distinction between what they may call the conditions of finitude on the one hand and sin and moral evil on the other. If the distinction is "ultimately" invalid, in this life it seems inscrutably so. Here and now it may be usefully drawn. Inequalities in the need for medical treatment are taken, it appears, to reflect the conditions of finitude more than anything else.

One can even go on to argue that among our basic or essential needs, the case of medical treatment is conspicuous in the following sense. While food and shelter are not matters about which we are at liberty to please ourselves, they are at least predictable. We can plan, for instance, to store up food and fuel for the winter. It may be held that responsibility increases along with the power to predict. If so, then many health crises seem peculiarly random and uncontrollable. Cancer, given the present state of knowledge at any rate, is a contingent diaster, whereas hunger is a steady threat. Who will needs serious medical care, and when, is then perhaps a classic example of uncertainty. . . .

. . . [J]ustice has properly to do with pleas of "But I could not help it." It seeks to distinguish such cases from those acknowledged to be praiseworthy or blameworthy. The formula "to each according to his needs" is one cogent way of identifying the moral relevance of these pleas. To ignore them may be thought to be unfair as well as unkind when they arise from the deprivation of some essential need. The move to confine the notion of justice wholly to desert considerations is thereby resisted as well. Hence we may say that sometimes "questions of social justice arise just because people are unequal in ways they can do very little to change and . . . only by attending to these inequalities can one be said to be giving their interests equal consideration."[23]

Similar Treatment for Similar Cases

This conception is perhaps the most familiar of all. Certainly it is the most formal and inclusive one. It is frequently taken as an elementary appeal to consistency and linked to the universalizability test. One should not make an arbitrary exception on one's own behalf, but rather should apply impartially whatever standards one accepts. The conception can be fruitfully applied to health care questions and I shall assume its relevance. Yet as literally interpreted, it is necessary but not sufficient. For rightly or not, it is often held to be as compatible with no positive treatment whatever as with active promotion of other peoples' interests, as long as all are equally and impartially included. Its exponents sometimes assume such active promotion without demonstrating clearly how this is built into the conception itself. Moreover, it may obscure a distinction that we have seen agapists and others make: between equal consideration and identical treatment. Needs may differ and so treatments must, if benefits are to be equalized.

I have placed this conception at the end of the list partly because it moves us, despite its formality, toward practice. Let me suggest briefly how it does so. Suppose first of all one agrees with the case so far offered. Suppose, that is, it has been shown convincingly that a need conception of justice applies with greater relevance than the earlier three when one reflects about the basic purpose of medical care. To treat one class of people differently from another because of income or geographic location should therefore be ruled out, because such reasons are irrelevant. (The irrelevance is conceptual, rather than always, unfortunately, causal.) In short, all persons should have equal access," as needed, without financial, geographic, or other barriers, to the whole spectrum of health services."[24]

Suppose however, second, that the goal of equal access collides on some occasions with the realities of finite medical resources and needs that prove to be insatiable. That such collisons occur in fact it would be idle to deny. And it is here that the practical bearing of the formula of similar treatment for similar cases should be noticed. Let us recall Williams's conclusion: "the proper ground of distribution of medical care is ill health: this is a necessary truth." While I agree with the essentials of his argument—for all the reasons above—I would prefer, for practical purposes, a slightly more modest formulation. Illness is the proper ground for the *receipt* of medical care. However, the *distribution* of medical care in less-than-optimal circumstances requires us to face the collisions. I would argue that in such circumstances the formula of similar treatment for similar cases may be constructed so as to guide actual choices in the way most compatible with the goal of equal access. The formula's allowance of no positive treatment whatever may justify exclusion of entire classes of cases from a priority list. Yet it forbids doing so for irrelevant or arbitrary reasons. So (1) if we accept the case for equal access, but (2) if we simply cannot, physically cannot, treat all who are in need, it seems more just to discriminate by virtue of categories of illness than, for example, between the rich ill and poor ill. All persons with a certain rare, noncommunicable disease would not receive priority, let us say, where the costs were inordinate, the prospects for rehabilitation remote, and for the sake of equalized benefits to many more. Or with Ramsey we may urge a policy of random patient selection when one must decide between claimants for a medical treatment unavailable to all. Or we may acknowledge that any notion of "comprehensive benefits" to which persons should have equal access is subject to practical restrictions that will vary from society to society depending on resources at a given time. Even in a country as affluent as the United States there will surely always be items excluded, e.g., perhaps over-the-counter drugs, some teenage orthodontia, cosmetic surgery, and the like.[25] Here, too, the formula of similar treatment for similar cases may serve to modify the application of a need conception of justice in order to address the insatiability problem and limit frivolous use. In all of the foregoing instances of restriction, however, the relevant feature remains the illness, discomfort, etc. itself. The goal of equal access then retains its prima facie authoritativeness. It is imperfectly realized rather than disregarded. . . .

Notes

1. Edward M. Kennedy, *In Critical Condition: The Crisis in America's Health Care* (New York: Simon and Schuster, 1972), pp. 234–52; Richard M. Nixon, "President's Message on Health Care System," Document No. 92–261 (March 2, 1972), House of Representatives, Washington, D.C., 1.

2. Hugo A. Bedau, "Radical Egalitarianism," pp. 168–80, in *Justice and Equality,* ed. by Hugo A. Bedau (Englewood Cliffs, N.J.: Prentice-Hall, 1971); John Hospers, *Human Conduct* (New York: Harcourt, Brace, 1961), 416–68; J. R. Lucas, "Justice," *Philosophy* 47, No. 181 (July 1972), 229–48; Ch. Perelman, *The Idea of Justice and the Problem of Argument,* trans. by John Petrie (London: Routledge and Kegan Paul, 1963); Nicholas Rescher, *Distributive Justice* (Indianapolis: Bobbs-Merrill, 1966); Gregory Vlastos, "Justice and Equality," pp. 31–72 in *Social Justice,* ed. by Richard B. Brandt (Englewood Cliffs, N.J.: Prentice-Hall, 1962).

3. Paul Ramsey, *The Patient as Person: Explorations in Medical Ethics* (New Haven: Yale University Press, 1970), p. 240.

4. Nancy Hicks, "Nation's Doctors Move to Police Medical Care," *New York Times* (Sunday, October 28, 1973), 52.

5. Lucas, 321.

6. Bernard A. O. Williams, "The Idea of Equality," pp. 116–37 in *Justice and Equality,* ed. by Hugo A. Bedau (Englewood Cliffs, N.J.: Prentice-Hall, 1971), pp. 126–37.

7. Ibid., 127.

8. See also Thomas Nagel, "Equal Treatment and Compensatory Discrimination," *Philosophy and Public Affairs* 2, No. 4, 348–63 (Summer 1973), 354.

9. Rescher, pp. 79–80.

10. Quoted in Ramsey, p. 248.

11. Ramsey, p. 256.

12. Ibid.

13. Ibid.

14. Anne R. Somers, *Health Care in Transition: Directions for the Future* (Chicago: Hospital Research and Educational Trust, 1971), p. 20.

15. Cf. Rescher, pp. 80–81.

16. Robert M. Sade, "Medical Care as a Right: A Refutation," *The New England Journal of Medicine* 285, 1288–92 (December 1971), 1289.

17. Ibid.

18. Ibid., p. 1291.

19. Charles L. Schultze, Edward R. Fried, Alice M. Rivlin, and Nancy H. Teeters, *Setting National Priorities: The 1973 Budget* (Washington, D.C.: The Brookings Institution, 1972), pp. 214–15.

20. Gene Outka, *Agape: An Ethical Analysis* (New Haven: Yale University Press, 1972), pp. 264–65.

21. Perelman, p. 22.

22. See John Rawls, *A Theory of Justice* (Cambridge: Harvard University Press, 1971), e.g. 104.

23. Benn, Stanley I., "Egalitarianism and the Equal Consideration of Interests," pp. 152–67 in *Justice and Equality,* ed. by Hugo A. Bedau (Englewood Cliffs, N.J.: Prentice-Hall, 1971), p. 164.

24. Anne R. Somers and Herman M. Somers, "The Organization and Financing of Health Care: Issues and Directions for the Future," *American Journal of Orthopsychiatry* 42, 119–36 (January 1972), 122.

25. Anne R. Somers and Herman M. Somers, "Major Issues in National Health Insurance," *Milbank Memorial Fund Quarterly 50,* No. 2, Part 1, 177–210 (April 1972), 182.

For Further Reading

Anderson, O. W. *Blue Cross Since 1929: Accountability and the Public Trust.* Cambridge, Mass.: Ballinger, 1975.

Anderson, R., Lion, J., and Anderson, O. W. *Equity in Health Services: Empirical Analyses in Social Policy.* Cambridge, Mass.: Ballinger, 1975.

Bullough, B., and Bullough, V. L. *Poverty, Ethnic Identity, and Health Care.* New York: Appleton-Century-Crofts, 1972.

Carlson, R. J. *Future Directions in Health Care: A New Public Policy*. Cambridge, Mass.: Ballinger, 1978.

Clark, K. B. *Dark Ghetto: Dilemmas of Social Power*. New York: Holt, Rinehart & Winston, 1965.

Clark, M. *Health in the Mexican-American Culture: A Community Study*. Berkeley, Calif.: University of California Press, 1970.

Davis, K., and Schoen, C. *Health and the War on Poverty: A Ten Year Appraisal*. Washington, DC: The Brookings Institute, 1978.

Fitzpatrick, M. S. *Environmental Health Planning*. Cambridge, Mass.: Ballinger, 1978.

Fox, R. C., and Swazey, J. P. *The Courage to Fail: A Social View of Organ Transplants and Dialysis*. Chicago: University of Chicago Press, 1974.

Friedson, E. *Professional Dominance: The Social Structure of Medical Care*. New York: Atherton Press, 1970.

Fuchs, V. R. *Who Shall Live?: Health Economics and Social Change*. New York: Basic Books, 1974.

Gray, E. *In Failing Health: The Medical Crisis and the A.M.A.* Indianapolis: Bobbs-Merrill, 1970.

Greenberg, S. *The Quality of Mercy: A Report on the Critical Condition of Hospitals and Medical Care in America*. New York: Atherton Press, 1971.

Kriesberg, L. *Mothers in Poverty: A Study of Fatherless Families*. Chicago: Aldine, 1970.

Miller, G. W. *Moral and Ethical Implications of Human Organ Transplants*. Springfield, Ill.: Charles C Thomas, 1971.

Pilisak, M., and Pilisak, P., eds. *Poor America: How the Poor White Live*. Chicago: Aldine, 1971.

Piven, F., and Cloward, R. A. *Regulating the Poor: The Functions of Public Welfare*. New York: Vintage Books, 1971.

Wertz, R. W. *Readings on Ethical and Social Issues in Biomedicine*. Englewood Cliffs, N.J.: Prentice-Hall, 1973.

Appendix A

THE OATH OF HIPPOCRATES

I swear by Apollo Physician and Asclepius and Hygieia and Panaceia and all the gods and goddesses, making them my witnesses, that I will fulfill according to my ability and judgment this oath and this covenant:

To hold him who has taught me this art as equal to my parents and to live my life in partnership with him, and if he is in need of money to give him a share of mine, and to regard his offspring as equal to my brothers in male lineage and to teach them this art—if they desire to learn it—without fee and covenant; to give a share of precepts and oral instruction and all the other learning to my sons and to the sons of him who has instructed me and to pupils who have signed the covenant and have taken an oath according to the medical law, but to no one else.

I will apply dietetic measures for the benefit of the sick according to my ability and judgment; I will keep them from harm and injustice.

I will neither give a deadly drug to anybody if asked for it, nor will I make a suggestion to this effect. Similarly I will not give to a woman an abortive remedy. In purity and holiness I will guard my life and my art.

I will not use the knife, not even on sufferers from stone, but will withdraw in favor of such men as are engaged in this work.

Whatever houses I may visit, I will come for the benefit of the sick, remaining free of all intentional injustice, of all mischief and in particular of sexual relations with both female and male persons, be they free or slaves.

What I may see or hear in the course of the treatment or even outside of the treatment in regard to the life of men, which on no account one must spread abroad, I will keep to myself holding such things shameful to be spoken about.

If I fulfill this oath and do not violate it, may it be granted to me to enjoy life and art, being honored with fame among all men for all time to come; if I transgress it and swear falsely, may the opposite of all this be my lot.

Reprinted with permission of the publisher from Ancient Medicine: Selected Papers of Ladwig Edelstein, *edited by Ousei Temkin and C. Lilian Temkin, p. 6. Copyright* © 1967 by The Johns Hopkins Press, Baltimore.

Appendix B

THE ETHICAL CODE OF THE AMERICAN MEDICAL ASSOCIATION

Preamble

These principles are intended to aid physicians individually and collectively in maintaining a high level of ethical conduct. They are not laws but standards by which a physician may determine the propriety of his conduct in his relationship with patients, with colleagues, with members of allied professions, and with the public.

Section 1

The principle objective of the medical profession is to render service to humanity with full respect for the dignity of man. Physicians should merit the confidence of patients entrusted to their care, rendering to each a full measure of service and devotion.

Section 2

Physicians should strive continually to improve medical knowledge and skill, and should make available to their patients and colleagues the benefits of their professional attainments.

Section 3

A physician should practice a method of healing founded on a scientific basis; and he should not voluntarily associate professionally with anyone who violates this principle.

Section 4

The medical profession should safeguard the public and itself against physicians deficient in moral character or professional competence. Physicians should observe all laws, uphold the dignity and honor of the profession, and accept its self-imposed disciplines. They should expose, without hesitation, illegal or unethical conduct of fellow members of the profession.

Section 5

A physician may choose whom he will serve. In an emergency, however, he should render service to the best of his ability. Having undertaken the care of a patient, he may not neglect him; and unless he has been discharged he may discontinue his services only after giving adequate notice. He should not solicit patients.

Section 6

A physician should not dispose of his services under terms or conditions that tend to interfere with or impair the free and complete exercise of his medical judgment and skill or tend to cause a deterioration of the quality of medical care.

Section 7

In the practice of medicine a physician should limit the source of his professional income to medical services actually rendered by him, or under his supervision, to his patients. His fee should be commensurate with the service rendered and the patient's ability to pay. He should neither pay nor

---

receive commission for referral of patients. Drugs, remedies, or appliances may be dispensed or supplied by the physician provided it is in the best interest of the patient.

Section 8

A physician should seek consultation upon request; in doubtful or difficult cases; or whenever it appears that the quality of medical service may be enhanced thereby.

Section 9

A physician may not reveal the confidences entrusted to him in the course of medical attendance, or the deficiencies he may observe in the character of patients, unless he is required to do so by law or unless it becomes necessary in order to protect the welfare of the individual or of the community.

Section 10

The honored ideals of the medical profession imply that the responsibility of the physician extend not only to the individual, but also to society, and these responsibilities deserve his interest and participation in activities that have the purpose of improving both the health and the well-being of the individual and the community.

Appendix C

A "LIVING WILL"

Directive to Physicians

Directive made this _____ day of _____ (month, year).

I_____, being of sound mind, willfully and voluntarily making known my desire that my life shall not be artificially prolonged under the circumstances set forth below, do hereby declare:

1. If at any time I should have an incurable injury, disease, or illness certified to be a terminal condition by two physicians, and where the application of life-sustaining procedures would serve only to artificially prolong the moment of my death and where my physician determines that my death is imminent whether or not life-sustaining procedures are utilized, I direct that such procedures be withheld or withdrawn and that I be permitted to die naturally.

2. In the absence of my ability to give directions regarding the use of such life-sustaining procedures, it is my intention that this directive shall be honored by my family and physician(s) as the final expression of my legal right to refuse medical or surgical treatment and shall accept the consequences from such refusal.

3. If I have been diagnosed as pregnant and that diagnosis is known to my physician, this directive shall have no force or effect during the course of my pregnancy.

4. I have been diagnosed and notified at least 14 days ago as having a terminal condition by _____, M.D., whose address is _____ and whose telephone number is _____.
I understand that if I have not filled in the physician's name and address, it shall be presumed that I did not have a terminal condition when I made out this directive.

5. This directive shall have no force or effect five years from the date filled in above.

6. I understand the full import of this directive, and I am emotionally and mentally competent to make this directive.

Signed _____

City, County, and State of Residence _____

The declarant has been personally known to me and I believe him or her to be of sound mind.

Witness _____

Witness _____

Index

Danbury 1980 by Houghton Mifflin Co.